T0202642

Lecture Notes in Computer Science   14331

Founding Editors

Gerhard Goos
Juris Hartmanis

Editorial Board Members

The series Lecture Notes in Computer Science (LNCS), including its subseries Lecture Notes in Artificial Intelligence (LNAI) and Lecture Notes in Bioinformatics (LNBI), has established itself as a medium for the publication of new developments in computer science and information technology research, teaching, and education.

LNCS enjoys close cooperation with the computer science R & D community, the series counts many renowned academics among its volume editors and paper authors, and collaborates with prestigious societies. Its mission is to serve this international community by providing an invaluable service, mainly focused on the publication of conference and workshop proceedings and postproceedings. LNCS commenced publication in 1973.

Xiangyu Song · Ruyi Feng · Yunliang Chen ·
Jianxin Li · Geyong Min
Editors

Web and Big Data

7th International Joint Conference, APWeb-WAIM 2023
Wuhan, China, October 6–8, 2023
Proceedings, Part I

 Springer

Editors
Xiangyu Song ⓘ
Peng Cheng Laboratory
Shenzhen, China

Ruyi Feng ⓘ
China University of Geosciences
Wuhan, China

Yunliang Chen ⓘ
China University of Geosciences
Wuhan, China

Jianxin Li ⓘ
Deakin University
Burwood, VIC, Australia

Geyong Min ⓘ
University of Exeter
Exeter, UK

ISSN 0302-9743   ISSN 1611-3349 (electronic)
Lecture Notes in Computer Science
ISBN 978-981-97-2302-7   ISBN 978-981-97-2303-4 (eBook)
https://doi.org/10.1007/978-981-97-2303-4

Preface

This volume (LNCS 14331) and its companion volumes (LNCS 14332, LNCS 14333, and LNCS 14334) contain the proceedings of the 7th Asia-Pacific Web (APWeb) and Web-Age Information Management (WAIM) Joint Conference on Web and Big Data, called APWeb-WAIM 2023. Researchers and practitioners from around the world came together at this leading international forum to share innovative ideas, original research findings, case study results, and experienced insights in the areas of the World Wide Web and big data, thus covering web technologies, database systems, information management, software engineering, knowledge graphs, recommend systems and big data.

The 7th APWeb-WAIM conference was held in Wuhan during 6–8 October 2023. As an Asia-Pacific flagship conference focusing on research, development, and applications in relation to Web information management, APWeb-WAIM builds on the successes of APWeb and WAIM. Previous APWeb conferences were held in Beijing (1998), Hong Kong (1999), Xi'an (2000), Changsha (2001), Xi'an (2003), Hangzhou (2004), Shanghai (2005), Harbin (2006), Huangshan (2007), Shenyang (2008), Suzhou (2009), Busan (2010), Beijing (2011), Kunming (2012), Sydney (2013), Changsha (2014), Guangzhou (2015), and Suzhou (2016); and WAIM was held in Shanghai (2000), Xi'an (2001), Beijing (2002), Chengdu (2003), Dalian (2004), Hangzhou (2005), Hong Kong (2006), Huangshan (2007), Zhangjiajie (2008), Suzhou (2009), Jiuzhaigou (2010), Wuhan (2011), Harbin (2012), Beidaihe (2013), Macau (2014), Qingdao (2015), and Nanchang (2016). The APWeb-WAIM conferences were held in Beijing (2017), Macau (2018), Chengdu (2019), Tianjin (2020), Guangzhou (2021), Nanjing (2022), and Wuhan (2023). With the ever-growing importance of appropriate methods in these data-rich times and the fast development of web-related technologies, APWeb-WAIM will become a flagship conference in this field.

The high-quality program documented in these proceedings would not have been possible without the authors who chose APWeb-WAIM for disseminating their findings. APWeb-WAIM 2023 received a total of 434 submissions and, after the double-blind review process (each paper received at least three review reports), the conference accepted 133 regular papers (including research and industry track) (acceptance rate 31.15%), and 6 demonstrations. The contributed papers address a wide range of topics, such as big data analytics, advanced database and web applications, data mining and applications, graph data and social networks, information extraction and retrieval, knowledge graphs, machine learning, recommender systems, security, privacy and trust, and spatial and multi-media data. The technical program also included keynotes by Jie Lu, Qing-Long Han, and Hai Jin. We are grateful to these distinguished scientists for their invaluable contributions to the conference program.

We would like to express our gratitude to all individuals, institutions, and sponsors that supported APWeb-WAIM2023. We are deeply thankful to the Program Committee members for lending their time and expertise to the conference. We also would like to acknowledge the support of the other members of the Organizing Committee. All of them helped to make APWeb-WAIM 2023 a success. We are grateful for the guidance of the honorary chair (Lizhe Wang), the steering committee representative (Yanchun Zhang) and the general co-chairs (Guoren Wang, Schahram Dustdar, Bruce Xuefeng Ling, and Hongyan Zhang) for their guidance and support. Thanks also go to the program committee chairs (Yunliang Chen, Jianxin Li, and Geyong Min), local co-chairs (Chengyu Hu, Tao Lu, and Jianga Shang), publicity co-chairs (Bohan Li, Chang Tang, and Xin Bi), proceedings co-chairs (David A. Yuen, Ruyi Feng, and Xiangyu Song), tutorial co-chairs (Ye Yuan and Rajiv Ranjan), CCF TCIS liaison (Xin Wang), CCF TCDB liaison (Yueguo Chen), Ph.D. consortium co-chairs (Pablo Casaseca, Xiaohui Huang, and Yanan Li), Web co-chairs (Wei Han, Huabing Zhou, and Wei Liu), and industry co-chairs (Jun Song, Wenjian Qin, and Tao Yu).

We hope you enjoyed the exciting program of APWeb-WAIM 2023 as documented in these proceedings.

October 2023

Yunliang Chen
Jianxin Li
Geyong Min
David A. Yuen
Ruyi Feng
Xiangyu Song

Organization

General Chairs

Guoren Wang   BIT, China
Schahram Dustdar   TU Wien, Austria
Bruce Xuefeng Ling   Stanford University, USA
Hongyan Zhang   China University of Geosciences, China

Program Committee Chairs

Yunliang Chen   China University of Geosciences, China
Jianxin Li   Deakin University, Australia
Geyong Min   University of Exeter, UK

Steering Committee Representative

Yanchun Zhang   Guangzhou University & Pengcheng Lab, China;
                                Victoria University, Australia

Local Co-chairs

Chengyu Hu   China University of Geosciences, China
Tao Lu   Wuhan Institute of Technology, China
Jianga Shang   China University of Geosciences, China

Publicity Co-chairs

Bohan Li   Nanjing University of Aeronautics and
                                Astronautics, China
Chang Tang   China University of Geosciences, China
Xin Bi   Northeastern University, China

Proceedings Co-chairs

David A. Yuen   Columbia University, USA
Ruyi Feng   China University of Geosciences, China
Xiangyu Song   Swinburne University of Technology, Australia

Tutorial Co-chairs

Ye Yuan   BIT, China
Rajiv Ranjan   Newcastle University, UK

CCF TCIS Liaison

Xin Wang   Tianjin University, China

CCF TCDB Liaison

Yueguo Chen   Renmin University of China, China

Ph.D. Consortium Co-chairs

Pablo Casaseca   University of Valladolid, Spain
Xiaohui Huang   China University of Geosciences, China
Yanan Li   Wuhan Institute of Technology, China

Web Co-chairs

Wei Han   China University of Geosciences, China
Huabing Zhou   Wuhan Institute of Technology, China
Wei Liu   Wuhan Institute of Technology, China

Industry Track Co-chairs

Jun Song	China University of Geosciences, China
Wenjian Qin	Shenzhen Institute of Advanced Technology CAS, China
Tao Yu	Tsinghua University, China

Program Committee Members

Alex Delis	University of Athens, Greece
Amr Ebaid	Google, USA
An Liu	Soochow University, China
Anko Fu	China University of Geosciences, China
Ao Long	China University of Geosciences, Wuhan, China
Aviv Segev	University of South Alabama, USA
Baoning Niu	Taiyuan University of Technology, China
Bin Zhao	Nanjing Normal University, China
Bo Tang	Southern University of Science and Technology, China
Bohan Li	Nanjing University of Aeronautics and Astronautics, China
Bolong Zheng	Huazhong University of Science and Technology, China
Cai Xu	Xidian University, China
Carson Leung	University of Manitoba, Canada
Chang Tang	China University of Geosciences, China
Chen Shaohao	China University of Geosciences, China
Cheqing Jin	East China Normal University, China
Chuanqi Tao	Nanjing University of Aeronautics and Astronautics, China
Dechang Pi	Nanjing University of Aeronautics and Astronautics, China
Dejun Teng	Shandong University, China
Derong Shen	Northeastern University, China
Dong Li	Liaoning University, China
Donghai Guan	Nanjing University of Aeronautics and Astronautics, China
Fang Wang	Hong Kong Polytechnic University, China
Feng Yaokai	Kyushu University, Japan
Giovanna Guerrini	University of Genoa, Italy
Guanfeng Liu	Macquarie University, Australia

Guoqiong Liao	Jiangxi University of Finance & Economics, China
Hailong Liu	Northwestern Polytechnical University, USA
Haipeng Dai	Nanjing University, China
Haiwei Pan	Harbin Engineering University, China
Haoran Xu	China University of Geosciences, Wuhan, China
Haozheng Ma	China University of Geosciences, Wuhan, China
Harry Kai-Ho Chan	University of Sheffield, UK
Hiroaki Ohshima	University of Hyogo, Japan
Hongzhi Wang	Harbin Institute of Technology, China
Hua Wang	Victoria University, Australia
Hui Li	Xidian University, China
Jiabao Li	China University of Geosciences, Wuhan, China
Jiajie Xu	Soochow University, China
Jiali Mao	East China Normal University, China
Jian Chen	South China University of Technology, China
Jian Yin	Sun Yat-sen University, China
Jianbin Qin	Shenzhen University, China
Jiannan Wang	Simon Fraser University, Canada
Jianqiu Xu	Nanjing University of Aeronautics and Astronautics, China
Jianxin Li	Deakin University, Australia
Jianzhong Qi	University of Melbourne, Australia
Jianzong Wang	Ping An Technology (Shenzhen) Co., Ltd., China
Jinguo You	Kunming University of Science and Technology, China
Jizhou Luo	Harbin Institute of Technology, China
Jun Gao	Peking University, China
Jun Wang	China University of Geosciences, Wuhan, China
Junhu Wang	Griffith University, Australia
K. Selçuk Candan	Arizona State University, USA
Krishna Reddy P.	IIIT Hyderabad, India
Ladjel Bellatreche	ISAE-ENSMA, France
Le Sun	Nanjing University of Information Science and Technology, China
Lei Duan	Sichuan University, China
Leong Hou U.	University of Macau, China
Li Jiajia	Shenyang Aerospace University, China
Liang Hong	Wuhan University, China
Lin Xiao	China University of Geosciences, Wuhan, China
Lin Yue	University of Newcastle, UK

Lisi Chen	University of Electronic Science and Technology of China, China
Lizhen Cui	Shandong University, China
Long Yuan	Nanjing University of Science and Technology, China
Lu Chen	Zhejiang University, China
Lu Qin	UTS, Australia
Luyi Bai	Northeastern University, China
Miaomiao Liu	Northeast Petroleum University, China
Min Jin	China University of Geosciences, Wuhan, China
Ming Zhong	Wuhan University, China
Mirco Nanni	CNR-ISTI Pisa, Italy
Mizuho Iwaihara	Waseda University, Japan
Nicolas Travers	Pôle Universitaire Léonard de Vinci, France
Peiquan Jin	University of Science and Technology of China, China
Peng Peng	Hunan University, China
Peng Wang	Fudan University, China
Philippe Fournier-Viger	Shenzhen University, China
Qiang Qu	SIAT, China
Qilong Han	Harbin Engineering University, China
Qing Xie	Wuhan University of Technology, China
Qiuyan Yan	China University of Mining and Technology, China
Qun Chen	Northwestern Polytechnical University, China
Rong-Hua Li	Beijing Institute of Technology, China
Rui Zhu	Shenyang Aerospace University, China
Runyu Fan	China University of Geosciences, China
Sanghyun Park	Yonsei University, South Korea
Sanjay Madria	Missouri University of Science & Technology, USA
Sara Comai	Politecnico di Milano, Italy
Shanshan Yao	Shanxi University, China
Shaofei Shen	University of Queensland, Australia
Shaoxu Song	Tsinghua University, China
Sheng Wang	China University of Geosciences, Wuhan, China
ShiJie Sun	Chang'an University, China
Shiyu Yang	Guangzhou University, China
Shuai Xu	Nanjing University of Aeronautics and Astronautics, China
Shuigeng Zhou	Fudan University, China
Tanzima Hashem	Bangladesh University of Engineering and Technology, Bangladesh

Tianrui Li	Southwest Jiaotong University, China
Tung Kieu	Aalborg University, Denmark
Vincent Oria	NJIT, USA
Wee Siong Ng	Institute for Infocomm Research, Singapore
Wei Chen	Hebei University of Environmental Engineering, China
Wei Han	China University of Geosciences, Wuhan, China
Wei Shen	Nankai University, China
Weiguo Zheng	Fudan University, China
Weiwei Sun	Fudan University, China
Wen Zhang	Wuhan University, China
Wolf-Tilo Balke	TU Braunschweig, Germany
Xiang Lian	Kent State University, USA
Xiang Zhao	National University of Defense Technology, China
Xiangfu Meng	Liaoning Technical University, China
Xiangguo Sun	Chinese University of Hong Kong, China
Xiangmin Zhou	RMIT University, Australia
Xiao Pan	Shijiazhuang Tiedao University, China
Xiao Zhang	Shandong University, China
Xiao Zheng	National University of Defense Technology, China
Xiaochun Yang	Northeastern University, China
Xiaofeng Ding	Huazhong University of Science and Technology, China
Xiaohan Zhang	China University of Geosciences, Wuhan, China
Xiaohui (Daniel) Tao	University of Southern Queensland, Australia
Xiaohui Huang	China University of Geosciences, Wuhan, China
Xiaowang Zhang	Tianjin University, China
Xie Xiaojun	Nanjing Agricultural University, China
Xin Bi	Northeastern University, China
Xin Cao	University of New South Wales, Australia
Xin Wang	Tianjin University, China
Xingquan Zhu	Florida Atlantic University, USA
Xinwei Jiang	China University of Geosciences, Wuhan, China
Xinya Lei	China University of Geosciences, Wuhan, China
Xinyu Zhang	China University of Geosciences, Wuhan, China
Xujian Zhao	Southwest University of Science and Technology, China
Xuyun Zhang	Macquarie University, Australia
Yajun Yang	Tianjin University, China
Yanfeng Zhang	Northeastern University, China

Yanghui Rao   Sun Yat-sen University, China
Yang-Sae Moon   Kangwon National University, South Korea
Yanhui Gu   Nanjing Normal University, China
Yanjun Zhang   University of Technology Sydney, Australia
Yaoshu Wang   Shenzhen University, China
Ye Yuan   China University of Geosciences, Wuhan, China
Yijie Wang   National University of Defense Technology,
                           China
Yinghui Shao   China University of Geosciences, Wuhan, China
Yong Tang   South China Normal University, China
Yong Zhang   Tsinghua University, China
Yongpan Sheng   Southwest University, China
Yongqing Zhang   Chengdu University of Information Technology,
                           China
Youwen Zhu   Nanjing University of Aeronautics and
                           Astronautics, China
Yu Liu   Huazhong University of Science and Technology,
                           China
Yuanbo Xu   Jilin University, China
Yue Lu   China University of Geosciences, Wuhan, China
Yuewei Wang   China University of Geosciences, Wuhan, China
Yunjun Gao   Zhejiang University, China
Yunliang Chen   China University of Geosciences, Wuhan, China
Yunpeng Chai   Renmin University of China, China
Yuwei Peng   Wuhan University, China
Yuxiang Zhang   Civil Aviation University of China, China
Zhaokang Wang   Nanjing University of Aeronautics and
                           Astronautics, China
Zhaonian Zou   Harbin Institute of Technology, China
Zhenying He   Fudan University, China
Zhi Cai   Beijing University of Technology, China
Zhiwei Zhang   Beijing Institute of Technology, China
Zhixu Li   Soochow University, China
Ziqiang Yu   Yantai University, China
Zouhaier Brahmia   University of Sfax, Tunisia

Contents – Part I

A BERT-Based Semantic Enhanced Model for COVID-19 Fake News Detection

Hui Yin[1]($\boxtimes$)(iD), Xiao Liu[2](iD), Yutao Wu[2](iD), Hilya Mudrika Arini[3](iD),
and Rami Mohawesh[4](iD)

[1] Social Innovation Research Institute, Swinburne University of Technology,
Melbourne, Australia
huiyin@swin.edu.au
[2] School of Information Technology, Deakin University, Geelong, Australia
{xiao.liu,yutao.wu}@deakin.edu.au
[3] Department of Mechanical and Industrial Engineering, Gadjah Mada University,
Yogyakarta, Indonesia
hilya.mudrika@ugm.ac.id
[4] College of Engineering, Al Ain University, Al Ain, UAE
rami.mohawesh@aau.ac.ae

Abstract. During the COVID-19 pandemic, COVID-19-related news keeps growing and spreading daily across social media platforms, including text, pictures, and videos. Meanwhile, fake news spreads widely on the Internet, preventing authoritative information from spreading and hindering the fight against the disease. To detect and recognize fake news, as well as to prevent its spread, effective detection models are urgently required. Text information is the most significant component of news content and is easy to be adopted by news consumers, so text-based fake news detection models are highly desirable. In this study, we propose a transformer-based semantic enhanced classification model for COVID-19 fake news detection. The model adds a semantic extraction module to the vanilla classifier to extract topic information from data samples as additional features to supplement text representations. Using k-fold cross-validation, we validate the model's performance on a publicly available COVID-19 fake news dataset, demonstrating its effectiveness and robustness. On evaluation metrics, the proposed model performs better than the vanilla model by more than 3%.

Keywords: Semantic enhanced · BERT · Fake news detection · COVID-19 News

1 Introduction

The unprecedented COVID-19 pandemic has raged around the world for more than two years. Countries have developed measures to flatten the curve, such as closing borders, banning people from gathering, working from home, severely

X. Song et al. (Eds.): APWeb-WAIM 2023, LNCS 14331, pp. 1–15, 2024.
https://doi.org/10.1007/978-981-97-2303-4_1

impacting people's daily lives. During the pandemic, people increasingly rely on the Internet to work, study, access the latest information, communicate with others, and more. A new report finds that people worldwide have been spending much more time online since the COVID-19 pandemic, and the trend shows no signs of slowing down[1]. People use the Internet to obtain the latest news, including the spread of the pandemic, measures issued by government departments, the latest policies, preventive measures announced by medical institutions. The fight against the pandemic is made more challenging by the fact that fake news about the pandemic is also spreading across different platforms at an astonishing speed, making it even harder to fight it in the first place.

Fake news has been causing severe consequences since the beginning of the pandemic. In the first three months of 2020, there may have been at least 800 deaths related to misinformation about Coronavirus[2]. Such as "Coronavirus is a BIOWEAPON and made in a lab", "The flu shot increases by 36% the risk of having the coronavirus", "A hospital says consuming alcohol kills the coronavirus". As fake news spreads globally, it is becoming increasingly clear that it has serious consequences at many levels. For example, rumors of drinking raw alcohol as a cure kills hundreds of Iranians [15], the conspiracy that "5G virus" in the UK has resulted in the burning of 77 cell phone towers as a result of the conspiracy[3], deaths caused by poisoning have increased since Trump endorsed bleach and disinfectants as treatments for COVID-19 in the US[4]. It has been revealed by Reuters that from January to March of 2020, there have been 900% more fact-checks in English than ever before.[5]. Fake news threatens social harmony, democracy, and public health, especially during pandemics, and is viewed by all as a public enemy.

Such information poses a severe threat to communities, misleads people into risky behaviors that may damage health, undermines public health responses, threatens social stability, and leads to distrust in health authorities and government agencies. Thus, it is urgent to develop a tool that can verify the authenticity of information to prevent fake news from further dissemination. This study aims to propose a novel BERT-based framework for misinformation detection coupled with an enhanced semantic module. The following are the main contributions of this work.

– We propose an approach based on the BERT algorithm for the detection of fake news, which uses text as input and a topic extraction module to complement the text representation to improve the accuracy of fake news detection.

---

[1] https://datareportal.com/reports/digital-2022-global-overview-report.

[2] https://www.bbc.com/news/world-53755067.

[3] https://www.businessinsider.com/77-phone-masts-fire-coronavirus-5g-conspiracy-theory-2020-5.

[4] https://www.indiatvnews.com/news/world/us-coronavirus-deaths-by-bleach-disinfectant-injection-major-rise-trump-covid-19-treatment-616708.

[5] https://reutersinstitute.politics.ox.ac.uk/types-sources-and-claims-covid-19-misinformation.

– We employ two strategies to learn the topic information of the text. The first is the Latent Dirichlet Allocation, which is based on bag-of-words and is a statistical method for identifying abstract "topics" that exist in a collection of documents. In the second technique, text topics are extracted using a clustering technique, which is based on sentence embedding.
– Extensive experiments show that extracting text topics as additional features can improve text representation ability, and the topics generated by the sentence embedding-based clustering algorithm (K-means clustering) are more effective.

Following is the structure of the remaining sections. Section 2 discusses some existing work related to COVID-19 fake news detection. Section 3 explains the proposed methodology this study follows. Results and parameter analysis are presented in Sect. 4. Finally, this work is concluded in Sect. 5 with an outlook on the future.

2   Related Work

Social media platforms have penetrated into the daily lives of people. As of January 2022, the top three social networks by active users are Facebook, YouTube and WhatsApp.[6]. It must be pointed out that while social media platforms facilitate access to information, they also greatly facilitate the spread of fake news [14,18,20,23]. False news travels faster than real stories[7], especially during the outbreak. Fake news has also hindered the fight against the outbreak during the COVID-19 pandemic, so in the following sections, we will discuss the data collection related to COVID-19, the current methods for fake news detection, and the working mechanism of the BERT model.

2.1   COVID-19 Fake News Collection

According to the sources, COVID-19 related information on the Internet falls into three categories. They are (1) fact-checking websites such as Poynter, Snopes; (2) authoritative websites (e.g., WHO, Reuters, BBC); (3) various social media websites (e.g., YouTube, Twitter, Facebook), the formats including text, videos, images, and URLs. Many researchers are committed to fighting the epidemic, there are already many multi-modal COVID-19 misinformation datasets based on different languages and platforms available for free use [19,21,22,24,25]. Such as Banda et al. [2] collected more than 1.12 billion tweets in 5 languages (English, French, Spanish, German) from January 1, 2020 to June 27, 2021, for extensive research by researchers around the world. Such as emotional and psychological responses to measures, identifying sources of misinformation, and more.

---

[6] https://datareportal.com/social-media-users.
[7] https://news.mit.edu/2018/study-twitter-false-news-travels-faster-true-stories-0308.

2.2 COVID-19 Fake News Detection

In existing studies, traditional machine learning algorithms (e.g., LR, SVM, RF) are always used as baseline methods or for the initial exploration of the collected datasets; deep learning algorithms have been extensively studied and deeply explored. Cui et al. [4] adopted different traditional machine learning methods (SVM, LR, RF) on the collected dataset CoAID for binary classification tasks (true or false). Kar et al. [8] proposed a multilingual BERT embedding model to detect fake tweets, then trained the model on multiple Indic-Languages fake news datasets. Kumari et al. [9] proposed a multitask learning misinformation detection framework (LSTM-based, BERT-based). The results revealed an improvement in the fake detection with two auxiliary tasks: emotion detection and novelty detection. Glazkova et al. [6] adopted an ensemble of three CT-BERT models and achieved the first place in the leaderboard at the Constraint@AAAI2021 shared task (English). Al-Rakhami et al. [1] collected over 400,000 tweets and created a stacking-based ensemble learning model. The varieties of transformers they use are: BERT, RoBERTa [11], DistilRoBERTa, ALBERT [10], DeBERTa [7].

2.3 BERT Model

BERT [5] stands for Bidirectional Encoder Representations from Transformers, which is a bidirectional training of the popular attention model transformer [17] for language modeling. It has demonstrated excellent results on a variety of NLP tasks [12,16], including question answering, natural language inference, and others. As we have mentioned in Sect. 2.2, BERT and its variants are widely used in detection tasks, such as RoBERTa, ALBERT, XLM-RoBERTa, mBERT. In the BERT framework, there are two steps: pre-training and fine-tuning. To fine-tune the model, the parameters are first initialized with the pre-trained parameters and then fine-tuned using labeled data from downstream tasks. Pre-trained parameters are obtained from various pre-training tasks and then fine-tuned using labeled data. A "masked language model" (MLM) is used to randomly mask some tokens in the input and predict the masked words only based on their context. This bidirectional strategy allows the MLM objective to fuse left and right contexts, therefore, the word embeddings generated by BERT contain contextual information.

3 Methodology

The proposed approach involves three steps: generating a textual representation of the samples using a language model, extracting the topics of the samples using topic extraction techniques (Semantic Enhancement Module), and concatenating the textual representation and the topic as the input of the vanilla classifier. In Fig. 1, a framework for addressing the three steps has been proposed. Using BERT as the backbone, the upper left part of the model is the language model that generates the embedding of a sentence. Using BERT's encoder mechanism and MLM strategy, individual sentence can be converted to sentence embeddings, preserving both semantic and positional information in the latent space. At the bottom is the semantic enhancement module, which is used to generate the topics of the data samples. In the module, two different strategies are employed to extract topics. The first is Latent Dirichlet Allocation (LDA), a statistical-based (bag of words) topic modeling technique. Another strategy is clustering techniques based on sentence embeddings. In the right part, we have a classifier with fully connected layers, followed by a softmax output layer which outputs the probability of the sample in each label.

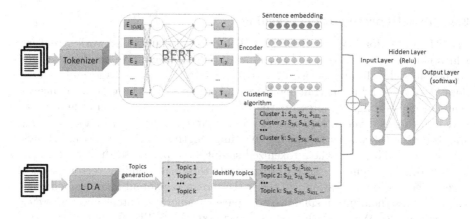

Fig. 1. The structure of the proposed framework.

3.1   Dataset

We select the Contraint@AAAI 2021 COVID-19 Fake news detection dataset in English [13] for the experiment. There are three groups in the original dataset- the training group, validation group, and test group. We combine these groups to get a dataset containing 5,600 real news articles and 5,100 fake news articles from various platforms such as Facebook, Twitter, Instagram. To prepare the data for input into the model, we perform general preprocessing on the raw data, such as lowering the case, and removing stopwords, punctuation, and URLs.

3.2   Problem Statement

This problem can be defined as a classification task. We define a set of news samples S, with a total number of $N \in \mathbb{N}$. Each sample has a label $y = (0,1)$ to indicate the context authenticity (true/false), that is, each sample $\{s_1 \cdots s_n\} \in S$ has a corresponding label $\{y_1 \cdots y_n\} \in Y$. We employ the language model M to produce the embedding of each sample $\{e_1 \cdots e_n\} \in E$, use the topic generation technology C to generate the topic of each sample $\{p_1 \cdots p_k\} \in P$, where k indicates the total number of topics in S and is generated by topic generation technology C, the value of k is different under different topic extraction techniques. The new text representation can be represented as: $E_{new} = E \oplus P$, E is the original sample vector generated by language model and P is the topic of sample, which is convert categorical data to numerical data, $\oplus$ means concatenation. Our purpose is to train a classifier with new text representation E_{new} and label Y and maximize the test accuracy on all test samples. The proposed method is shown in Fig. 1 and detailed in following sections.

3.3   Text Representation Learning

SBERT stands for Sentence-BERT, a BERT based siamese network structure which can generate fixed-sized vectors for input sentences. We employ the pre-trained model "all-mpnet-base-v2" which was trained on a large and diverse dataset of over 1 billion training pairs and achieved the best performance in sentence embedding. S represents the raw dataset, we first convert the original text into tokens, refer to Equation (1). w_{ij} represents the word in a sample, t_{ij} stands for the token after tokenizing (padding length is 512), the "arrow" stands for tokenizer. Then the whole tokenized samples are passed to the transformer-based model M (Sentence BERT) to generate the corresponding vector for each token. The pooling operation will then take the average of all token embeddings and compress them into a single vector space to generate sentence vectors, refer to Equation (2). We input the preprocessed text E into SBERT and obtain the embedding for each sample (denoted as e_n) with a dimension of $1 \times H$ ($H = 768$), for the whole dataset E is an $N \times H$ matrix, N indicates the total number of samples, n represents the n_{th} sample.

$$
S = \begin{bmatrix} w_{11} & w_{12} & \cdots & w_{1k} \\ w_{21} & w_{22} & \cdots & w_{2k} \\ \cdots & \cdots & \cdots & \cdots \\ w_{(n-1)1} & w_{(n-1)2} & \cdots & w_{(n-1)k} \\ w_{n1} & w_{n2} & \cdots & w_{nk} \end{bmatrix} \Rightarrow \begin{bmatrix} t_{11} & t_{12} & \cdots & t_{1k} \\ t_{21} & t_{22} & \cdots & t_{2k} \\ \cdots & \cdots & \cdots & \cdots \\ t_{(n-1)1} & t_{(n-1)2} & \cdots & t_{(n-1)k} \\ t_{n1} & t_{n2} & \cdots & t_{nk} \end{bmatrix} \tag{1}
$$

$$
E = M \begin{bmatrix} t_{11} & t_{12} & \cdots & t_{1k} \\ t_{21} & t_{22} & \cdots & t_{2k} \\ \cdots & \cdots & \cdots & \cdots \\ t_{(n-1)1} & t_{(n-1)2} & \cdots & t_{(n-1)k} \\ t_{n1} & t_{n2} & \cdots & t_{nk} \end{bmatrix} = \begin{bmatrix} e_1 \\ e_2 \\ \cdots \\ e_{n-1} \\ e_n \end{bmatrix} = \begin{bmatrix} 0.5808 & \cdots \cdots & 0.504 \\ 0.823 & \cdots \cdots & 0.419 \\ \cdots & \cdots \cdots & \cdots \\ -0.557 & \cdots \cdots & 0.169 \\ 1.284 & \cdots \cdots & -206 \end{bmatrix} \tag{2}
$$

3.4   Topic Generation

Each piece of news seems to be independent, but in fact, they are related and under a certain topic. This distinguishes it from other texts, especially in terms of the authenticity of its content. For this reason, we extract topic features from samples according to semantics by using topic extraction and clustering technology. The topic features in this study are generated using two strategies. One is the traditional topic modeling LDA [3] based on statistical model. The implementation of this technology is shown in the lower part of Fig. 1, and we classify the samples into different topics according to their probability of topic distribution, such as $\{s_2, s_{85}, s_{2065}\} \in p_5$. The other is sentence embedding-based clustering technique that utilizes the spatial location information contained in sentence vectors. Under this strategy, we cluster the samples using the density-based clustering algorithm HDBSCAN and the K-means clustering technique. The clustering algorithm is applied after the language model generates the text representation, i.e. E, we cluster the samples in the latent space, thinking that the samples in each cluster belong to the same topic, such as $\{s_9, s_{104}, s_{465}\} \in p_{12}$.

Latent Dirichlet Allocation. Latent Dirichlet Allocation (LDA) is a classical topic model used to classify text into categories in a document. LDA is a probabilistic method that first learns the number k of topics present in all documents, then computes the probability distribution of each document over k topics, and categories the document into topic with the highest probability. It creates a topic per document and word per topic model based on Dirichlet distributions. We use the coherence score (Equation (3)) to determine the optimal number of topics:

$$Coherence = \sum_{i<j} score(w_i, w_j). \tag{3}$$

The coherence selects top n frequently occurring words in each topic, then aggregates all the pairwise scores of the top n words $w_i, \cdots, w_n$ of the topic. We set the number of topics in range (1–30), and use the "coherencemodel" in the open source library "gensim" to calculate the coherence score, and we get the highest coherence score when the number of topics is 20 (Alpha = 0.6, Beta = 0.9). We set the number of topic to 20 and employ LDA model to generate topic for each samples, $\{p_1 \cdots p_k\} \in P$ where k represents the topic number. After using a one-hot encoder, we get a topic matrix P of data samples of dimension $N * 20$, where N is the number of samples.

HDBSCAN. HDBSCAN stands for Hierarchical Density-Based Spatial Clustering of Applications with Noise. Its prototype uses a density-based clustering algorithm, DBSCAN, to find any arbitrary spatial database buried within noisy data by taking the largest set of points connected by density and grouping them into clusters. A hierarchical clustering algorithm called HDBSCAN extends DBSCAN by creating a flat clustering system based on cluster stability and will not force outliers into any clusters. Compared with the DBSCAN algorithm, HDBSCAN defines a way to measure the distance between two points mutual reachability distance, which is defined as follows:

$$d_{mreach-k}(a, b) = max\, \{core_k(a), core_k(b), distance(a, b)\} \tag{4}$$

where parameter k is the core distance and $core_k(a)$ represents the distance between core and point a, $distance(a, b)$ is the original metric distance between a and b. Under this metric, dense points close to the core maintain the original distance, but sparse points farther away from the core are pushed away, ensuring their core distance from any other points. This way makes the boundaries of clusters in low-dimensional spaces clearer.

With the same parameter setting with BERTopic, HDBSCAN generates 85 topics and the visualization results are shown in Fig. 2. We can see that some outlines (grey dots) do not belong to any topic, increasing the parameter in HDBSCAN min_cluster_size can get fewer topics. After using a one-hot encoder, we get a topic matrix P of data samples of dimension $N*85$, where N represents the total number of samples.

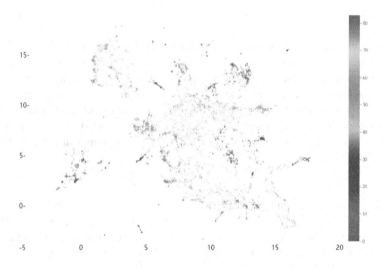

Fig. 2. Visualization of clustering results under HDBSCAN.

K-Means Clustering. As a method of vector quantization, K-means clustering offers a way to partition a set of observations $(x_1, x_2, ..., x_n)$ into $k(\leq n)$ clusters $S = \{S_1, S_2, ..., S_k\}$ in which each observation belongs to the cluster with the nearest mean (cluster centers or cluster centroid), which serves as a prototype for a cluster[8].

$$argmin \sum_{i=1}^{k} \sum_{x \in S_i} \| x - u_i \|^2 = argmin \sum_{i=1}^{k} |S_i| Var S_i \qquad (5)$$

where μ_i is the mean of points in S_i. In K-means, a cluster refers to a collection of data points aggregated together because of certain similarities. In this study, the

---

[8] https://en.wikipedia.org/wiki/K-means_clustering.

sentence embeddings contain both semantic and spatial location information of sentences, thus the samples aggregated together because they have high semantic similarity. We employ Elbow Method[9] in Yellowbrick to select the optimal k in this algorithm. As shown in Fig. 3, the optimal number of topic is 14, so we set k to 14 and then apply "MiniBatchKMeans" in sklearn to finish the K-means clustering and generate topic for each sample. After using a one-hot encoder, we get a topic matrix P of data samples of dimension $N * 14$, where N represents the total number of samples.

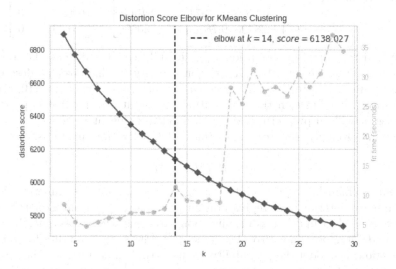

Fig. 3. Distortion score elbow for K-means clustering to find the optimal the number of topics.

The advantage of the proposed model is that we do not have to dive into what each topic is talking about, but we simply need to identify that cluster to which the data sample belongs and distinguish it from the other clusters. That is, we only need to know the topic/cluster to which the sample belongs and use the topic/cluster number as the semantic enhancement of the sample.

3.5 Classifier Design

In Sect. 3.3, we employ the SBERT with the pretrained model "all-mpnet-base-v2" to generate sentence embeddings E of dimension $N \times H$ (in this case, $H = 768$). In Sect. 3.4, We adopt three methods to generate the topic for each sentence $\{p_1 \cdots p_k\} \in P$ and the number of topics k under the three methods is different, LDA categories the samples into 20 topics, K-means clustering into 14 topics and HDBSCAN into 85 topics. Then we convert topic numbers to vectors using

---

[9] https://www.scikit-yb.org/en/latest/api/cluster/elbow.html?highlight=elbow.

one-hot encoding, $p_n = [0, 0, 1, 0, ..., 0]$, n represents the n_{th} sample's topic. We concatenate the topic feature P with the original text representation E as a new text representation E_{new}. Where E is the $N \times H$ matrix, P is the $N \times k$ matrix and the value of k is different under the three topic generation methods, thus the dimension of new text representation E_{new} is $N \times (H + k)$.

$$E_{new} = E \ concate \ T = \begin{bmatrix} 0.58108 & -2.7709 & \cdots & 0.50074 \\ 0.8123 & -1.2817 & \cdots & 0.41439 \\ \cdots & \cdots & \cdots & \cdots \\ -0.5578 & 0.4974 & \cdots & 0.1069 \\ 1.28458 & 0.78436 & \cdots & -20606 \end{bmatrix} \oplus \begin{bmatrix} 0 & 0 & 0 & \cdots & 1 \\ 0 & 0 & 0 & \cdots & 0 \\ & & \cdots & \\ 0 & 1 & 0 & \cdots & 0 \\ 1 & 0 & 0 & \cdots & 0 \end{bmatrix} \tag{6}$$

We feed the text representation E_{new} in fully connected dense layer with 512 hidden units, we introduce ReLU non-linearity to this layer as described in Eq. (7). ReLU, or rectified linear activation function ($f(x) = max(0, x)$), is a piecewise linear function that will output the input directly if it is positive; otherwise, it will output zero. Many neural networks use it because it is easier to train and often gives better results than other activation functions.

$$Q = ReLU(E_{new}) \tag{7}$$

We compute final label probabilities using a standard softmax function using the non-linear intermediate representation (denoted as Q) from Equation (7). Softmax is used in multi-class classification problems where class membership is required for more than two class labels, but it is also suitable for binary classification, to model the probability of the output in each class. The output of this softmax layer P has a dimension of $N \times 2$, N being the total number of samples, and 2 representing the number of labels (true/false).

$$P = Softmax(Q) \tag{8}$$

For the model training, we adopt the binary cross-entropy loss function to update the classifier and the learning rate is set to 0.005 and the epochs is set to 10. We employ Accuracy, Precision, Recall and F1 score to evaluate the performance of the models.

$$Loss = -\frac{1}{N} \sum_{i=1}^{N} y_i \cdot log(p(y_i)) + (1 - y_i) \cdot log(1 - p(y_i)) \tag{9}$$

where y_i ($i \in [0, 1]$)is the label and $p(y_i)$ is the predicted probability of the sample belongs to y_i for all N samples. In this study, the output is two values, corresponding to the possibility that the sample belongs to the true and false categories. For multi-class classification, we adopt the categorical crossentropy:

$$Loss = -\frac{1}{N} \sum_{i=1}^{N} \sum_{c=1}^{C} 1_{y_i \in C_c} log p_{model}[y_i \in C_c] \tag{10}$$

4   Experimental Results and Parameter Analysis

We discussed the proposed method and the specific implementation steps in the previous section. Below we will present the experimental results for each method as well as the effect of parameter settings on classification.

4.1   Experimental Results

We employ evaluation metrics to determine the performance of a binary or multi-class classification task, accuracy, precision, recall and F1 score. We choose Decision Tree(DT) as baseline methods with TF-IDF transformer, and set vocabulary size to 10,000. We utilize ELMo (Embedding from Language Model) and Long Short Term Memory (LSTM) as our deep learning model baseline and the early preprocessing stage is the same as BERT. We add a semantic enhancement module to the top-performing SBERT classifier, using three different topic extraction strategies to find new semantic feature, they are SBERT_L, SBERT_H and SBERT_K, respectively. The three strategies are based on different mechanisms and generate different numbers of topics across the dataset, 20, 85 and 14, respectively. According to the experimental results, semantically enhanced text representation can be useful in improving the accuracy of classification. The K-means clustering technique gave the best results among the three strategies that were proposed for extracting topics (Table 1).

Table 1. Experiment results (5 Fold cross-validation) in terms of accuracy, precision, recall and F1 score (%), L stands for LDA, K stands for K-means, H stands for HDB-SCAN.

Model	Accuracy	Precision(Macro)	Recall(Macro)	F1 score(Macro)
DT	81.38	81.44	80.41	80.92
LSTM	87.22	82.00	86.55	83.51
ELMo	90.05	90.25	90.13	90.17
SBERT	90.58	90.51	90.62	90.55
SBERT_L	91.07	91.25	91.16	91.07
SBERT_H	91.12	91.23	91.33	91.15
SBERT_K	**93.50**	**93.48**	**93.50**	**93.48**

When considering the complexity of the model, SBERT_H is the most complicated because we need to reduce the dimensionality before using HDBSCAN. UMAP has two parameters that need to be adjusted, n_neighbours determines the size of the local neighbors, and n_components, which determines the target dimension. Information is lost when the target dimension is too low, and clustering results are worse when the target dimension is too high. In addition, we need to set min_cluster_size in HDBSCAN to determine the minimum size of the cluster. The algorithm can't use a perfect scientific standard as a reference, which makes human interference very significant, and even the slightest deviation from the standard parameter may affect the clustering result.

4.2 Parameter Analysis

Since the proposed SBERT_K method performs the best, we further investigate this method to understand the impact of parameters on model performance.

Dense Layer Setting. In the proposed structure shown in Fig. 1, the classifier is composed of one fully connected hidden layers with 512 hidden units with "ReLU" activation, and the last dense layer is a softmax layer for generating class probabilities with the number of categories (2 in this study, fake or real). As part of our testing of the dense layer settings, we set various layer sizes and the number of units to determine how they affect classification results. The results are shown in Fig. 4, and we can see that the changes in layers did not cause significant fluctuations in the results, the values of all metrics still fluctuate between 92% and 94%.

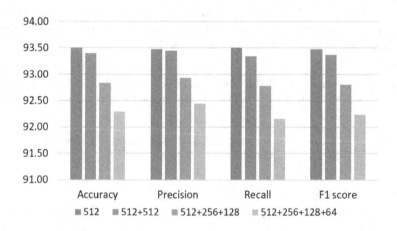

Fig. 4. Performance comparison under different layer setting.

Hyper-parameter Analysis. The learning rate is a hyper-parameter that controls how much the model should change when its weights are updated. In selecting the learning rate, a small value could result in a long training process that gets stuck, while a large value may lead to an unstable training program or learning of a sub-optimal set of weights. We set different learning rates to see the change in running time and the impact on the performance. In table 2, we can see that the model achieves the best performance when the learning rate is 0.01.

Table 2. Comparison of experimental results (5 fold cross-validation) under different learning rates in the SBERT_K model, including accuracy, precision, recall, F1 score (%) and running time.

Learning rate	Accuracy	Precision	Recall	F1 score	Time
0.0001	91.31	91.30	91.28	91.28	57.115 s
0.0005	92.66	92.71	92.67	92.65	57.763 s
0.001	92.98	93.00	92.97	92.96	60.652 s
0.005	93.46	93.44	93.50	93.45	77.955 s
0.01	93.77	93.75	93.76	93.75	58.581 s
0.05	92.64	92.64	92.64	92.62	67.169 s

5 Conclusion

In this work, we propose a BERT-based semantic enhancement model for COVID-19 fake news detection. The model consists of three parts: text representation module, topic generation module and MLP classifier. The text representation module process raw text into vectors, the semantic enhancement module uses three different techniques to generate topics for the text, then use topic features to supplement text representations. The new representation will then be fed into the classifier. The experiment results haven demonstrated that topics generated by K-means clustering algorithm is more effective in short text semantic enhancement, and the new representation improves classifier performance by more than 3% in four evaluation matrices while with less complexity and running time. For deep neural network-based classifiers, a simple dense layer structure and a few units can achieve better results with less structural complexity and running time in NLP tasks. The designed model can enhance the short text representation and is equally applicable to other short text classification tasks.

Considering the importance of this work, we are hoping to extend it in the following directions in the future:

- How to optimize the model so that the deep learning-based classifier can achieve optimal results even with small datasets?
- Are there other modalities of data that can improve classifier performance without increasing model complexity, such as publisher, publication time, and image in the data sample?
- Is the proposed method suitable for multilingual fake news detection and does it require special modifications to the model?

Acknowledgments. This work is supported by the Study Melbourne Research Partnership program which has been made possible by funding from the Government of Victoria through Study Melbourne (Project ID: veski-SMRP#1906).

References

1. Al-Rakhami, M., Alamri, A.: Lies kill, facts save: detecting COVID-19 misinformation in twitter. IEEE Access **8**, 155961–155970 (2020). https://doi.org/10.1109/ACCESS.2020.3019600
2. Banda, J.M., et al.: A large-scale COVID-19 twitter chatter dataset for open scientific research - an international collaboration. CoRR **abs/2004.03688** (2020). https://arxiv.org/abs/2004.03688
3. Blei, D.M., Ng, A.Y., Jordan, M.I.: Latent dirichlet allocation. J. Mach. Learn. Res. **3**, 993–1022 (2003)
4. Cui, L., Lee, D.: Coaid: COVID-19 healthcare misinformation dataset. CoRR abs/2006.00885 (2020). https://arxiv.org/abs/2006.00885
5. Devlin, J., Chang, M., Lee, K., Toutanova, K.: BERT: pre-training of deep bidirectional transformers for language understanding. In: Burstein, J., Doran, C., Solorio, T. (eds.) Proceedings of the 2019 Conference of the North American Chapter of the Association for Computational Linguistics: Human Language Technologies, NAACL-HLT 2019, Minneapolis, MN, USA, June 2–7, 2019, Volume 1 (Long and Short Papers). pp. 4171–4186. Association for Computational Linguistics (2019). https://doi.org/10.18653/v1/n19-1423
6. Glazkova, A., Glazkov, M., Trifonov, T.: g2tmn at constraint@aaai2021: Exploiting CT-BERT and ensembling learning for COVID-19 fake news detection. CoRR abs/2012.11967 (2020). https://arxiv.org/abs/2012.11967
7. He, P., Liu, X., Gao, J., Chen, W.: Deberta: decoding-enhanced BERT with disentangled attention. CoRR abs/2006.03654 (2020). https://arxiv.org/abs/2006.03654
8. Kar, D., Bhardwaj, M., Samanta, S., Azad, A.P.: No rumours please! A multi-indiclingual approach for COVID fake-tweet detection. CoRR abs/2010.06906 (2020). https://arxiv.org/abs/2010.06906
9. Kumari, R., Ashok, N., Ghosal, T., Ekbal, A.: Misinformation detection using multitask learning with mutual learning for novelty detection and emotion recognition. Inf. Process. Manag. **58**(5), 102631 (2021). https://doi.org/10.1016/j.ipm.2021.102631
10. Lan, Z., Chen, M., Goodman, S., Gimpel, K., Sharma, P., Soricut, R.: ALBERT: a lite BERT for self-supervised learning of language representations. In: 8th International Conference on Learning Representations, ICLR 2020, Addis Ababa, Ethiopia, April 26–30, 2020. OpenReview.net (2020). https://openreview.net/forum?id=H1eA7AEtvS
11. Liu, Y., et al.: Roberta: a robustly optimized BERT pretraining approach. CoRR abs/1907.11692 (2019). http://arxiv.org/abs/1907.11692
12. Mitra, S., Banerjee, S., Naskar, M.K.: Remodelling correlation: a fault resilient technique of correlation sensitive stochastic designs. Array **15**, 100219 (2022). https://doi.org/10.1016/j.array.2022.100219
13. Patwa, P., et al.: Fighting an infodemic: COVID-19 fake news dataset. In: Chakraborty, T., Shu, K., Bernard, H.R., Liu, H., Akhtar, M.S. (eds.) Combating Online Hostile Posts in Regional Languages during Emergency Situation - First International Workshop, CONSTRAINT 2021, Collocated with AAAI 2021, Virtual Event, February 8, 2021, Revised Selected Papers. Communications in Computer and Information Science, vol. 1402, pp. 21–29. Springer (2021https://doi.org/10.1007/978-3-030-73696-5_3

14. Song, X., Li, J., Lei, Q., Zhao, W., Chen, Y., Mian, A.: Bi-clkt: Bi-graph contrastive learning based knowledge tracing. Knowl. Based Syst. **241**, 108274 (2022). https://doi.org/10.1016/j.knosys.2022.108274

15. Tanne, J.H., Hayasaki, E., Zastrow, M., Pulla, P., Smith, P., Rada, A.G.: Covid-19: how doctors and healthcare systems are tackling coronavirus worldwide. BMJ **368** (2020)

16. Ueda, I., Shishido, H., Kitahara, I.: Spatio-temporal aggregation of skeletal motion features for human motion prediction. Array **15**, 100212 (2022). https://doi.org/10.1016/j.array.2022.100212

17. Vaswani, A., et al.: Attention is all you need. In: Advances in Neural Information Processing Systems, vol. 30 (2017)

18. Xia, J., Li, M., Tang, Y., Yang, S.: Course map learning with graph convolutional network based on AUCM. World Wide Web 1–20 (2023).https://doi.org/10.1007/s11280-023-01194-8

19. Yang, S., Jiang, J., Pal, A., Yu, K., Chen, F., Yu, S.: Analysis and insights for myths circulating on twitter during the COVID-19 pandemic. IEEE Open J. Comput. Soc. **1**, 209–219 (2020). https://doi.org/10.1109/OJCS.2020.3028573

20. Yin, H., Song, X., Yang, S., Li, J.: Sentiment analysis and topic modeling for COVID-19 vaccine discussions. World Wide Web **25**(3), 1067–1083 (2022). https://doi.org/10.1007/s11280-022-01029-y

21. Yin, H., Yang, S., Li, J.: Detecting topic and sentiment dynamics due to COVID-19 pandemic using social media. In: Yang, X., Wang, C.-D., Islam, M.S., Zhang, Z. (eds.) ADMA 2020. LNCS (LNAI), vol. 12447, pp. 610–623. Springer, Cham (2020). https://doi.org/10.1007/978-3-030-65390-3_46

22. Zhang, F., Wang, X., Li, Z., Li, J.: Transrhs: a representation learning method for knowledge graphs with relation hierarchical structure (2020). https://doi.org/10.24963/ijcai.2020/413

23. Zhang, M., Wang, G., Ren, L., Li, J., Deng, K., Zhang, B.: Metonr: a meta explanation triplet oriented news recommendation model. Knowl. Based Syst. **238**, 107922 (2022). https://doi.org/10.1016/j.knosys.2021.107922

24. Zhou, J., Yang, S., Xiao, C., Chen, F.: Examination of community sentiment dynamics due to COVID-19 pandemic: a case study from a state in Australia. SN Comput. Sci. **2**(3), 201 (2021). https://doi.org/10.1007/s42979-021-00596-7

25. Zhou, J., Zogan, H., Yang, S., Jameel, S., Xu, G., Chen, F.: Detecting community depression dynamics due to COVID-19 pandemic in Australia. IEEE Trans. Comput. Soc. Syst. **8**(4), 982–991 (2021). https://doi.org/10.1109/TCSS.2020.3047604

Mining Frequent Geo-Subgraphs in a Knowledge Graph

Yixin Wu[1], Jingyan Huang[1], Dingming Wu[1($\boxtimes$)], Christian S. Jensen[2], and Kezhong Lu[1]

[1] College of Computer Science and Software Engineering, Shenzhen University, Shenzhen, China
{2070276039,2210274051}@email.szu.edu.cn, {dingming,kzlu}@szu.edu.cn
[2] Department of Computer Science, Aalborg University, Aalborg, Denmark
csj@cs.aau.dk

Abstract. Frequent subgraph mining aims to find all subgraphs that occur frequently in a graph database or in a single large graph. It finds applications in social networks, citation networks, protein interaction networks, etc. This paper studies frequent subgraph mining in a knowledge graph, where some vertices are associated with geo-coordinates, called geo-vertices. We introduce geo-constraints and propose the problem of frequent geo-subgraph (FGS) mining to find each subgraph in a knowledge graph that (i) contains at least two geo-vertices, (ii) has no pairs of geo-vertices with a distance that exceeds τ, and (iii) has support at least δ. Such frequent geo-subgraphs can be used to analyze the relationships among geo-entities in spatial regions, enabling downstream applications such as classification, clustering, and recommendation of geo-entities. To solve the FGS problem in a large knowledge graph efficiently, we propose algorithm FreGeoSubgraphAlg and three optimizations. Extensive evaluations on real datasets demonstrate that the proposed algorithms can solve the FGS problem efficiently on large graphs and outperform the state-of-the-art algorithm by several orders of magnitude.

1 Introduction

Frequent subgraph mining aims to find all subgraphs that occur frequently in a graph database or in a single large graph. This functionality is used in protein function discovery [19], game recommendation [3], software engineering [11], chemical compound classification [8], etc.

A knowledge graph consists of a set of triples of the form (h, r, t), indicating that relationship r holds between entities h and t. A knowledge graph is thus a graph where the vertices denote entities and the edges represent relationships between entities. Existing knowledge graphs such as DBpedia [5], Yago [20], and Freebase [6] contain millions to billions of triples. Frequent subgraph mining has been applied to knowledge graphs. For instance, kCP [14] finds top-k frequent subgraphs based on a given core pattern.

© The Author(s), under exclusive license to Springer Nature Singapore Pte Ltd. 2024
X. Song et al. (Eds.): APWeb-WAIM 2023, LNCS 14331, pp. 16–31, 2024.
https://doi.org/10.1007/978-981-97-2303-4_2

Knowledge graphs often contain large numbers of vertices with associated geo-coordinates, called geo-vertices. This paper extends the definition of a frequent subgraph by integrating a geo-constraint and proposes the problem of frequent geo-subgraph (FGS) mining: finding each subgraph in a knowledge graph that satisfies three conditions: (i) it contains at least two geo-vertices, (ii) the distances between pairs of geo-vertices are at most τ, and (iii) the support is at least δ. Frequent geo-subgraph mining can be used to analyze relationships among multiple geo-entities in spatial regions and enables downstream applications such as classification, clustering, and recommendation of geo-entities.

A straightforward solution to the FGS problem is to first use an existing frequent subgraph mining algorithm, e.g., GRAMI [9], to find all subgraphs with support at least δ and then check whether each satisfies the other two conditions. However, this is inefficient and fails to scale to large graphs (shown in the experiments). The key idea in most frequent subgraph mining algorithms is as follows. Based on the anti-monotone property, first find small frequent subgraphs and then consider extensions of these subgraphs as candidates. The next step is to compute the isomorphisms of the candidates and to determine whether the candidates are frequent. The resulting frequent subgraphs are then extended into larger candidates. This procedure is repeated until no new frequent subgraphs can be found. This idea has two limitations: (i) generated candidates may not be frequent subgraphs, making it a waste to compute their isomorphisms; and (ii) given a candidate subgraph, it is expensive to find isomorphisms that satisfy both the geo-constraint and the support threshold. We propose algorithms that tackle these limitations and support efficient computation of frequent geo-subgraphs in a large knowledge graph.

The main contributions are summarized as follows. We propose the problem of frequent geo-subgraph mining in a knowledge graph, which is a variant of frequent subgraph mining. Based on the anti-monotone property, we propose algorithm FreGeoSubgraphAlg that is able to compute frequent geo-subgraphs in a large knowledge graph. To tackle limitation (i), we propose arc consistency based candidate generation to reduce unnecessarily generated candidates. To tackle limitation (ii), we propose image vertex reuse and geo-grid based vertex ordering that are able to quickly find isomorphisms satisfying thresholds τ and δ. Extensive experiments on real datasets demonstrate the efficiency of the proposed algorithms and offer evidence that the algorithms can outperform the state-of-the-art algorithm by several orders of magnitude.

The rest of the paper is organized as follows. Section 2 formalizes the frequent geo-subgraph mining problem. Section 3 presents the FreGeoSubgraphAlg algorithm, and Sect. 4 proposes the three optimizations. Section 5 covers the experimental evaluation. Section 6 reviews related work, and Sect. 7 concludes.

2   Problem Definition

This section presents the preliminaries and defines the problem of frequent geo-subgraph mining.

A knowledge graph can be modeled as a labeled graph $G = (V, E, L)$ where each vertex $v \in V$ represents an entity, each edge $e = (u, v) \in E$ represents a relationship between a pair of entities, and function $L()$ assigns labels to vertices and edges. The label $L(v)$ of a vertex v indicates the type or category of v. The label $L(u, v)$ of an edge (u, v) is a predicate satisfied by u and v. Some vertices have associated geo-coordinates and are called *geo-vertices*. A graph $S = (V_S, E_S, L_S)$ is a subgraph of G iff $V_S \subseteq V$, $E_S \subseteq E$, $\forall v \in V_S$ $(L_S(v) = L(v))$, and $\forall e \in E_S$ $(L_S(e) = L(e))$.

A subgraph isomorphism of S to G is an injective function $f : V_S \to V$, such that (i) $\forall v \in V_S$ $(L_S(v) = L(f(v)))$ and (ii) $\forall (u, v) \in E_S$ $((f(u), f(v)) \in E \wedge L_S(u, v) = L(f(u), f(v)))$. Given an isomorphism f_i of a subgraph S, $f_i(v), v \in V_S$ is called the *image vertex* of v, and $f_i(S)$ is called the *image graph* of S. The terms image graph and isomorphism are used interchangeably in the rest of the paper. Let $\{f_1, \cdots, f_m\}$ be the set of isomorphisms of a subgraph $S = (V_S, E_S, L_S)$ in a graph G and let $F(v) = \{f_1(v), \cdots, f_m(v)\}$ be the image vertex set of v. The minimum image based support (MNI) of S in G, denoted by $\mathsf{MNI}(S, G)$, is defined as $\mathsf{MNI}(S, G) = \min_{v \in V_S}\{|F(v)|\}$.

Example 1. Figure 1 shows a knowledge graph G with eight vertices $v_1, v_2, \cdots, v_8$ and seven edges. There are two vertex labels (CO and CY) and three edge labels (RI, HB, and HI). Vertices v_1, v_3, v_6, and v_8 are geo-vertices. $S_1 = (u_1 \overset{\text{HB}}{——} u_2)$ is a subgraph of G. There are two subgraph isomorphisms of S_1 to G: $f_1(S_1) = (v_5 \overset{\text{HB}}{——} v_1)$ and $f_2(S_1) = (v_5 \overset{\text{HB}}{——} v_6)$. The image vertex set of u_1 in S_1 is $F(u_1) = \{v_5\}$. The image vertex set of u_2 in S_1 is $F(u_2) = \{v_1, v_6\}$. Thus, we have $\mathsf{MNI}(S_1, G) = 1$.

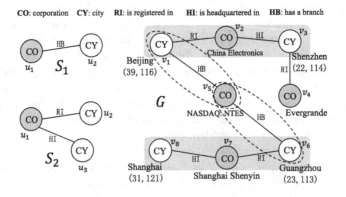

Fig. 1. Frequent geo-subgraphs.

Definition 1. *Given a subgraph $S = (V_S, E_S, L_S)$, let $\Lambda \subseteq V_S$ be the set of geo-vertices in S. Subgraph S is called a **geo-subgraph** if $|\Lambda| \geq 2$. The **diameter** of geo-subgraph S is defined as $\mathsf{Dia}(S) = \max_{u,v \in \Lambda}\{\|u - v\|_2\}$, where $\|u - v\|_2$ is the L_2-norm applied to the geo-coordinates of u and v.*

Definition 2. *Given a geo-subgraph S, a subgraph isomorphism f_i of S in a graph G is called a* **geo-constrained subgraph isomorphism** *if* $\text{Dia}(f_i(S)) \leq \tau$.

Definition 3. *Given a geo-subgraph S, let $\{f_1, \cdots, f_m\}$ be the set of geo-constrained subgraph isomorphisms of S in a graph G and let $F_\tau(v) = \{f_1(v), \cdots, f_m(v)\}$ be the image vertex set of v. The* **geo-constrained MNI** *of S in G is defined as follows:* $\text{MNI}_\tau(S, G) = \min_{v \in V_S}\{|F_\tau(v)|\}$.

Definition 4. *Given a graph G, a minimum support threshold δ, and a geo-constraint threshold τ, the* **frequent geo-subgraph isomorphism mining** *problem.is to find all geo-subgraphs S in G such that $\text{MNI}_\tau(S, G) \geq \delta$.*

Different support metrics have been proposed, including MNI [7], MIS [22], HO [10], MI, and MVC [16]. We follow GRAMI [9] and choose to extend MNI that can be computed efficiently and provides a superset of the the results obtained when using other metrics.

Lemma 1. MNI_τ *is an anti-monotone support metric.*

Proof. The MNI is an anti-monotone support metric, meaning that any super-graph of an infrequent graph cannot be frequent. The geo-constrained MNI MNI_τ extends the MNI by imposing a geo-constraint on each isomorphism. MNI_τ is also an anti-monotone support metric. The reason is as follows. Consider a geo-subgraph S that is not a frequent geo-subgraph. Then, for any supergraph S^s of S, if assuming that S^s is a frequent geo-subgraph, we have a set of isomorphisms $\{f_1, \cdots, f_m\}$, where $\forall f_i$ $(\text{Dia}(f_i(S^s)) \leq \tau)$, such that $\min_{v \in V_{S^s}}\{|F_\tau(v)|\} \geq \delta$. Since S is a subgraph of S^s, we can construct m image graphs $\{f_i(S) \mid 1 \leq i \leq m\}$ of S by removing the vertices in $V_{S^s} \setminus V_S$. Then, we have $\forall f_i$ $(\text{Dia}(f_i(S)) \leq \tau)$ and $\min_{v \in V_S}\{|F_\tau(v)|\} \geq \delta$. This means that S is a frequent geo-subgraph, which contradicts the assumption that S is not frequent and thus proves the lemma by contradiction.

Example 2. Consider the knowledge graph G in Fig. 1. $S_2 = (u_2 \xrightarrow{\text{RI}} u_1 \xrightarrow{\text{HI}} u_3)$ is a geo-subgraph of G because it contains geo-vertices u_2 and u_3. Given $\tau = 20$, there are two geo-constrained subgraph isomorphisms of S_2 in G: $f_1(S_2) = (v_1 \xrightarrow{\text{RI}} v_2 \xrightarrow{\text{HI}} v_3)$ and $f_2(S_2) = (v_6 \xrightarrow{\text{RI}} v_7 \xrightarrow{\text{HI}} v_8)$. Their diameters are $\text{Dia}(f_1(S_2)) = 17 < \tau$ and $\text{Dia}(f_2(S_2)) = 11 < \tau$. The image vertex sets of the vertices in S_2 are $F_\tau(u_1) = \{v_2, v_7\}$, $F_\tau(u_2) = \{v_1, v_6\}$, and $F_\tau(u_3) = \{v_3, v_8\}$. Thus, we have $\text{MNI}_\tau(S_2, G) = 2$.

3   Frequent Geo-Subgraph Mining

This section presents algorithm FreGeoSubgraphAlg that solves the problem of frequent geo-subgraph isomorphism mining. It first computes small frequent geo-subgraphs and then utilizes the anti-monotone property of MNI_τ to prune supergraphs of infrequent geo-subgraphs. It adopts the CSP model [9] so that

infrequent geo-subgraphs can be identified quickly without enumerating all isomorphisms. To determine whether a geo-subgraph S is frequent, we need to find
a set of isomorphisms (image graphs) such that (i) the diameter of each isomorphism is at most τ and (ii) the number of image vertices of each vertex in S is
at least δ. FreGeoSubgraphAlg (Algorithm 1) integrates the geo-constraint check
into the process of finding an image graph, so that image graphs that violate
the geo-constraint are detected early. We proceed to present the details of the
algorithm.

FreGeoSubgraphAlg first obtains edge set E^{fre}, where the support of each
edge is no less than δ in graph G. Then, the edges without any geo-vertex and
whose diameters exceed τ are removed from E^{fre}. The remaining edges in E^{fre}
are taken as the initial frequent subgraphs. Using the anti-monotone property,
any frequent geo-subgraph must consist of some of the edges in E^{fre}. Next,
ExtendSubgraph recursively tries to extend each frequent subgraph by adding an
edge in E^{fre}, thus generating a set of candidates. ExtendSubgraph is similar to
function SubgraphExtension in GRAMI [9], but calls IsFreGeo (Algorithm 2) to
determine whether a candidate subgraph S is frequent and does not violate the
geo-constraint. IsFreGeo adopts the CSP model [9] and calls DFS (Algorithm 3) to
find image graphs of S without violating the diameter constraint. DFS conducts
a depth-first search on graph G starting from a given vertex u, which is an image
vertex of v in candidate subgraph S. An image graph sol is initialized as vertex
u. During the traversal, if an encountered vertex can be an image vertex, it is
added to the image graph sol, and the diameter of sol is checked. If the geo-
constraint is not violated, the traversal proceeds. Otherwise, the encountered
vertex is discarded. When each vertex in candidate subgraph S finds an image
vertex, DFS returns TRUE. Otherwise, no image graph containing vertex u exists
in G, and DFS returns FALSE.

Algorithm 1. FreGeoSubgraphAlg(G, δ, τ)

1: result $\leftarrow \emptyset$; $E^{\text{fre}} \leftarrow$ get all frequent edges in G;
2: **for each** edge $e \in E^{\text{fre}}$ **do**
3:     **if** neither of the vertices of e is a geo-vertex **then**
4:         Remove e from E^{fre};
5:     **if** Dia(e) $> \tau$ **then**
6:         Remove e from E^{fre};
7: **for each** edge $e \in E^{\text{fre}}$ **do**
8:     result $\leftarrow$ result $\cup$ ExtendSubgraph($e, G, \delta, \tau, E^{\text{fre}}$);
9:     Remove e from G and E^{fre};
10: Remove non-geo-subgraphs from result;
11: **Return** result;

Example 3. Consider the knowledge graph G in Fig. 1. Given $\tau = 15$ and
$\delta = 1$, we illustrate the process of determining whether S_2 is a frequent
geo-subgraph. The domains of the vertices in S_2 are $D_{u_1} = \{v_2, v_4, v_5, v_7\}$
and $D_{u_2} = D_{u_3} = \{v_1, v_3, v_6, v_8\}$. After applying arc consistency, we have
$D_{u_1} = \{v_2, v_7\}$ and $D_{u_2} = D_{u_3} = \{v_1, v_3, v_6, v_8\}$. Next, IsFreGeo tries to find

isomorphisms of S_2 and updates the image vertex sets of all the vertices in S_2. It first considers v_2 in the domain of u_1. DFS is called to compute an isomorphism that sets $f(u_1) = v_2$, $f(u_2) = v_1$, and $f(u_3) = v_3$. However, the diameter of this image graph is $\mathsf{Dia}(f(S_2)) = \|v_1 - v_3\|_2 = 17 > \tau$. Thus, no isomorphism that maps v_2 to u_1 exists and v_2 is removed from the domain of u_1. Then, consider v_7 in the domain of u_1. DFS is called to compute an isomorphism that sets $f(u_1) = v_7$, $f(u_2) = v_6$, and $f(u_3) = v_8$. The diameter of this image graph is $\mathsf{Dia}(f(S_2)) = \|v_6 - v_8\|_2 = 11 < \tau$. Thus, we have $\mathsf{MNI}_\tau(S_2, G) = 1 = \delta$ and S_2 is a frequent geo-subgraph.

Algorithm 2. IsFreGeo(S, G, δ, τ)

1: Consider S to G CSP; Apply node and arc consistency;
2: **for each** vertex v in subgraph S, whose domain is D_v **do**
3:     count $\leftarrow 0$; Apply arc consistency;
4:     **if** the size of any domain is less than δ **then**
5:         **Return** FALSE;
6:     **for each** vertex $u \in D_v$ **do**
7:         Add u to sol;
8:         **if** u is already marked **then**
9:             count $+ +$;
10:        **else if** DFS$(u, S, G, \tau, \text{sol})$ **then**
11:           Mark all values of sol in the corresponding domains; count(v)++;
12:        **else**
13:           Remove u from D_v;
14:        **if** count $= \delta$ **then**
15:           **Break**;   ▷ Process the next vertex in S at line 3.
16:     **Return** FALSE;   ▷ Cannot find δ image vertices for v.
17: **Return** TRUE;

Algorithm 3. DFS$(u, S, G, \tau, \text{sol})$

1: **if** sol contains an image vertex of each vertex in S **then**
2:     **Return** TRUE;
3: **for each** neighbor vertex u' of u in G **do**
4:     **if** u' can be an image vertex **then**
5:         Add u' to sol;
6:         **if** Dia(sol) $> \tau$ **then**
7:            Remove u' from sol;
8:            **Continue**;
9:         **Return** DFS$(u', S, G, \tau, \text{sol})$;
10: **Return** FALSE;

---

4 Optimizations

Here, we present three optimizations that improve the performance of algorithm FreGeoSubgraphAlg by (i) reducing the number of generated candidate subgraphs and (ii) speeding up the computations of geo-constrained isomorphisms.

4.1   Arc Consistency Based Candidate Generation

ExtendSubgraph generates candidates by adding frequent edges to frequent sub-graphs. These candidates are passed to IsFreGeo to eliminate infrequent geo-subgraphs. Figure 2 illustrates the process of the candidate generation. Given graph G and $\delta = 2$, S_1 is a frequent subgraph and S_2 is a frequent single-edge graph. The vertices $\{v_i\}$ under each vertex u_i are possible image vertices. They make up the domain of u_i in the CSP model [9]. Subgraph S_3 is a candidate generated by combining S_2 and S_1 that share a vertex with label C. In S_3, the domain of u_3 is the union of the domains of u_3 in S_1 and u_4 in S_2, i.e., $D_{u_3} = \{v_3, v_6, v_8\}$. Then, the domain of each vertex in S_3 is no smaller than δ, so S_3 is passed to IsFreGeo to find its isomorphisms. However, it is observed that S_3 is not frequent in G. This shows that ExtendSubgraph is inefficient because it generates candidates that are not frequent meaning that computations for finding isomorphisms of such candidates are wasted.

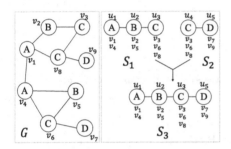

Fig. 2.   Candidate   generation   in
FreGeoSubgraphAlg.

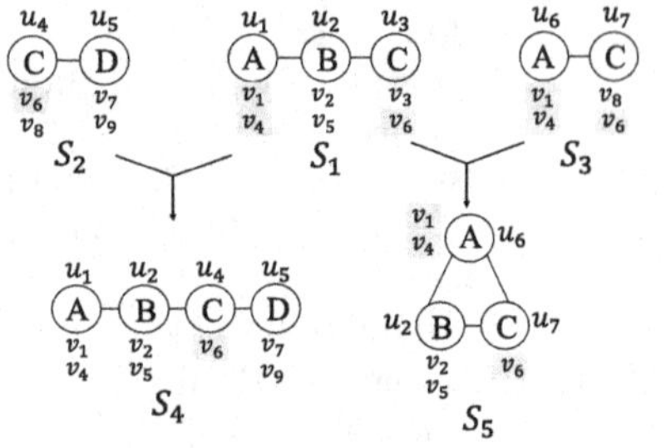

Fig. 3. Arc consistency based candidate generation.

We propose to generate candidates based on arc consistency, aiming to detect infrequent candidates at generation time so that fewer candidates are passed to isomorphism computation. Specifically, given a frequent subgraph $S = (V_S, E_S, L_S)$ and a frequent edge $e = (u, v)$ in graph $G = (V, E, L)$, if there exist $u', v' \in V$ such that $L_S(u) = L(u')$ and $L_S(v) = L(v')$, when adding e to S, the domains of u and v are updated to be $D_u = D_u \cap D_{u'}$ and $D_v = D_v \cap D_{v'}$. If there exists only one vertex in S whose label matches that of one vertex in edge e, e.g., $L_S(u) = L(u')$, when adding e to S, the domain of u is updated to be $D_u = D_u \cap D_{u'}$. After adding the frequent edge to the frequent subgraph, if $|D_u| \geq \delta$ and $|D_v| \geq \delta$, S is a candidate. Otherwise, S is discarded.

Example 4. Figure 3 illustrates the process of arc consistency based candidate generation, where the set of vertices $\{v_i\}$ under each vertex u_i is the domain. Given a frequent subgraph S_1, frequent edges S_2 and S_3, and $\delta = 2$, if adding S_2 to S_1, we may obtain S_4 where the domain of u_4 is updated to be $D_{u_4} = D_{u_4} \cap D_{u_3} = \{v_6\}$. Since $|D_{u_4}| < \delta$, S_4 is infrequent and is not passed to isomorphism computation. If adding S_3 to S_1, we may obtain S_5 where the domain of u_6 is updated to be $D_{u_6} = D_{u_6} \cap D_{u_1} = \{v_1, v_4\}$ and the domain of u_7 is updated to be $D_{u_7} = D_{u_7} \cap D_{u_3} = \{v_6\}$. Since $|D_{u_7}| < \delta$, S_5 is infrequent and is not passed to isomorphism computation.

4.2   Image Vertex Reusage

Given a candidate subgraph S, DFS (Algorithm 3) is used to find image graphs of S from scratch. However, candidate S is constructed by adding a frequent single-edge graph to a frequent subgraph and the image graphs of this frequent subgraph have already been found. Based on this observation, we propose algorithm ReuseImages (Algorithm 4) that reuses the image vertices of the frequent subgraph found before computing the image graphs of a candidate so that the cost of determining whether a candidate is a frequent geo-subgraph is reduced.

Algorithm 4. ReuseImages(S, S^*, G, τ, δ)

1: ▷ S is generated by adding edge (u, u') at u to S^*.
2: **for each** vertex $v \in F(u) \setminus D_u$ **do**
3:      Remove the isomorphisms that map v to u;
4: $F(u) \leftarrow F(u) \cap D_u$;
5: **for each** vertex $v \in F(u)$ **do**
6:      $\mathcal{F} \leftarrow$ the set of isomorphisms that map v to u;
7:      **if** $\exists v' \in D_{u'}$ such that v' is a neighbor of $v \wedge v'$ does not belong to the image vertex sets in $\mathcal{F} \wedge$ adding v' does not violate the geo-constraint **then**
8:          Add v' to $F(u')$;
9:      **else**
10:         Remove all the isomorphisms in $\mathcal{F}$;
11: **while** $\exists u \in S$ $(|F(u)| < \delta)$ **do**
12:     **for each** vertex $v \in D_u \setminus F(u)$ **do**
13:         DFS($v, S, G, \tau, \text{sol}$);
14:         **if** $|F(u)| = \delta$ **then**
15:             **Break;**
16:     **if** $|F(u)| < \delta$ **then**
17:         **Return** FALSE;
18: **Return** TRUE;

Specifically, let S be a candidate generated by adding single-edge graph (u, u') at vertex u in a frequent subgraph S^*. The set of isomorphisms of S^* is stored. Let $F(u)$ be the image vertex set of u in S^* and let D_u be the domain of vertex u in S. Obviously, the vertices in set $F(u) \setminus D_u$ cannot be reused, and thus, the isomorphisms that map these vertices to u cannot be reused. Then, for each image vertex v in $F(u) \cap D_u$, let $\mathcal{F}$ be the set of isomorphisms that map v to u. If a vertex v' exists in the domain of u' such that (i) v' is a neighbor of v, (ii) v' does not belong to the image vertex sets in $\mathcal{F}$, and (iii) adding v' does not violate the geo-constraint then an isomorphism in $\mathcal{F}$ can be reused, and v' can be taken as an image vertex of u'. If none of the vertices in the domain of u' satisfies the above three conditions, no isomorphism in $\mathcal{F}$ can be reused.

After considering all the isomorphisms of frequent subgraph S^*, if the sizes of the image vertex sets of some vertices in S are still below δ, ReuseImages calls DFS to find isomorphisms from scratch. Finally, when the vertices in all domains are exhausted and $\mathsf{MNI}_\tau(S, G) < \delta$, candidate S is infrequent and ReuseImages returns FALSE.

Example 5. Figures 4 and 5 illustrate how to reuse image vertices to compute isomorphisms of a candidate subgraph. Given $\delta = 2$, Fig. 4 shows four frequent edges in graph G. In Fig. 5, S_1 is a frequent subgraph that has two isomorphisms. The image vertices are shown under each vertex, i.e., $F(u_1) = \{v_1, v_6\}$ and $F(u_2) = \{v_3, v_9\}$. Subgraph S_2 is a candidate generated by adding edge (u_3, u_4) to S_1, and we have $F(u_3) = D_{u_3} = \{v_3, v_9\}$. Then, for each vertex in $F(u_3)$, we can find one image vertex for u_4 and derive $F(u_4) = \{v_4, v_7\}$. Thus, two isomorphisms of S_2 are found based on the isomorphisms of S_1, and candidate S_2 is frequent. Next, subgraph S_3 is a candidate generated by adding edge (u_5, u_6) to S_2, and we have $F(u_5) = D_{u_5} = \{v_1, v_6\}$. Then, for each vertex in $F(u_5)$, we can find one image vertex for u_6 and derive $F(u_6) = \{v_4, v_7\}$. Thus, two isomorphisms of S_3 are found based on the isomorphisms of S_2, and candidate S_3 is frequent. Next, S_4 is a candidate generated by adding edge (u_7, u_8) to S_3, and we have $F(u_7) = \{v_4, v_7\}$ and $D_{u_7} = \{v_4, v_7, v_{11}\}$. Then, for the vertices in $F(u_7)$, we can find only one image vertex v_5 for u_8. For the vertices in $D_{u_7} \setminus F(u_7)$, no image vertex can be found for u_8. Hence, candidate S_4 is not frequent.

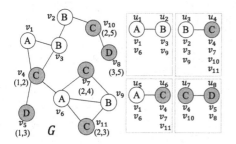

Fig. 4. Frequent edges in graph G.

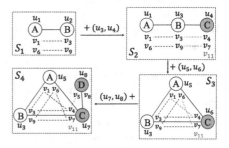

Fig. 5. Image vertex reusage.

4.3   Geo-Grid Based Vertex Ordering

Given a candidate subgraph, the domains of some geo-vertices may be large, and the vertices in these domains may spread over a large spatial region. Algorithm FreGeoSubgraphAlg ignores the spatial distribution of the vertices when computing isomorphisms, making it vulnerable to spending considerable time on computing isomorphisms whose diameters exceed τ. We propose a geo-grid based vertex ordering that prioritizes vertices close to the image vertices found already when computing isomorphisms. Thus, the isomorphisms that satisfy the geo-constraint can be found quickly.

First, an $h \times h$ grid is constructed on the spatial region that contains all geo-vertices in the graph. Each grid cell is indexed by (column id, row id) and each geo-vertex is assigned to a grid cell. For each vertex in graph G, its neighboring geo-vertices are indexed based on the corresponding grid cells. For example in Fig. 6, a 7×7 grid is constructed, and seven geo-vertices $v_3, v_4, \cdots, v_9$ are assigned to grid cells. For instance, v_3 is assigned to cell $(3,3)$. Considering vertex v_1, suppose that it has five neighboring geo-vertices in graph G, i.e., v_4, v_6, v_7, v_8, v_9. These geo-vertices are indexed based on their corresponding grid cells, as shown in Fig. 6.

Given a candidate subgraph S, the process of computing its isomorphisms is to find an image vertex for each vertex in S one by one. If a geo-vertex v exists among the image vertices already found, for the next vertex u to be considered, the geo-vertices in the domain of u are processed in ascending order of the distances between their corresponding grid cells and v. For example in Fig. 6, given $\tau = 2$, let the side length of the grid cell be 1. Consider the candidate subgraph S, where both u_3 and u_4 are geo-vertices. Suppose that we are computing an isomorphism that already has mappings: $f(u_1) = v_1$, $f(u_2) = v_2$, and $f(u_3) = v_3$. Then, the domain of u_4 is the intersection of the neighbor vertex sets of v_1 and v_3, i.e., $D_{u_4} = \{((2,4) : v_9), ((5,5) : v_8), ((6,1) : v_6, v_7)\}$. The vertices in D_{u_4} are indexed by their corresponding grid cells. Since v_3 is a geo-vertex already found in this isomorphism, when determining the image vertex of u_4, we first consider the closest grid cell to v_3 in D_{u_4}, i.e., $(2,4) : v_9$ and set $f(u_4) = v_9$. Thus, a geo-constrained subgraph isomorphism is found, and $\mathsf{Dia}(f(S)) = \|v_3 - v_9\|_2 < 2$. In contrast, algorithm FreGeoSubgraphAlg simply considers the geo-vertices in D_{u_4} one by one. It may spend computational effort on vertices v_6, v_7, and v_8 before finding $f(u_4) = v_9$.

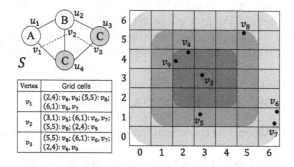

Fig. 6. Geo-grid based vertex ordering.

5   Experimental Study

We study the proposed algorithms by experiments on real datasets, and we compare them with the state-of-the-art algorithm.

5.1   Setup

Datasets. Five datasets are used. Table 1 shows statistics of these datasets. Dataset *Email*[1] is an email network, where vertices represent email accounts and edges represent communications between email accounts. Dataset *MiCo* [9] is a collaboration network where vertices refer to authors and edges refer to collaboration relationships between authors. There are no geo-coordinates in *Email* and *MiCo*. Thus, we generate random coordinates in the unit square for each vertex. Datasets *US* and *UK* are extracted from DBpedia[2] and contain geo-entities in the US and the UK, respectively. Dataset *World* is the largest connected component extracted from DBpedia. The geo-coordinates in datasets *US*, *UK*, and *World* are normalized to the unit square.

Table 1. Statistics of datasets.

Dataset	#Vertices	#Vertex labels	#Geo vertices	#Edges	#Edge labels	Density
Email	224,832	29	181,808	394,400	1	Sparse
MiCo	100,000	30	96,547	1,080,298	100	Dense
UK	165,925	5,533	80,765	878,608	434	Medium
US	582,970	8,892	228,034	3,272,992	537	Medium
World	7,987,748	240,217	882,920	39,749,264	653	Medium

Algorithms. To assess the efficiency of the proposed algorithms, five algorithms are compared in the experiments.

- GRAMI: The state-of-the-art GRAMI [9] is extended to solve the FGS problem in a two-step manner. We first adopt GRAMI to find all the subgraphs with support at least δ. Then, we filter out the subgraphs that satisfy either of the two conditions: (i) the number of geo-vertices is below 2 and (ii) the diameter exceeds τ.
- FreGeoSubgraphAlg: Algorithm 1 presented in Sect. 3.
- FreGeoSubgraphAlg*: FreGeoSubgraphAlg using the three optimizations introduced in Sects. 4.1–4.3.
- FreGeoSubgraphAlgOpt1: FreGeoSubgraphAlg using the optimization introduced in Sect. 4.1.
- FreGeoSubgraphAlgOpt2: FreGeoSubgraphAlg using the two optimizations introduced in Sects. 4.1 and 4.2.

---

[1] https://networkrepository.com/networks.php.
[2] https://www.dbpedia.org.

Settings. We implement all the algorithms in C++ and compile them using GNU G++ 5.4.0. Experiments are conducted on a server with an Intel Xeon 2.67 GHz CPU, 384 GB main memory, and running Linux (ubuntu Linux 16.04, 64 bits). On each dataset and for each algorithm, we report the running time and the number of generated candidate subgraphs. The effects of varying parameters τ and δ are studied.

5.2 Performance Evaluations

Comparisons of Algorithms. Table 2 shows the running times of the algorithms on the five datasets. GRAMI cannot finish the computations within 4 h on all datasets, while FreGeoSubgraphAlg is able to report results within reasonable time. Thus, the state-of-the-art algorithm is not applicable to the FGS problem. On UK, the running time of FreGeoSubgraphAlgOpt1 is around 70% below that of FreGeoSubgraphAlg. On the other four datasets, the running time of FreGeoSubgraphAlgOpt1 is 90% below that of FreGeoSubgraphAlg. So arc consistency based candidate generation is effective at reducing the number of candidate subgraphs. Comparing with FreGeoSubgraphAlgOpt1, the running time of FreGeoSubgraphAlgOpt2 is around 90% less on UK, around 60% less on MiCo, and 10%–30% less on the other three datasets. This result verifies that reusing image vertices reduces costs compared to computing isomorphisms from scratch. Although the running time of FreGeoSubgraphAlg* is the same as that of FreGeoSubgraphAlgOpt2 on UK, the running time of FreGeoSubgraphAlg* is around 5%–20% below that of FreGeoSubgraphAlgOpt2 on the other four datasets. These results offer evidence that the geo-grid based vertex ordering makes it possible to quickly find geo-constraint isomorphisms. Overall, this experiment offers evidence of the high efficiency of the proposed algorithm and optimizations.

Table 2. Running time (seconds) of five algorithms.

Algorithm	Dataset				
	Email $\delta = 130$ $\tau = 0.42$	MiCo $\delta = 3400$ $\tau = 0.4$	UK $\delta = 840$ $\tau = 0.25$	US $\delta = 2080$ $\tau = 0.4$	World $\delta = 4000$ $\tau = 0.4$
GRAMI	4+ hours	4+ hours	4+ hours	4+ hours	4+ hours
FreGeoSubgraphAlg	526	1,152	2,421	464	949
FreGeoSubgraphAlgOpt1	14	116	617	28	23
FreGeoSubgraphAlgOpt2	13	39	63	21	17
FreGeoSubgraphAlg*	12	34	63	19	15
#frequent geo-subgraphs	150	14	12	22	13

Effect of Varying δ. Figure 7 shows the effect of varying δ on the performance of GRAMI, FreGeoSubgraphAlg, and FreGeoSubgraphAlg* when fixing the value of τ on the five datasets. Except on dataset UK, GRAMI fails to terminate within 4 h. In terms of the running time, FreGeoSubgraphAlg* is faster than GRAMI by two orders of magnitude and is faster than FreGeoSubgraphAlg by one order of magnitude. As δ increases, the running times of the algorithms decrease, and the numbers of candidates generated decrease. This is because when δ increases, the number of frequent edges decreases, so that the number of candidate subgraphs decreases and the costs of computing isomorphisms is reduced. On all the datasets, when δ is small, the improvements in running times of FreGeoSubgraphAlg* over FreGeoSubgraphAlg are large.

Effect of Varying τ. Figure 8 shows the effect of varying τ on the performance of GRAMI, FreGeoSubgraphAlg, and FreGeoSubgraphAlg* when fixing the value of δ on the five datasets. As τ increases, the running times of the algorithms increase and the numbers of candidate subgraphs increase. This is because when τ is large, the number of geo-subgraphs that satisfy the geo-constraint is large, so that the costs of computing isomorphisms is high. On all datasets, when τ is large, FreGeoSubgraphAlg* outperforms FreGeoSubgraphAlg substantially.

6   Related Work

We review relevant studies on frequent subgraph mining in graph databases and single graphs.

Frequent Subgraph Mining in Graph Databases. Given a graph database that consists of a set of small data graphs, the support of a subgraph S is defined as the number of data graphs that contain an isomorphism of S. FSG [13] adopts the idea of the Apriori algorithm [2] to generate candidate subgraphs and compute isomorphisms. gSpan [25] encodes edges using the DFS approach and finds frequent subgraphs consisting of $k + 1$ edges based on frequent subgraphs consisting of k edges. LEAP [24] prunes the search space according to similar substructures. Later, to tackle the problem of large numbers of frequent subgraphs, constraints are introduced to mine variants of frequent subgraphs, such as maximal frequent subgraphs [21], closed frequent subgraphs [26], and significant subgraphs [18].

In the setting of graph databases, individual data graphs are usually small, so that few candidates are generated from each graph. However, we generate candidates from a single large and complex graph. In this setting, the algorithms in this category are inefficient.

Frequent Subgraph Mining in Single Graph. Given a single large graph, different definitions of support of a subgraph have been investigated, such as MNI [7], MIS [22], HO [10], MI, and MVC [16]. The computational cost of frequent subgraph mining is dominated by two parts: generating candidates and computing isomorphisms. Most existing studies adopt the anti-monotone property to reduce the number of candidate subgraphs. SIGRAM [27] generates a candidate with $k + 1$ edges by combining two k-edge frequent subgraphs that share $k - 1$ common edges. It caches all isomorphisms of frequent subgraphs to

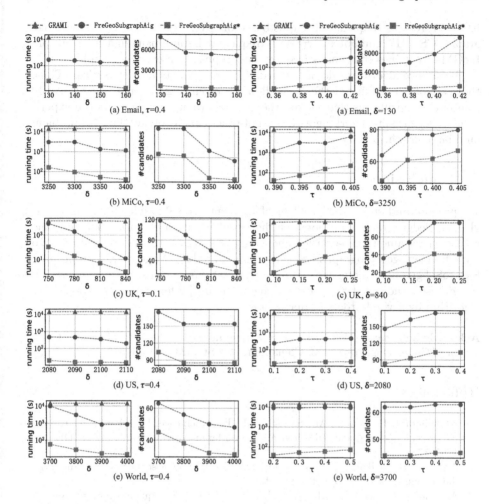

Fig. 7. Varying δ.   **Fig. 8.** Varying τ.

speed up isomorphism computation of candidates, which requires a large amount of memory and renders it inapplicable to large graphs. GRAMI [9] adopts a similar way of generating candidates as SIGRAM, but solves the problem of frequent subgraph mining differently by considering it as a CSP problem. The benefit is that there is no need to find all isomorphisms but only to find enough image vertices. FSSG [12] identifies symmetric parts in the graph and partitions the vertices. In frequent components, it generates candidates by enumerating all relevant graphs.

Frequent subgraph mining in a single graph has been applied in various applications. WeFreS [4] mines frequent subgraphs in weighted graphs. WeGraMi [15] considers the weights of vertices when computing frequent subgraphs. DTop-kPM [23] finds top-k frequent subgraphs that are diverse in terms of structure.

KCP [14] returns frequent subgraphs based on given core patterns in knowledge graphs. CSM [17] computes correlated frequent subgraph pairs in a single graph. IncGM+ [1] mines continuous frequent subgraphs in an evolving graph.

The above algorithms do not consider geo-constraints among vertices. Thus, they cannot determine whether a candidate is promising, and may unnecessarily compute isomorphisms that violate the geo-constraints. As shown in the experiments, the state-of-the-art algorithm is not practical for solving the FGS problem on a large graph.

7   Conclusion

We propose and study the problem of frequent geo-subgraph mining in a single knowledge graph. Solutions to the problem are useful for analyzing the relationships among geo-entities and can enable downstream applications. To mine frequent geo-subgraphs efficiently on a large knowledge graph, we propose FreGeoSubgraphAlg along with three optimizations. Extensive experiments on real datasets show that the proposed algorithm outperforms the state-of-the-art algorithm by several orders of magnitude.

Acknowledgements. This work was supported in part by the Natural Science Foundation of Guangdong Province of China under Grant 2023A1515011619.

References

1. Abdelhamid, E., Canim, M., Sadoghi, M., Bhattacharjee, B., Chang, Y., Kalnis, P.: Incremental frequent subgraph mining on large evolving graphs. In: ICDE, pp. 1767–1768 (2018)
2. Agrawal, R., Srikant, R.: Fast algorithms for mining association rules in large databases. In: VLDB, pp. 487–499 (1994)
3. Alobaidi, I.A., Leopold, J.L., Allami, A.A.: The use of frequent subgraph mining to develop a recommender system for playing real-time strategy games. In: ICDM, pp. 146–160 (2019)
4. Ashraf, N., et al.: Wefres: weighted frequent subgraph mining in a single large graph. In: ICDM, pp. 201–215 (2019)
5. Auer, S., Bizer, C., Kobilarov, G., Lehmann, J., Cyganiak, R., Ives, Z.G.: Dbpedia: a nucleus for a web of open data. In: ISWC, vol. 4825, pp. 722–735 (2007)
6. Bollacker, K.D., Evans, C., Paritosh, P.K., Sturge, T., Taylor, J.: Freebase: a collaboratively created graph database for structuring human knowledge. In: SIGMOD, pp. 1247–1250 (2008)
7. Bringmann, B., Nijssen, S.: What is frequent in a single graph? In: PAKDD, pp. 858–863 (2008)
8. Deshpande, M., Kuramochi, M., Wale, N., Karypis, G.: Frequent substructure-based approaches for classifying chemical compounds. IEEE Trans. Knowl. Data Eng. **17**(8), 1036–1050 (2005)
9. Elseidy, M., Abdelhamid, E., Skiadopoulos, S., Kalnis, P.: GRAMI: frequent subgraph and pattern mining in a single large graph. Proc. VLDB Endow. **7**(7), 517–528 (2014)

10. Fiedler, M., Borgelt, C.: Support computation for mining frequent subgraphs in a single graph. In: MLG (2007)
11. Henderson, T.A.: Frequent subgraph analysis and its software engineering applications. Case Western Reserve University (2017)
12. Kavitha, D., Haritha, D., Padma, Y.: Optimized candidate generation for frequent subgraph mining in a single graph. In: ICDE, pp. 259–272 (2021)
13. Kuramochi, M., Karypis, G.: Frequent subgraph discovery. In: ICDM, pp. 313–320 (2001)
14. Kuramochi, M., Karypis, G.: Finding frequent patterns in a large sparse graph. Data Min. Knowl. Discov. **11**(3), 243–271 (2005)
15. Le, N., Vo, B., Nguyen, L.B.Q., Fujita, H., Le, B.: Mining weighted subgraphs in a single large graph. Inf. Sci. **514**, 149–165 (2020)
16. Meng, J., Pitaksirianan, N., Tu, Y.: Counting frequent patterns in large labeled graphs: a hypergraph-based approach. Data Min. Knowl. Discov. **34**(4), 980–1021 (2020)
17. Prateek, A., Khan, A., Goyal, A., Ranu, S.: Mining top-k pairs of correlated subgraphs in a large network. Proc. VLDB Endow. **13**(9), 1511–1524 (2020)
18. Ranu, S., Singh, A.K.: Graphsig: a scalable approach to mining significant subgraphs in large graph databases. In: ICDE, pp. 844–855 (2009)
19. Saha, T.K., Katebi, A., Dhifli, W., Hasan, M.A.: Discovery of functional motifs from the interface region of oligomeric proteins using frequent subgraph mining. IEEE ACM Trans. Comput. Biol. Bioinform. **16**(5), 1537–1549 (2019)
20. Suchanek, F.M., Kasneci, G., Weikum, G.: YAGO: a core of semantic knowledge. In: WWW, pp. 697–706 (2007)
21. Thomas, L.T., Valluri, S.R., Karlapalem, K.: MARGIN: maximal frequent subgraph mining. In: ICDM, pp. 1097–1101 (2006)
22. Vanetik, N., Gudes, E., Shimony, S.E.: Computing frequent graph patterns from semistructured data. In: ICDM, pp. 458–465 (2002)
23. Wang, X., Tang, L., Liu, Y., Zhan, H., Feng, X.: Diversified pattern mining on large graphs. In: DEXA, pp. 171–184 (2021)
24. Yan, X., Cheng, H., Han, J., Yu, P.S.: Mining significant graph patterns by leap search. In: SIGMOD, pp. 433–444 (2008)
25. Yan, X., Han, J.: GSPAN: graph-based substructure pattern mining. In: ICDM, pp. 721–724 (2002)
26. Yan, X., Han, J.: Closegraph: mining closed frequent graph patterns. In: SIGKDD, pp. 286–295 (2003)
27. Zeng, J., U, L.H., Yan, X., Han, M., Tang, B.: Fast core-based top-k frequent pattern discovery in knowledge graphs. In: ICDE, pp. 936–947 (2021)

Locality Sensitive Hashing for Data Placement to Optimize Parallel Subgraph Query Evaluation

Mingdao Li[1], Bo Zhai[2], Yuntao Jiang[1], Yunjian Li[3], Zheng Qin[1], and Peng Peng[1(✉)]

[1] Hunan University, Changsha, China
{limingdao,jiangyuntao,zqin,hnu16pp}@hnu.edu.cn
[2] Beijing Institute of Astronautical Systems Engineering, Beijing, China
[3] Institute of Land Aviation, Beijing, China

Abstract. Recently, parallel computing systems composed of interconnected workers through a high-speed network have become readily available, thereby presenting an opportunity for parallelizing subgraph queries in large graphs. To effectively evaluate these subgraph queries, it is crucial to place vertices among different workers. In contrast to widely used hash-based techniques, our approach leverages the utilization of locality sensitive hashing methods for data placement. This paper introduces a novel graph locality sensitive hashing method named VMH, which is specifically designed for data placement by considering the labels of vertices. By employing VMH, we can effectively place similar vertices to the same worker while considering the labels of vertices, thereby reducing redundant communication and computation across multiple workers during parallel subgraph query evaluation. Extensive experimental studies conducted on both large real and synthetic graphs demonstrate that our proposed techniques lead to significant improvements in query performance compared to existing methods.

1 Introduction

Graphs have become increasingly significant in various domains in recent years, and subgraph queries represent a fundamental category of queries in graph analysis. In this context, the task of evaluating subgraph queries involves the identification of all unique instances of a query subgraph Q within a given data graph G. Subgraph queries find practical applications across a diverse range of fields, including protein interaction analysis [24], social network analysis [7], among others. However, due to its well-established NP-hard complexity [10], the processing of subgraph queries often poses a significant performance bottleneck in these applications.

With the rapid growth in graph sizes, the demand for extending subgraph query evaluation to parallel computing systems has been significantly amplified. A parallel computing system refers to a distributed system comprising multiple workers interconnected by a high-speed network [22]. These workers can

X. Song et al. (Eds.): APWeb-WAIM 2023, LNCS 14331, pp. 32–47, 2024.
https://doi.org/10.1007/978-981-97-2303-4_3

encompass processors within mainframe computers or supercomputers, as well as commodity computers within a local network. The high-speed network can manifest as the bus within a mainframe computer or supercomputer or as a Myrinet network within the local network context. Parallel computing systems harness the abundant computing nodes to collaboratively execute application programs to enhance overall performance.

The core concept behind parallelizing subgraph query evaluation in parallel computing systems involves breaking down the query evaluation into independent units of sequential evaluation. Each independent unit, referred to as a *task*, can be executed by any worker within the system. However, a critical concern arises regarding the optimal placement of different vertices among the available workers. Existing methods typically employ a simple hash function to partition the data graph into subgraphs distributed across multiple workers in a fine-grained manner. However, these methods often overlook the importance of considering the locality aspect in subgraph query evaluation.

In light of these considerations, this paper explores the extension of locality sensitive hashing methods to facilitate the partitioning of the data graph among different workers, adopting a coarser granularity. This approach enables the utilization of locality and allows for the reuse of common computations during parallel subgraph query evaluation. In order to achieve these objectives, we must address the following challenging problem.

Challenge: Extending Locality Sensitive Hashing Methods for Data Placement. Existing systems often employ a simple hash function to assign vertices to workers, disregarding the computational cost of data placement [1]. Unfortunately, this approach may result in suboptimal data locality. However, vertices sharing the same label and residing in close proximity are more likely to be involved in the same query and share common computations. Therefore, when distributing vertices among workers, it is preferable to place vertices with the same label and proximity in the same worker. In this paper, we propose a novel data placement approach named $Vertex\ MinHash$ (VMH), which utilizes a graph-based locality sensitive hashing method. VMH incorporates both vertex labels and neighboring vertices in the hashing process. Vertices sharing the same label and residing close to each other exhibit a higher probability of having the same hash value in VMH, thereby favoring their placement in the same worker.

Contributions. To summarize, we make the following contributions in the paper.

- We propose a novel data placement approach based on a new locality sensitive hashing method named VMH that integrates both labels and neighbors of vertices.
- The proposed VMH locality sensitive hashing method is mathematically proven to guarantee that the probability of two vertices having the same hash value in VMH is equal to their vertex similarity (Theorem 1).
- We conduct comprehensive experiments on both real and synthetic graphs to evaluate the effectiveness of our proposed approach. The results demonstrate

that our approach can significantly surpass existing methods in terms of query performance for subgraph queries.

2 Background

2.1 Preliminaries

In this section, we review the terminology used in this paper, and formally define our problem. In this work, we focus on undirected graphs with vertex labels (Definition 1). Note that, our method can be easily extended to handle "directed" and "edge-labeled" graphs.

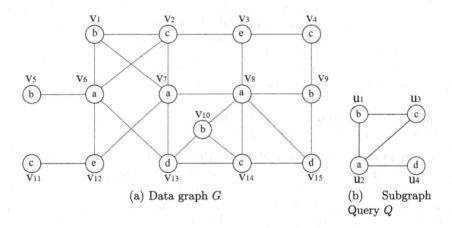

(a) Data graph G (b) Subgraph Query Q

Fig. 1. A running example of a data graph and a subgraph query

Definition 1. (Graph) *A graph is denoted as* (V, E, L, f), *where* V *is a set of vertices;* $E \subseteq V \times V$ *is a set of edges and an edge connecting* v_1 *and* v_2 *is denoted as* $\overline{v_1 v_2}$, *where* $v_1, v_2 \in V$; L *is a set of vertex labels; and* f *defines the label mapping function* $f : V \to L$.

Given a vertex v, we use $N(v)$ to denote the neighbors of v, i.e., $N(v) = \{v' \in V | \overline{vv'} \in E\}$.

Given a large data graph G and a query graph Q, where $|V^Q| \ll |V^G|$, the problem of this paper is to find all *matches* of Q in G, where matches are defined in Definition 2.

Definition 2. (Match) *Given a data graph* $G = (V^G, E^G, L^G, f^G)$ *and a query graph* $Q = (V^Q, E^Q, L^Q, f^Q)$, Q *is isomorphic to* G *if and only if there exists an injective mapping* $\mu : V^Q \to V^G$ *meeting the following constraints: 1)* $\forall u \in V^Q, f^Q(u) = f^G(\mu(u))$; *2)* $\forall \overline{u_1 u_2} \in E^Q, \overline{\mu(u_1)\mu(u_2)} \in E^G$.

A mapping μ *is called a* match *of* Q *over* G *in this paper.*

Example 1. Figure 1 shows an example data graph and two subgraph queries, where *vertex IDs* are assigned beside the vertices and the letters inside the vertices represent *vertex labels*. For the data graph G and query graph Q in Fig. 1, there are three matches:

- $\mu_1 = \{(u_1, v_1), (u_2, v_6), (u_3, v_2), (u_4, v_{12})\}$;
- $\mu_2 = \{(u_1, v_1), (u_2, v_7), (u_3, v_2), (u_4, v_{12})\}$;
- $\mu_3 = \{(u_1, v_{10}), (u_2, v_8), (u_3, v_{14}), (u_4, v_{15})\}$.   □

In this paper, we mainly focus on how to parallelize the subgraph query processing algorithm, and our proposed parallel optimization does not highly depend on any particular subgraph query processing algorithm.

2.2 Parallel Execution Model

Parallel computing systems leverage multiple computing nodes to collaboratively execute application programs, aiming to enhance overall performance. In the context of subgraph query processing, the architecture of a parallel system is depicted in Fig. 2. The data graph G is partitioned into a collection of subgraphs, which are subsequently distributed across a cluster of nodes. The fundamental processing unit is referred to as a *worker*, which can be conceptualized as a process that occupies a CPU core. Typically, a single physical machine with multiple cores can accommodate several workers. The *task*, serving as the basic scheduling unit, corresponds to a specific subset of data where a query can be independently evaluated. To achieve load balancing, numerous tasks are distributed among different workers. Each worker maintains its own task queue to ensure continuous processing and mitigate contention. Given the potentially large number of tasks, it is infeasible to retain all tasks in memory and some tasks towards the end of the task queue are stored in the file systems.

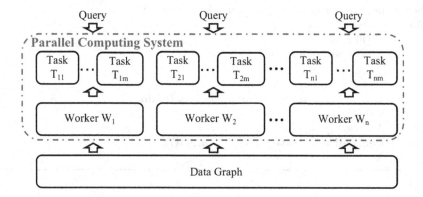

Fig. 2. Execution Model of Subgraph Query Processing in Parallel Computing Systems

The fundamental idea is that the query processing is decomposed into some tasks of sequential processing. In this paper, our decomposition of the query processing is based on the same hash function applied to the candidate set of the first vertex in the search order. This is because different candidates of the first vertex in the search order can result in different matches and the subgraph query processing starting from different candidates of the first vertex in the search order can be executed independently.

Algorithm 1 presents the framework of parallel subgraph query processing algorithm used in this paper. In particular, similar to [5,11], we first generates the *candidate set* $C(u)$ of query vertex u based on its label and degree (Line 1), and define the search order φ as the BFS traversal order starting from $u_r = argmin_{u \in V^Q} \frac{|C(u)|}{d(u)}$ (Line 2). We denote the ith vertex in φ as $\varphi[i]$. Then, we distribute the candidate set of the first vertex u_r in the search order among different workers by using a hash function H, and initialize a task for each candidate in $C(u_r)$ (Lines 3–5). Last, all workers call function *ParallelEnumerate* to enumerate the matches in parallel (Lines 7–11).

For Function *ParallelEnumerate*, if all query vertices have been matched, the mapping μ is added into the result set of matches (Lines 1–3). Otherwise, we select the next vertex u in φ to match (Line 4). We first compute a local candidate vertex set $LC(u)$ that is around $\mu(u_r)$ and can match u (Line 5), and then we loop over $LC(u)$ to extend the matches, and recursively call function *ParallelEnumerate* (Lines 6–11). If a vertex in the local candidate set $LC(u)$ is not in the current worker we send it and its neighbors into the current worker (Lines 7–8).

Algorithm 1: Parallel Hash Subgraph Query Processing Algorithm

---

Input: A data graph G and a query graph Q
Output: The set M of all matches of Q over G
1 Generate candidate set $C(u)$ for $u \in V^Q$;
2 Select the start vertex u_r and generate the search order φ;
3 **for** *each vertex* $v \in C(u_r)$ **do**
4     $i \leftarrow H(v) \bmod n$;
5   Initialize a task t_v for v in W_i;
6 $M \leftarrow \emptyset$;
7 **for** *each worker* W_i **in parallel do**
8   **for** *each task* t_v *initialized from vertex* v *in* W_i **do**
9        $\mu \leftarrow \emptyset$;
10   Add (u_r, v) in μ;
11   **ParallelEnumerate**$(Q, G, \varphi, M, \mu, 2, W_i)$;
12 Return M;

Example 2. Figure 3 shows how our parallel hash subgraph query processing algorithm works for Q in Fig. 1(b). Here, for Q, $C(u_2) = \{v_6, v_7, v_8\}$ and the

Function. ParallelEnumerate($Q, G, \varphi, M, \mu, i, W$)

1 if $i = |\varphi| + 1$ then
2 | Add μ in M;
3 | Return;
4 $u \leftarrow \varphi[i]$;
5 $LC(u) \leftarrow \bigcap_{u' \in N(u)} N(\mu(u'), f^Q(u))$;
6 for each vertex $v \in LC(u)$ do
7 | if v is not in W then
8 | | Send v and its neighbors to W;
9 | Add (u, v) in μ;
10 | ParallelEnumerate($Q, G, \varphi, M, \mu, i + 1, W$);
11 | Remove (u, v) from μ;

search order is $\varphi_1 = [u_2, u_1, u_3, u_4]$. We assume there are two workers W_0 and W_1 and we just use the modulus operator on the vertex identifier as the hash function. Thus, $H(v_6) = H(v_8) = 0$ and $H(v_7) = 1$. Therefore, v_6 and v_8 are placed at worker W_0, while v_7 is placed at worker W_1.*Then*,   □

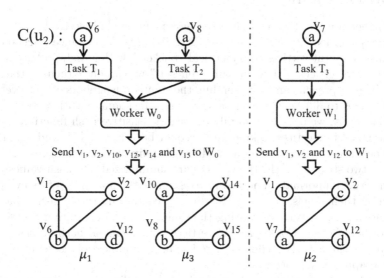

Fig. 3. Example of Parallel Hash Subgraph Query Processing

3   Locality Sensitive Hashing for Data Placement

In the aforementioned parallel subgraph query execution model, a crucial challenge is to reduce the communication and computation overhead associated with multiple tasks. Given that different tasks across various workers may share common subgraphs, there is an opportunity to reuse communication and computation for these shared subgraphs. Consequently, it is essential to initialize tasks that share common subgraphs within the same worker to reduce redundant communication and computation. As outlined in Algorithm 1 (Line 4), achieving this requires close candidates to have identical hash values and be assigned to the same worker. Subsequently, their corresponding tasks can be initialized within the worker.

Example 3. Let us consider the parallel hash subgraph query processing in Example 2. In this case, v_6 and v_7 are placed at different workers and initialized as separate tasks. Consequently, both workers W_0 and W_1 receive duplicates of v_1, v_2, and v_{12} for match formation, which is redundant. However, it is noteworthy that v_6 and v_7 share the same label and have some common neighbors. This suggests a strong likelihood that they are often the candidates of the same query vertices and their corresponding matches share substantial common subgraphs. Therefore, to eliminate redundancy, v_6 and v_7 should have the same hash value and be placed at the same worker.

3.1 Vertex Similarity

In this paper, we aim to accomplish the aforementioned objective by extending the technique of locality sensitive hashing (LSH) [20]. We enhance LSH to effectively hash adjacent vertices into identical "buckets" with a high probability. By grouping close vertices within the same "bucket," we can place them to the same worker. This approach can avoid sending the common neighbors of close vertices multiple times to different workers and reduces the communication cost.

In this study, we develop a family of locality sensitive hash functions specifically designed for vertices, taking into account both their labels and neighbors. This procedure can be viewed as a dimensionality reduction technique, performed in two stages. Firstly, the vertices are summarized into hash values while retaining crucial information necessary to estimate the proximity of two vertices, considering their labels. Secondly, by directly comparing these hash values, we can identify pairs of vertices sharing the same label and exhibiting a high likelihood of being close. Consequently, vertices with identical hash values can be placed to the same worker, effectively reducing communication costs during parallel subgraph query processing.

In our locality sensitive hashing method, we utilize the similarity between hash values as an estimate for the proximity of vertices, taking into account their labels. Specifically, vertices with the same label and close proximity are likely to exhibit similar hash values, while vertices with different labels and significant distance between them tend to have dissimilar hash values. As a result, it is crucial to establish the definition of similarity based on the labels and closeness of vertices, serving as the foundation for the design of our LSH method. By incorporating these considerations, we can effectively capture the relationships between vertices and create meaningful hash values.

Thus, we define the similarity of two vertices based on their neighbor sets while considering their labels. In particular, there are two factors that we should consider. First, the two vertices should have the same label, since they are more likely to be the candidates of the same query vertex, which means that their corresponding matches can overlap. Second, these two vertices should have as many common neighbors as possible. If two vertices have more common neighbors, matches containing them are more likely to overlap.

Example 4. Let us consider the graph in Fig. 1(a). v_6, v_7 and v_8 have the same label a, so all of them are possible to be candidates of a query vertex. If we just use the modulus operator on the vertex identifier as the hash function to assign the vertex to two workers, v_6 and v_8 are placed into the same worker, while v_7 is placed in another worker. However, we should note that v_6 and v_7 share many neighbors. Thus, if v_6 and v_7 are put in the same worker, they can share more communication and computation, when they are candidates of a query vertex. □

To achieve the above goal, given two vertices v_1 and v_2, we define a binary variable $\mathbb{I}_G(v_1, v_2)$ to record whether v_1 and v_2 have the same label as follows.

$$\mathbb{I}_G(v_1, v_2) = \begin{cases} 1 & \textit{if } f^G(v_1) = f^G(v_2); \\ 0 & \textit{otherwise.} \end{cases} \tag{1}$$

Further, given two vertices v_1 and v_2, we use the Jaccard similarity between the neighbor sets $N(v_1)$ and $N(v_2)$ of v_1 and v_2 in the definition of their similarity. Then, we can formally define the vertex similarity between v_1 and v_2 as follows.

Definition 3. *(Vertex Similarity) Given two vertices v_1 and v_2, the vertex similarity between v_1 and v_2 is as follows.*

$$VS(v_1, v_2) = \mathbb{I}_G(v_1, v_2) \times \frac{N(v_1) \cap N(v_2)}{N(v_1) \cup N(v_2)} \tag{2}$$

3.2   Vertex MinHash

Based on the above definition of vertex similarity, we present a vertex locality sensitive hashing method, called Vertex MinHash (VMH). Given a vertex v, we summarize the label and neighbors of v into a sketch $\vec{s}$. First, we always add the label of v into a sketch $\vec{s}$. Second, we run the original MinHash algorithm on the neighbor set $N(v)$ of v and also add the result into the sketch $\vec{s}$.

Function VMH shows the procedure. We also first initialize an empty sketch $\vec{s}$ of length $m' + 1$, where m' is an user-defined parameter (Line 1 in Function VMH). Then, we add the label of v into $\vec{s}$ (Line 2 in Function VMH). We use a hash function h_1 to accept each neighbor of v and add the smallest m' hash values into $\vec{s}$ (Lines 3–7 in Function VMH). Last, we return $\vec{s}$ (Line 8 in Function VMH).

Function. VMH(v)

1 Initialize an empty sketch $\vec{s}$ of length $m' + 1$ with $+\infty$;
2 $\vec{s}[1] \leftarrow f^G(v)$;
3 **for** $v' \in N(v)$ **do**
4   **if** $\exists i > 1, \vec{s}[i] = +\infty$ **then**
5          $|\quad s[i] \leftarrow h_1(v')$;
6   **else if** $\exists i > 1, \vec{s}[i] > h_3(l, k)$ **then**
7          $|\quad s[i] \leftarrow h_1(v')$;
8 Return $\vec{s}$;

Using the VMH function, we can map vertices to unique sketches. Then, a consistent hashing function can be used to assign sketches with unique identifiers to a physical partition index uniformly at random. Due to the definition of VMH, if two vertices have the same label and many common neighbors, then their VMH sketches are likely to the same. This indicates that they are likely be assigned to the same hash value using the consistent hashing function, so they are likely to be placed into the same worker. Then, when we initialize the tasks for two vertices having the same sketches (Line 5 in Algorithm 2), they are likely in the same worker which can avoid redundant communication.

Example 5. Let us consider v_6 and v_7 in Fig. 1(a). The labels of v_6 and v_7 are a, so $f^G(v_6) = f^G(v_7)$, which indicates $\mathbb{I}_G(v_6, v_7) = 1$. In addition, there are 4 common neighbors in $N(v_6) \cap N(v_7)$ and 6 in $N(v_6) \cup N(v_7)$, so the Jaccard similarity between $N(v_6)$ and $N(v_7)$ is $\frac{2}{3}$. Therefore, the vertex similarity between v_6 and v_7 is $\frac{2}{3}$, and the probability of v_6 and v_7 having the same hash value in VMH is $\frac{2}{3}$ which means it is high probabile that v_6 and v_7 are put into the same worker. □

Theorem 1. *Given two vertices v_1 and v_2, $Pr(VMH(v_1) = VMH(v_2)) = VS(v_1, v_2)$.*

Proof. If $f^G(v_1) \neq f^G(v_2)$, $\mathbb{I}_G(v_1, v_2) = 0$. Thus, $VS(v_1, v_2) = 0$. Meanwhile, since the labels of v_1 and v_2 are different, the first element in the sketches of v_1 and v_2 should be different according to Line 2 in Function VMH, which means $Pr(VMH(v_1) = VMH(v_2)) = 0$.

On the other hand, if $f^G(v_1) = f^G(v_2)$, $\mathbb{I}_Q(v_1, v_2) = 1$. Then, $VS(v_1, v_2) = \frac{N(v_1) \cap N(v_2)}{N(v_1) \cup N(v_2)}$, which is the Jaccard similarity between the neighbor sets of v_1 and v_2. The proposed technique is equivalent to incorporating MinHash proposed in [6]. Hence, it has been proven that $Pr(VMH(v_1) = VMH(v_2)) = VS(v_1, v_2)$ [6].

4   System Implementation

In this section, we use the above algorithm to implement a prototype system named PSP (*P*arallel *S*ubgraph query *P*rocessing), whose codes are released at

GitHub[1]. The graph is stored at the Hadoop Distributed File System (HDFS). We use MPICH-3.3.2 running on C++ for communication.

Algorithm 2: Parallel Hash Subgraph Query Processing Algorithm based on VMH

Input: A data graph G and a query graph Q
Output: The set M of all matches of Q over G

1 Generate candidate set $C(u)$ for $u \in V^Q$;
2 Select the start vertex u_r and generate the search order φ;
3 **for** *each vertex* $v \in C(u_r)$ **do**
4      $i \leftarrow VMH(v) \ mod \ n$;
5   Initialize a task t_v for v in W_i;
6 $M \leftarrow \emptyset$;
7 **for** *each worker* W_i ***in parallel*** **do**
8   **for** *each task* t_v *initialized from vertex* v *in* W_i **do**
9         $\mu \leftarrow \emptyset$;
10   Add (u_r, v) in μ;
11   **ParallelEnumerate**$(Q, G, \varphi, M, \mu, 2, W_i)$;
12 Return M;

In PSP, the data graph is stored as a set of vertices, where each vertex is stored with its adjacency list. Each machine in the parallel computing system only loads a fraction of vertices along with their adjacency lists into its memory, kept in a local vertex table. Vertices are assigned to machines by using our proposed locality sensitive hashing on their vertex IDs, and the aggregate memory of all machines is used to keep a big graph. The local vertex tables of all machines form a distributed key-value store where any task can request for the adjacency list using the ID of v.

5   Experiments

5.1   Experimental Setting

In this section, we conducted extensive experiments to test the effectiveness of our proposed techniques on real and synthetic graphs. Table 1 lists the statistics.

Table 1. Datasets

	Dataset	$|V^G|$	$|E^G|$	$|L^G|$
Real Graphs	DBLP	7,706,505	56,754,876	501
	Twitter	41,652,230	1,202,513,195	900
Synthetic Graphs	WatDiv	25,548,374	580,949,687	195

---

[1] https://github.com/bnu05pp/PSP.

For real graphs, DBLP [25] is a citation network of papers, where the venues are used as the labels. Twitter [18] is a social network containing users and social relations, where we randomly assign a label to each vertex. For synthetic graphs, we use WatDiv [2], which is a customizable synthetic knowledge graph generator that models an e-commerce platform. Entities in WatDiv are vertices and relations between two entities are edges. Each entity has some properties, and we randomly select one of them as the label. We generate a WatDiv graph of 100 million edges.

We design a generator to generate the query graphs based on the datasets. Given the number of subgraph queries N, we randomly chooses N data vertices. For each vertex, it generates a subgraph around this vertex by random walk. The number of edges ranges from 3 to 12, while the number of vertices is from 4 to 10. Each subgraph query generated this way can be guaranteed to have at least one match. By default, we generate 100 queries for each dataset.

We also use our proposed techniques to implement a prototype system named PSP (Parallel Subgraph query Processing), which is detailed in Sect. 4. We conduct all experiments on a cluster of 11 machines running Linux in Alibaba, each of which has two CPUs with six cores of 1.2 GHz. Each machine works as one worker, and has 32 GB memory and 200 GB disk storage. We select one of these machines as the coordinator machine.

We compare our method with two recent parallel subgraph query processing systems, G-thinker and Patmat. G-thinker [26] is a distributed framework that adopts a subgraph-centric vertex-pulling API and supports distributed subgraph query processing. Patmat is a prototype system [19] that covers representative distriubted subgraph query algorithms. Patmat implements subgraph query processing using different kinds of joins, including binary join and worst-case optimal generic join. Patmat also support two data placement methods, Hash and Triangle, where the former places vertices among workers via a hash function and the latter further incorporates the edges among the neighbors in each worker after hash placement.

5.2   Effect of Our Proposed Techniques

In this experiment, we leverage various datasets alongside a suite of 100 queries to assess the performance of the technique proposed in this study. We establish a baseline scenario (referred to as 'Basic'), devoid of any implemented optimization techniques. As depicted in Fig. 4, the PSP method that incorporates VMH noticeably outperforms the Basic scenario. This superior performance is attributable to PSP's strategic vertex placement, wherein vertices frequently qualifying as candidates for the same query vertices are co-located within the same workers. This tactic of vertex placement mitigates redundant communication of common neighbors during task initialization, thus enhancing overall performance efficiency.

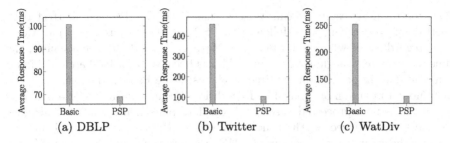

Fig. 4. Average Query Response Time

5.3   Comparison with Other Parallel Subgraph Query Systems

In this experiment, we use 100 queries on different graphs to compare the query performance of PSP against the four competitors. Figure 5 shows the experimental results. Here, Patmat-BJ and Patmat-GJ denote Patmat using "binary join" and Patmat using "worst-case optimal generic join". Since Patmat using triangle partitioning always achieves better performance than using hash partitioning, both Patmat-BJ and Patmat-GJ in Fig. 5 use triangle partitioning, Patmat fails over the largest graph Twitter. Generally, the query performances of PSP over different graphs can outperform others greatly, since PSP can avoid lots of intraquery communication.

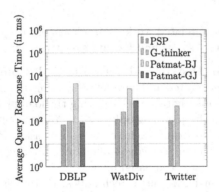

Fig. 5. Query Performance Comparison

5.4   Data Placement Performance

In this experiment, we evaluate the data placement performance over different graphs. Here, PSP in this paper uses VMH for data placement, G-thinker uses hash data placement, and Patmat uses Hash and Triangle data placement. The corresponding Patmats are denoted as Patmat-H and Patmat-T respectively. The data placement processing time of each method is shown in Table 2. From the experimental results, it can be seen that G-thinker has the shortest data partitioning processing time, followed by PSP, and Patmat is the worst. Moreover, regardless of whether Patmat uses Hash or Triangle data placement, the time it consumes is significantly more. This is because both PSP and G-thinker can read data in parallel on multiple machines, while Patmat only supports the main node to read on one machine. In addition, because Patmat's Triangle data placement is post-processed based on Hash data placement, Triangle consumes the most time. Moreover, the time consumed by PSP is more than G-thinker, because the computing operations of VMH are more complex than those of Hash. However, as can be seen from the experimental results in Fig. 5, the VMH data placement used by PSP can significantly reduce the query response time, so the additional overhead of VMH data placement is valuable.

Table 2. Data Placement Performance (in s)

Dataset	PSP	G-thinker	Patmat-H	Patmat-T
DBLP	19.99	11.31	177.37	218.84
Twitter	224.52	146.39	3,234.19	–
WatDiv	30.70	20.96	400.26	479.16

6   Related Work

There are two threads of related work: data placement of graphs in distributed and parallel computing systems and locality sensitive hashing.

Data Placement in Distributed and Parallel Computing Systems. Within the realm of distributed and parallel computing systems, the partitioning of graph structures commonly takes place during the load phase. The hash-based partitioning approach is frequently employed [19,26], leveraging a uniformly distributed hash function to assign each vertex to a worker. Patmat [19] extends this methodology by including the edges among neighboring vertices within each worker subsequent to hash-based assignment.

Beyond the scope of hash-centric techniques, there exist alternative graph partitioning strategies aimed at optimizing data placement. Application-centric graph partitioning [8,9] adapts to varying graph algorithms by learning computational and communicational cost functions, thereby refining either an edge-cut or a vertex-cut partition into a hybrid partition predicated on these cost functions. In the context of distributed RDF systems, H-RDF-3X [13] deploys METIS [16] to segregate RDF graphs while mitigating the number of edge crossings between distinct worker nodes. The MPC methodology [23] introduces an innovative partitioning scheme that aims to minimize the count of unique crossing properties in RDF graphs.

Locality Sensitive Hashing (LSH). Indyk [14] proposes the fast nearest neighbor search algorithm for massive high-dimensional data, which has been used in various ways.

LSH for Similarity. The work by Marçais et al. [21] introduces an LSH method for edit distance computations known as Order Min Hash (OMH), an advanced iteration of MinHash. Ji et al. [15] propose SBLSH, a data-independent locality-sensitive hashing technique, which they demonstrate to offer an unbiased estimate of angular similarity. Additionally, Hu et al. [12] put forward a nearly load-optimal LSH-centric join algorithm, accompanied by rigorous experimental validations in a distributed setting. Building on this, Aumüller et al. [4] provide a more profound analysis of the techniques suggested in [12], intending to reconcile their theoretical considerations with the practical aspects of a robust distributed similarity joining algorithm rooted in LSH.

LSH for Graph Database. Aluç et al. [3] formulate a novel locality-sensitive hashing (LSH) scheme named Tunable-LSH, crafted to cluster records within an RDF graph system. Remarkably, Tunable-LSH incorporates the latest query access patterns from the database for automatic tuning. Kiran et al. [17] devise a nearest-neighbor search algorithm that enables the exploration of graph collections for graphs bearing similarity to a designated query graph, utilizing the LSH method to construct the nearest-neighbor search data infrastructure. Furthermore, [27] offers a swift algorithm reliant on local sensitive hashing for large-scale graph similarity searches. This technique initiates by transforming the complex graph into a vector representation based on prototypes present in the database. Consequently, similarity searches within the Euclidean space are expedited by leveraging locally sensitive hashing, ensuring search accuracy.

7 Conclusion

In this paper, for data placement, we propose a vertex locality sensitive hashing methods, named vertex MinHash (VMH), which can place similar vertices into the same worker to reduce communication during query processing. Extensive experiments show that the proposed techniques can highly improve the query performance.

Acknowledgement. This work was supported by NSFC under grants (U20A20174), Science and Technology Major Projects of Changsha City (No. kh2205032), and Hunan Provincial Natural Science Foundation of China under grant 2022JJ30165.

References

1. Abbas, Z., Kalavri, V., Carbone, P., Vlassov, V.: Streaming graph partitioning: an experimental study. Proc. VLDB Endow. **11**(11), 1590–1603 (2018)
2. Aluç, G., Hartig, O., Özsu, M.T., Daudjee, K.: Diversified stress testing of RDF data management systems. In: Mika, P., et al. (eds.) ISWC 2014. LNCS, vol. 8796, pp. 197–212. Springer, Cham (2014). https://doi.org/10.1007/978-3-319-11964-9_13

3. Aluç, G., Özsu, M.T., Daudjee, K.: Building self-clustering RDF databases using tunable-LSH. VLDB J. **28**(2), 173–195 (2019)
4. Aumüller, M., Ceccarello, M.: Implementing distributed similarity joins using locality sensitive hashing. In: EDBT, pp. 1:78–1:90. OpenProceedings.org (2022)
5. Bi, F., Chang, L., Lin, X., Qin, L., Zhang, W.: Efficient subgraph matching by postponing cartesian products. In: SIGMOD, New York, NY, USA, pp. 1199–1214. Association for Computing Machinery (2016)
6. Broder, A.: On the resemblance and containment of documents. In: SEQUENCES, USA, p. 21. IEEE Computer Society (1997)
7. Fan, W.: Graph pattern matching revised for social network analysis. In: ICDT, New York, NY, USA, pp. 8–21. Association for Computing Machinery (2012)
8. Fan, W., et al.: Application driven graph partitioning. In: SIGMOD, New York, NY, USA, pp. 1765–1779. Association for Computing Machinery (2020)
9. Fan, W., Xu, R., Yin, Q., Yu, W., Zhou, J.: Application-driven graph partitioning. VLDB J. **32**(1), 149–172 (2023)
10. Garey, M.R., Johnson, D.S.: Computers and Intractability: A Guide to the Theory of NP-Completeness. W. H. Freeman & Co., USA (1979)
11. Han, W.-S., Lee, J., Lee, J.-H.: Turbo$_{iso}$: towards ultrafast and robust subgraph isomorphism search in large graph databases. In: SIGMOD, New York, NY, USA, pp. 337–348. Association for Computing Machinery (2013)
12. Hu, X., Yi, K., Tao, Y.: Output-optimal massively parallel algorithms for similarity joins. ACM Trans. Database Syst. **44**(2), 61–636 (2019)
13. Huang, J., Abadi, D.J., Ren, K.: Scalable SPARQL querying of large RDF graphs. PVLDB **4**(11), 1123–1134 (2011)
14. Indyk, P.: Nearest neighbors in high-dimensional spaces. In: Handbook of Discrete and Computational Geometry, 2nd edn., pp. 877–892. Chapman and Hall/CRC (2004)
15. Ji, J., Li, J., Yan, S., Zhang, B., Tian, Q.: Super-bit locality-sensitive hashing. In: NIPS, NIPS 2012, Red Hook, NY, USA, pp. 108–116. Curran Associates Inc. (2012)
16. Karypis, G., Kumar, V.: A fast and high quality multilevel scheme for partitioning irregular graphs. SIAM J. Sci. Comput. **20**(1), 359–392 (1998)
17. Kiran, P., Sivadasan, N.: Scalable graph similarity search in large graph databases. In: 2015 IEEE Recent Advances in Intelligent Computational Systems (RAICS), pp. 207–211 (2015)
18. Kwak, H., Lee, C., Park, H., Moon, S.: What is Twitter, a social network or a news media? In: WWW, New York, NY, USA, pp. 591–600. Association for Computing Machinery (2010)
19. Lai, L., Qing, Z., Yang, Z., Jin, X., Lai, Z., Wang, R., Hao, K., Lin, X., Qin, L., Zhang, W., Zhang, Y., Qian, Z., Zhou, J.: Distributed Subgraph Matching on Timely Dataflow. Proc. VLDB Endow. **12**(10), 1099–1112 (2019)
20. Leskovec, J., Rajaraman, A., Ullman, J.D.: Mining of Massive Datasets, 2nd edn. Cambridge University Press, Cambridge (2014)
21. Marçais, G., DeBlasio, D.F., Pandey, P., Kingsford, C.: Locality-sensitive hashing for the edit distance. Bioinform. **35**(14), i127–i135 (2019)
22. Özsu, M.T., Valduriez, P.: Principles of Distributed Database Systems, 4th edn. Springer, Cham (2020). https://doi.org/10.1007/978-3-030-26253-2
23. Peng, P., Ozsu, M., Zou, L., Yan, C., Liu, C.: MPC: minimum property-cut RDF graph partitioning. In: ICDE, Los Alamitos, CA, USA, pp. 192–204. IEEE Computer Society (2022)

24. Pržulj, N., Corneil, D.G., Jurisica, I.: Efficient estimation of graphlet frequency distributions in protein-protein interaction networks. Bioinformatics **22**(8), 974–980 (2006)
25. Tang, J., Zhang, J., Yao, L., Li, J., Zhang, L., Su, Z.: Arnetminer: extraction and mining of academic social networks. In: SIGKDD, New York, NY, USA, pp. 990–998. Association for Computing Machinery (2008)
26. Yan, D., Guo, G., Chowdhury, M.M.R., Özsu, M.T., Ku, W., Lui, J.C.S.: G-thinker: a distributed framework for mining subgraphs in a big graph. In: ICDE, pp. 1369–1380, Dallas, TX, USA. IEEE (2020)
27. Zhang, B., Liu, X., Lang, B.: Fast graph similarity search via locality sensitive hashing. In: Ho, Y.-S., Sang, J., Ro, Y.M., Kim, J., Wu, F. (eds.) PCM 2015. LNCS, vol. 9314, pp. 623–633. Springer, Cham (2015). https://doi.org/10.1007/978-3-319-24075-6_60

DUTD: A Deeper Understanding of Trajectory Data for User Identity Linkage

Qian Li, Qian Zhou, Wei Chen$^{(\boxtimes)}$, and Lei Zhao

Shool of Computer Science and Technology, Soochow University, Suzhou, China
20204227004@stu.suda.edu.cn,
{qzhou0,robertchen,zhaol}@suda.edu.cn

Abstract. The widespread adoption of mobile devices and sensing technologies has significantly boosted the availability of trajectory data, thereby facilitating studies on users' spatio-temporal behaviors. Among these studies, the task of trajectory-based user identity linkage (UIL) has attracted extensive attention recently. Existing efforts for the task mainly rely on data mining methods and sequence models to extract features and measure user identity similarity. Most of them commonly regard trajectories as mere sequences, thus failing to take full advantage of the hidden spatio-temporal characteristics within trajectories, such as temporal regularities and mobility patterns. To fill the gap, we propose a novel model namely **DUTD** (A **D**eeper **U**nderstanding of **T**rajectory **D**ata for User Identity Linkage). Specifically, we first adopt the skip-gram module on historical trajectories to generate grid representations of each trajectory. Subsequently, a transformer-based encoder is designed to capture long-term temporal regularities and mobility patterns. Particularly, to enhance the effectiveness and efficiency of the encoder, we introduce two self-supervised learning tasks: trajectory recovery and trajectory contrastive learning. Ultimately, to capture the inter-trajectory correlations between users, a matcher comprising a multi-interaction module and a prediction layer is designed. The extensive experiments conducted on two real-world datasets demonstrate the superiority of our proposed model compared to the state-of-the-art methods.

Keywords: Trajectory · User identity linkage · Spatio-temporal characteristics · Pre-training

1 Introduction

The ubiquity of location-capturing devices and the flourishing of location-based services have resulted in the exponential growth of GPS-recorded trajectories and bring great convenience for a deeper understanding of users' behaviors [20]. Among existing studies, the task of user identity linkage (UIL), which is proposed by Zafarani et al. [23] and aims to identify the same user across different platforms, has attracted increasing attention, due to its wide-ranging applications, such as location-ware maps and cross-domain recommendation [4].

X. Song et al. (Eds.): APWeb-WAIM 2023, LNCS 14331, pp. 48–62, 2024.
https://doi.org/10.1007/978-981-97-2303-4_4

As an important branch of UIL, the trajectory-based user identity linkage is to identify the same user across different platforms based on the trajectories sampled by GPS-enabled devices. The cornerstone of realizing the UIL task is to explore the spatio-temporal characteristics of trajectories. Traditional studies adopt different data mining methods to model human mobility, such as DBSCAN [21] and kernel density estimation (KDE) [4]. However, these methods rely on presumed parameters or manually designed features. On the contrary, the deep learning approaches can adaptively adjust model parameters according to the target task, and automatically extract significant features from historical data. Recent work has been inspired to encode trajectories for feature extraction [8] by recurrent neural networks (RNN). Despite the promising results achieved by the study, it falls short in capturing the long-term dependency of sampled points in a trajectory, and ignores the users' periodic behaviors hidden in trajectories.

Intuitively, the GPS-recorded trajectories are usually data-intensive and exhibit periodic repetitions, and contain rich spatio-temporal characteristics. Specifically, on the one hand, the users' behaviors may present temporal regularities. For example, a commuter may arrive at work region around 9:00 am and leave for home around 6:00 pm on each weekday. On the other hand, the mobility patterns can be extracted from historical trajectories. By way of illustration, the above-mentioned office worker may follow a fixed route from home to work region, then to the gym, and finally back home. These features are crucial for achieving more accurate UIL, while neither traditional machine learning methods nor sequence model-based methods can make full use of these spatio-temporal characteristics.

To tackle the above-discussed defects that lie in existing methods, we propose a novel model DUTD to extract the spatio-temporal characteristics from trajectories for high-performance UIL. In detail, the model is composed of the following three components. 1) **Grid feature extractor**, as the GPS-enabled devices usually sample points with a fixed and short interval [3], we first divide the entire space into non-overlapping grids to reduce data redundancy. Then, the skip-gram model [16], which has proven to be successful on predicting context words [1], is applied to obtain initial grid embeddings that are inputs of the following encoder. 2) **Transformer-based encoder**, to further capture the mobility patterns and temporal regularities, the trajectory of each user in an identity pair is fed into a transformer for encoding, where a stacked self-attention mechanism enables it to model long-term dependencies and extract global trajectory features. Particularly, to improve the learning effectiveness and efficiency of the encoder, we introduce two self-supervised learning tasks, namely trajectory recovery and trajectory contrastive learning, which directly take mobility patterns and trajectory semantics as training targets without task-specific labels. The former is to capture local patterns and regularities by masking continuous subsequences in trajectories, and the latter is to capture the correlations of the same user and the difference between different users in terms of spatio-temporal behaviors by contrastive learning. 3) **Matcher**, to capture the inter-trajectory relationships for measuring the similarity of different users, we design a matcher consisting of a multi-interaction module and a prediction layer. Particularly, the multi-interaction module captures the correlations of user trajectories through linear interaction and sparse co-attention mechanism, and a probability distribution is generated by the prediction layer.

In summary, the main contributions of this paper are summarized as follows:

- To further improve the performance of existing work in the task of trajectory-based UIL, we develop a novel model entitled DUTD, where the deep spatio-temporal characteristics involved in historical trajectories are fully explored.
- In our model, a transformer-based encoder is introduced to capture the long-term temporal regularities and mobility patterns in trajectories. We pre-train the encoder with two self-supervised learning tasks to improve the representation learning capability and design a matcher to improve model performance by capturing inter-trajectory correlations between two users.
- We conduct extensive experiments on two real-world datasets. The experiment results demonstrate that our proposed model significantly outperforms the state-of-the-art baselines.

2   Related Work

Early studies have explored different ways to link user identities by using service-specific data such as user profile attributes and social graphs. For instance, Zafarani et al. [23] investigate the possibility of linking identities across various communities on the web. Mu et al. [17] propose a framework based on latent user space modeling by utilizing the basic user profiles. Unfortunately, the user profiles may be ambiguous and unreliable because users may fill in fake information (e.g., name, gender) in their profiles. Vosecky et al. [19] introduce a method to match web profiles from different domains and extend its effectiveness by incorporating the user's social network. However, different friendship networks tend to be heterogeneous and may vary substantially in terms of scale, which poses a challenge to measuring their similarity accurately.

Apart from the above work, recent advances pay attention to location-based UIL. For example, Jin et al. [13] formalize the UIL task as a k-nearest neighbor (k-NN) query based on signature similarity and develop a dimensionality reduction strategy with WR-tree indexing to speed up the search. Feng et al. [8] extract features from trajectories through a recurrent neural network (RNN) and introduce a co-attention mechanism to capture correlations between trajectories. Chen et al. [3] propose a grid-based KDE method for computing user similarity, and design an entropy-based weighting scheme to alleviate the challenge of negative coincidence. As the trajectory data usually contain rich information about users' behaviors, various studies on spatio-temporal data have emerged in recent years, such as trajectory-user linking (TUL), and trajectory similarity computation. Gao et al. [9] utilize RNN-based models to learn mobility patterns from trajectories and link them to users. However, the standard RNN-based models lack the ability to model long-term dependencies of trajectories. Deng et al. [5] employ contrastive learning to learn the latent representations of trajectories and calculate the similarity between sub-trajectories generated from one dense trajectory, which fails to model the relationship of trajectories from different datasets. Due to the inherent limitations of these models, they are not effective in solving the UIL task. In contrast, our model can not only extract spatio-temporal features from trajectories but also model the inter-trajectory correlations between users.

3   Preliminary

In this section, we introduce some notations used throughout this paper and define the trajectory-based UIL problem.

Definition 1. *Trajectory.* *A trajectory is a sequence of points recorded by GPS-enabled devices for a moving object. Each trajectory $T = \{p_1, \cdots, p_m\}$ is a sequence of tuples where $p_i = (lat, lon, t)$ stands for the longitude and latitude information of a sampled point at timestamp t.*

Definition 2. *Trajectory-based User Identity Linkage.* *Given a user identity pair (u_1^i, u_2^j) where $u_1^i \in U_1$ and $u_2^j \in U_2$, U_1 and U_2 denote the set of user identities on two different platforms, each user identity is associated with a set of trajectories, i.e., T_1^i and T_2^j. Our goal is to learn a binary classifier f that infers whether u_1^i and u_2^j refer to the same natural person or not, i.e.,*

$$f(u_1^i, u_2^j) = \begin{cases} 1, & u_1^i \text{ and } u_2^j \text{ belong to the same user} \\ 0, & \text{otherwise} \end{cases} \tag{1}$$

4   Proposed Model

In this section, we introduce the details of our proposed model DUTD. As shown in Fig. 1, DUTD consists of three key components: a grid feature extractor, a transformer-based encoder, and a matcher. We first employ a skip-gram module on historical trajectories to extract features of trajectory units. The trajectories of a given user identity pair are then fed into a transformer-based encoder to extract spatio-temporal characteristics. In particular, we design two self-supervised learning tasks to pre-train the encoder aiming to improve the learning capability. After obtaining the trajectory representations, we design a matcher to capture the inter-trajectory correlations between two users and perform the UIL task.

4.1   Grid Feature Extractor

Similar to the strategy commonly used in spatio-temporal data analysis [24], we divide the entire space into grids of equal size and map each point of a trajectory into a specific grid that is called unit throughout the paper. In this manner, each

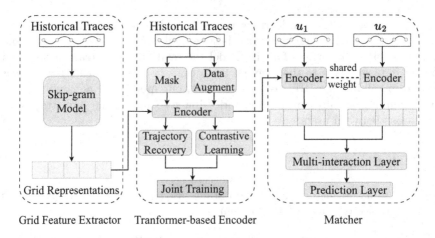

Fig. 1. Overview of the proposed model DUTD

trajectory in the given dataset is transformed into a sequence of grids. In order to obtain the contextual information of the trajectory T, we adopt the skip-gram model to obtain the grid embedding by maximizing the log probability:

$$Loss = -\frac{1}{N}\sum_{i=1}^{n}\sum_{-w<j<w,j\neq0} logP(g_j|g_i) \tag{2}$$

$$P(g_j|g_i) = \frac{exp(v_{g_j}^T v_{g_i})}{\sum_{k=1}^{w} exp(v_{g_k}^T v_{g_i})} \tag{3}$$

where g_i represents current grid and v_{g_i} represents its corresponding representation. g_j stands for the context neighbor of g_i and v_{g_j} stands for its representation. N is the number of grids and w is the window size.

4.2   Tranformer-Based Encoder

After obtaining the initial grid embedding, the trajectory of each user is then modeled by the transformer-based encoder. We introduce this module with three parts as illustrated in Fig. 2. The first is a temporal embedding layer, which captures the periodicity of user trajectory in terms of day and week, respectively. The second is a self-attention module for modeling long-term temporal dependencies in trajectories. The last is the self-supervised learning tasks, which are designed to improve the learning effectiveness and efficiency of the encoder.

Temporal Embedding Layer. To encode the timestamps of grids corresponding to trajectory T, we first convert their timestamps to minute and day-of-week indices. Note that there are 1440 min in a day and seven days in a week. Next, we use two embedding layers to generate temporal embeddings at two levels of

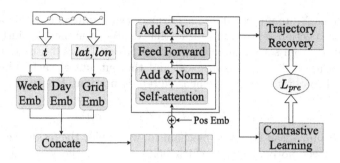

Fig. 2. The structure of transformer-based encoder

time information, respectively. By encoding the timestamps in this way, we can capture the periodicities in user behaviors. Details are as below:

$$te_{m_i} = Emb(m_i), \ te_{d_i} = Emb(d_i) \tag{4}$$

where $m_i \in \{0, 1, \cdots, 1439\}$ and $d_i \in \{0, 1, \cdots, 6\}$ are the minute index and day-of-week index of the grid g_i, $te_{m_i} \in \mathbb{R}^d$ and $te_{d_i} \in \mathbb{R}^d$ are both temporal embeddings of g_i.

Subsequently, we fuse these embeddings with the position encoding to obtain the comprehensive representation $x_i \in R^d$ of the grid g_i:

$$x_i = v_{g_i} + te_{m_i} + te_{d_i} + pe_i \tag{5}$$

where v_{g_i} represents the initial grid representation obtained by the skip-gram model, te_{m_i} and te_{d_i} are the corresponding temporal embeddings, and pe_i represents the position encoding used in the transformer [18]. In this way, the following self-attention mechanism can not only discover temporal regularity but also be aware of the position information of the input trajectory.

Self-attention Module. After obtaining the grid embedding matrix $X \in \mathbb{R}^{|T| \times d}$ of the trajectory T, we feed it into the self-attention module which is the core mechanism of the transformer. The details are as follows:

$$(Q, K, V) = X(W^Q, W^K, W^V) \tag{6}$$

$$A = softmax(\frac{QK^T}{\sqrt{d}})V \tag{7}$$

where W^Q, W^K, $W^V \in \mathbb{R}^{d \times d}$ are learnable projection matrices. Q, K, and V are the query matrix, key matrix, and value matrix respectively, which are transformed from the input representations X. The attention scores are calculated via scaled inner-dot products, $softmax(\cdot)$ is the activation function, and A is the output of the self-attention operation.

Furthermore, we employ a multi-head attention mechanism to model the dependency in the trajectory from multiple perspectives. Specifically, the input representations X are projected into h subspaces with different learnable parameters, each projection represents a head that enables the model jointly focuses on information from several independent subspaces. The self-attention mechanism is then applied to each subspace, and the outputs of each subspace are concatenated to obtain the comprehensive representations $E \in \mathbb{R}^{|T| \times d}$:

$$E = FC(concat(A_1, A_2, \cdots, A_h)) \tag{8}$$

After the multi-head attention mechanism, we employ the residual connection followed by layer normalization, which helps to avoid the vanishing gradient problem. Finally, the output representations E are passed through a feed-forward network (FFN) consisting of two layers of linear transformations and a ReLU activation function. Next, the residual connection and layer normalization are applied. Details are as follows:

$$\tilde{E} = ReLU(EW_1 + b_1)W_2 + b_2 \tag{9}$$

$$\tilde{X} = LayerNorm(E + \tilde{E}) \tag{10}$$

where W_1, $W_2 \in \mathbb{R}^{d \times d}$, b_1, and b_2 are learnable parameters. $\tilde{E}$ is the output of FFN. By stacking multiple self-attention modules, we can effectively capture complex temporal dependencies in the trajectory. $\tilde{X}$ is the final output matrix stacked by grid embeddings. To get a vector representation of the trajectory T, we add a [CLS] token at the first position and use the token's embedding vector e as the representation of the entire trajectory.

Self-supervised Learning Tasks. Inspired by the classic transformer-based model BERT [6], we introduce two self-supervised learning tasks for the target UIL task to pre-train our model for the purpose of maximizing the utilization of the available data. The masked language modeling (MLM) and next sentence prediction (NSP) proposed in BERT are designed to improve the representation ability of words and sentences, which are different tasks compared with our study, thus we cannot directly apply the original self-supervised learning tasks. Therefore, we adapt and design two new self-supervised tasks, which are described below.

1) Trajectory Recovery. In the original MLM task, each word in a sentence is randomly masked with a certain probability and is predicted by the model, which forces the model to learn the relationship between words in the sentence. However, applying this method to masked individual points in a trajectory does not work well, since the model can easily infer the value of the masked point, due to the low task difficulty caused by people's travel habits. In general, people tend to visit nearby locations [7], which leads to the phenomenon that temporally consecutive points in a trajectory also be close in location. To guide the model to detect this pattern for better feature representation, we mask several contiguous subsequences of length l for a trajectory, whose total length is $p\%$ of the trajectory. The task compels the model to capture the local pattern of the trajectory in order to accurately predict the masked subsequences.

2) Trajectory Contrastive Learning. To perceive the correlations between users' trajectories, we introduce a contrastive learning method. The principle is to guide the model to understand the semantics of trajectories and obtain distinguishable representations by making positive samples with similar semantics closer and negative samples farther apart. The key to contrastive learning lies in which data augmentation method is used to construct different views. We employ an effective method namely dropout, which has been demonstrated to be feasible in SimCSE [10]. The essence is to randomly discard some elements in the embedding layer with a certain probability and set their values to zero.

Training. We first feed each trajectory into the transformer twice to get two views, which are different due to the built-in dropout of the transformer. Then we take the two views as positive samples, and other trajectories in the batch as negative samples. Following the contrastive framework [2], we employ an intra-batch negative cross-entropy objective called *InfoNCE* loss for training, which maximizes the agreement between positive samples and minimizes the agreement between negative samples. Formally, the loss function for a positive pair (t_i, t_j) is defined as:

$$L_{i,j} = -log\frac{exp(sim(t_i, t_j)/\tau)}{\sum_{k=1}^{2N} \mathbb{1}_{[k\neq i]}exp(sim(t_i, t_k)/\tau)} \quad (11)$$

where $sim(t_i, t_j)$ is the cosine similarity. τ is the temperature hyperparameter that controls the contribution of the negative samples, and $\mathbb{1}$ is the indicator that equals to 1 if $k \neq i$, and zero otherwise.

Overall, we jointly pre-train our model with the two self-supervised learning tasks. The pre-training loss L^{pre} is defined as:

$$L^{pre} = \lambda L^{mask} + (1 - \lambda)L^{cl} \quad (12)$$

where L^{mask} and L^{cl} are the loss of trajectory recovery and trajectory contrastive learning, respectively. λ is the hyperparameter to balance the loss of two tasks.

4.3   Matcher

After encoding trajectories with the above encoder, we obtain the respective trajectory representations e_1 and e_2 of u_1 and u_2 in a given user pair. To capture the correlations between user behaviors for better UIL, we design a matcher, which is composed of a multi-interaction module and a prediction layer.

Multi-interaction Module. In fact, the multi-interaction module consists of two parts, one is the linear interaction and the other is the sparse co-attention mechanism.

Specifically, linear interaction is a collective term for subtraction and dot product between trajectory representations, which is used to capture low-order correlations between users. Through the subtraction and dot product, we can obtain the trajectory-level correlation vectors l_1 and l_2.

As a complement to the linear interaction, we further introduce a sparse co-attention mechanism to capture high-order correlations. The sparse co-attention mechanism can dynamically adjust the interaction weight between the trajectory representations of two users, thus providing a fine-grained similarity measure. Specifically, we first generate an affinity matrix C based on the grid embedding matrixs $\tilde{X}_1$ and $\tilde{X}_2$ of the two users:

$$C = tanh(\tilde{X}_1^T W_c \tilde{X}_2) \tag{13}$$

where $W_c \in \mathbb{R}^{d \times d}$ indicates the weights. After this, we compute the co-attention scores via the affinity matrix:

$$M_1 = tanh(W_{m_1} \tilde{X}_1 + (W_{m_2} \tilde{X}_2)C), \quad a_1 = sparsemax(w_{a_1}^T M^1) \tag{14}$$

$$M_2 = tanh(W_{m_2} \tilde{X}_2 + (w_{m_1} \tilde{X}_1)C^T), \quad a_2 = sparsemax(w_{a_2}^T M^2) \tag{15}$$

where $W_{m_1}, W_{m_2} \in R^{k \times d}$, $w_{a_1}, w_{a_2} \in R^k$ are the weight parameters. M_1 and M_2 are the outputs after co-attention interaction. a_1 and a_2 are co-attention scores, essentially denoting the probability that a grid of one user's trajectory appears in another user's trajectory for an identity pair. Sparsemax [15] is an alternative to the softmax function that makes useful information more compact. It tends to yield sparse probability distributions and the formula is as follows:

$$sparsemax(x) = \underset{p \in \delta^d}{\arg\min} \|p - x\|^2 \tag{16}$$

where x is the input vector and p is the output vector. Compared with $softmax(\cdot)$ that may assign weights to useless data, $sparsemax(\cdot)$ tends to produce zeros for low scores in vectors, which retains the most important factors.

Finally, we compute the weighted sum of trajectory representations based on the co-attention scores and the respective trajectory representations of two users to obtain the grid-level correlation vectors s_1 and s_2:

$$s_1 = \sum_{m=1}^{|T_1|} a_1^m \tilde{e}_1, \quad s_2 = \sum_{n=1}^{|T_2|} a_2^n \tilde{e}_2 \tag{17}$$

Prediction Layer. The prediction layer based on MLP (multi-layer perceptron) aims to convert the UIL task into a binary classification task. For the two users of an identity pair, we first concatenate their trajectory-level and grid-level correlation vectors obtained from the above multi-interaction module to generate a fused vector z, which is used as the input of the prediction layer. Finally, the probability $\tilde{p}$ that indicates two identities belong to the same user is generated through the sigmoid function. Details are as follows:

$$\tilde{z} = tanh(w_3 z + b_3), \quad \tilde{p} = sigmoid(w_4 \tilde{z} + b_4) \tag{18}$$

where w_3, w_4, b_3, and b_4 denote the learnable parameters. Correspondingly, we train the model by minimizing the binary cross-entropy loss function.

5   Experiment

5.1   Datasets

We briefly describe the two datasets used in the paper as follows: 1) **GeoLife** [25] is a GPS-recorded dataset collected in Beijing, which captures the trajectory information of 182 users with 17,621 trajectories between April 2007 and August 2012. This dataset covers various outdoor activities of users, ranging from mundane routines like commuting to work or school to leisure activities like shopping, sightseeing, hiking, and cycling. Notably, 91% of the trajectories are densely logged, with points recorded every 1–5 s or every 5–10 m. 2) **BJ taxi** contains 201,328 trajectories collected by devices embedded in taxis, which is generated by 12,577 taxis in Beijing from October 1, 2012 to October 15, 2012. Each taxi is associated with a set of historical trajectories and the time interval of most points in a trajectory is about one minute. We randomly select 400 taxis with 7200 trajectories as our research objects.

During data processing, we uniformly resample both datasets every 5 min, which reduces redundancy while preserving the main information of the trajectories. In addition, we adopt the same strategy as [3,12], i.e., randomly divide each original dataset into two parts D_1 and D_2 to simulate two platforms. Each user exists in both sub-datasets, whose trajectories are divided into two parts accordingly. In our experiments, we use the first 60% of each user's trajectories as the training set on the dataset it belongs to, the following 20% as the validation set, and the remaining 20% as the test set.

5.2   Baselines

We compare DUTD with the following state-of-the-art baselines. Among them, T3S and CL-TSim are models for trajectory similarity calculation. HFUL and DPLink are models of UIL based on spatio-temporal data. Note that we use the source codes released by the authors and the parameter settings recommended in the original papers which are fine-tuned on each dataset to be optimal.

TULER [9]: The model leverages recurrent networks to encode trajectories in a single mobility dataset and identify the user who produced the trajectories.

T3S [22]: The method preserves the spatial and structural information of trajectories for similarity computation by applying LSTM and self-attention based networks.

CL-TSim [5]: The approach employs contrastive learning to learn the latent representations of trajectories and calculate the similarity between trajectories.

HFUL [3]: The method measures users' similarity based on KDE and designs an entropy-based weight scheme to reduce the impact of negative coincidence.

DPLink [8]: The paper pre-trains the model by a simple task to address the heterogeneity of data and introduces a co-attention mechanism to capture the potential correlations between trajectories.

5.3   Parameter Setting and Evaluation Metrics

Following previous work [5], we set the width of each grid to 100 m. Similar to [11], our encoder is a 6-layer transformer with a hidden size of 256, and the number of self-attention heads is 8. Besides, we set the dropout ratio to 0.1 for all layers. During the pre-training stage of the encoder, we set the mask length l as 2 and the mask ratio $p\%$ is 15% in the trajectory recovery task. λ is set to 0.6 to balance the pre-training loss. During the fine-tuning phase, we choose AdamW [14] as our optimizer with a learning rate of 0.0002. We evaluate the model performance by metrics commonly used in classification tasks, including precision, recall, and F1 score. Precision represents the prediction accuracy of positive sample results. Recall is the ratio of the number of correctly classified positive samples to the number of true samples. F1 score is the harmonic average of precision and recall to comprehensively consider the classification performance. In particular, we measure the model performance based on the F1 score in the training phase and select the optimal parameters in the validation step.

5.4   Performance Comparison

We conduct experiments on both datasets and the results are given in Table 1, with the best in bold and the second best underlined. From this table, we have the following observations based on the most important metric F1 score.

Table 1. Performance comparison of DUTD with baselines on two datasets.

Method	Geolife			BJ taxi		
	Precision	Recall	F1	Precision	Recall	F1
TULER	0.8429	0.5552	0.6694	0.4480	0.6594	0.5335
CL-TSim	0.6410	0.6897	0.6644	0.3737	0.6825	0.4830
T3S	0.8684	0.5690	0.6875	0.3971	0.5619	0.4653
HFUL	0.6923	0.7500	0.7199	0.6418	0.5375	0.5850
DPLink	<u>0.9357</u>	<u>0.8034</u>	<u>0.8646</u>	<u>0.6963</u>	<u>0.7106</u>	<u>0.7034</u>
DUTD	**0.9533**	**0.8448**	**0.8958**	**0.7810**	**0.8356**	**0.8073**

DUTD achieves superior performance on all metrics on both datasets. This demonstrates the effectiveness of the transformer for exploring temporal regularities and mobility patterns in trajectory and the necessity of pre-training with self-supervised learning tasks. Meanwhile, we notice that DPLink performs best among all baselines, which owes to its design of an attention-based selector to capture the correlations between features of two trajectories. Likewise, DPLink introduces a pre-training mechanism, but the performance is still worse than our model. This is because we introduce two self-supervised tasks that are closer to the target task, and the transformer-based encoder can capture long-term dependencies compared to RNN. As the only method based on data mining techniques in these baselines, HFUL is second only to DPLink in terms of F1 score. This can be attributed to two reasons: 1) HFUL employs a KDE-based algorithm to measure the similarity. 2) An entropy-based weighting scheme is designed to reduce negative coincidences. These techniques enable HFUL to achieve UIL with high performance. However, the data mining method is still inferior to deep learning technology in terms of the depth of feature extraction, so the room for improvement is limited. Besides, we compare our model with three other state-of-the-art trajectory representation methods: TULER, T3S and CL-TSim. As expected, our model significantly outperforms these models. Because TULER focuses on the trajectory user linking on the single mobility dataset while ignoring the potential correlations between two different mobility datasets. T3S and CL-TSim obtain trajectory representations by measuring the similarity of subtrajectories generated from one dense trajectory but fall short in modeling the relationship between trajectories from different datasets.

Overall, all methods perform better on the GeoLife dataset than on the BJ taxi dataset. This can be attributed to two reasons: 1) The time span of the GeoLife dataset is larger than that of the BJ taxi dataset, which means that we can extract the long-term regularities of user behavior. 2) The GeoLife dataset captures diverse activities of users, including daily routines such as commuting as well as recreational and sports activities. While the BJ taxi dataset is limited to street routes and does not involve private areas, resulting in lower discriminability for identification.

5.5 Ablation Study

We conduct the ablation study by independently removing different key components of DUTD to verify their impacts on model performance. Specifically, we design the following variants and conduct experiments on GeoLife, and the experiments on BJ taxi are omitted here due to the space limitation. 1) *w/o skipgram*: The model has no skip-gram module to extract grid features. 2) *w/o pretrain*: The encoder of the model is trained from scratch. 3) *w/o TR*: The trajectory recovery task is absent in the pre-training phase. 4) *w/o CL*: The contrastive learning task is absent in the pre-training phase. 5) *w/o linear-IR*: The linear interaction is removed from the matcher. 6) *w/o SA*: The sparse co-attention is removed from the matcher. 7) *w/softmax*: Sparsemax is applied as the transformation function for the sparse co-attention mechanism. We divide the experiment results into four groups for analysis, and our observations are drawn from Fig. 3. Note that all analyses are based on F1 score.

Impact of Skip-Gram Model. According to the results in Fig. 3(a), the variant *w/o skipgram*, which does not extract grid features through the skip-gram module, has a degradation in model performance compared to the original model. This is because the skip-gram model learns grids' embeddings by capturing the relationship between them, thereby helping the model understand the context of the trajectory. Moreover, the grid representation provides meaningful initialization for the embedding layer in the encoder, which helps the model better understand the trajectory semantics, thereby improving the accuracy of the UIL task.

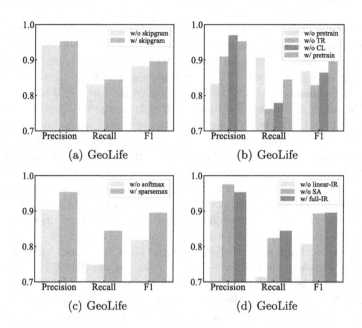

Fig. 3. Results of ablation study on variants. *w/ skipgram* in 3(a), *w/ pretrain* in 3(b), *w/ sparsemax* in 3(c) and *w/ full-IR* in 3(d) all refer to the original model DUTD.

Impact of Pertaining Strategy. It can be seen from Fig. 3(b) that the model trained from scratch performs worse than the original model. This demonstrates that pre-training can help the model fully utilize the spatio-temporal characteristics in the trajectory data to learn better feature representations and achieve more accurate UIL. Specifically, the trajectory recovery task help the model learn the context dependencies between grids, thereby extracting regularity in user behavior. The trajectory contrastive learning task guides the model to learn how to link the same user through similar trajectories. In particular, *w/o TR* is worse than *w/o CL*, which indicates that the trajectory recovery task can extract more sufficient and effective spatio-temporal characteristics than the contrastive learning task. By designing suitable self-supervised tasks, the pre-trained model can obtain sufficient prior knowledge from trajectories.

Impact of Transformation Functions in Sparse Co-attention Module. We can observe from Fig. 3(c) that the performance of the model drops significantly after replacing *sparsemax* in the sparse co-attention mechanism with *softmax*. This is because *softmax* still assigns weights to less important output nodes, which can easily cause the output of important features and unimportant features to be indistinguishable. While *sparsemax* discretizes the output by assigning zero probability to low-scoring choices, and finally only outputs more important features, which helps the co-attention mechanism selectively filter information. Furthermore, *sparsemax* prevents the over-learning problem of *softmax*, which makes it more suitable for pre-training scenarios.

Impact of the Multi-interaction Module in Matcher. As can be seen from Fig. 3(d) that whether the sparse co-attention module or the linear interaction module is removed, the performance of the model shows a decline in varying degrees. Not only that, the performance drops the most after removing the linear interaction module, which can reach about 10%. This can be attributed to the small size and single type of our datasets, which results in the complex co-attention mechanism tends to overfit and cannot bring expected performance improvement. Instead, simple linear interactions (subtraction and dot product) are less prone to overfitting. In a word, capturing the correlations between the trajectories of different users is an extremely important step in the UIL task, and even the simplest interactions can handle such task well.

Notably, the parameter analysis about embedding size, learning rate, dropout, and so on are omitted due to the space limitation.

6 Conclusion

We propose a novel trajectory-based UIL model DUTD to explore the spatio-temporal characteristics in trajectories and capture the inter-trajectory correlations. Specifically, we first utilize the skip-gram model on the historical data to obtain the initial embeddings of grids. Then a transformer-based encoder is designed to extract spatio-temporal characteristics by capturing long-term dependencies. Particularly, we introduce two self-supervised learning tasks to improve the effectiveness of learning. Moreover, a matcher is designed to capture inter-trajectory correlations for more accurate UIL. The extensive experiments conducted on two real-world datasets demonstrate the superiority of our proposed model DUTD compared to the state-of-the-art methods.

Acknowledgments. This work is supported by the National Natural Science Foundation of China No. 62272332, the Major Program of the Natural Science Foundation of Jiangsu Higher Education Institutions of China No. 22KJA520006.

References

1. Chen, M., Zhao, Y., Liu, Y., Yu, X., Zheng, K.: Modeling spatial trajectories with attribute representation learning. IEEE Trans. Knowl. Data Eng. **34**(4), 1902–1914 (2020)
2. Chen, T., Kornblith, S., Norouzi, M., Hinton, G.E.: A simple framework for contrastive learning of visual representations. In: ICML, pp. 1597–1607 (2020)
3. Chen, W., Wang, W., Yin, H., Zhao, L., Zhou, X.: HFUL: a hybrid framework for user account linkage across location-aware social networks. VLDB J. **32**(1), 1–22 (2023)
4. Chen, W., Yin, H., Wang, W., Zhao, L., Zhou, X.: Effective and efficient user account linkage across location based social networks. In: ICDE, pp. 1085–1096 (2018)
5. Deng, L., Zhao, Y., Fu, Z., Sun, H., Liu, S., Zheng, K.: Efficient trajectory similarity computation with contrastive learning. In: CIKM, pp. 365–374 (2022)

6. Devlin, J., Chang, M., Lee, K., Toutanova, K.: BERT: pre-training of deep bidirectional transformers for language understanding. In: NAACL-HLT, pp. 4171–4186 (2019)
7. Feng, J., et al.: Deepmove: predicting human mobility with attentional recurrent networks. In: WWW, pp. 1459–1468 (2018)
8. Feng, J., et al.: Dplink: user identity linkage via deep neural network from heterogeneous mobility data. In: WWW, pp. 459–469 (2019)
9. Gao, Q., Zhou, F., Zhang, K., Trajcevski, G., Luo, X., Zhang, F.: Identifying human mobility via trajectory embeddings. In: IJCAI, pp. 1689–1695 (2017)
10. Gao, T., Yao, X., Chen, D.: Simcse: simple contrastive learning of sentence embeddings. In: EMNLP, pp. 6894–6910 (2021)
11. Jiang, J., Pan, D., Ren, H., Jiang, X., Li, C., Wang, J.: Self-supervised trajectory representation learning with temporal regularities and travel semantics. CoRR abs/2211.09510, 1–13 (2022)
12. Jin, F., Hua, W., Xu, J., Zhou, X.: Moving object linking based on historical trace. In: ICDE, pp. 1058–1069 (2019)
13. Jin, F., Hua, W., Zhou, T., Xu, J., Francia, M., Orlowska, M.E., Zhou, X.: Trajectory-based spatiotemporal entity linking. IEEE Trans. Knowl. Data Eng. **34**(9), 4499–4513 (2022)
14. Loshchilov, I., Hutter, F.: Decoupled weight decay regularization. In: ICLR, pp. 1–19 (2019)
15. Martins, A., Astudillo, R.: From softmax to sparsemax: a sparse model of attention and multi-label classification. In: ICML, pp. 1614–1623 (2016)
16. Mikolov, T., Sutskever, I., Chen, K., Corrado, G.S., Dean, J.: Distributed representations of words and phrases and their compositionality. In: NIPS, pp. 3111–3119 (2013)
17. Mu, X., Zhu, F., Lim, E.P., Xiao, J., Wang, J., Zhou, Z.H.: User identity linkage by latent user space modelling. In: SIGKDD, pp. 1775–1784 (2016)
18. Vaswani, A., et al.: Attention is all you need. In: NIPS, pp. 5998–6008 (2017)
19. Vosecky, J., Hong, D., Shen, V.Y.: User identification across social networks using the web profile and friend network. Int. J. Web Appl. **2**(1), 23–34 (2010)
20. Wang, S., Cao, J., Yu, P.: Deep learning for spatio-temporal data mining: a survey. IEEE Trans. Knowl. Data Eng. **34**(8), 3681–3700 (2022)
21. Xue, H., Sun, B., Si, C., Zhang, W., Fang, J.: DBUL: a user identity linkage method across social networks based on spatiotemporal data. In: ICTAI, pp. 1461–1465 (2021)
22. Yang, P., Wang, H., Zhang, Y., Qin, L., Zhang, W., Lin, X.: T3s: effective representation learning for trajectory similarity computation. In: ICDE, pp. 2183–2188 (2021)
23. Zafarani, R., Liu, H.: Connecting corresponding identities across communities. In: ICWSM, pp. 354–357 (2009)
24. Zhang, Y., Liu, A., Liu, G., Li, Z., Li, Q.: Deep representation learning of activity trajectory similarity computation. In: ICWS, pp. 312–319 (2019)
25. Zheng, Y., Li, Q., Chen, Y., Xie, X., Ma, W.Y.: Understanding mobility based on GPS data. In: UbiComp, pp. 312–321 (2008)

Large-Scale Rank Aggregation from Multiple Data Sources Based D^3MOPSO Method

Xian Tan[1] 🆔, Wei Yu[2], and Li Tan[3(✉)]

[1] College of Literature and Communication, Hubei Minzu University, Enshi 445000, China
[2] Computer School, Wuhan University, Wuhan 430072, China
[3] International School Longpanhu, Yichang 443007, China
richardleetanli@gmail.com

Abstract. Aggregating the search result from multiple data sources is a challenging problem in the metasearch engines. However, the ordinary methods do not have enough ability to deal with these tremendous data. Aiming at this issue, based on the big data we aim to harness the efficiency and effectiveness of the aggregation result from multiple data sources and propose an aggregation framework, which consists of aggregating the multi-users requirements and preferences ranking lists and modelling discrete multi-objective evolutionary model. Based on the DPSO algorithm, we improve and optimize its encoding scheme, initialization methods, position and velocity definition, integrating updating, turbulence operator, external archive updating strategy and leaders selection, which could address the problem of low efficiency on the large scale data sources. Extensive experiments on the public datasets, real-world datasets and synthetic simulation datasets demonstrate that our method outperforms existing state-of-the-art ranking aggregation method and multi-objective evolutionary method by the efficiency, performance and convergence.

Keywords: Rank Aggregating · metasearch · user preference · multi-objective optimization · particle

1 Introduction

Metasearch engines do not maintain their own indexes on Web pages. After accepting a query from the end user, they simply forward that query to several second-level search engines (i.e. data sources), collect the returned search results and reorder them, then produce an aggregate ranking list. This aggregate list is displayed to the user finally as a response to the query. In this process, the key issue is the rank aggregation algorithm that retrieves the ranked results from selected data sources and then combines and reorders those results. If different rank aggregation algorithms are used to collate search results, the results of metasearching for the same query may vary for the same set of participating search engines. Therefore, the actual success of a meta-search engine directly depends on the aggregation technique underlying it.

In metasearch, as the aggregation is performed in runtime when the user is waiting for the final result for his query, it is essential for the algorithm to be of low complexity. If we directly use multi-objective particle swarm optimization (i.e. MOPSO) algorithm for rank aggregation, the final order of all results will be computed in $O(N^M)$ time if there are N search results to sort and M programming objectives. This complexity is unacceptably high. In this paper, we wish to develop an approximation of MOPSO called D^3MOPSO (Discrete Multi-Objective Particle Swarm Optimizer based on Decomposition and Dominance) to reduce the aggregation complexity and maintain the effectiveness. Here, we also discuss a user preference based metasearch system that uses D^3MOPSO as its rank aggregation method. In this, user preference is obtained and organized as a multi-component objective vector for each query, and then the D^3MOPSO is performed to search for the multi-dimensional optimum solution efficiently.

Also, in metasearch, qualities of the candidate search results given by individual search engines are not equal, because the attributes of a search engine, such as indexed set of web-pages, ranking features, ranking function etc., often affect its ranking. Due to this fact, there is a need to identify the qualities of the candidate results and use that quality information while performing the aggregation. So, the data sources that we adopt here are ranked by the measures of their accuracy, reliability, freshness, access cost, and relevance to the query.

The specific contributions of this work are three-fold:

(1) We take the rank aggregation in metasearch as a multi-objective programming problem on ranking all search results, and thus propose a user preference based rank aggregation algorithm for metasearch called D3MOPSO, which enables the aggregation of search results meet the multi-objective demands of end users.
(2) The algorithm is performed in a parallel pattern of seeking the optimum results for those large-scale data sources. This brings the time complexity down and promotes scalability of the algorithm impressively.
(3) We improved multi-objective particle swarm optimization algorithm to fit for the discrete data objects in the rank aggregation task. Experimental results based on benchmark functions indicate that this improved algorithm has superiority in convergence and search efficiency of solution space when dealing with discrete objects.

2   Related Work

Metasearching is widely discussed in literature. There are a good number of studies that have been performed on metasearching [1–7].

Amin and Emrouznejad [8] present a positional score based algorithm(a method base on positional score) that assigns weights to each position dynamically, by looking at the data. However, if there are N items in the set and the length of the longest list is L, then the method needs to solve N linear programming problems, each with $O(N + L)$ constraints. Several Markov chain based methods were used for rank aggregation in Dworketal. [9]. For each pair of items i_1 and i_2, the corresponding state transition probability depends on what fraction of the input rankers have ranked both the items, and also how many lists have i_1 before i_2. Four methods were proposed, MC1, MC2, MC3, and MC4. The methods are different from each other in the ways the transition probabilities are calculated. Coppersmith et al. [10] have posed the task of optimizing average Kendall-Tau distance or Kemeny distance as the minimum feedback arc set problem and proposed algorithms for the task. Montogue and Aslam [11] proposed a *Condorcet method* called Condorcet-fuse format searching, the aggregation is then performed in a number of steps, by selecting one Condorcet winner (which corresponds to an item from the input ranked lists) in each step. Wu et al. [12] proposed a data fusion approaches using linear aggregation of scores are suggested in literature for the Learning to Rank task, which weighted sum or its variations are used as aggregation operators. Klementievetal. [13] proposed a Mallowsmodel which has also been used to aggregate "typed rankings" coming from domain-specific experts–where the type or domain of the candidate items are known.

We now give a formal definition of the rank aggregation problem and explain our solution for approaching the problem.

3 Definitions and Problem Formulation

Before describing our proposed algorithm for solving the multi-objective rank aggregation problem, we formally introduce the concept of "user preference based rank aggregation" as follows.

Definition 1 [Data Source, DS]. A data source here is a participating search engine to which the metasearch engine forward its query and from which the metasearch engine retrieve the search results. It can be characterized by a four-tuple $< DS, IP, OP, DQ >$.

Definition 2 [Retrieval Request with User Preference, RRUP]. A tuple $< IPs, UPs >$ is presented here to characterize the query and the user's preference. *IPs* is the set of input queries, and *UPs* is the vector of user preferences. Each element in this tuple, either a query or a preference, stands for a demand of the user. A typical user preference vector is like $(c_1, c_2, c_4, c_3, c_5)$, in which each element represents one preference and the position of an element represents the precedence of the corresponding preference. In this paper, the preference information of a metasearch user is acquired by mining his/her features and history searching behaviors. The mining algorithm is discussed in [15].

Definition 3 [Set of Candidate Item, SCI]. In a rank aggregation process, the candidate items are those search results retrieved from the participating search engines, i.e., the data sources. The set of these candidate items can be characterized as $SCI = \Gamma_{DS_1, DS_2, ..., DS_n}^{\sigma_{IPs}} = \bigcup_{1 \leq i \leq n} \Gamma_{DS_i}^{\sigma_{IPs}}$, if the candidate items are retrieved from n data sources $DS_1, DS_2, ..., DS_n$ after the queries *IPs* are submitted. Here $\Gamma_{DS_i}^{\sigma_{IPs}}$ refers to the dataset of all the candidate items returned by data source DS_i for queries *IPs*. It can be expressed in the form of a vector, $\Gamma_{DS_i}^{\sigma_{IPs}} = (d_{i1}, d_{i2}, \cdots, d_{im_i})$, in which d_{ij} is the j^{-th} candidate item returned by DS_i for *IPs* and m_i is the number of candidate items from data source DS_i. Then we can calculate the size of this set of candidate items by the following:

$$N = \prod_{i=1}^{n} m_i.$$

Definition 4 [Quality Measures of Search Result, QMSR]. For each item in the set of candidate items, the quality features (accuracy, reliability, freshness, access cost, and relevance to the query) can be measured with the method mentioned in [14]. Thus we define that the t^{-th} quality feature of the candidate item d_{ij} has the value of $f_t^{\sigma_{IPs}}(d_{ij})$, shorted as $f_t(d_{ij})$.

Definition 5 [Partial Y-Dominance (PY-Dominance)]. Let the objective vector:
$f(d) = (f_1(d), f_2(d), \cdots, f_M(d))$ consists of M components, among which $f_i(d)$ measures an item d against the i^{-th} demand of user in the QUP. Consider two candidate items d_1 and d_2, item d_1 is said to PY-dominate d_2(written as $d_1 \succ_{PY} d_2$) if the following two conditions hold (in a maximum programming problems):

(1) $\forall i \in \{1, 2, \cdots, M\} : f_i(d_1) \geq f_i(d_2)$
(2) $\exists S = \{s_1, s_2, \cdots, s_y\}, s_j \in \{1, 2, \cdots, M\} : f_{s_j}(d_1) > f_{s_j}(d_2)$

where y is the size of Set S. Particularly, when $y = 1$, PY-Dominance is exactly the pareto dominance. Only if $y \leq [M/2]$, PY-Dominance describes a partial order of two items, otherwise it makes no sense.

Definition 6[Optimal Feasible Set, OFS]. Optimal Feasible Set is a complete set that any item in the set PY-Dominance any other item, the set OFS can be expressed as:

$$OFS \triangleq \{d^* | \neg \exists d \in \mathbf{SCI} : d \succ_{PY} d^*\}$$

Problem Formulation [Rank Aggregation with User Preferences, RAUP]. With the definitions above, we formally introduce the task of rank aggregation. Rank Aggregation with User Preferences is a vector which satisfies multi-objective optimization and user ordinal preferences. Let the candidate items in SCI are search results retrieved from n data sources for query IPs. Assume the user who input query IPs has the preferences of UPs. The rank aggregation algorithm is to find the top k optimal items $\mathbf{RAUP} = (d_1, d_2, \cdots, d_k)$, where $\forall i, d_i \in OFS$, and with the following rule: $\forall d_i \in \mathbf{RAUP}, \forall d_j \in \Gamma_{DS_i}^{IPs} - \mathbf{RAUP} : d_i \succ_{PY} d_j$, and di is superior to d_j in user preferences UPs.

   Each of the k items should be globally PY optimal in the multiple objectives assigned by UPs.

   Then the rank aggregation task can be formulated as the following multi-objective programming problem:

$$Max \quad \mathbf{F}(d) = (f_1(d), f_2(d), \cdots, f_M(d))^T$$
$$d \in \Gamma_{DS_1, DS_2, \dots, DS_n}^{\sigma_{IPs}} \tag{1}$$

   The final aggregate list is the set of PY optimal solutions. The optimization process is to search for solutions that are optimal in M objectives from N candidate items, in general case, the complexity of this optimization is $O(N^M)$.

   The participating search engines of metasearch can return massive search results from diverse data sources in Web. As a result, the set of candidate items is usually of huge size. Rank aggregation of these candidate items becomes a problem of searching the optimal solutions in large-scale dataset. Aiming at this problem, a lot of researches have been done already which cannot effectively resolve the accuracy and efficiency problem created by high dimensional and huge amounts of data. In this paper, we propose an improved MOPSO algorithm.

4   Proposed Method

In this section, the proposed D^3MOPSO (Discrete Multi-Objective Particle Swarm Optimizer based on Decomposition and Dominance) method for retrieval result integration problem is described.

4.1   Strategy on Encoding Scheme and Multi-directional Search

In the traditional discrete multi-objective particle swarm method, all discrete points are sorted in a certain direction by linear results on a certain dimension in order to ensure random walk search in both direction. However, because of the large-scale characteristics of discrete points, this search method is inefficient. Therefore, according to the characteristics of data sources and result set, the data on the *SCI* is divided into a grid graph in accordance with one feature of *DQ*, and the structure of search graph is defined, as shown in Fig. 1.

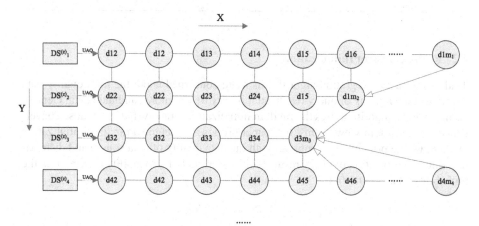

Fig. 1. Schematic diagram about search path of Particles on objective dimension z.

n data sources in the target dimension z, which is in descending order according to their data source quality'sV_z of *DQ*, formed a sequence $(DS_1^{(z)}, DS_2^{(z)}, \cdots, DS_n^{(z)})$.The result collection of each data source $DS_i^{(z)}$ which meets *RRUP* is $(d_{i1}, d_{i2}, \cdots, d_{im_i})$, the size of the collection is m_i. Therefore, each dimension of the particles can move back and forth as shown in the both X and Y directions. In order to facilitate understanding, we virtually define four pointers to each data d in the search graph:*d- > Xnext,d- > Xprior, d- > Ynext,d- > Yprior*(in fact, this algorithm use continuous two dimensional matrix to store and search the address by the movement of memory in order to improve the search efficiency), it is respectively defined as:

$$d_{ij} \to Xnext = \begin{cases} d_{i(j+1)}, j < m_i \\ null, j \geq m_i \end{cases} \qquad d_{ij} \to Xprior = \begin{cases} null, j \leq 1 \\ d_{i(j-1)}, j > 1 \end{cases}$$

$$d_{ij} \to Ypnext = \begin{cases} d_{(i+1)j} & i < n, m_{i+1} \geq j \\ d_{(i+1)m_{i+1}} & i < n, m_{i+1} < j \\ null & i \geq n \end{cases} \quad d_{ij} \to Yprior = \begin{cases} null & i \leq 1 \\ d_{(i-1)j} & i > 1, m_{i-1} \geq j \\ d_{(i-1)m_{i-1}} & i > 1, m_{i-1} < j \end{cases} \quad (2)$$

Due to the ordered in X and Y directions, so by this distribution, the target value on the z-dimension of each particle could keep relatively smooth in a small field, thus facilitating particles to find extreme value in local searches. What more important is that particles can walk in multiple directions in the diagram structure, improving the search efficiency greatly. Algorithm 1 describes a corresponding grid chart for the M targets.

Algorithm 1 Encoding Scheme

Input: n datasets from each data source by UAQ as $DataS_1, DataS_2, \ldots DataS_n$
Output: M matrix to each object
1. $Maxitem = MAX(m_i)$
2. $ParticleGraph(z) = $ new matrix$(Maxitem, n)$ //$z=1,2,\ldots,M$
3. $SortIndex(z) = $ new matrix$(1, n)$ //storage index of data sources by z- dimension descending order
4.   **for** $z=1:M$
5. $SortIndex(z) = SORT(DS1, DS2, \ldots, DSn)$ by $DQ.v_z$
6. $ParticleGraph(z) = (DataS_{SortIndex(z)(1)}, DataS_{SortIndex(z)(2)}, \ldots)$ // reorder data sources
7.   **end for**

4.2   Particle Swarm Initialization

Traditional query integration algorithm is more concerned about the high quality data source and the ranking of query result, so it is easy to lose more candidate data. General particle swarm algorithm uses the random initialization strategy, likely to cause convergence difficulties. In view of this, this paper proposes a weighted random initialization method, whose principle is combined with the data quality of data source and distribution rule of the position of query result position for random probability distribution, the probability of any data d_{ij} is:

$$P(d_{ij}, z) = P(DS = i|z)P(Rank = j)$$

where $P(DS = i|z) = DQ_i(v_z) / \sum_{k=1:n} DQ_k(v_z)$, $P(Rank = j) = cj^{-\beta}$ follow Zipf's Law Distribution, and $\sum_{j=1:m_1} hj^{-\beta} = 1$, $h(m_i)^{-\beta} = a$, where a is the minimum probability in the end of the query results, it is get by tuning parameters of experiment. Parameters h and β can be solved by the numerical approximation method.

Algorithm 2 describes the particle initialization process on the target dimension z, when generating particle swarm, algorithm 2 will be called for many times according to the size of the particle swarm.

Algorithm 2 Initialization of Population

Input: z//the objective dimension which want to generate initial particle
Output: *Particle* //a initialized particle

1. *rand1=Random()* //a random number in $[0,1)$
2. ***SortIndex*= SORT(DS1,DS2,...,DSn) by DQ.v_z** //vector to storage index of data sources which
    descending order by z-dimension (DQ.v_z)
3. *TotalScore* $= \sum\limits_{i=1:n} DS_i.DQ.v_z$ //accumulation of prior value z-dimension in all data sources
4. $s_i = 0$ //the index of random selection data source (new order)
5. **for** $i=1$:n
6.   **if** *rand1*$<DS_{\text{SortIndex}(i)}.DQ.v_z$ / *TotalScore*
7. $s_i = i$
8.   **break**
9.   **else**
10.   *rand1*$-= DS_{\text{SortIndex}(i)}.DQ.v_z$/ *TotalScore*
11.   **end if**
12. **end for**
13. $t_i = 0$ //the index of random selection item
14. DataNumber=m_{si} //the number of results from selected data source s_j
15. *rand2=Random()* // a random number in $[0,1)$
16. **for** $i=1$: DataNumber
17.   **if** *rand2*$< hi^{-beta}$
18. $t_i = i$
19.   **break**
20.   **else**
21.   *rand2*$-= hi^{-beta}$
22.   **end if**
23. **end for**
24. **return** d(s_i,t_i) //the t_i^{th} particle has been selected in s_i^{th} data source

---

4.3 Definition of Discrete Position and Velocity

In order to solve the large-scale retrieval result integration problem, in this paper, we redefine the term position and velocity used in PSO in discrete form. The definitions are as following:

(1) Definition of position: In PSO, each vertex in Fig. 1 represents a solution to the optimized problem. For the retrieval result integration problem, each particle p_k corresponds to an item (d_{ij}, vertex in Fig. 1), thus s-size particle swarm can be expressed as $P = \{p_1, p_2, \cdots, p_s\}$.

(2) Definition of velocity: Velocity works on the position sequence and it is rather crucial. A good velocity gives the particle guidance and determines whether the particle can reach its destination and by how fast it could. The discrete velocity of

particle I is defined as $V_i =< v_X, v_Y >$, where v_X, v_Y are respectively the particles velocity component in the X and Y direction as shown in Fig. 2, its value is an integer. If v_X is positive, then the pointer to move for v_X time along the direction of $d_{ij} \rightarrow$ Xnext, if v_X is negative, then the pointer moves for $-v_X$ time along the direction of $d_{ij} \rightarrow$ Xprior. If v_Y is positive, then the pointer to move for v_Y time along the direction of $d_{ij} \rightarrow$ Ynext, if v_Y is negative, then the pointer moves for $-v_Y$ time along the direction of $d_{ij} \rightarrow$ Yprior.

4.4 Discrete Particle Statue Updating

In the proposed DPSO, a velocity provides a particle with the moving direction and tendency. After updating the velocity, one particle makes use of the new velocity to build new position. Since the position and velocity, in our approach, are all integer vectors, the mathematical updating rules in continuous PSO no longer fit the discrete situation; therefore, we define them to meet the requirements of retrieval result integration problem. We first redefine the velocity updating rule in discrete form as

$$
\begin{aligned}
v_{iX} &= [\omega_X v_{iX} + c_{X1} r_{X1} DX (Pbest_i - p_i) + c_{X2} r_{X2} DX (Obest_i - p_i) + c_{X3} r_{X3} DX (Gbest_i - p_i)] \\
v_{iY} &= [\omega_Y v_{iY} + c_{Y1} r_{Y1} DY (Pbest_i - p_i) + c_{Y2} r_{Y2} DY (Obest_i - p_i) + c_{Y3} r_{Y3} DY (Gbest_i - p_i)]
\end{aligned} \tag{3}
$$

where *Pbest* is the optimal position in the individual history of particle *I*, *Obest* is the local optimal location of particle swarm on the target dimension; *Gbest* are the optimal location of global history on various dimensions, the get method see the *selection 4.6*.

ω_X is the inertia weight in X-axis; c_{X1}, c_{X2} and c_{X3} are the cognitive and social components respectively in x-axis; And ω_Y is the inertia weight in Y-axis; c_{Y1}, c_{Y2} and c_{Y3} are the cognitive and social components respectively in y-axis; r_{X1}, r_{X2}, r_{X3}, r_{Y1}, r_{Y2} and r_{Y3} are the two random numbers with range [0,1].

In formula (3), DX, DY respectively represent a function of the distance in the X and Y directions, for example $pbest_i$ is $d_{23,17}$, and p_i is $d_{15,32}$, then $DX (Pbest_i - p_i) = -15$, $DY (Pbest_i - p_i) = 8$. [] is operator for the floor value of integer.

In our algorithm, to promote exploration and exploitation, the inertia weight ω_X and ω_Y and randomly generated between [0.1], and the cognitive and social components c_{X1}, c_{X2}, c_{X3}, c_{Y1}, c_{Y2} and c_{Y3} are set by experimental parameters tuning.

Based on the newly defined discrete velocity updating rule, we now redefine the position updating rule as the following discrete form:

$$
x_i = x_i \odot V_i = x_i \odot (v_{Xi}, v_{Yi}) \tag{4}
$$

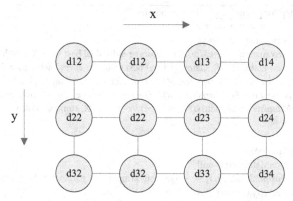

Fig. 2. Schematic diagram about search direction of particles

In formula (4), the operator ⊙ is the key procedure in the particle status updating process. It directly affects the performance of the algorithm since the sight operand of ⊙ determines the direction to which the particle flies. A good operator ⊙ should help to guide the particle to better place that is much closer to the food rather than a bad position that is far away from it.

Given a position x_i as d_{kj} vertex and a velocity $V_i = < v_X, v_Y >$, position ⊙ velocity generates a new position that corresponds to a new solution to the optimization problem, i.e., $x_1 ⊙ V_1 = x_2$, the position of the x_2 is represented by the vertex $d_{k'j'}$. The element of x_2 is defined as

$$\begin{cases} k' = k + v_y & 1 \le k + v_y \le n \\ k' = n + v_y - k & k + v_y \ge n \\ k' = 2 - v_y - k & k + v_y < 1 \end{cases} \begin{cases} j' = j + v_x & 1 \le j + v_x \le m_{k'} \\ j' = m_{k'} + v_x - j & j + v_x \ge m_{k'} \\ j' = 2 - v_x - j & j + v_x < 1 \end{cases} \quad (5)$$

where n is the number of data source, and $m_{k'}$ is the record number of query result of data source k'. And if $\neg \exists d_{k'j'}$, then search for the nearest vertex in the neighborhood of the coordinate randomly. If $d_{k'j'} \in history$, then select a vertex in the global randomly to avoid falling into local search. As shown in Fig. 2, if from d_{14}, $v_x = -2$, $v_y = 1$, the particle moves to the point vertices of the field points which closest to d_{22}.

72 X. Tan et al.

Algorithm 3 A step for position and velocity updating

Input: *SwarmSize* //the size of particle swarm

Output: *Particle* // generated particles per step

1. *PBestMatrix*=new matrix(*SwarmSize*, *M*) //storage individual best value for each particle
2. *OBestMatrix*=new matrix(*1*, *M*) //storage local best value for every each dimension
3. PositionMatrix=new matrix(*SwarmSize*, *M*) // storage current position <k,j> for each particle
4. VelocityMatrix=new matrix(*SwarmSize*, *M*) // storage current velocity <v_x,v_y> for each particle
5. *ParticleGraphStatus*(z)=new matrix(*Maxitem*, *n*) //z=1,2,...,M, storage Boolean value about whether visited
6. **for** i=1 : M
7.   **for** j= 1 : *SwarmSize*
8.   **if** PositionMatrix[j,i] is null
9.   PositionMatrix[j, i] ← Algorithm 2(input *i* for parameter *z*) // initialization
10.   VelocityMatrix[j, i] ← <0, 0>
11.   **else**
12.   Set random for each r_{X1}, r_{X2}, r_{X3}, r_{Y1}, r_{Y2}, r_{Y3}
13.   VelocityMatrix[j, i].vx ← fix(ω_x VelocityMatrix[j, i].vx+ $c_{x1}r_{x1}$ (PBestMatrix[j, i].j- PositionMatrix[j, i].j) + $c_{x2}r_{x2}$ (*OBestMatrix* [1, i].j- PositionMatrix[j, i].j) + $c_{x3}r_{x3}$ (GBest.j- PositionMatrix[j, i].j)
14.   VelocityMatrix[j, i].vy ← fix(ω_y VelocityMatrix[j, i].vy+ $c_{y1}r_{y1}$ (PBestMatrix[j, i].k- PositionMatrix[j, i].k) + $c_{y2}r_{y2}$ (*OBestMatrix* [1, i].k- PositionMatrix[j, i].k) + $c_{y3}r_{y3}$ (GBest.k- PositionMatrix[j, i].k)
15.   PositionMatrix[j, i] ← PositionMatrix[j, i] ⊙ VelocityMatrix[j, i] //by formula(5)
16.   **if** *ParticleGraphStatus(i,* PositionMatrix[j, i]*)*
17.   PositionMatrix[j, i] ← Algorithm 2(input *i* for parameter *z*) //visited, flyout
18.   **end if**
19.   **end if**
20.   *ParticleGraphStatus(i,* PositionMatrix[j, i]*)* ← true //record the visited status
21.   **if** *ParticleGraph* (i, PositionMatrix[j, i]) ≻ *ParticleGraph* (i, *PBestMatrix*[j, i])
22.   Update(*PBestMatrix*[j, i]) ← PositionMatrix[j, i] //Update new *Pbest* if domination
23.   **end if**
24.   **if** *ParticleGraph* (i, PositionMatrix[j, i]) ≺ *ParticleGraph* (i, *OBestMatrix*[1, i])
25.   Update(*OBestMatrix*[1,i]) ← PositionMatrix[j,i] // Update new *Obest* if domination
26.   **end if**
27.   Update(GBest) //Update global optimal solution *GBest* as selection 4.6
28.   **end for**
29. **end for**

4.5 Framework of the Proposed Algorithm

The procedure of the proposed D^{3}MOPSO is illustrated by Algorithm 4.

Algorithm 4 Framework of the proposed D³MOPSO
Input: n datasets from each data source by UAQ as DataS$_1$, DataS$_2$,...DataS$_n$

Output: Top-k optimal solution in order
1. **Step 1) Initialization**
2. **Step 1.1)** recode all items for each objective dimension according to selection 4.1
3. **Step 1.2)** generate initialized particle swarm($M*$ *SwarmSize* in size) according to selection 4.2
4. **Step 1.3)** Define architecture, position, velocity for each particle according to selection 4.3
5. **Step 1.4)** Generate a well-distributed weighted vectors: $W = \{\omega_{1_X}, \omega_{1_Y}, \cdots\}$
6. **Step 1.5)** Initialization *PBest* matrix, *OBest* vector in Algorithm 3
7. **Step 1.6)** Initialization *GBest* Function
8. **Step 2) Cyling**
9. **Step 2.1)** iterative execute Algorithm 3 and update *PBest*, *OBest* 和 *GBest*
10. **Step 2.2)** if a particle which is domination all particles has been generated, judge and join in EA
11. **Step 2.3)** Update lifecycle of *GBest* or renewal it after each iterative according to selection 4.6
12. **Step 3 Stopping criteria**: if iterations $t<maxgen$, then $t++$ and go to **Step 2**, otherwise, stop the algorithm and output EA.

From the above algorithm, we know that although the D³MOPSO algorithms adopts the framework of PSO with the ideas of the crossover and mutation, it improves PSO by encoding scheme, initialization methods, position and velocity definition, integrating updating, turbulence operator, external archive updating strategy and leaders selection.

4.6 Complexity Analysis

(1) Space Complexity: In our algorithm, two main memorizers are needed. The first one is all original data in Algorithm 1, which needs a complexity of $M \cdot n \cdot Maxitem \approx M \cdot N$, i.e. $O(M \cdot N)$. The second memorizer is for the particles in Algorithm 3, say there are *SwarmSize* particles, then the complexity is $3SwarmSize \cdot M$, i.e. $O(SwarmSize \cdot M)$. Thus, the total space complexity of our algorithm is $O(N)$.

(2) Computational Complexity: The main times complexity lies in Step 2 of our algorithm since Step 1 can be accomplished in once Initialization and linear time. Here, Step 3.1 needs $O(SwarmSize \cdot M)$ basic operations, Step 3.2 requires $O(MAXP)$ where *MAXP* is the volume of external archive. Step 3.3 needs $O(SwarmSize \cdot M)$ basic operations for *GBest* selection for each particle. According the operational rules of the symbol O, the worst case time complexity for D³MOPSO can be simplified as $O(maxgen \cdot SwarmSize \cdot M)$, where *maxgen* is the iteration number.

5 Experimental Studies

In this section, the experiments carried out are summarized.

5.1 Comparison Algorithms

In this paper, 4 algorithms named as ANN, WT-INDEG, NPDA, BHTPSO are chosen to compare with our algorithm named as D³MOPSO.

User feedback based Metasearching by Artificial neural network [37] (ANN, in short) is a combined method which use different rank aggregation techniques to aggregate the various rankings of the search results to generate an overall ranking.

Weighted aggregation Rank Algorithm [38] (named WT-INDEG) is an implement and efficient algorithm for unsupervised rank aggregation. It assigns varying weights to the input graphs to reduce the influence of the bad rankers on the aggregation process.

Nadir Point Determination Algorithm [39] (NPDA, in short) is based on an exhaustive search of the (p-2)-dimensional space. NPDA guarantees to find the nadir point exactly in a finite number of steps. In the paper, the algorithm has been proved superior to the discrete multi-objective evolutionary algorithms such as Ehrgott and Tenfelde-Podehl's algorithm (ETPA), Jorge's algorithm (JA) and so on. When compared with our method, we applied the above mentioned algorithm to the problem of formula (1), and then the maximization problem can be transformed into a minimization problem.

Binary hybrid topology particle swarm optimization [40] (BHTPSO, in short) algorithm has been introduced to extend the binary search capability and to improve convergent efficiency, which can solve memetic binary particle swarm optimization for discrete optimization problems. The paper has been proved that the algorithm is superior to the discrete Particle swarm algorithms such as BPSO, NBPSO, BGSA and so on. In comparisons with our method, the aforementioned algorithms were applied to the problem of formula (1) and transformed the maximization problem into minimization problem.

5.2 Experimental Settings

Our experiments use 3 types of datasets[1] to measure the algorithm performance.

(1) LETOR Datasets

We used 3 data sources included Yahoo Learning to Rank Challenge (LTRC) dataset, Microsoft Learning to Rank (MSLR-WEB10K) dataset and LETOR 4.0 rank aggregation datasets (MQ2008-agg) for our experiments. The LETOR datasets are specifically for rank aggregation and we use them directly.

(2) Real-world Datasets

Another dataset in this paper came from the retrieval results of twenty different web data sources. Moreover, it was separately from thirty different retrieval requirements. There were about 200 records of each retrieval requirement for one data source. Add up to 129853 records. The retrieval requirement format is as <"keywords", value > shows. The main format of the retrieval result is OPs = { <"title", $value_1$ >, <"price", $value_2$ >, <"Merchant", $value_3$ >, <"date",$value_4$ >, <"AccessTime", $value_5$ > }. On the basis of previous studies, the diverse estimate values of definition 4 for each record were performed on the semantic computations of the data results.

(3) Synthetic Simulation Datasets

There was also a simulation dataset that has the same distribution features as real dataset. It includes 1000 data sources. For each data source there were average 150 records. We have simulated about 100 retrieval requirements. Add up to 15837493 records. And for each record it had been numerically processed. Consequently, each record can be expressed as the values of M objective dimensions such as accuracy, reliability, efficiency, access speed, result relevance and so on. It can be represented as $\{f_1, f_2, f_3, f_4, f_5\}$.

---

[1] All datasets have been published in http://www.whudml.com.

All experiments are performed on a PC cluster with Intel(R) Core(TM) CPU i7–4790 3.60 GHz and 8 GB memory, running on a Windows 2008 operating system. Our algorithms are implemented using the C++ language and each execution is performed as a single process (i.e., no parallel processes), where very minor simplifications are done in the implemented versions.

5.3 Evaluation Metrics

(1) Kullback–Tau Distance (KTD)

In order to express the bias between the retrieval result and the expected result, we use Kullback–Tau distance to measure the distances of corresponding results.

(2) Discounted Cumulative Gain (NDCG)

The goal is to find, for each query in the web data sources, the best possible rank of aggregation results, ranking accuracy being assessed by the means of the Normalized Discounted Cumulative Gain (NDCG) measure.

(3) Distance metric (DM)

Due to the particularity of discrete data and data integration, the ordinary spacing metric cannot efficiently describe the distances among the solutions in the approximate optimal solution set. Taking the location distance, semantic distance and sequence distance into consideration, *RAUP* express the approximate optimal solution vector.

(4) Convergencemetric (CM)

For the ordinary continuous multi-objective optimization problem, the convergence metric describes the normalized euclidean distance of the approximate optimal complete solution set. While in the discrete multi-objective optimization problem, it has arrange optimal solution denoted as $\mathbf{OFS} = (d_1, d_2, \cdots, d_{|\mathbf{OFS}|})$.

(5) Iteration Convergence Generations Metric (ICGM)

When the multi-objective evolutionary algorithm was applied to solve this problem, the numerical convergence value of the iteration process has reflected its convergence speed. IC_i represents that *when* the i^{-th} iteration finished we will update the number of records for document EA.

(6) Times

In order to compare the efficiency of different methods, under the same dataset we directly use execution time (measurement unit: second) to compare them. The lower the execution time is, the higher the efficiency of the corresponding algorithm is.

5.4 The Results

5.4.1 Comparing the Performance

The indictors used to evaluate the performance mainly include KTD, NDCG, DM, CM and so on. The three multi-objective evolutionary algorithms such as D³MOPSO 、NPDA and BHTPSO all set *maxgen* = 100. The mean results of every metric at the three category dataset are as follows:

(1) mean in LETOR Datasets

In Table 1 it shown the performance comparison results on LETOR Datasets of different methods. In which we can see that for KTD and NDCG indicator D^3MOPSO were close to ANN and WT-INDEG. This was due to that the sorting algorithm ANN and WT-INDEG used had better comparison feature. Meanwhile much more efficient than NPDA and BHTPSO. Because the optimal solutions of NPDA and BHTPSO was an unordered set. Which gave a poor account in the sorting process. For DM indicator D^3MOPSO was a little worse than NPDA and BHTPSO. It was because for the variety selection NPDA and BHTPSO mainly distinguished particle location which result in while the location of the optimal solution set was big the DM indicator was well. At the same time much more efficient than ANN and WT-INDEG. Because for variety ANN and WT-INDEG had been considered inadequate. For CM indicator D^3MOPSO was close to NPDA and BHTPSO. And for these five methods all play well at indicator CM. It proved that the results from every method were all close to optimal solution set. Therefore, the experiment on LETOR Datasets simplified that D^3MOPSO play well on performance indicator.

Table 1. Mean results of performance metric in LETOR Datasets

	KTD	NDCG	DM	CM
D^3MOPSO	0.189	0.673	4.897	0.021
ANN	0.181	0.618	2.298	0.112
WT-INDEG	0.192	0.682	2.489	0.078
NPDA	0.372	0.387	6.982	0.019
BHTPSO	0.312	0.412	5.498	0.018

(2) Real-world Datasets

With regard to real-world Datasets there was no optimal expectation sort result as Ground truth. Consequently, for this dataset we mainly discussed its performance on indicator DM and CM. in Table 2 it shown the performance of every method on Real-world Datasets. The experimental results demonstrated that for Real-world Datasets the performance indicator of D^3MOPSO play similarity as on LETOR Datasets.

(3) Synthetic Simulation Datasets

The optimal expectation sort result for Synthetic Simulation Datasets came from the optimal automatically creation process. In Table 3 it shown the performance of every method on Synthetic Simulation Datasets. The experimental results demonstrated that in the integration of large scale data result ANN and WT-INDEG gave a poor account. Especially for CM indicator it was long away from the optimal solution set. It was because the large solution space. Although NPDA and BHTPSO play well on diversity indicator DM at the expense of decreasing CM indicator. The main

Table 2. Mean results of performance metric in Real-world Datasets

	DM	CM
D^3MOPSO	4.613	0.019
ANN	2.497	0.315
WT-INDEG	2.721	0.253
NPDA	6.395	0.017
BHTPSO	5.956	0.016

reason were that the solution space was large and its hardness to converge while the step velocity of particle was too fast. Consequently for performance indicator on Synthetic Simulation Datasets D^3MOPSO play better than other methods under the circumstances of large scale solution space.

Table 3. Mean results of performance metric in Synthetic Simulation Datasets

	KTD	NDCG	DM	CM
D^3MOPSO	0.265	0.694	4.411	0.035
ANN	0.398	0.572	2.072	0.626
WT-INDEG	0.339	0.458	2.503	0.575
NPDA	0.367	0.353	8.804	0.053
BHTPSO	0.309	0.452	6.225	0.038

5.4.2 Comparing the Efficiency

Each method has different time complexity and efficiency. The time complexity of WT-INDEG was $O(Nm^2)$.where Nis the number of input rankings, and the total number of distinct items was m transformed into the measure symbol defined in our paper, then the time complexity expressed as $O(N \cdot MAXP^2)$. The time complexity of BHTPSO was $O(maxgen \cdot M \cdot MAXP^2)$.through the time complexity analysis, our method was superior to them. Then the experiment results were analyzed and discussed.

(1) Mean in LETOR Datasets

The characteristics of LETOR Datasets were its limited data source and the larger results from a single query. The experiment results on LETOR Datasets were shown as Fig. 3. The time complexity of ANN and WT-INDEG was the lowest, while D^3MOPSO was a litter less than NPDA and BHTPSO. When the amount of data was small the iteration step was large, the time complexity of these three evolutionary algorithms were higher than ANN and WT-INDEG. On the iteration convergence aspect, as Fig. 6 shown the convergence step of D^3MOPSO was a litter small than

NPDA and BHTPSO. So on LETOR Datasets the efficiency of D^{3}MOPSO was at medium levels.

(2) Mean in Real-world Datasets

While the characteristics of Real-world Datasets were its limited data source and the proper results from a single query. The experimental results on Real-world Datasets were as Fig. 4 shown. It was similar as that of LETOR Datasets. On the iteration convergence aspect, as Fig. 6 shown the same as that of LETOR Datasets. We had almost identical result that on Real-world Datasets the efficiency of D^{3}MOPSO was also at medium levels.

(3) Mean in Synthetic Simulation Datasets

The characteristics of Synthetic Simulation Datasets were its variety data source and the proper results from a single query. The experimental results on Synthetic Simulation Datasets were as Fig. 5 shown. In which the time consuming of ANN and WT-INDEG were extremely undesirable, due to its big candidate data amount. On the iteration convergence aspect, as Fig. 6 shown, the convergence step ICGM of D^{3}MOPSO was far less than NPDA and BHTPSO. So on Simulation Datasets the efficiency of D^{3}MOPSO was much better than other methods.

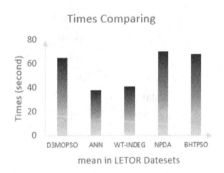

Fig. 3. Times Comparing in LETOR Datesets

Fig. 4. Times Comparing in Real-world Datesets

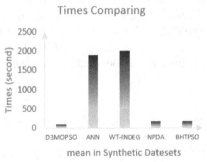

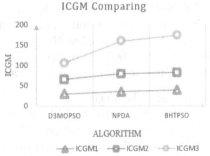

Fig. 5. Times Comparing in Synthetic Datesets

Fig. 6. ICGM Comparing among 3 MOEA

6 Conclusion

In this paper, we propose a fast, simple, easy to implement and efficient algorithm for Large-scale Rank Aggregation from multiple data sources based User Preference. The method to solve rank aggregation problem by Multi-objective Evolutionary Algorithm (MOEA) is an original work. Nice rank reflects use demand in different objective and preference, so how to find the best rank can be formalized a multi-objective optimization problem, which is possible in theory. However, conventional Multi-objective Disperse Algorithm (MODA) can't be applied in rank aggregation directly, which the reason is lack of mapping from data item to particle. So multi-strategy have been proposed in this paper, named D^3MOPSO, which improve PSO by encoding scheme, initialization methods, position and velocity definition, integrating updating, turbulence operator, external archive updating strategy and leaders selection.

Also, our experiments provide concrete evidence that the performance and efficiency of D^3MOPSO as well as current rank aggregation method approximately in small-scale data, and the performance and efficiency of D^3MOPSO are superior to current rank aggregation method and other MOEA/D in large-scale data.

As future work, we plan to research the inner mechanism that how the best results been found by MOEA, and improve the stability and performance. And a useful metasearch will be published by our research team.

References

1. Desarkar, M.S., Sarkar, S., Mitra, P.: Preference relations based unsupervised rank aggregation for metasearch. Expert Syst. Appl. **49**, 86–98 (2016)
2. Ozdemiray, A.M., Altingovde, I.S.: Explicit search result diversification using score and rank aggregation methods. J. Am. Soc. Inf. Sci. **66**(6), 1212–1228 (2015)
3. Ali, R., Naim, I.: User feedback based metasearching using neural network. Int. J. Mach. Learn. Cybern. **6**(2), 265–275 (2015)
4. Li, L., Xu, G., Zhang, Y., Kitsuregawa, M.: Random walk based rank aggregation to improving web search. Knowl.-Based Syst. **24**(7), 943–951 (2011)
5. Keyhanipour, A.H., Moshiri, B., Kazemian, M., Piroozmand, M., Lucas, C.: Aggregation of web search engines based on users' preferences in WebFusion. Knowl.-Based Syst. **20**(4), 321–328 (2007)
6. Amin, G.R., Emrouznejad, A., Sadeghi, H.: Metasearch information fusion using linear programming. Rairo-Oper. Res. **46**(04), 289–303 (2012)
7. Meng, W., Wu, Z., Yu, C., Li, Z.: A highly scalable and effective method for metasearch. ACM Trans. Inf. Syst. (TOIS) **19**(3), 310–335 (2001)
8. Amin, G.R., Emrouznejad, A.: Optimizing search engines results using linear programming. Expert Syst. Appl. **38**(9), 11534–11537 (2011)
9. Dwork, C., Kumar, R., Naor, M., Sivakumar, D.: Rank aggregation methods for the web, pp. 613–622. ACM (2001)
10. Coppersmith, D., Fleischer, L.K., Rurda, A.: Ordering by weighted number of wins gives a good ranking for weighted tournaments. ACM Trans. Algorithms (TALG) **6**(3), 1–13 (2010)
11. Montague, M., Aslam, J.A.: Condorcet fusion for improved retrieval, pp. 538–548. CIKM (2002)
12. Wu, S., Li, J., Zeng, X., Bi, Y.: Adaptive data fusion methods in information retrieval. J. Am. Soc. Inf. Sci. **65**(10), 2048–2061 (2014)

13. Klementiev, A., Roth, D., Small, K., Titov, I.: Unsupervised rank aggregation with domain-specific expertise, pp.1101–1106. IJCAI (2009)
14. Wei, Y., Shijun, L.: Automatically discovering of inconsistency among cross-source data based on Web big data. Comput. Res. Dev. **52**(2), 295–308 (2015)
15. Zhao, Y., Shen, B.: Empirical study of user preferences based on rating data of movies. PloS One **11**(1), e0146541 (2016)

Hierarchically Delegatable and Revocable Access Control for Large-Scale IoT Devices with Tradability Based on Blockchain

Liang Zhang[1,2] , Haibin Kan[2,3(✉)] , Jinrong Huang[2,3],
and Zhanpeng Zhang[2,3]

[1] School of Computer Science, Hainan University, Haikou 570228, China
zhangliang@hainanu.edu.cn
[2] School of Computer Science, Fudan University, Shanghai 200433, China
hbkan@fudan.edu.cn
[3] Shanghai Engineering Research Center of Blockchain, Shanghai 200433, China

Abstract. Access control is deemed a practical approach for managing the allowed list of who can access IoT devices. Different IoT use cases have shown requirements for various functionalities in access control. Hierarchical delegation on access control allows granted users to authorize other users. Revocability demonstrates the ability of the IoT owner to cancel access to her device. Tradability, meaning changing IoT device ownership, provides a solution for the IoT devices ownership trading market. However, the existing access control approaches do not implement these functionalities simultaneously or efficiently. This paper fills the gap by proposing a blockchain-based access control framework, leveraging the hierarchical deterministic (HD) wallet address technique. We achieve hierarchical access control of IoT devices using HD wallet. A signature scheme is incorporated to prove knowledge of HD child key pairs in the proposed framework. Further, we conduct concrete experiments on Ethereum and evaluate the gas consumption. The functionalities of the framework cost constant time or gas, implying that the framework scales well.

Keywords: Ethereum · HD wallet · IoT · access control ·
delegatability · revocability · tradability

1 Introduction

The internet of things (IoT) constructs a network of digital objects, embedded with sensors and software, to exchange data with other devices and systems over the internet. Smart IoT devices in offices, hospitals, vehicles, homes, retail stores, and smart buildings have brought conveniences to people's modern life. A promising solution to guarantee effective and safe access is through an access control mechanism, where access to an IoT device is restricted according to a specific rule. Access control of IoT devices contains two implications, i.e.,

X. Song et al. (Eds.): APWeb-WAIM 2023, LNCS 14331, pp. 81–95, 2024.
https://doi.org/10.1007/978-981-97-2303-4_6

authentication and authorization. Authentication [1] is a procedure that checks the authenticity of a user, while authorization [5] is a process that allows a user to access a resource.

In recent years, blockchain prevails in IoT access control systems [2–7], since it provides decentralized storage and a transparent computation environment. The programs of blockchain smart contract are stored and invoked among distributed nodes, indicating transparency and tamper-resistance. Intuitively, a smart contract could record access control policies, thereby enabling it to authorize users automatically. Besides, digital currency is available in blockchain smart contract, enabling users to trade digital assets. Moreover, blockchain incorporates cryptographic key pairs as pseudonymity, which is conducive to identifying distributed roles and protecting privacy in applications. With a cryptographic key pair as a user's identity, a smart contract can check whether the user is authenticated or not.

Nevertheless, none of the previous IoT access control literature considers functionalities of hierarchical delegation, revocation and tradability simultaneously. We aim at filling the gap by proposing an access control framework that leverages the hierarchical deterministic (HD) wallet address technique [8] and Ethereum smart contract [9]. Notably, hierarchical deterministic (HD) wallet address is a technique for users to manage multiple key pairs (identities) hierarchically. All key pairs generated by HD wallet address can be employed as independent blockchain identities. Parent private keys have higher permissions, making access control revocation possible in applications. Thus, we achieve hierarchical delegatable and revocable access control for IoT owners with HD wallet. Also, we implement a deposit smart contract on Ethereum, thereby quickly tackling the publicly known problem—fair exchange or fair trade [15,16].

Our contributions are fourfold:

1. We propose a novel access control framework for IoT devices, where hierarchical delegation, revocation and tradability are achieved. Besides, the framework guarantees decentralization, pseudonymity, non-repudiation and auditability. The idea of the framework differs significantly from previous approaches due to the use of HD wallet technique.
2. We leverage signature scheme to prove knowledge of HD wallet secret keys, guaranteeing the correctness of hierarchical delegation and revocation. Also, we have considered the replay attack and impersonate attack to the proposed framework by a granting user.
3. We achieve fair tradability for ownership of IoT devices without a trusted third party through a deposit smart contractWe consider all occasions that may break the fairness based on the smart contract, guaranteeing the feasibility of ownership transfer.
4. We conduct concrete experiments for the access control framework. Moreover, compared with the most relevant literature [18] that supports hierarchical delegation, our framework costs $O(1)$ for both time and space complexity in adding and deleting delegation. Consequently, our framework is well suited for the large-scale IoT environment.

2   Building Blocks

2.1   Blockchain and Ethereum

A blockchain is a transparent machine with public available computation and storage services, relying on decentralized consensus algorithms. As the fundamental ingredients of the blockchain blocks, transactions are linked with cryptographic hashes. Transactions are executed sequentially, in an order set by the block miner, resolving the concurrency problem. A blockchain is decentralized, implying that no single point of failure problem occurs or no DoS attack succeeds. As a successor of the first blockchain system—Bitcoin [12], Ethereum [9] brings significant evolution. Unlike Bitcoin, Ethereum not only allows for transferring digital currency (Ether), but also enables programmers to develop complex decentralized applications (DAPPs). Transactions or DAPPs are executed and verified in the Ethereum virtual machine (EVM). Every transaction invoker needs to pay for the corresponding computation and storage. The payment is calculated in gas. Ethereum address $addr_K$ is widely used as the unique identifier of an entity in a smart contract. For the sake of brevity, we use the public key K, rather than $addr_K$ which can be derived from K, as pseudonymity throughout the paper.

2.2   Digital Signature

Digital signature is used to prove ownership of specific digital assets or provide authentication of transactions. Signature schemes have the properties of public verifiability and non-repudiation. Each transaction on Ethereum is signed by default, using elliptic curve digital signature algorithm (ECDSA) [13,14]. In our framework, digital signatures are also implemented to prove knowledge of BIP-32 secret keys. A signature is generated using a private key k, and it is verified using the corresponding public key K, where $K = kG$ and G is a generator of group $\mathbb{G}$ based on elliptic curve "Secp256k1". The BIP-32 key pairs are compatible with ECDSA, which includes two algorithms:

 - $\mathsf{Sign}(m, k) \to \sigma$: outputs the signature σ of message m, using secret key k.
 - $\mathsf{Verify}(m, \sigma, K) \to$ True or False: outputs True if the signature σ is verified successfully with the public key K, otherwise outputs False.

2.3   BIP-32 Standard

HD wallets manage the keys in a tree structure, hence a private key can derive a sequence of children keys and grandchildren keys. Users can hide the relation (i.e., achieve unlinkability [10]) between parent key pair and child key pairs when necessary. The most extensively employed HD wallet technique is the one defined by the BIP-32 standard [11]. BIP is short for Bitcoin [12] improvement proposals. As a standard to generate cryptographic key pairs, the BIP-32 standard applies the elliptic curve "Secp256k1". BIP-32 allows each parent key to generate 2^{32} child keys and allows to generate child keys of any depth. The BIP-32 standard is realized applying extended keys, which is denoted as (k, c), where k is the normal private key, and c the called chain code. (K, c) is the corresponding extended public keys, where K is the corresponding public key generated from the private key k. The BIP-32 provides child key derivation (CKD) functions, which are described as follows:

- CKDpriv$(k_{par}, c_{par}, i) \rightarrow k_i, c_i$: calculates the ith child private key from the parent private key k_{par}.
- CKDpub$(K_{par}, c_{par}, i) \rightarrow K_i, c_i$: calculates the ith child public key from the parent public key K_{par}.

BIP-32 is prone to privilege escalation attack [17], in which an attacker can calculate k_{par} if given (k_i, i, c_{par}). Thus, CKDpriv and CKDpub should be invoked off-chain. To simplify the presentation, we omit the chain code c_{par} in the remainder of the paper. Therefore, we obtain:

- CKDpriv$(k_{par}, i) \rightarrow k_i$
- CKDpub$(K_{par}, i) \rightarrow K_i$

3   System Assumption and Requirements

3.1   System Entities

- IoT device: Each IoT device has a unique identity IoT_{id}, which is initialized by the corresponding manufacturer. Cars, refrigerators and smart locks in buildings are examples of IoT devices.
- Owner: Each IoT device belongs to an owner. Each owner has a key pair as its unique identity. With the private key, the owner takes full control of her IoT devices.
- Buyer: A buyer is someone who deposits digital currency using her private key to bid for an owner's IoT device ownership. If the trade is accomplished, the buyer becomes the new owner of the IoT device.
- Granting user: A granting user refers to a privileged user who is able to authorize another user. It can be inferred that an owner is always a granting user.
- IoT user: An IoT user, with key pair (k_{user}, k_{user}), is someone who wants to use the IoT device. A user is given a key pair (k_i, K_i) related to the device, generated using the BIP-32 standard by a granting user with (k_{par}, K_{par}).

3.2   System Assumption

Each smart IoT device is assumed to be under full control of its unique owner. However, we do not consider the occasion of asking an owner to exhibit a physical

IoT device. Also, the case that a user damages the device is also out of the discussion.

If an IoT user's key pair is valid, which is judged by Verify, the key pair is assumed to be generated using the BIP-32 standard. The assumption is reasonable, because BIP-32 standard enables everyone to derive child keys with her parent private key. Hence, no granting user wants to hold extra key pairs without any benefits. We allow adversaries to start replay attack on the signature schemes, where a valid signature is used multiple times. Due to the use of BIP-32 standard, a granting user with k_{par} needs to prove possession of child private key k_i when delegation, where $k_i = \mathsf{CKDpriv}(k_{par}, i)$. However, a dishonest granting user may also initiate impersonate attack when accessing IoT devices.

Blockchain provide reliable storage and computation services, which are available for all system entities. We assume the off-chain communication channel is secure when a granting user sends child key pairs to users. Transactions are the interfaces to invoke smart contract functions. Usually, the transaction parameter msg contains the invoker's public key $msg.sender$ and some digital currency $(msg.value)$. Each transaction is authenticated in blockchain by default, guaranteeing the authenticity of msg. Also, double-spending attacks (or variable consistency problems) via multiple transactions are avoided in blockchain. "Read-only" operations on blockchain are prone to replay attack by malicious blockchain nodes.[1]

3.3   System Requirements

- **Decentralization**: To avoid single-point of failure problem and guarantee trustworthiness, the proposed framework should eliminate trusted third parties. Even the smart contract deployer does not have any advantage over IoT owners or users.
- **Pseudonymity**: To protect privacy of personal information, pseudonymous identifiers are used to represent entities in the proposed framework.
- **Tradability**: The ownership of an IoT device can be transferred from an owner to another one (buyer). The framework should allow the owner to choose a buyer when there are multiple buyers.
- **Free/Fair trade agreement**: The trade between the owner and the buyer should guarantee fairness, i.e., either the ownership is transferred and the original owner obtains some digital currency, or nobody loses anything.
- **Hierarchical delegation**: A granting user can authorize access permissions to other users. The authorized users can also be granting users, demonstrating the functionality of hierarchical delegation for an IoT device.
- **Revocation**: An IoT device owner has the privilege to revoke the access permission for an authorized user.
- **Auditability**: The activities should be publicly auditable.
- **Non-repudiation**: No one can deny its activities in the proposed framework and no impersonate attack occurs.

---

[1] Read-only operations output calculation results instantly, since no new record is written to blockchain and no consensus algorithm is required.

– **Scalability**: The proposed framework supports numerous IoT users and heterogeneous distributed IoT devices.

4   The Proposed Framework

4.1   High-Level Overview

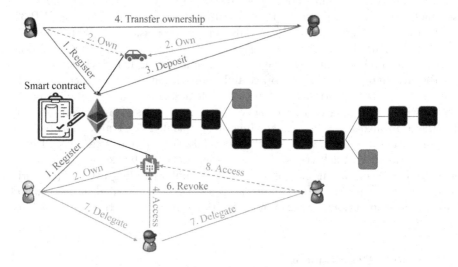

Fig. 1. High-level overview of the proposed framework

Figure 1 presents a high-level overview of the proposed framework. The purple and brown arrows demonstrate IoT devices registration and ownership, respectively. The teal and blue arrows depict depositing digital currency and ownership transfer, respectively. The green and red arrows show the delegation and revocation of access control, respectively. The cyan arrow represents that users are trying to access IoT devices.

Ethereum inherits the implementation of BIP-32 protocol. Hence, IoT device owners or users could generate their key pairs with the Ethereum account interface. Then, the framework works once the smart contract is deployed onto Ethereum. In the proposed framework, the smart contract is implemented to manage the registered IoT devices and the access control list. With the smart contract, a granting user can authorize for other users; an owner can transfer ownership to a buyer or revoke a user's permission after IoT registration.

The smart contracts can be deployed to Ethereum by any user, thereby guaranteeing that no trusted third party is responsible for our proposed framework. The global variables in the smart contracts are *deposits*, *IoTDevices*, *accList*, which are dictionaries. *deposits* represents the deposited Ether of a buyer for some IoT device. *IoTDevices* records all the IoT devices in the system. *accList* maintains all the authorization list for all IoT devices. The related functionalities in smart contracts will be introduced in the following subsections.

4.2   IoT Device Registration

The registration phase is interactively between an IoT device owner and the smart contract. The Algorithm 1 demonstrates how an owner registers an IoT device (with identifier IoT_{id}) to claim its ownership. The owner invokes the algorithm to record its ownership of IoT_{id} in *IoTDevices*. The registration information includes the device name ("name"), the value of the device ("price"), and the register's public key ("owner"). The name is a short description of the device, and it does not necessarily be unique in our framework. The algorithm is executed with constant gas consumption since *IoTDevices* is a dictionary. The information can be modified only by the owner, which is not detailed in this paper.

Algorithm 1. IoT device registeration

%Invoked by an IoT owner
function Register(IoT_{id}, name, price):
   global *IoTDevices*;
   temp={};
   temp["name"]=name;
   temp["price"]=price;
   temp["owner"]=$msg.sender$;
   IoTDevices[IoT_{id}]=temp;
   return True;

4.3   Ownership Transfer/Trading of IoT Device

As introduced in Sect. 3.1, an IoT device has a unique owner. In real life, an IoT device is usually required to be tradable, i.e., the ownership should be transferred from an owner to another. Moreover, an IoT owner is allowed to choose the highest bidder to maximize the benefit.

Algorithm 2 provides functionalities of how buyers deposit and withdraw Ether on Ethereum, as well as how the owner chooses a buyer and transfers the ownership. The ownership transfer functionality is interactively accomplished between an owner and multiple buyers. Multiple buyers could bid for the same IoT device (IoT_{id}) with different offers by invoking Deposit. The owner of IoT_{id} chooses a buyer, to whom he would like to transfer the ownership through Transfer. Buyers who fail to obtain the ownership of IoT_{id} could withdraw their deposited Ether using Withdraw.

In the Transfer function, the smart contract first checks the target buyer's deposit; second, it checks whether the *accList* is empty; third, it verifies the identity of the invoker; fourth, it changes the ownership and transfers the deposited Ether from the buyer to the original owner. We require *accList* to be empty in the functionality Transfer to avoid disputes between the new buyer and old users.

Algorithm 2. Deposit contract

%Invoked by an IoT buyer
function Deposit(IoT_{id}):

    global *deposits, IoTDevices*;
    assert($msg.value \geq IoTDevices[\text{IoT}_{id}][\text{"price"}]$);
    $deposits[msg.sender][\text{IoT}_{id}] = msg.value$;
    return True;

%Invoked by an IoT buyer
function Withdraw(IoT_{id}):

    global *deposits*;
    assert($deposits[msg.sender][\text{IoT}_{id}] > 0$);
    amount $= deposits[msg.sender][\text{IoT}_{id}]$;
    $deposits[msg.sender][\text{IoT}_{id}] = 0$;
    $msg.sender$.transfer(amount); %Transfer the deposit back to the sender
    return True;

%Invoked by the IoT owner
function Transfer(K_{buyer}, IoT_{id}):

    global *deposits, IoTDevices, accList*;
    assert($deposits[K_{buyer}][\text{IoT}_{id}] > 0$); %Ensure the deposit is not withdrawed
    assert($accList[\text{IoT}_{id}].\text{length} == 0$); %Ensure the IoT device is not in use
    assert($IoTDevices[\text{IoT}_{id}][\text{"owner"}] == msg.sender$); %Ensure the original ownership
    $IoTDevices[\text{IoT}_{id}][\text{"owner"}] = K_{buyer}$;
    amount $=$ deposits[buyer][IoT_{id}];
    deposits[buyer][IoT_{id}] $= 0$;
    K_{owner}.transfer(amount); %Transfer the deposit to the IoT owner
    return True;

As is known to all, it is not easy to achieve fair trade or fair exchange [15, 16]. One of the most challenging obstacle is obtaining fairness (or atomicity), which means no honest user losses digital currency or IoT ownership simultaneously. We achieve fairness through the smart contract by considering the following four requirements:

1. An owner transfers the ownership to a buyer if he confirms the buyer's bid;
2. An owner cannot transfer the ownership of an IoT device to multiple buyers simultaneously;
3. A buyer who fails to bid for the ownership could withdraw its deposit;
4. A buyer cannot withdraw deposit if ownership is transferred to him.

We prove that our framework satisfies the above four requirements as follows. The requirement 1) is straightforward if an owner obtains a buyer's bid, where Transfer algorithm is executed without revert. The requirement 2) is satisfied due to the blockchain virtual machine, which resolves the double-spending problem by executing transactions one after another. We provide Withdraw function to enable a buyer to withdraw its deposit before the deposit is maintained in *deposits.* Thus, the requirement 3) is guaranteed. Since the owner's private key has total control over an IoT device, the original owner has no privilege to manage the IoT device if the ownership is transferred. Therefore, the requirement 4) is guaranteed.

As a dictionary, operations on *IoTDevices* cost constant time. Thus, functionalities (Deposit, Withdraw and Transfer) in Algorithm 2 cost constant gas.

4.4   (Hierarchical) Delegation of Access Control

We leverage BIP-32 key pairs to manage keys and enable hierarchical delegation of access control for IoT devices. A granting user generates a child key pair, i.e., $k_i \leftarrow$ CKDpriv(k_{par}, i) and $K_i \leftarrow$ CKDpub(K_{par}, i), and securely sends them to the target user (with key pairs (k_{user}, K_{user})). The variable i is an increasing number held by the granting user. If a user is given permissions to access multiple IoT devices, he manages all the corresponding key pairs in a table. We do not go into details about key management off-chain in this paper.

It can be inferred that digital signature schemes can be used to prove knowledge of a BIP-32 standard secret key. Thus, we leverage Verify to prove knowledge of child private keys for a granting user.

We describe the functionality of hierarchical delegation of access control by Algorithm 3. To call Delegate algorithm, a granting user with (k_{par}, K_{par}) prepares the parameters: K_{user}, IoT$_{id}$, "expire", "right", K_i, σ_i. K_{user} is the target user's public key and (k_i, K_i) is also sent to the target user, where (k_i, K_i) is generated using CKDpriv(k_{par}, i) and CKDpub(K_{par}, i) by the granting user. "expire" represents the expiration time of the authorized access. Only when "right" is True will the target user have permission to authorize another user, demonstrating the functionality of hierarchical delegation. K_{user} is used as the message to produce the signature $\sigma_i =$ Sign(K_{user}, k_i) by the granting user.

Algorithm 3. Delegation of access control

%Invoked by an IoT owner or a granting user with k_{par}
function Delegate(K_{user}, IoT$_{id}$, expire, right, K_i, σ_i):
    global *IoTDevices, accList*;
    assert(Verify(K_{user}, σ_i, K_i) == True); %Check whether the invoker has knowledge of k_i
    assert(*IoTDevices*[IoT$_{id}$]["owner"]==*msg.sender*   or
    accList[IoT$_{id}$][*msg.sender*]["right"]); %Ensure the invoker is the owner or a granting user
    temp={};
    temp["from"]=*msg.sender*;
    temp["expire"]=expire;
    temp["right"]=right;
    temp["kuser"]=K_{user};
    accList[IoT$_{id}$][K_i]=temp;
    return True;

In Algorithm 3, the smart contract first checks the invoker has knowledge of private key k_i using $\mathsf{Verify}(K_{user}, \sigma_i, K_i)$. Then, the smart contract checks whether the invoker is the owner or a granting user of IoT_{id}. Finally, the targeted user is appended to the table $accList$. Here, replay attack of signature scheme is under consideration—The target user K_{user} may invoke $\mathsf{Delegate}$ with the tuple $(K_{user}, \sigma_i, K_i)$ to gain access permission of other IoT devices. We eliminate the replay attack by checking whether invoker of $\mathsf{Delegate}$ algorithm is a valid granting user to an IoT device.

Therefore, all authorization records in $accList$ are valid, as expected. Since both $accList$ and $IoTDevices$ are dictionaries, the functionality $\mathsf{Delegate}$ also costs constant gas. In addition, since the BIP-32 standard allows for generating multiple levels of child key pairs, our framework enables multi-level and hierarchical delegation of access control.

4.5   Access an IoT Device

By Algorithm 3, we have demonstrated how a granting user delegates the access control permission through the smart contract. In this section, we introduce the process of accessing an IoT device. Once a user is granted to access an IoT device, he could access or use the IoT device with a "sessionID" generated by $\mathsf{genSessionID}$ in Algorithm 4. In function $\mathsf{genSessionID}$, the smart contract first checks whether the user is in the authenticated list (i.e., $\mathrm{assert}(accList[\mathrm{IoT}_{id}][K_i]$ ["kuser"]$== msg.sender$)). If the user is granted, the smart contract generates a "sessionID" as $K_{user}\|\mathrm{IoT}_{id}\|\mathrm{user}[\text{"expire"}]$ for the user.

Due to resource limitations, IoT devices should conduct computation as little as possible. When the user tries to access an IoT device with the "sessionID", the IoT device first verifies the user's identity via $\mathsf{Verify}(\mathrm{sessionID}, \sigma, K_{user})$. If Verify returns True, σ must be generated by $\mathsf{Sign}(\mathrm{sessionID}, k_{user})$. Then, the IoT device outsources the judgement to blockchain, introduced as $\mathsf{allowAccess}$.[2] In $\mathsf{allowAccess}$, the smart contract checks whether K_i corresponds to K_{user} correctly (i.e., $accList[\mathrm{IoT}_{id}][K_i][\text{"kuser"}]==K_{user}$). Next, the smart contract checks whether the "sessionID" is expired. Finally, the IoT device make decision according to the response of the smart contract.

Obviously, the $\mathsf{genSessionID}$ function, executed in smart contract, costs constant gas and the $\mathsf{JudgeAccess}$ function, executed in IoT devices, takes constant time. Besides, the outsourced computation of $\mathsf{allowAccess}$ costs no gas, because it is a read-only operation on Ethereum.

4.6   Revocation

To revoke the privileges of accessing the IoT device IoT_{id} for a user K_i, our framework provides two methods to do this. One implicit method is to set an

---

[2] "sessionID" is reusable in the proposed framework. To prevent malicious blockchain nodes eavesdropping the tuple (sessionID, σ, K_{user}) and initiating replay attack to IoT devices, the Verify is not outsourced to Ethereum.

Algorithm 4. Access an IoT device

%Invoked by an IoT user with k_{user}
function genSessionID(K_i, IoT$_{id}$):

    global *accList*;
    assert(*accList*[IoT$_{id}$][K_i]["kuser"]==$msg.sender$);
    expire=*accList*[IoT$_{id}$][K_i]["expire"];
    return K_i∥IoT$_{id}$∥expire;

%Invoked by an IoT device
function allowAccess(sessionID, K_{user}) *readonly*:

    global *accList*;
    K_i, IoT$_{id}$, expr=sessionID.split("∥");
    assert(*accList*[IoT$_{id}$][K_i]["kuser"]==K_{user});
    assert(expr > current time);
    return True;

%Invoked by a user and executed in an IoT device
function JudgeAccess(sessionID, σ, K_{user}):

    assert(Verify(sessionID, σ, K_{user})==True);
    allowed=allowAccess(sessionID, K_{user});
    if allowed **then**
        allow K_{user} to access;
    else
        reject K_{user} to access;
    end if
    return;

---

expiration time when delegating, as depicted by Algorithm 3. When the expiration time is exceeded, the IoT device will reject a user's access, as the JudgeAccess function in Algorithm 4 depicts. The other method is explicit, i.e., the IoT device owner removes a user from the authorization list, as Algorithm 5 shows.

Algorithm 5. Revocation

%Invoked by an IoT owner
function Revoke(K_i, IoT$_{id}$):

    global *accList, IoTDevices*;
    assert(*IoTDevices*[IoT$_{id}$]["owner"] == $msg.sender$);
    delete *accList*[IoT$_{id}$][K_i];
    return;

The smart contract first verifies whether the invoker is the owner of an IoT device. Then, the target user K_i is removed from authorized list *accList*. Once the user is removed, it will fail to generate a "sessionID" with genSessionID function. Further, our framework can be easily extended so that all granting users are also able to revoke permissions. It is easy to conclude that the Revoke algorithm costs constant gas.

5 Experimental Results

The experiments are conducted on MacBook Air macOS 11.3.1, 1.6 GHz Dual-Core Intel Core i5 CPU, 16 GB RAM. We develop the proposed access control for IoT framework with nodejs 14.16.1, a runtime built on Chrome's V8 JavaScript engine and solidity, an Ethereum smart contract programming language. We use Ganache 2.3.0 as an Ethereum full node to run the smart contract, which is compiled by solidity codes compiler v0.5.17 in Truffle.

The smart contracts, concerning the proposed functionalities, are deployed to Ethereum. It can be inferred that the deployer has no privilege of gaining private information in the proposed framework.

The functions related to private keys should be invoked off-chain. We evaluate these functions by giving the time costs (ms) on a personal computer (PC). These functions includes CKDpriv, CKDpub, Sign. Other functions (i.e., Verify, Register, Deposit, Transfer, Withdraw, Delegate, genSessionID and Revoke) are executed on-chain and they are publicly verifiable.

Table 1. Gas or time cost of functions in the framework

Function	Environment	Gas/Time
CKDpriv	PC	0.057 ms
CKDpub	PC	0.093 ms
Sign	PC	15 ms
deployment	Ethereum	0.1Ether
Verify	Ethereum	30820wei
Register	Ethereum	128539wei
Deposit	Ethereum	44038wei
Transfer	Ethereum	42791wei
Withdraw	Ethereum	21523wei
allowAccess	Ethereum	0wei
Delegate	Ethereum	175004wei
genSessionID	Ethereum	30766wei
Revoke	Ethereum	32620wei

Table 1 summarizes the time cost (on PC, measured by ms) and gas consumption (on Ethereum, measured by wei and 1wei $= 10^{-18}$Ether) of the functions in the framework. The deployment of the smart contract costs about 0.1Ether. The gas cost of allowAccess is 0, since it is a read-only function. By comparing the cost of function Register and genSessionID in Table 1, it can be inferred that the storage on Ethereum is quite costly.

6   Security Analysis

In this section, we prove that the proposed framework satisfies the security requirements defined in Sect. 3.3.

- **Decentralization**: As is known to all, Ethereum is totally decentralized and the smart contract deployment is a transaction. The smart contract source code of our framework is totally public and the corresponding deployer has no control over the data, as Sect. 4 demonstrates. The core of the proposed framework is the smart contract, which can be deployed by any third party. IoT owners and users maintain their respective BIP-32 standard key pairs, which only imply their privileges of corresponding IoT devices. Therefore, the proposed framework keeps decentralization, as well as availability.
- **Pseudonymity**: The framework reveals no information about the real-world identity of an IoT owner or a user, since we use BIP-32 standard public keys as the identifiers. Moreover, BIP-32 standard achieves unlinkability between parent keys and child keys. Hence, we protect the privacy of users through pseudonymity.
- **Tradability**: We design a deposit contract to enable IoT device owners to transfer their ownership with corresponding digital rewards, as introduced by Algorithm 2. The smart contract allows multiple potential users to bid for the ownership of an IoT device and the owner can choose a buyer as he/she likes. We have assumed each IoT device has a unique identity and it is under full control of its owner's private key. Once the ownership is transferred, the ability to control the device is shifted from the original owner to the new owner.
- **Free/Fair trade agreement**: The deposit contract has three independent functionalities: Transfer, Withdraw and Deposit. The Transfer guarantees that an owner exchanges ownership of an IoT device with a buyer's deposited digital currency. Other buyers can withdraw his/her deposited money via Withdraw. Moreover, we have assumed that double-spending attacks (or variable consistency problems) are avoided in blockchain. Hence, no one can invoke Transfer or Withdraw for an IoT device twice. Thus, no buyer suffers a loss, guaranteeing fairness.
- **Hierarchical delegation**: We leverage BIP-32 standard key pairs in our design, hence each child key pair can generate a grandchild key pair. In the Delegate function, the smart contract record a child public key for each new granted user. If the record mark "right" as True, the user is authorized as a granting user, as introduced in Algorithm 3. Thus, our framework enables hierarchical delegation. Moreover, only public keys are recorded on blockchain, implying the security and feasibility of hierarchical delegation.
- **Revocation**: Smart contract records the ownership of each IoT device. In our design, only the device owners are allowed to remove authorized users through the functionality Revoke, demonstrating that ownership means full control. The identities of invokers are verified to guarantee the requirement, as introduced in Algorithm 5.

- **Auditability**: Storage on blockchain is tamper-resistant and publicly available. In addition, transactions are authenticated. Thus, everyone can verify whether an activity to invoke functionalities is valid or not. Thus, all operations of the proposed framework are auditable.
- **Non-repudiation**: This property extends from **Auditability**. The on-chain operations are executed and verified publicly in the smart contract. Thus, no user can alter the global variables to remove its activity records. In BIP-32 standard, a granting user has knowledge of child private keys of granted users. This fact brings possibility of impersonate attack by granting users. We resolve this kind of attack by authenticating the invoker in genSessionID and allowAccess. Moreover, we have considered replay attacks when leveraging signatures as identity proofs in functionality Delegate and when designing the read-only functionality allowAccess. Thus, no one can deny its activities in the framework.
- **Scalability**: All the functionalities (introduced in Sect. 4) of the proposed framework cost constant time or gas, regardless of the number of registered devices or delegations. Thus, our protocol scales well.

7   Conclusions

In this paper, we introduce a new framework to enhance the properties of access control in the industrial IoT environment. By incorporating HD wallet address, we achieve hierarchical delegation and revocation in authorizing access permission of IoT devices. The HD wallet address enables an IoT owner to manage (delegate or revoke access permissions and trade) her device only with the master secret key. In addition, we achieve fair ownership transfer between different users, which is implemented through a deposit contract on Ethereum. Ethereum is publicly available and tamper-resistant with digital currency, guaranteeing the proposed framework to obtain desirable properties, such as non-repudiation, auditability, and tradability. The compatibility of the HD wallet address in Ethereum facilitates authentication in the framework with pseu-donymity. Moreover, all operations in our framework cost constant time or gas, leading to scalability. Comparison with related works has shown the distinctions and contributions of the framework.

Acknowledgements. This work was supported by National Key R & D Program of China (No. 2019YFB2101703), National Natural Science Foundation of China (Grant No. 62272107 and U19A2066), Foundation of Hainan University (No. KYQD22168) and the Key Laboratory of Internet Information Retrieval of Hainan Province Research Found (No. 2022KY01).

References

1. Yu, Y., Zhao, Y., Li, Y., Du, X., Wang, L., Guizani, M.: Blockchain-based anonymous authentication with selective revocation for smart industrial applications. IEEE Trans. Industr. Inf. **16**(5), 3290–3300 (2020)

2. Pal, S., Dorri, A., Jurdak, R.: Blockchain for IoT access control: recent trends and future research directions. J. Netw. Comput. Appl. **203**, 103371 (2022)
3. Han, P., Zhang, Z., Ji, S., Wang, X., Liu, L., Ren, Y.: Access control mechanism for the Internet of Things based on blockchain and inner product encryption. J. Inf. Secur. Appl. **74**, 103446 (2023)
4. Wu, N., Xu, L., Zhu, L.: A blockchain based access control scheme with hidden policy and attribute. Futur. Gener. Comput. Syst. **141**, 186–196 (2023)
5. Siris, V.A., Dimopoulos, D., Fotiou, N., Voulgaris, S., Polyzos, G.C.: Decentralized authorization in constrained IoT environments exploiting interledger mechanisms. Comput. Commun. **152**, 243–251 (2020)
6. Ouaddah, A., Kalam, A.A.E., Ouahman, A.A.: Fairaccess: a new blockchain-based access control framework for the internet of things. Secur. Commun. Networks **9**, 5943–5964 (2016)
7. Novo, O.: Blockchain meets IoT: an architecture for scalable access management in IoT. IEEE Internet Things J. **5**(2), 1184–1195 (2018)
8. Maxwell, G., Bentov, I.: Deterministic Wallets. https://www.cs.cornell.edu/~iddo/detwal.pdf. Accessed 25 Mar 2023
9. Wood, G.: Ethereum: a secure decentralised generalised transaction ledger. Ethereum project yellow paper, pp. 1–32 (2014)
10. Das, P., Faust, S., Loss, J.: A formal treatment of deterministic wallets. In: CCS 2019, pp. 651–668 (2019)
11. Wuille, P.: BIP 32: Hierarchical deterministic wallets. https://github.com/bitcoin/bips/blob/master/bip-0032.mediawiki. Accessed 25 Mar 2023
12. Nakamoto, S.: Bitcoin: a peer-to-peer electronic cash system (2008)
13. Blake-Wilson, S., Bolyard, N., Gupta, V., Hawk, C., Moeller, B.: Elliptic Curve Cryptography (ECC) Cipher Suites for Transport Layer Security (TLS). RFC rfc4492 (2006)
14. Paar, C., Pelzl, J.: Understanding Cryptography: A Textbook for Students and Practitioners. Springer, Cham (2010). https://doi.org/10.1007/978-3-642-04101-3
15. Bentov, I., Kumaresan, R.: How to use bitcoin to design fair protocols. In: Garay, J.A., Gennaro, R. (eds.) CRYPTO 2014. LNCS, vol. 8617, pp. 421–439. Springer, Heidelberg (2014). https://doi.org/10.1007/978-3-662-44381-1_24
16. Dziembowski, S., Eckey, L., Faust, S.: Fairswap: how to fairly exchange digital goods. In: Proceedings of the 2018 ACM SIGSAC Conference on Computer and Communications Security, pp. 967–984 (2018)
17. Fan, C., Tseng, Y., Su, H., Hsu, R., Kikuchi, H.: Secure hierarchical bitcoin wallet scheme against privilege escalation attacks. In: IEEE Conference on Dependable and Secure Computing, DSC, pp. 1–8 (2018)
18. Tapas, N., Longo, F., Merlino, G., Puliafito, A.: Experimenting with smart contracts for access control and delegation in IoT. Futur. Gener. Comput. Syst. **111**, 324–338 (2020)

Distributed Deep Learning for Big Remote Sensing Data Processing on Apache Spark: Geological Remote Sensing Interpretation as a Case Study

Ao Long[1], Wei Han[1], Xiaohui Huang[1,2]✉, Jiabao Li[1], Yuewei Wang[1], and Jia Chen[1]

[1] School of Computer Science, China University of Geosciences, Wuhan, China
[2] Hubei Key Laboratory of Intelligent Geo-Information Processing, Wuhan, China
xhhuang@cug.edu.cn

Abstract. The advent of sensor technologies has led to the sheer amount volume of remote sensing data containing fruitful spatial and spectral information. Insights into Earth's surface's objects are gained with the help of remote sensing processing methods and techniques and are applied in various applications. Recently, deep-learning-based methods are widely used in remote sensing data processing due to their ability to mine relationships using multiple layers. However, the time spent by deep learning-based methods with numerous layers and large parameter sizes in processing remote sensing data with "big data" characteristics is unacceptable in real-time applications. Combining deep learning with distributed computing namely distributed deep learning, has become an emerging topic in deep learning-based remote sensing processing. This paper first surveys recent methods and open-source solutions of Apache Spark-based distributed deep learning. Then, the pros and cons of each distributed deep learning open-source solution in processing remote sensing data are summarized. Later, the geological remote sensing interpretation is chosen as the case study by implementing the online training of a deep learning-based interpretation model called D-AMSDFNet for geological environments on Apache Spark. Experiments on Landsat 8 and Sentinel 2 satellite images investigate the effectiveness of the proposed D-AMSDFNet, which also indicates the promising development of distributed deep learning in processing remote sensing data.

Keywords: Remote sensing · Distributed deep learning · Apache spark · Distributed geological remote sensing interpretation

1 Introduction

Remote sensing is a way of collecting information on electromagnetic waves from targets through non-contact methods [1]. Remote sensing technologies can provide long-term and large-scale continuous observation of regions, especially

X. Song et al. (Eds.): APWeb-WAIM 2023, LNCS 14331, pp. 96–110, 2024.
https://doi.org/10.1007/978-981-97-2303-4_7

restricted areas, and obtain spatial and spectral information of surface objects in the region. By using remote sensing data processing techniques such as manual or intelligent interpretation, hidden relationships between elements can be analyzed to serve various remote sensing applications [2]. However, complex scenes of object classification and detection exist in remote sensing data interpretation. Recently, deep learning (DL) methods that can automatically extract deep semantic information have been widely used in remote sensing data interpretation [3], such as satellite image classification [4], semantic segmentation [5], and object detection [6]. However, the training of deep neural networks is time-consuming. The reason is two folds. At first, the extremely deep layers of the networks result in large model parameters (e.g., ResNet can achieve 100 or even 1000 layers by stacking residual modules [7]). In addition, training a DL-based model generally requires multiple iterations until it converges. Consequently, using DL methods to process remote sensing data consume a lot of time difficult to meet application needs [8]. At the same time, remote sensing data have entered the "big data era", with the improvement of spatial, spectral, and temporal resolution of sensors carried by remote sensing platforms, remote sensing data types are becoming diverse and their volume is also growing rapidly. Processing and analyzing such a sheer amount of remote sensing data is also time-consuming [9]. However, a portion of remote sensing applications has real-time demands. For example, multi-source remote sensing data need to be processed in reasonable time to accomplish the mapping tasks for natural disasters (e.g., earthquakes, floods, and landslides) [10]. Therefore, with advanced distributed computing at researchers' disposal, combining DL methods and distributed computing techniques to efficiently process massive remote sensing data has gradually become a research hot topic in the field of remote sensing recently.

Distributed deep learning (DDL) is a parallel computing strategy that decomposes a deep learning task into several sub-tasks, each of which is distributed to computing nodes in a distributed cluster. Each computing node is responsible for processing a portion of data or model parameters, thus speeding up the training process. On the one hand, considering that DL frameworks (e.g., PyTorch [11], Tensorflow [12]) are mainly used to train neural networks in remote sensing data interpretation [5, 8]. On the other hand, big data frameworks (e.g., Hadoop [13], Spark [14]) are generally used for processing and storing remote sensing data. Letting the training of DL models near the location where the data is stored significantly improves the efficiency of the processing of remote sensing. [15]. Currently, methods implementing DDL on the Apache Spark platform can be divided into two categories:

(1) Offline training and online inference. This category first process data on a Spark cluster and then export the processed data to a single machine for training. Finally, the trained model is deployed to the Spark cluster for distributed reasoning. In this mode, when new remote sensing data is coming, the trained model needs to be trained offline again and then to be redeployed, resulting in lower efficiency.

(2) Online training and inference. This type of method implements model training and inference on a Spark cluster. Several representative libraries support the training of DL models on the Spark cluster. MLlib is a distributed
machine learning library bound to Spark [16], but it can only build multilayer
perceptron (MLP). Another method uses a "connector" (e.g., SparkTorch,
TensorflowOnSpark [17]) that provides network building using Pytorch or
Tensorflow. However, these frameworks may result in significant overhead
due to Apache Spark's limited support for DL models at the algorithm level
[18]. Alternatively, other frameworks such as BigDL [15] and Horovod [19]
can be used to directly implement DDL on the Spark cluster, making it
easier to construct end-to-end deep learning pipelines following a unified
programming paradigm, which results in higher efficiency. Currently, there
are many open source libraries available for researchers, but there has been
little in-depth research, analysis, and summary of DL models for remote
sensing data processing with these libraries.

To solve the issues mentioned above, this paper first summarizes methods,
and open source implementations of current Apache Spark-based distributed
deep learning. In addition, recent works utilizing DDL to process remote sensing
data are also surveyed by summarizing their pros and cons. Then, our previous
work that adopts DL methods to conduct geological remote sensing interpretation is chosen as a case study to indicate the efficiency of DDL on remote sensing
processing. In conclusion, this paper makes the following contributions.

- We summarize existing open source frameworks, including MLlib, Spark-
  Torch, TensorflowOnSpark, DeepLearning4Java [20], BigDL, and Horovod,
  that implement DDL on Spark. We analyze the demands for DL-based
  remote sensing data processing, summarize the advantages and disadvantages of existing DDL libraries, and summarize the feasibility of their
  implementation.
- We choose geological remote sensing interpretation as a case, we modify the
  AMSDFNet, which is proposed in our previous work, into a distributed structure, and implement online training with data parallelism on an Apache Spark
  cluster. In experiments, we compare the performance between the traditional
  Spark cluster utilizing only the CPU and the Spark cluster adopting both
  CPU and GPU. In addition, we investigate the acceleration effect of DDL
  and the scalability of nodes.

2   Related Works

This section introduces the current research status of DDL's development status
and DDL-based remote sensing data processing.

2.1   Distributed Deep Learning's Development Status

The common approaches of DDL consist of data parallelism and model parallelism, which are depicted in Fig. 1. Data parallelism divides data into several

parts, each of which is processed by a computing node. Data parallelism, as shown in Fig. 1(a), is generally used for situations where the data volume is large. It can accelerate model training by multiple computing nodes. Because the data are divided into several subsets, each computing node uses one subset to train the model. Sun et al. [21] used data parallelism to train DNN with large batches. Model parallelism divides the model into multiple submodels and computes them in parallel on different computing nodes. Figure 1(b) shows how the model parallelism works. It is typically used when the model is too large to fit into a single computing node. Hong et al. [22] applied a Transformer network whose training needs model parallelism to hyperspectral classification.

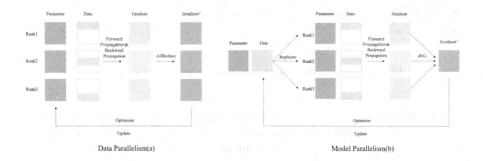

Fig. 1. Two methods of DDL.

2.2   DDL-Based Remote Sensing Data Processing

Data parallelism of DDL is mainly applied in remote sensing. Because convolutional neural networks (CNN) are currently mainly used in remote sensing data interpretation [23]. The volume of the CNN model is not large, but the amount of remote sensing data is usually huge, so data parallelism is suitable for accelerating remote sensing interpretation. However, DDL is mainly based on big data ecosystems to process remote sensing data which is stored in distributed file systems (e.g., HDFS, Ceph, etc.). Mainstream methods implementing DL models for remote sensing data processing are based on distributed computing frameworks (e.g., Apache Spark). As is shown in Fig. 2, The methods can be divided into two categories, i.e., offline training & online inference and online training & online inference.

Offline Training and Online Inference. This method completes the storage, processing, and deployment inference of remote sensing data online, while the training of the model is completed offline. Liu et al. [24] trained a high-precision remote sensing object detection model offline using PyTorch and deployed the trained model on Spark to support distributed inference on large-scale remote sensing images. Zhong et al. [25] built a data-driven framework with offline training and online inference, aiming to achieve efficient and robust airport positioning and aircraft detection.

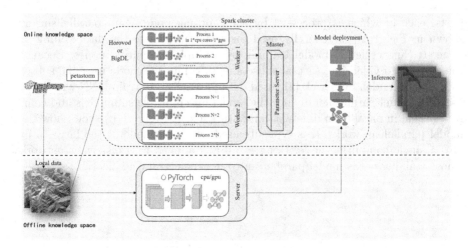

Fig. 2. Online or Offline training and inference architecture.

For offline training and online inference, the big data platform and the DL processing are separated. When new remote sensing data is coming, the training data is updated, which requires re-training the existing DL model to ensure its accuracy. Re-training the existing DL model is generally performed on a single machine and thus is a complex process and time-consuming. The reason is that the new remote sensing data is exported from big data platforms to a single machine. Then, the DL model is re-trained and deployed into the big data platform.

Online Training and Inference. This method directly implements model training and inference in the Spark cluster. Haut et al. [26] used MLlib to build a multilayer perceptron and conducted distributed training directly in Spark, which greatly improves the efficiency. However, MLlib cannot implement complex networks built with Pytorch or Tensorflow. Sun et al. [21] applied the TensorflowOnSpark to train a DNN built with Tensorflow with remote sensing data on Spark. Frameworks such as TensorflowOnSpark and SparkTorch are "connectors", and the training process is not integrated into Spark pipelines, so they are not fast and efficient enough. Boulila et al. [27] used BigDL to build RS-DCNN and implemented pixel-level classification for big remote sensing data. BigDL supports neural network models built using Pytorch or Tensorflow and directly implements DDL in the Spark big data system. However, BigDL does not support GPU and coarse-grained operations, which makes it difficult to support high-precision remote sensing interpretation models.

3   Distributed Deep Learning Frameworks

In this section, we investigate existing open source frameworks that support DDL on Spark. Considering the characteristics of remote sensing data and the demand for interpreting remote sensing data based on DL, the advantages and disadvantages of these libraries and the feasibility of their implementation for remote sensing data processing are summarized in Table 1.

Table 1. Summarization of DL frameworks.

Frameworks	Whether support Pytorch or Tensorflow	Whether support GPU
MLlib	No	No
SparkTorch,TensorflowOnSpark	Yes,coarse-grained	Yes
DeepLearning4Java	Partial compatibility	Yes
BigDL	Yes,coarse-grained	No
Horovod	Yes,fine-grained	Yes

3.1   MLlib

It is a machine-learning library that is natively supported by Spark. It is designed for the DataFrame data structure, so MLlib includes algorithms (e.g., regression analysis, principal component analysis, and MLP) designed for tabular data. Each input sample can be flattened as a vector. However, MLP is too simple to achieve high-precision remote sensing image interpretation, which requires deeper neural networks (e.g., CNN).

3.2   SparkTorch and TensorflowOnSpark

These frameworks adopt a "connector approach" which develops interfaces to connect different data processing and DL components. However, the adaptation between different frameworks can pose very large overheads in practice. Specifically, they suffer from impedance mismatches, which arise from crossing boundaries between heterogeneous components [15]. For example, SparkTorch first adopts Spark to allocate resources and then runs DL tasks based on the allocated resources. However, big data and DL systems have different execution models - Spark is parallel and independent of each other, while DL tasks need to coordinate with others. For instance, the Spark cluster just relaunches the worker, when a Spark worker fails, but this is incompatible with the Pytorch execution model and can cause the entire workflow to block indefinitely.

3.3   DeepLearning4Java

DeepLearning4Java is developed with Java and also implements DDL on Spark. However, it builds neural networks following a declarative, layer-by-layer construction pattern. It is relatively outdated compared to Tensorflow and Pytorch frameworks because it appeared earlier. Therefore, the pre-trained models provided by the framework are not as rich as those of Pytorch. Therefore, models such as ResNet and the semantic segmentation network DeeplabV3, etc., cannot be implemented with DeepLearning4Java. So this framework is rarely used for high-precision geological remote sensing interpretation models.

3.4   BigDL

BigDL directly implements DDL on Spark by providing support to build neural networks using PyTorch or TensorFlow. Therefore, it offers the ability to conveniently construct end-to-end DL pipelines in a unified programming paradigm. The pipelines can then be executed as standard Spark jobs to process and train large-scale datasets stored on the Spark cluster within the Spark framework. This approach fully eliminates impedance mismatch problems and significantly enhances the efficiency of developing and executing DL models for big data. However, BigDL is designed for traditional CPU clusters and is also limited to providing coarse-grained operations for functional computation models. Specifically, it does not support customizing the training process (e.g., the hard sample mining algorithm can not be realized after one epoch).

3.5   Horovod

In 2017, Uber released Horovod, which is a lightweight, high-performance DDL training framework. The module running Horovod on Spark aims to integrate Horovod with a Spark cluster. This integration allows the entire machine learning loop, including data processing, model training, and model evaluation, to reside within Spark. Horovod initially integrates DL frameworks (e.g., Pytorch, Tensorflow) with MPI. MPI is a library focused on portable performance, but it disregards programmer productivity because its low-level primitives require large development code. Moreover, the rate of success of MPI clusters is not satisfactory since the failure of a single task calls for restarting the entire cluster. Due to this poor fault tolerance, Horovod integrates with Spark, aiming to use Spark's fault tolerance mechanism. Specifically, Horovod relies on Spark as an underlying mechanism to distribute PyTorch. Therefore, invoking PyTorch processes through Spark executors is possible, and realizing a more finely-grained training process compared to BigDL. In addition, it requires minimal changes to the training scripts developed for PyTorch to the Spark cluster using CPUs or GPUs.

In conclusion, our analyses demonstrate that Horovod is the most suitable framework to achieve distributed training of high-precision DL models used in remote sensing data interpretation on Spark.

4   D-AMSDFNet: Distributed Deep Learning-Based AMSDFNet for Geological Remote Sensing Interpretation

As a case study, this section focuses on the geological remote sensing interpretation and proposes a modification to our proposed AMSDFNet model [8] by switching it to a data parallelism distributed model.

4.1   AMSDFNet

AMSDFNet utilizes multi-source data and DL features to interpret multiple geological elements. Figure 3 shows the structure of the AMSDFNet model. The characteristics of the AMSDFNet model are as follows:

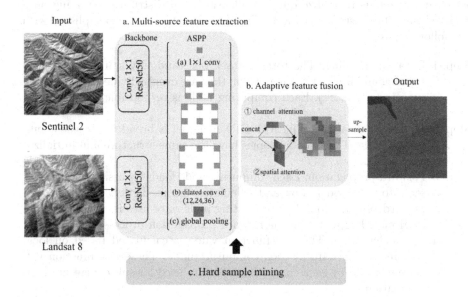

Fig. 3. The overall architecture of the proposed AMSDFNet [8].

(1) Multi-source feature extraction [28]. AMSDFNet has two parallel DeeplabV3 networks, each of which includes a ResNet50 as a backbone pre-trained on ImageNet and an atrous spatial pyramid pooling (ASPP) module [29]. The above-mentioned two branches can simultaneously extract visual and spectral features from Sentinel-2 and Landsat-8 satellite images. The extracted features are then stacked together.

(2) Adaptive feature fusion. The stacked features are fused using a convolutional block attention module (CBAM) [30], an attention mechanism module. Valuable information may be affected because the previous step involves large-scale feature connections. Therefore, an adaptive feature fusion module is adopted to increase the weight of key features and enhance valuable information while suppressing redundant information.

(3) Hard sample mining. To address the issue of sample imbalance in geological elements, a hard sample mining algorithm is used to automatically select key samples and further improve model accuracy. This algorithm classifies pixel-level samples based on updated gradients, and AMSDFNet only learns pixels with gradients that exceed a predefined threshold.

4.2   Design of Distributed AMSDFNet

AMSDFNet improved the accuracy of geological remote sensing interpretation. But considering its time cost, AMSDFNet took 171.3 min to complete the full training when processing 2.1 GB remote sensing data on an RTX3090 GPU using Pytorch. The whole training process is time-consuming. When applying AMSDFNet in applications with real-time requirements, its time cost is unacceptable, which motivates us to realize data parallelism-based distributed training on a Spark cluster. The distributed AMSDFNet (D-AMSDFNet) is accomplished with the following steps.

Step. 1 Data partitioning: The total dataset was randomly split into 80% for training and 20% for testing. Then, they are evenly divided into multiple subsets $(x^1, ..., x^n)$. Each computing node is responsible for processing one subset.

Step. 2 Model broadcasting: The AMSDFNet model is broadcast to each computing node. These model copies have the same structure and initialization parameters.

Step. 3 Forward propagation: Each computing node loads its assigned data subset into the model copy and calculates the prediction result through forward propagation. The computing node only uses its local data and local model copy during the forward propagation process.

Step. 4 Loss calculation: The loss function value is calculated based on local remote sensing data at each computing node. These loss function values generated by each computing node are used for subsequent gradient calculation.

Step. 5 Backward propagation and gradient calculation: Each computing node calculates gradients through backward propagation. During this process, the computing node adopts its local data and local model copy to calculate the gradients by $\frac{\partial L(x^i;\omega)}{\partial \omega}(i = 1, ..., n)$, where ω refers to weights of the neural network.

Step. 6 Gradient aggregation: The computing nodes send the locally calculated gradients to the master node. The master node is responsible for collecting and aggregating the gradients of all computing nodes, this process is called all-reduce. Specifically, the sum of per-parameter gradients that are computed using subsets $(x^1, ..., x^n)$ of a mini-batch x matches the per-parameter gradients for the entire input batch (i.e., $\frac{\partial L(x;\omega)}{\partial \omega} = \frac{\partial L(x^1;\omega)}{\partial \omega} + ... + \frac{\partial L(x^n;\omega)}{\partial \omega}$).

Step. 7 Model update: The master node sends the aggregated gradient to each computing node, and each computing node updates the model parameters based on the aggregated gradient. Since each computing node receives the same gradient, the model remains the same after the update.

Step. 8 Repeat iteration: Repeating Step. 3 to Step. 7 until the model is converged.

5 Experiments

This section evaluates the efficiency of the proposed D-AMSDFNet using Horovod to implement online training of AMSDFNet on Spark clusters using CPUs and GPUs. And we discuss DDL's acceleration effect and node scalability.

5.1 Settings

Hardware and Software. All experiments were performed on a cluster equipped with two servers, each of which has a 64-cores Intel(R) Xeon(R) Gold 6346 CPU @ 3.10GHz, 503GB of memory, and two RTX 3090 GPUs. In the experiments, we used the pre-trained models provided by Pytorch 1.12, Horovod 0.28.0, and Spark 3.2.1.

Experimental Data. The Sentinel 2, Landsat 8, and their annotation images were aligned and segmented into 1101 image patches. The Landsat 8 and annotation image patches were with 224×224 pixels, while the Sentinel 2 image patches were cropped and resized to 448×448 pixels to facilitate feature fusion.

According to the original paper, when training AMSDFNet on an RTX 3090 GPU, the batch size was set to eight, and the initial learning rate was set to 0.003. The entire training process for AMSDFNet was 120 epochs [8].

When training AMSDFNet on a Spark cluster using GPUs (termed as Spark-GPU), hyper-parameters would be changed. The epochs remained constant, but the batch size and initial learning rate were multiplied by the number of GPUs. This is because the data parallelism approach increases the batch size. To ensure the model can be converged, it's necessary to multiply the learning rate or increase the number of iterations.

When training AMSDFNet on a Spark cluster using CPUs (termed as Spark-CPU), there are 128 CPU cores across two servers (each core can run one training process). To fully utilize the computing resources, we used 128 processes, each of which has a batch size of 2. The initial learning rate was adjusted to 0.03 (the original learning rate was multiplied by 10), and the epochs were set to 150.

5.2 Analysis of Experimental Results

We first compare the efficiency between a single machine with one GPU and the Spark-GPU cluster. The evaluating indicators are the running time and acceleration ratio, which is calculated by Eq. 1.

$$AccelerationRatio = \frac{T_0}{T} \qquad (1)$$

where T_0 refers to the complete training time on a single machine with a single GPU, and T is the training time under other circumstances.

Analysis of Spark-GPU Cluster's Results. This experiment aims to compare the time taken for training AMSDFNet using a single GPU on a Spark-GPU cluster, as well as to examine the acceleration effect of using the data parallelism approach on the Spark-GPU cluster and the scalability of the nodes.

Table 2 shows that using a single machine with one GPU took 127.0 minutes to complete training. While training on Spark single node with one GPU took 120.1 minutes, and two GPUs took 101.0 minutes.

The introduction of Spark slightly reduced training time with the same GPU. In Spark, the dataset is loaded into the memory as a resilient distributed dataset before it is processed. During the training of DL models, the data will be copied from the memory to GPU for matrix computation. Additional time is required to load the data from the local disk into memory as well as from memory to GPU.

Table 2. Spark-GPU cluster's experiment results.

	Local machine (T_0)	Spark-GPU cluster (T)		
	Single machine with one GPU	Spark single node with one GPU	Spark single node with two GPUs	Spark two nodes with four GPUs
Runtime(min)	127.0	120.1	73.7	2400.8
AccelerationRatio	1.0	1.1	1.7	0.1

On a single node, the training time with two GPUs is just over half the training time of one GPU. Data parallelism makes it possible to double the batch size of a single training session, but because of parameter synchronization between the two processes, which results in extra communication costs between GPUs. That is to say, the acceleration ratio is less than 2. Nonetheless, data parallelism can still effectively accelerate the model training process on a single node.

Unfortunately, in the case of two nodes with four GPUs on the Spark cluster, the training time is longer than a single node with only one GPU, indicating that the communication loss between devices on the Spark-GPU cluster during the training of AMSDFNet is serious, and the scalability of the nodes is poor. The reason is that copying data between different GPUs on the same node requires extra time overhead, and data transmission between different nodes on the cluster also requires additional time overhead. Moreover, the speed of data transmission through the network is one-tenth or even one-hundredth of the speed of transmission between different GPUs on the same node [21].

Analysis of Spark-CPU Cluster's Results. This experiment aims to compare the performance of training DL models between traditional Spark-CPU clusters and Spark-GPU clusters, as well as the acceleration effect of data parallelism training on Spark-CPU clusters and the scalability of nodes.

Experimental results can be seen in Table 3, when using a high number of parallel processes and a large batch size for training on a two-node Spark-CPU cluster, the training time is still eight times longer than training on a single GPU. This is because GPU has a smaller volume but a larger number of cores, while CPU has a larger volume but a smaller number of cores. So GPU is more suitable for matrix operations, which makes it faster than CPU for training.

Table 3. AMSDFNet training time using Spark-CPU cluster and Spark-GPU cluster

	Spark-GPU(single node with one GPU)	Spark-CPU
Runtime(min)	120.1	1066.5

According to Fig. 4, when training on a Spark-CPU cluster with data parallelism, the training time can be reduced by approximately half when the number of processes doubles. Although communication losses become increasingly severe with an increase in parallel processes, the acceleration effect of distributed train-

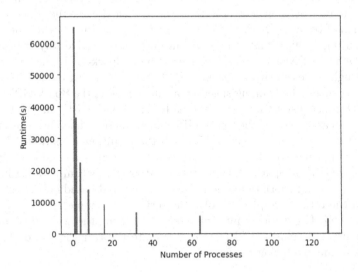

Fig. 4. The number of processes was 1, 2, 4, 8, 16, 32, 64, 128 (When the number of processes was small, i.e., using fewer CPU cores, the complete training time for AMSDFNet was very long. Here, we just verified the relationship between the number of processes and the acceleration effect. Therefore, the number of epochs was changed to 10 and other hyperparameters remained unchanged)

ing is still evident, and this acceleration is cross-node, indicating the good scalability of the Spark-CPU cluster.

In summary, the most suitable approach for the AMSDFNet model is to use a Spark-GPU cluster for distributed data parallelism training on a single node with multiple GPU cards. The reason is that the training is pixel-level training and has large model parameters. However, the scalability of the current Spark-GPU cluster is not good enough.

Although traditional CPU clusters have slower training speeds for this model, they exhibit good scalability in multi-node scenarios. Additionally, two servers each with 128 CPU cores can execute up to 128 processes, while a GPU can only support one process at a time. The Spark-CPU cluster has a high data throughput, allowing for the training of significantly larger batch sizes. Based on the good scalability and multi-parallel processes of the CPU cluster, as well as the fast calculation speed of the Spark-GPU cluster on a single node but poor scalability, we make the following judgment: The use of Spark-CPU cluster data parallelism is sufficient when the model comprises of a small size, but there's a considerable amount of data. For larger model sizes, training with the Spark-GPU clusters on a single node is preferable.

6   Conclusions

This paper searches the issue of implementing DDL on the big data platform. By investigating relevant DDL frameworks and combining them with distributed geological remote sensing interpretation as a practical case, the paper achieves acceleration of training D-AMSDFNet on Spark solving the real-time demand of remote sensing applications. There are currently two feasible ways to complete training D-AMSDFNet on a Spark cluster: using a Spark-GPU cluster for training on a single node or using a Spark-CPU cluster for training on multiple nodes. In the experiment, the training speed on a single node in the Spark-GPU cluster is at least 8 times faster than using two nodes in a Spark-CPU cluster. However, the communication loss of the Spark-GPU cluster across machines is extremely severe and currently not scalable. Although the traditional Spark-CPU cluster has slow computing speed, it has large data throughput and good scalability. Using Spark-GPU or Spark-CPU clusters for data parallelism training has pros and cons, so the appropriate method should be selected based on the actual situation. In the future, we plan to solve the serious communication loss problem of the Spark-GPU cluster on multiple nodes. Realizing the scalability of nodes on the Spark-GPU cluster will further accelerate the training speed of remote sensing data interpretation models.

Acknowledgements. This work is supported by the National Natural Science Foundation of China (U21A2013, 41925007, 42201415) and the Open Research Project of the Hubei Key Laboratory of Intelligent Geo-Information Processing (KLIGIP-2022-B16).

References

1. Weiss, M., Jacob, F., Duveiller, G.: Remote sensing for agricultural applications: a meta-review. Remote Sens. Environ. **236**, 111402 (2020)
2. ElGharbawi, T., Zarzoura, F.: Damage detection using SAR coherence statistical analysis, application to beirut, lebanon. ISPRS J. Photogramm. Remote. Sens. **173**, 1–9 (2021)
3. Han, W., et al.: A survey of machine learning and deep learning in remote sensing of geological environment: challenges, advances, and opportunities. ISPRS J. Photogrammetry Remote Sens. **202**, 87–113 (2023)
4. Kussul, N., Lavreniuk, M., Skakun, S., Shelestov, A.: Deep learning classification of land cover and crop types using remote sensing data. Geosci. Remote Sens. Lett. **14**(5), 778–782 (2017)
5. Zhang, X., et al.: Complex mountain road extraction in high-resolution remote sensing images via a light roadformer and a new benchmark. Remote Sens. **14**(19), 4729 (2022)
6. Khelifi, L., Mignotte, M.: Deep learning for change detection in remote sensing images: comprehensive review and meta-analysis. IEEE Access **8**, 126385–126400 (2020)
7. He, K., Zhang, X., Ren, S., Sun, J.: Deep residual learning for image recognition. In: 2016 IEEE Conference on Computer Vision and Pattern Recognition, CVPR 2016, Las Vegas, NV, USA, June 27-30, 2016, pp. 770–778. IEEE Computer Society (2016)
8. Han, W., et al.: Geological remote sensing interpretation using deep learning feature and an adaptive multisource data fusion network. IEEE Trans. Geosci. Remote Sens. **60**, 1–14 (2022)
9. Ma, Y., et al.: Remote sensing big data computing: challenges and opportunities. Futur. Gener. Comput. Syst. **51**, 47–60 (2015)
10. Bi, C., et al.: Machine learning based fast multi-layer liquefaction disaster assessment. World Wide Web **22**(5), 1935–1950 (2019)
11. Paszke, A., et al.: Pytorch: an imperative style, high-performance deep learning library. In: Wallach, H.M., Larochelle, H., Beygelzimer, A., d'Alché-Buc, F., Fox, E.B., Garnett, R. (eds.) Advances in Neural Information Processing Systems 32: Annual Conference on Neural Information Processing Systems 2019, NeurIPS 2019, December 8-14, 2019, Vancouver, BC, Canada, pp. 8024–8035 (2019)
12. Abadi, M., et al.: Tensorflow: a system for large-scale machine learning. In: Keeton, K., Roscoe, T. (eds.) 12th USENIX Symposium on Operating Systems Design and Implementation, OSDI 2016, Savannah, GA, USA, November 2-4, 2016, pp. 265–283. USENIX Association (2016)
13. Ji, H., Wu, G., Zhao, Y., Wei, L., Wang, G., Fan, Y.: A fault-tolerant optimization mechanism for spatiotemporal data analysis in flink. World Wide Web (WWW) **26**(3), 867–887 (2023)
14. Imran, M., Gévay, G.E., Quiané-Ruiz, J., Markl, V.: Fast datalog evaluation for batch and stream graph processing. World Wide Web **25**(2), 971–1003 (2022)
15. Dai, J.J., et al.: BigDL: a distributed deep learning framework for big data. In: Proceedings of the ACM Symposium on Cloud Computing, SoCC 2019, Santa Cruz, CA, USA, November 20-23, 2019, pp. 50–60. ACM (2019)
16. Meng, X., et al.: MLlib: machine learning in apache spark. CoRR arXiv:1505.06807 (2015)

17. Lu, X., Shi, H., Biswas, R., Javed, M.H., Panda, D.K.: DLoBD: a comprehensive study of deep learning over big data stacks on HPC clusters. IEEE Trans. Multi-Scale Comput. Syst. **4**(4), 635–648 (2018)
18. Lin, J., Ryaboy, D.V.: Scaling big data mining infrastructure: the twitter experience. SIGKDD Explor. **14**(2), 6–19 (2012)
19. Sergeev, A., Balso, M.D.: Horovod: fast and easy distributed deep learning in tensorflow. CoRR arXiv:1802.05799 (2018)
20. Caner, K., Gerdan, D., EmİNoGLu, M.B., YegüL, U., Bulent, K., VatandaŞ, M.: Classification of hazelnut cultivars: comparison of dl4j and ensemble learning algorithms. Notulae Botanicae Horti Agrobotanici Cluj-Napoca **48**(4), 2316–2327 (2020)
21. Sun, P., Wen, Y., Han, R., Feng, W., Yan, S.: Gradientflow: Optimizing network performance for large-scale distributed DNN training. IEEE Trans. Big Data **8**(2), 495–507 (2022)
22. Hong, D., et al.: Spectralformer: rethinking hyperspectral image classification with transformers. IEEE Trans. Geosci. Remote Sens. **60**, 1–15 (2022)
23. Zhu, X.X., et al.: Deep learning in remote sensing: a comprehensive review and list of resources. IEEE Geosci. Remote Sens. Mag. **5**(4), 8–36 (2017)
24. Liu, L., et al.: Object detection in large-scale remote sensing images with a distributed deep learning framework. IEEE J. Sel. Top. Appl. Earth Observations Remote Sens. **15**, 8142–8154 (2022)
25. Zhong, Y., Zheng, Z., Ma, A., Lu, X., Zhang, L.: COLOR: cycling, offline learning, and online representation framework for airport and airplane detection using GF-2 satellite images. IEEE Trans. Geosci. Remote Sens. **58**(12), 8438–8449 (2020)
26. Haut, J.M., et al.: Cloud deep networks for hyperspectral image analysis. IEEE Trans. Geosci. Remote Sens. **57**(12), 9832–9848 (2019)
27. Boulila, W., Sellami, M., Driss, M., Al-Sarem, M., Safaei, M., Ghaleb, F.A.: RS-DCNN: a novel distributed convolutional-neural-networks based-approach for big remote-sensing image classification. Comput. Electron. Agric. **182**, 106014 (2021)
28. Wu, S., Li, X., Dong, W., Wang, S., Zhang, X., Xu, Z.: Multi-source and heterogeneous marine hydrometeorology spatio-temporal data analysis with machine learning: a survey. World Wide Web (WWW) **26**(3), 1115–1156 (2023)
29. Chen, L., Papandreou, G., Schroff, F., Adam, H.: Rethinking atrous convolution for semantic image segmentation. CoRR arXiv:1706.05587 (2017)
30. Woo, S., Park, J., Lee, J., Kweon, I.S.: CBAM: convolutional block attention module. CoRR arXiv:1807.06521 (2018)

Graph-Enforced Neural Network for Attributed Graph Clustering

Zeang Sheng[1], Wentao Zhang[2(✉)], Wen Ouyang[3], Yangyu Tao[3], Zhi Yang[1], and Bin Cui[1(✉)]

[1] School of CS and Key Laboratory of High Confidence Software Technologies, Peking University, Beijing, China
{shengzeang18,zhiyang,bin.cui}@pku.edu.cn
[2] Mila - Québec AI Institute, HEC Montréal, Montreal, Canada
wentao.zhang@mila.quebec
[3] Tencent Inc., Shenzhen, China
{gdpouyang,brucetao}@tencent.com

Abstract. Graph clustering aims to discover cluster structures in graphs. This task becomes more challenging when each node in the graph is associated with an attribute vector (i.e., the attributed graph). Recently, methods built on Graph Auto-Encoder (GAE) have achieved state-of-the-art performance on the attributed graph clustering task. The performance gain mainly comes from GAE's ability to capture knowledge from graph structures and node attributes simultaneously. However, there is limited understanding of the critical issues that hinder the clustering performance of current GAE-based methods. To bridge this gap, we present a detailed empirical analysis and find that existing GAE-based methods suffer from graph information degradation issues of intra-cluster estrangement, attribute similarity neglection, and blurred cluster boundaries. Based on the observations, we design corresponding graph-enforcement tasks to address these degradation issues and include them in a unified multi-task learning framework called Graph-Enforced Neural Network (GENN). Extensive experimental results on four popular graph benchmark datasets illustrate that GENN consistently outperforms state-of-the-art attributed graph clustering methods.

Keywords: Graph Neural Network · Graph Clustering

1 Introduction

Graph clustering, a fundamental task in graph analysis, aims to partition the nodes in the graph into different clusters so that nodes within the same cluster are more similar than others. The attributed graph is a special kind of graph where each node inside the graph is associated with an attribute vector. Graph clustering on attributed graphs is more challenging than on graphs without node attributes since the method should capture knowledge from both graph structures and node attributes rather than only one of the two. Traditional clustering

X. Song et al. (Eds.): APWeb-WAIM 2023, LNCS 14331, pp. 111–126, 2024.
https://doi.org/10.1007/978-981-97-2303-4_8

methods like K-Means [5] are widely adopted as they can capture the knowledge within attributes. On the other hand, graph-based clustering methods like spectral clustering [16] are proposed to leverage the graph structural information. However, the performance of these two kinds of methods on the attributed graph clustering task is unsatisfactory since they cannot capture knowledge from graph structures and node attributes simultaneously.

Graph Neural Networks (GNNs) have recently shown incredible performance on various graph learning tasks, such as node classification [7,27], link prediction [19,23], self-supervised learning [3,24], active learning [26], recommendation systems [20], combinatorial optimization [13], and geometric learning [1]. Compared to traditional methods (e.g., matrix factorization [4]), the performance gain of GNNs mainly comes from the fact that they can effectively take advantage of both graph structures and node attributes during training. GNN-based methods have also been exploited for the attributed graph clustering task. Most existing methods follow the framework proposed by Graph Auto-Encoder (GAE) [9], which is composed of a GCN [8] encoder and an inner-product decoder. The training objective of the framework forces the learned node embeddings to recover the one-hop connection relationship between nodes, i.e., the graph's adjacency matrix.

Key Observations. However, based on a detailed empirical analysis, we find that the following graph information degradation issues severely limit the clustering performance of most existing GAE-based methods:

a) Intra-cluster estrangement: Most existing GAE-based attributed graph clustering methods force the learned node embeddings to recover only the one-hop connection relationship by reconstructing the graph's adjacency matrix. As a result, the high-order structural knowledge is neglected. Our empirical analysis finds that even within the same cluster, existing GAE-based methods would continuously push node pairs with a large distance away during training, which leads to sub-optimal clustering performance.

b) Attribute similarity neglection: In attributed graphs, nodes with more similar attributes are more likely to belong to the same cluster. Thus, attribute similarity is another critical graph information for the attributed graph clustering task. Most existing GAE-based graph clustering methods only implicitly utilize node attributes by treating them as input features. However, these methods only emphasize recovering the one-hop connection relationship during training and thus fail to preserve the corresponding attribute similarity information in their learned node embeddings. Our empirical analysis shows that the attribute similarity information between nodes gradually dissipates during the training process of existing GAE-based attributed graph clustering methods.

c) Blurred cluster boundaries: Cluster boundaries are hyperplanes in the high-dimensional space that separate the nodes into different clusters. An intuition is that indistinguishable cluster boundaries imply unsatisfactory clustering performance. Since both intra-cluster edges and inter-cluster edges exist in the graph, optimizing these two kinds of edges indiscriminately like most existing GAE-based methods may blur the cluster boundaries. In the empirical analysis, we remove all the inter-cluster edges in the graph and find that the performance of GAE-based methods is significantly boosted. This observation suggests that optimizing the inter-cluster edges without specific designs would result in blurred cluster boundaries and degrade the final clustering performance.

Graph-Enforced Neural Network. Inspired by the above observations, we propose a new multi-task learning framework, Graph-Enforced Neural Network (GENN), for the attributed graph clustering task. GENN introduces three carefully-designed tasks corresponding to the three graph information degradation issues. Specifically, the first two tasks enforce the learned node embeddings to preserve high-order structural proximities and attribute similarities between nodes. To keep the cluster boundaries distinguishable, the last task continuously pulls nodes belonging to the same cluster to be close and pushes nodes belonging to different clusters away. These three tasks are optimized jointly under a unified multi-task learning framework in the end-to-end manner. Experimental results on four popular graph benchmark datasets show that GENN consistently outperforms state-of-the-art methods on the attributed graph clustering task.

Contributions. The major contributions of this paper can be summarized as following three points:

1. *Analysis:* We conduct a detailed empirical analysis and find that most existing GAE-based attributed graph clustering methods suffer from three graph information degradation issues, which points out new directions for improvements on the attributed graph clustering task.
2. *Framework:* We design three enforcement tasks corresponding to the degradation issues and propose a multi-task learning framework, GENN, to jointly optimize the three different training objectives in an end-to-end manner.
3. *Performance:* The evaluation results on four popular graph datasets illustrate that GENN outperforms state-of-the-art attributed graph clustering methods and achieves comparable or better runtime efficiency.

2 Related Works

GAE-based attributed graph clustering methods have achieved state-of-the-art performance recently. Unlike traditional clustering methods, GAE-based methods can simultaneously utilize knowledge within graph structures and node attributes. Graph Auto-Encoder (GAE) [9] learns node embeddings by GCN [8] and decodes the learned node embeddings to recover the localized neighborhood structures. Specifically, an inner-product decoder is adopted to reconstruct the graph's original adjacency matrix:

$$\mathbf{A}_{pred} = sigmoid(\mathbf{Z}\mathbf{Z}^\mathsf{T}), \tag{1}$$

where $\mathbf{Z}$ is the learned node embedding matrix by GCN. Then, a binary cross entropy loss is adopted to minimize the difference between $\mathbf{A}_{pred}$ and the original adjacency matrix of the graph.

Many GAE-based graph clustering methods have been proposed in recent years. ARGA and ARVGA [12] introduce adversarial training into GAE by assuming that the learning node embeddings fit a Gaussian distribution. MGAE [18] reconstructs the node attribute matrix instead of the adjacency matrix using a denoising marginalized auto-encoder. DAEGC [17] explicitly incorporates a clustering-specific self-supervised module to boost its clustering performance. AGC [28] proposes an improved filter matrix to filter out high-frequency information better. AGE [2] reconstructs the adjacency matrix using cosine similarity between node embeddings, and the ground-truth labels are updated iteratively by ranking the values of the similarity matrix. ACMin [22] proposes a linear-time solver to optimize cluster assignment and model training jointly. GIC [11] presents an end-to-end framework that maximizes the mutual information between nodes and their assigned clusters. S-DAGC [6] contains one additional auto-encoder to estimate the natural cluster number of the input data.

3   Notations and Problem Formulation

An attributed graph can be represented as $\mathcal{G} = (\mathcal{V}, \mathcal{E}, \mathbf{X})$. $\mathcal{V}$ is the node set where $v_i \in \mathcal{V}$ represents a node of the graph. The total number of nodes in the graph is denoted as N. $\mathcal{E}$ is the edge set where $(v_i, v_j) \in \mathcal{E}$ denotes an edge connecting node v_i and v_j. We denote the adjacency matrix of $\mathcal{G}$ by a binary-valued matrix $\mathbf{A} \in \{0,1\}^{N \times N}$. And $\mathbf{A}_{ij}$ equals 1 if and only if $(v_i, v_j) \in \mathcal{E}$. $\mathbf{X}$ is the node attribute matrix and $\mathbf{X} = \{\mathbf{x}_1, \mathbf{x}_2..., \mathbf{x}_N\}$ where $\mathbf{x}_i \in \mathbb{R}^d$ for $i \in \{1, 2, ..., N\}$.

The main goal of the attributed graph clustering task is to partition the nodes in $\mathcal{G} = (\mathcal{V}, \mathcal{E}, \mathbf{X})$ into m disjoint clusters $\{\mathcal{G}_1, \mathcal{G}_2, ..., \mathcal{G}_m\}$ where nodes within the same cluster are more similar in terms of both graph structures and node attributes than others. In this paper, we concentrate only on homophily graphs where connected nodes are more probable to belong to the same cluster.

4   Degradation Analysis

This section provides an empirical analysis of existing GAE-based attributed graph clustering methods. This analysis explores whether existing GAE-based methods suffer from the three graph information degradation issues proposed in Sect. 1: intra-cluster estrangement, attribute similarity neglection, and blurred cluster boundaries. For conciseness, we only evaluate the original 2-layer GAE [9] here since most existing GAE-based methods share the same framework and training objective. The experiments are conducted on two popular graph datasets: Cora and Citeseer [8]. The label of each node is treated as the ground-truth cluster each node belongs to.

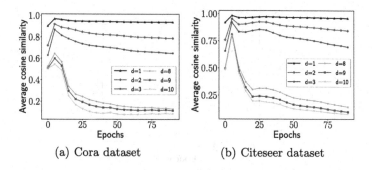

(a) Cora dataset   (b) Citeseer dataset

Fig. 1. Intra-cluster estrangement

4.1   Intra-cluster Estrangement

Node pairs within the same cluster may not be directly connected. For example, two nodes within the same cluster may be four or five hops away. Most GAE-based methods only focus on reconstructing the one-hop connection relationship. Thus, they may neglect the deeper structural knowledge and generate highly different embeddings for two nodes belong to the same cluster.

Here, we evaluate the variation trend of the cosine similarities of the embeddings of node pairs that consist of nodes within the same cluster during training. The node pairs are grouped by their shortest-path length d, and the corresponding average cosine similarities of different groups are shown in Fig. 1. For a more precise presentation, we only draw the curves of node pairs with shortest-path lengths being 1, 2, 3, 8, 9, and 10 in the figure. The evaluation results illustrate that the cosine similarities of node pairs all rise at the initial training period. Then, the similarities of node pairs with large shortest-path lengths (i.e., d=8, 9, 10) decrease rapidly and finally become close to zero. On the other hand, the similarities of node pairs with small shortest-path lengths only encounter slight drops and remain steady.

Although it is reasonable to generate more similar embeddings for node pairs with small distances, the attributed graph clustering task also requires the embeddings of nodes within the same cluster to be similar. Thus, the desired method should consider deeper structural knowledge and identify node pairs in the same cluster with large distances, which may help to boost the clustering performance.

4.2   Attribute Similarity Neglection

The attribute graph clustering task is different from the traditional graph clustering task in that the nodes provided by the former are further associated with attribute vectors. Thus, the method needs to consider both graph structures and node attributes when learning node embeddings. Most existing GAE-based methods only implicitly utilize the node attributes by treating them as inputs to the GNN encoders. And their training objectives only reconstruct the adjacency

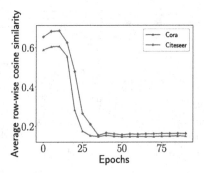

Fig. 2. Attribute similarity neglection

Table 1. Blurred cluster boundaries

Variants	Cora			Citeseer		
	ACC	NMI	ARI	ACC	NMI	ARI
Original $\mathcal{G}$	61.1	48.2	35.2	54.6	26.3	24.1
Modified $\mathcal{G}$	**70.2**	**60.3**	**51.0**	**63.2**	**39.9**	**37.1**
Δ	+9.1	+12.1	+15.8	+8.6	+13.6	+13.0

matrix of the graph. Therefore, we assume that these methods cannot sufficiently capture the knowledge within the node attributes.

A straightforward intuition concerning node attributes is that nodes with more similar attributes are more likely to belong to the same cluster. To validate our assumption, we calculate the average row-wise cosine similarity between the pair-wise similarity matrices of the embeddings learned by GAE ($\mathbf{Z}$) and the node attributes ($\mathbf{X}$), which are denoted by $\mathbf{S}_Z$ and $\mathbf{S}_X$, respectively.

$$\mathbf{S}_Z = \frac{\mathbf{Z}\mathbf{Z}^\top}{\|\mathbf{Z}\|_2^2}, \quad \mathbf{S}_X = \frac{\mathbf{X}\mathbf{X}^\top}{\|\mathbf{X}\|_2^2}, \tag{2}$$

As shown in Fig. 2, the average row-wise cosine similarity between $\mathbf{S}_Z$ and $\mathbf{S}_X$ drops rapidly during the initial period of training and then remains below 0.2 on both two datasets. Although GAE treats the node attributes as input, the evaluation results illustrate that its optimization direction is inconsistent with preserving pair-wise attribute similarities between nodes, which may degrade its clustering performance on the attributed graph clustering task.

4.3   Blurred Cluster Boundaries

Cluster boundaries [15] are hyperplanes that separate the data points into different groups according to node-to-cluster assignments. Moreover, more distinguishable cluster boundaries often indicate better clustering performance. As introduced in Sect. 2, most existing GAE-based clustering methods adopt the training objective of reconstructing the adjacency matrix, which pulls connected node pairs to be close. However, there are also inter-cluster edges in the graphs.

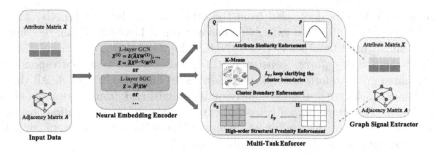

Fig. 3. Overview of the GENN framework.

Forcing nodes connected by inter-cluster edges to be close may blur the cluster boundaries and thus degrade the clustering performance.

We go through all the edges in the two datasets and find approximately 19% and 26% edges connecting nodes from different clusters in Cora and Citeseer, respectively. To evaluate the impacts these inter-cluster edges have on the clustering performance, we run GAE on the modified $\mathcal{G}$ where all the inter-cluster edges are removed. The evaluation results are shown in Table 1 along with the performance of GAE on the original $\mathcal{G}$. It is observed that GAE achieves significantly better clustering performance on the modified $\mathcal{G}$, up to 15.8% on specific performance metrics. We speculate that the huge performance gap comes from that GAE stops forcing connected node pairs from different clusters to be close. Thus, we argue that continuously clarifying the cluster boundaries during training would help to boost the clustering performance.

5   The Proposed Method

This section introduces Graph-Enforced Neural Network (GENN), a unified multi-task learning framework. The content of this section is organized as follows: Sect. 5.1 introduces the GENN framework in general, and Sects. 5.2 to 5.4 explain the detailed designs of the three graph enforcement tasks. Section 5.5 describes how GENN jointly optimizes the training objectives of different tasks.

5.1   Multi-task Learning Framework

GENN is composed of three modules: the *neural embedding encoder*, the *multi-task enforcer*, and the *graph signal extractor*. The overview of the GENN framework is shown in Fig. 3.

The *neural embedding encoder* is a graph neural network that takes in the adjacency matrix ($\mathbf{A}$) and the node attribute matrix ($\mathbf{X}$) as input and encodes them to the node embeddings ($\mathbf{Z}$). The default option of the *neural embedding encoder* in GENN is a L-layer Graph Convolutional Network (GCN) [8]:

$$\mathbf{X}^{(1)} = \delta\left(\hat{\mathbf{A}}\mathbf{X}\mathbf{W}^{(1)}\right), ..., \mathbf{Z} = \hat{\mathbf{A}}\mathbf{X}^{(L-1)}\mathbf{W}^{(L)}. \tag{3}$$

This module is fully decoupled from the other two modules of GENN. Thus, the default option of *neural embedding encoder* (an L-layer GCN) can be conveniently substituted by other GNNs — NDLS [25], GAMLP [27], etc.

The *multi-task enforcer* consists of three enforcement tasks corresponding to the three graph information degradation issues analyzed in Sect. 4. All three enforcement tasks are optimized jointly under a multi-task framework in an end-to-end way. The *graph signal extractor* extracts graph signals required by the enforcement tasks in the *multi-task enforcer*. The options for the graph signals are flexible and can even include auxiliary models.

5.2 High-Order Structural Proximity Enforcement

The empirical analysis in Sect. 4 illustrates that disconnected nodes within the same cluster may be assigned with highly dissimilar embeddings when the method only focuses on reconstructing the one-hop connection relationship. To address this issue, GENN proposes a high-order structural proximity enforcement task to enforce the learned node embeddings to reflect the high-order structural proximity knowledge.

The *graph signal extractor* in GENN is responsible for providing the high-order structural proximity matrix $\mathbf{H}$, where entry $\mathbf{H}_{ij}$ measures the structural proximity between node v_i and v_j. The default option for $\mathbf{H}$ is the weighted sum of the powers of the normalized adjacency matrix $\hat{\mathbf{A}}$, where the weights are the inverse of the powers:

$$\mathbf{H} = \sum_{i=1}^{K} \frac{1}{i} \hat{\mathbf{A}}^i = \sum_{i=1}^{K} \frac{1}{i} (\widetilde{\mathbf{D}}^{-\frac{1}{2}} \widetilde{\mathbf{A}} \widetilde{\mathbf{D}}^{-\frac{1}{2}})^i, \text{ where } \widetilde{\mathbf{A}} = \mathbf{I}_N + \mathbf{A}. \tag{4}$$

It is proposed in [21] that the number of different simple paths can indicate the high-order structural proximity between two nodes.

During training, the *multi-task enforcer* in GENN first calculates the pairwise cosine similarity matrix $\mathbf{S}_Z$ with the learned node embeddings $\mathbf{Z}$ as in Eq. 2. Then, it minimizes distance between $\mathbf{S}_Z$ and $\mathbf{H}$ as follows:

$$L_p = \frac{1}{2N} \|\mathbf{S}_Z - \mathbf{H}\|_F^2. \tag{5}$$

By enforcing the pair-wise similarity matrix of learned node embeddings ($\mathbf{S}_Z$) to be close to the high-order structural proximity matrix ($\mathbf{H}$) provided by the *graph signal extractor*, GENN can capture and preserve the high-order structural knowledge within the graph and achieves better clustering performance.

Furthermore, GENN allows the *graph signal extractor* to incorporate auxiliary models for extracting more sophisticated high-order structural knowledge. For example, sequence-based graph learning methods (e.g., DeepWalk [14]) can be utilized to provide supervision for the enforcement task. With the node embeddings $\mathbf{Z}_{\mathrm{DW}}$ learned by DeepWalk, one can set $\mathbf{H}$ to $\mathbf{S}_{\mathrm{DW}} = \mathbf{Z}_{\mathrm{DW}} \mathbf{Z}_{\mathrm{DW}}^{\mathsf{T}} / \|\mathbf{Z}_{\mathrm{DW}}\|_2^2$. The model-augmented graph signal provides GENN with more accurate supervision at the additional cost of obtaining the auxiliary model during pre-processing.

5.3 Attribute Similarity Enforcement

Most existing GAE-based attributed graph clustering methods only implicitly utilize the knowledge within the node attributes by treating them as input to their GNN encoders, yet lack explicit exploitation during the training process. On the contrary, GENN proposes an attribute similarity enforcement task to explicitly enforce the model to preserve pair-wise similarities with respect to node attributes in the learned node embeddings.

Similar to the high-order structural proximity enforcement task, the *graph signal extractor* is responsible for providing the pair-wise node attribute similarity matrix $\mathbf{P}$, where entry $\mathbf{P}_{ij}$ measures the attribute similarity between node v_i and v_j. Motivated by t-SNE [10], the *graph signal extractor* in GENN adopts a Gaussian distribution for encoding the pair-wise distances for the high-dimensional node attributes and constructs the pair-wise similarity matrix $\mathbf{P}$. On the other hand, a heavy-tailed student distribution is utilized for encoding the pair-wise distances for the low-dimensional node embeddings.

The heavy-tailed distribution would assign the node pairs with moderate distances in high-dimensional spaces with greater pulling force instead of treating them as distant node pairs in low-dimensional spaces. Thus, the "crowding problem" caused by the huge volume gap between high- and low-dimensional spaces can be alleviated. The corresponding similarity matrix is denoted as $\mathbf{Q}$. Let p_{ij} and q_{ij} be the (i, j)-th entry of $\mathbf{P}$ and $\mathbf{Q}$, respectively, we have:

$$p_{ij} = \frac{\exp\left(-\|\mathbf{X}_i - \mathbf{X}_j\|_2^2 / 2\sigma^2\right)}{\sum_{k \neq l} \exp\left(-\|\mathbf{X}_k - \mathbf{X}_l\|_2^2 / 2\sigma^2\right)}, \tag{6}$$

$$q_{ij} = \frac{\left(1 + \|\mathbf{Z}_i - \mathbf{Z}_j\|_2^2\right)^{-1}}{\sum_{k \neq l} \left(1 + \|\mathbf{Z}_k - \mathbf{Z}_l\|_2^2\right)^{-1}}. \tag{7}$$

Then, the *multi-task enforcer* adopts the KL divergence between these two distributions as the training objective, which forces the pair-wise similarity distribution $\mathbf{Q}$ extracted from node embeddings to be close to the distribution $\mathbf{P}$ extracted from node attributes:

$$L_s = \mathrm{KL}(\mathbf{P} \,\|\, \mathbf{Q}). \tag{8}$$

As a result, GENN can preserve the knowledge of pair-wise attribute similarities in the learned node embeddings. By contrast, most existing GAE-based attributed graph clustering methods only constrain the learned node embeddings to reconstruct the graph adjacency matrix, which neglects the knowledge within node attributes and leads to sub-optimal performance.

5.4 Cluster Boundary Enforcement

The empirical analysis in Sect. 4 shows that the clustering performance of existing GAE-based methods will be significantly boosted if all the inter-cluster edges are removed. We assume the reason is that forcing nodes connected by inter-cluster edges to be close blurs the cluster boundaries and degrades the clustering

Table 2. Overview of the four benchmark datasets.

Dataset	#Nodes	#Edges	#Attributes	#Clusters
Cora	2708	5429	1433	7
Citeseer	3327	4732	3703	6
Pubmed	19717	44338	500	3
Wiki	2405	17981	4973	17

performance. However, it is impossible to know whether certain edges connect nodes from different clusters in advance. Thus, we design a cluster boundary enforcement task that keeps clarifying the cluster boundaries during training.

The *multi-task enforcer* in GENN performs the K-Means algorithm on the embedding matrix $\mathbf{Z}$ every t (t is set to 10 by default) epochs and reports the cluster centroid C_j for each cluster $\mathcal{G}_j$. Then, the *multi-task enforcer* continuously pulls each node to be close to the centroid of the cluster it belongs to and pushes each node away from other centroids during training. The training objective of the cluster boundary enforcement task is defined as follows:

$$ L_c = \sum_{i=1}^{N} \|\mathbf{Z}_i - \mathbf{Z}_{Y_i}\|_2 - \gamma \sum_{i=1}^{N} \left(\frac{1}{m-1} \sum_{C_j \neq Y_i} \|\mathbf{Z}_i - \mathbf{Z}_{C_j}\|_2 \right), \tag{9} $$

where coefficient $\gamma > 0$, and m is the number of clusters. Y_i is the centroid of the cluster that node v_i belongs to, and $\mathbf{Z}_{C_j}$ is calculated as the average embeddings of all the nodes assigned to cluster $\mathcal{G}_j$. By pulling each node close to its assigned cluster and pushing each node away from other clusters, GENN can maintain distinguishable cluster boundaries throughout the training process, which helps to boost the clustering performance.

5.5   Joint Objective Optimization

Based on the idea of multi-task learning, GENN jointly optimizes the three training objectives described in the previous three subsections and formulates the final training objective as follows:

$$ L = L_s + \alpha \cdot L_p + \beta \cdot L_c, \tag{10} $$

where coefficients $\alpha > 0$ and $\beta > 0$ are the balancing factors for the three different enforcement tasks. The three enforcement tasks share the same *neural embedding encoder*, yet have their task-dependent *multi-task enforcers* and *graph signal extractors* when optimizing the final training objective cooperatively.

6   Experiments

In this section, we evaluate GENN against existing attributed graph clustering methods on four popular graph benchmark datasets. Code and data are available at https://github.com/shengzeang/GENN.

Table 3. Evaluation results on attributed graph clustering.

Methods	Cora			Citeseer			Pubmed			Wiki		
	ACC	NMI	ARI	ACC	NMI	ARI	ACC	NMI	ARI	ACC	NMI	ARI
GAE	61.1 ± 0.2	48.2 ± 0.3	35.2 ± 0.2	54.6 ± 0.4	26.3 ± 0.3	24.1 ± 0.3	63.1 ± 0.4	24.9 ± 0.3	21.7 ± 0.2	37.9 ± 0.2	34.5 ± 0.3	18.9 ± 0.2
VGAE	59.2 ± 0.3	40.8 ± 0.4	34.7 ± 0.3	55.7 ± 0.2	26.1 ± 0.3	25.6 ± 0.3	65.5 ± 0.2	25.0 ± 0.4	20.3 ± 0.4	45.1 ± 0.4	46.8 ± 0.3	26.3 ± 0.4
MGAE	63.4 ± 0.5	45.6 ± 0.3	43.6 ± 0.4	63.5 ± 0.4	39.7 ± 0.4	42.5 ± 0.5	59.3 ± 0.5	28.2 ± 0.2	24.8 ± 0.4	50.1 ± 0.3	48.0 ± 0.4	34.9 ± 0.5
ARGA	63.9 ± 0.4	45.1 ± 0.3	35.1 ± 0.5	57.3 ± 0.5	35.2 ± 0.3	34.0 ± 0.4	68.0 ± 0.5	27.6 ± 0.4	29.0 ± 0.4	38.1 ± 0.5	34.5 ± 0.3	11.2 ± 0.4
ARVGA	64.0 ± 0.5	44.9 ± 0.4	37.4 ± 0.5	54.4 ± 0.5	25.9 ± 0.5	24.5 ± 0.3	51.3 ± 0.4	11.7 ± 0.3	7.8 ± 0.2	38.7 ± 0.4	33.9 ± 0.4	10.7 ± 0.2
AGC	68.9 ± 0.5	53.7 ± 0.3	48.6 ± 0.3	66.9 ± 0.5	41.1 ± 0.4	41.9 ± 0.5	69.8 ± 0.4	31.6 ± 0.3	31.8 ± 0.4	47.7 ± 0.3	45.3 ± 0.5	34.3 ± 0.4
DAEGC	70.2 ± 0.4	52.6 ± 0.3	49.7 ± 0.4	67.2 ± 0.5	39.7 ± 0.5	41.1 ± 0.4	66.8 ± 0.5	26.6 ± 0.2	27.7 ± 0.3	48.2 ± 0.4	44.8 ± 0.4	33.1 ± 0.3
ACMin	65.6 ± 0.4	49.8 ± 0.3	45.3 ± 0.3	68.0 ± 0.4	42.2 ± 0.3	42.4 ± 0.5	69.1 ± 0.4	30.8 ± 0.5	31.2 ± 0.4	49.2 ± 0.5	48.6 ± 0.3	31.7 ± 0.3
GIC	70.4 ± 0.4	52.5 ± 0.5	49.5 ± 0.4	68.8 ± 0.3	43.8 ± 0.6	43.7 ± 0.4	64.3 ± 0.4	26.0 ± 0.5	21.6 ± 0.6	46.5 ± 0.3	48.2 ± 0.4	30.4 ± 0.2
S-DAGC	70.8 ± 0.3	54.4 ± 0.4	51.0 ± 0.3	69.3 ± 0.5	44.2 ± 0.5	44.7 ± 0.4	69.8 ± 0.3	33.3 ± 0.2	30.6 ± 0.4	53.0 ± 0.4	48.7 ± 0.3	32.7 ± 0.5
AGE	72.8 ± 0.5	58.1 ± 0.6	56.3 ± 0.4	70.0 ± 0.3	44.6 ± 0.4	45.4 ± 0.5	69.9 ± 0.5	30.1 ± 0.4	31.4 ± 0.6	51.1 ± 0.6	53.9 ± 0.4	32.4 ± 0.5
GENN	75.3 ± 0.5	56.5 ± 0.5	56.5 ± 0.4	70.3 ± 0.3	45.0 ± 0.5	46.7 ± 0.3	70.6 ± 0.4	33.5 ± 0.5	33.5 ± 0.4	51.4 ± 0.4	48.4 ± 0.5	32.2 ± 0.4
GENN-Augmented	76.6 ± 0.3	59.8 ± 0.3	59.9 ± 0.4	70.7 ± 0.4	45.5 ± 0.4	47.3 ± 0.3	71.5 ± 0.5	34.1 ± 0.4	34.5 ± 0.4	51.9 ± 0.5	51.6 ± 0.3	35.1 ± 0.4

6.1 Experiment Settings

Datasets. We evaluate GENN on four popular graph datasets. Cora, Citeseer, and Pubmed are three citation networks, and Wiki is a webpage network. The node attributes in Cora and Citeseer are binary word vectors; while in Pubmed and Wiki, each node is associated with a TF-IDF weighted word vector. The detailed statistics of the four graph datasets are summarized in Table 2.

Baselines and Evaluation Metrics. We evaluate two GENN variants in the experiments: **GENN**, the default version of GENN; **GENN-Augmented**, which sets $\mathbf{H}$ in the high-order structural proximity enforcement task to $\mathbf{S}_{DW} = \mathbf{Z}_{DW}\mathbf{Z}_{DW}^{T}/\|\mathbf{Z}_{DW}\|_2^2$ where $\mathbf{Z}_{DW}$ is the node embeddings generated by Deep-Walk. We compare GENN against a wide range of baseline methods: **GAE** and **VGAE** [9], **ARGA** and **ARVGA** [12], **AGC** [28], **DAEGC** [17], **ACMin** [22], **GIC** [11], **S-DAGC** [6], and **AGE** [2]. We use three widely-used performance metrics for evaluation on the attributed graph clustering task: Accuracy (**ACC**), Normalized Mutual Information (**NMI**), and Adjusted Rand Index (**ARI**).

Hyperparameter Settings. For GENN, we adopt a 2-layer GCN as the *neural embedding encoder* and set the hidden dimension to 256 and final embedding dimension to 64. The k in Eq. 4 is set to 5 and 2 for Cora and Wiki, respectively. Moreover, we set k to 4 for the other two datasets. For the DeepWalk algorithm, the window size, the walk length, and the number of walks from each node are set to 5, 10, and 10, respectively. The learning rate is set to 1e-3 for Cora and 5e-3 for the other three datasets. Hyperparameters for the baseline methods are all set according to their original papers. All the methods are trained for 200 epochs and optimized by the Adam optimizer. We run all the experiments 10 times and report the average value and the standard deviation of each metric. Considering that most baselines adopt K-Means as the clustering algorithm, we replace spectral clustering in AGE with K-Means for a fair comparison.

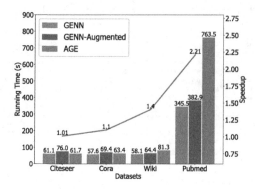

Fig. 4. Efficiency comparison among GENN, GENN-Augmented, and AGE.

6.2 Performance Comparison

The evaluation results of GENN and all the baselines on the attributed graph clustering task are shown in Table 3. The **bold** and <u>underlined</u> values indicate the highest and the second-highest scores among all the compared methods.

As Table 3 shows, GENN-Augmented consistently outperforms the strongest baseline AGE on all four graph datasets. For example, the accuracy of GENN-Augmented exceeds AGE by 3.8%, 0.7%, 1.6%, and 0.8% on Cora, Citeseer, Pubmed, and Wiki, respectively. AGE exploits knowledge inside graph structures and node attributes by pulling similar node pairs close and updates the chosen similar node pairs iteratively during training. However, AGE still adopts a similar training objective with the one of GAE and partially suffers from the same degradation issues analyzed in Sect. 4. The performance priority of GENN-Augmented over AGE mainly comes from GENN-Augmented's ability to better address these three graph information degradation issues with the help of the three enforcement tasks. On the other hand, GENN-Augmented outperforms GENN on all four datasets, which illustrates that the auxiliary DeepWalk module helps to better capture high-order structural knowledge within the graph.

6.3 Efficiency Comparison

In this subsection, we evaluate the runtime efficiency of GENN. Table 3 shows that AGE is the strongest method among all the baselines. Thus, we compare the runtime efficiency of GENN and GENN-Augmented with AGE. We train all three methods for 200 epochs, and the total runtime are shown in Fig. 4. We also draw the speedup of GENN against AGE as a red curve in the figure.

Figure 4 shows that GENN has higher runtime efficiency than AGE. Notably, GENN achieves 2.21× speedup against AGE on the largest dataset, Pubmed. Moreover, GENN-Augmented runs faster than AGE on the Wiki and Pubmed dataset, although it has to train an auxiliary DeepWalk model during pre-processing. The low efficiency of AGE comes from the design that it has to

Table 4. Ablation study on task effectiveness.

Model Variants	Cora			Citeseer		
	ACC	NMI	ARI	ACC	NMI	ARI
GENN-w/o-L_p	48.6	29.9	23.1	52.5	29.2	25.0
GENN-w/o-L_s	69.6	49.0	47.0	55.8	30.4	26.9
GENN-w/o-L_c	70.3	50.3	49.8	63.5	39.5	40.0
GENN	**75.3**	**56.5**	**56.5**	**70.3**	**45.0**	**46.7**

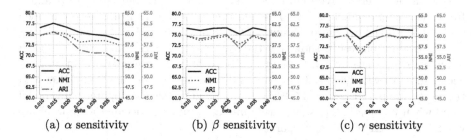

(a) α sensitivity   (b) β sensitivity   (c) γ sensitivity

Fig. 5. Hyperparameter sensitivity analysis on the Cora dataset

periodically rank N^2 similarities of node pairs, which incurs significant run-time overhead. The runtime of GENN-Augmented is 14.94 s, 11.82 s, 6.28 s, and 37.41 s longer than the runtime of GENN on Citeseer, Cora, Wiki, and Pubmed, respectively. The runtime overhead of GENN-Augmented is acceptable since GENN-Augmented consistently outperforms GENN in Table 3 For example, the overhead of GENN-Augmented on the Pubmed dataset consists of only 9.77% of its total runtime; and it outperforms GENN by 0.9% on clustering accuracy.

6.4   Ablation Study

To measure the impact each enforcement task has on the performance of GENN, we evaluate the performance of GENN: (1) without the high-order structural proximity enforcement task (denoted as "**GENN-w/o-L_p**"), (2) without the attribute similarity enforcement task (denoted as "**GENN-w/o-L_s**"), and (3) without the cluster boundary enforcement task (denoted as "**GENN-w/o-L_c**"). We report the clustering performance of these three GENN variants and the original GENN on the Cora and Citeseer dataset in Table 4.

The results in Table 4 illustrate that all three enforcement tasks are helpful to the attributed graph clustering task as GENN outperforms all the three variants. Among the three tasks, the high-order structural proximity enforcement task plays the most crucial role as the performance of GENN-no-L_p is lower than the other two variants on both datasets. We also find that the performance gap between GENN-no-L_s and GENN-no-L_p on the Cora dataset is more significant than on the Citeseer dataset. This observation is in accord with the characteristics of these two datasets since Cora has 41% more edges per node than Citeseer, implying that Cora contains richer structural information than Citeseer.

6.5   Hyperparameter Sensitivity Analysis

GENN proposes three enforcement tasks to alleviate the three graph information degradation issues analyzed in Sect. 4. These three enforcement tasks are optimized jointly in a multi-task learning framework to incorporate different aspects of graph knowledge. As a result, GENN has several hyperparameters (i.e., α, β, and γ) inside to balance the weights of different tasks. In this subsection, we evaluate GENN-Augmented on the attributed graph clustering task on the Cora dataset with different hyperparameter settings. We vary the values of α and β from the recommended 0.01 to 0.04 and the value of γ from the recommended 0.1 to 0.5. All the evaluation results of GENN-Augmented are shown in Fig. 5.

The evaluation results show that the hyperparameter α is more sensitive than β and γ as the variations of α result in observable clustering performance differences. In contrast, the variations of β and γ do not have such effects. However, the clustering performance of GENN-Augmented reflects an apparent uni-modal trend when varying α. Thus, it is still quite convenient to decide the appropriate value for α when tuning hyperparameters.

7   Conclusion

In this paper, we present a detailed empirical analysis of existing GAE-based attributed graph clustering methods, and find that they suffer from three graph information degradation issues: intra-cluster estrangement, attribute similarity neglection, and blurred cluster boundaries. Based on the empirical analysis, we propose three corresponding graph enforcement tasks and incorporate them in a unified multi-task learning framework, GENN. Extensive experimental results illustrate that GENN consistently outperforms state-of-the-art attributed graph clustering methods on the four popular graph datasets and achieves comparable or better runtime efficiency.

Acknowledgements. The work is supported by the National Natural Science Foundation of China under Grant (61832001, U22B2037), and PKU-Tencent joint research Lab.

References

1. Cao, W., Zheng, C., Yan, Z., Xie, W.: Geometric deep learning: progress, applications and challenges. Sci. China Inf. Sci. **65**(2), 126101 (2022)
2. Cui, G., Zhou, J., Yang, C., Liu, Z.: Adaptive graph encoder for attributed graph embedding. In: Proceedings of the 26th ACM SIGKDD International Conference on Knowledge Discovery & Data Mining, pp. 976–985 (2020)
3. Fang, P.F., et al.: Connecting the dots in self-supervised learning: a brief survey for beginners. J. Comput. Sci. Technol. **37**(3), 507–526 (2022)
4. Gao, L.G., Yang, M.Y., Wang, J.X.: Collaborative matrix factorization with soft regularization for drug-target interaction prediction. J. Comput. Sci. Technol. **36**, 310–322 (2021)

5. Hartigan, J.A., Wong, M.A.: Algorithm as 136: a k-means clustering algorithm. J. Roy. Stat. Soc. Ser. C **28**(1), 100–108 (1979)
6. Ji, J., Liang, Y., Lei, M.: Deep attributed graph clustering with self-separation regularization and parameter-free cluster estimation. Neural Netw. **142**, 522–533 (2021)
7. Jin, T., et al.: Deepwalk-aware graph convolutional networks. Sci. China Inf. Sci. **65**(5), 152104 (2022)
8. Kipf, T.N., Welling, M.: Semi-supervised classification with graph convolutional networks. arXiv preprint arXiv:1609.02907 (2016)
9. Kipf, T.N., Welling, M.: Variational graph auto-encoders. arXiv preprint arXiv:1611.07308 (2016)
10. Maaten, L.V.D., Hinton, G.: Visualizing data using t-SNE. J. Mach. Learn. Res. **9**(Nov), 2579–2605 (2008)
11. Mavromatis, C., Karypis, G.: Graph InfoClust: maximizing coarse-grain mutual information in graphs. In: Karlapalem, K., et al. (eds.) PAKDD 2021. LNCS (LNAI), vol. 12712, pp. 541–553. Springer, Cham (2021). https://doi.org/10.1007/978-3-030-75762-5_43
12. Pan, S., Hu, R., Long, G., Jiang, J., Yao, L., Zhang, C.: Adversarially regularized graph autoencoder for graph embedding. arXiv preprint arXiv:1802.04407 (2018)
13. Peng, Y., Choi, B., Xu, J.: Graph learning for combinatorial optimization: a survey of state-of-the-art. Data Sci. Eng. **6**(2), 119–141 (2021)
14. Perozzi, B., Al-Rfou, R., Skiena, S.: Deepwalk: Online learning of social representations. In: Proceedings of the 20th ACM SIGKDD international conference on Knowledge discovery and data mining, pp. 701–710 (2014)
15. Sisodia, D., Singh, L., Sisodia, S., Saxena, K.: Clustering techniques: a brief survey of different clustering algorithms. Int. J. Latest Trends Eng. Technol. **1**(3), 82–87 (2012)
16. Von Luxburg, U.: A tutorial on spectral clustering. Stat. Comput. **17**(4), 395–416 (2007)
17. Wang, C., Pan, S., Hu, R., Long, G., Jiang, J., Zhang, C.: Attributed graph clustering: a deep attentional embedding approach. arXiv preprint arXiv:1906.06532 (2019)
18. Wang, C., Pan, S., Long, G., Zhu, X., Jiang, J.: MGAE: marginalized graph autoencoder for graph clustering. In: Proceedings of the 2017 ACM on Conference on Information and Knowledge Management, pp. 889–898 (2017)
19. Wu, H., Song, C., Ge, Y., Ge, T.: Link prediction on complex networks: an experimental survey. Data Sci. Eng. **7**(3), 253–278 (2022)
20. Wu, S., Sun, F., Zhang, W., Xie, X., Cui, B.: Graph neural networks in recommender systems: a survey. ACM Comput. Surv. **55**(5), 1–37 (2022)
21. Yang, G., Zhan, Y., Li, J., Yu, B., Liu, L., He, F.: Exploring high-order structure for robust graph structure learning. arXiv preprint arXiv:2203.11492 (2022)
22. Yang, R., Shi, J., Yang, Y., Huang, K., Zhang, S., Xiao, X.: Effective and scalable clustering on massive attributed graphs. In: Proceedings of the Web Conference 2021, pp. 3675–3687 (2021)
23. Zhang, M., Li, P., Xia, Y., Wang, K., Jin, L.: Labeling trick: a theory of using graph neural networks for multi-node representation learning. Adv. Neural. Inf. Process. Syst. **34**, 9061–9073 (2021)
24. Zhang, W., et al.: NAFS: a simple yet tough-to-beat baseline for graph representation learning. In: International Conference on Machine Learning, pp. 26467–26483. PMLR (2022)

25. Zhang, W., et al.: Node dependent local smoothing for scalable graph learning. Adv. Neural. Inf. Process. Syst. **34**, 20321–20332 (2021)
26. Zhang, W., et al.: Grain: improving data efficiency of graph neural networks via diversified influence maximization. arXiv preprint arXiv:2108.00219 (2021)
27. Zhang, W., et al.: Graph attention multi-layer perceptron. In: Proceedings of the 28th ACM SIGKDD Conference on Knowledge Discovery and Data Mining, pp. 4560–4570 (2022)
28. Zhang, X., Liu, H., Li, Q., Wu, X.M.: Attributed graph clustering via adaptive graph convolution. arXiv preprint arXiv:1906.01210 (2019)

MacGAN: A Moment-Actor-Critic Reinforcement Learning-Based Generative Adversarial Network for Molecular Generation

Huidong Tang[1]([✉]) [iD], Chen Li[2]([✉]) [iD], Shuai Jiang[1] [iD], Huachong Yu[1] [iD], Sayaka Kamei[1] [iD], Yoshihiro Yamanishi[2] [iD], and Yasuhiko Morimoto[1] [iD]

[1] Graduate School of Advanced Science and Engineering, Hiroshima University, Kagamiyama 1-7-1, Higashi-Hiroshima, Hiroshima 739-8521, Japan
{d216083,jiangshuai,huachongyu,s10kamei,morimo}@hiroshima-u.ac.jp
[2] Graduate School of Informatics, Nagoya University, Chikusa, Nagoya 464-8602, Japan
li.chen.z2@a.mail.nagoya-u.ac.jp, yamanishi@i.nagoya-u.ac.jp

Abstract. Deep generative models such as variational autoencoders (VAEs) and generative adversarial networks (GANs) have demonstrated significant efficacy in drug discovery. However, GANs are typically employed to process continuous data such as images and are unstable in performance for discrete molecular graphs and simplified molecular-input line-entry system (SMILES) strings. Most previous studies use reinforcement learning (RL) methods (e.g., Monte Carlo tree search) to solve the above issues. However, the generation task is time-consuming and cannot be applied to large chemical datasets due to the extensive sampling required to generate each atomic token. This study introduces a moment-actor-critic RL-based GAN (MacGAN) for the novel molecular generation with SMILES strings. MacGAN leverages the robust architecture of GAN while incorporating a simple reward mechanism, making it suitable for larger datasets compared to computationally intensive Monte Carlo-based methods. Experimental results show effectiveness of MacGAN.

Keywords: Molecular generation · Generative adversarial networks · Moment-actor-critic reinforcement learning

1 Introduction

Drug discovery represents a crucial endeavor to identify novel chemical compositions that exhibit efficacy in disease treatment [12]. Its significance in modern healthcare cannot be overstated. However, the intricacy of this undertaking lies in the formidable challenge of discerning desirable properties from an extensive array of possibilities, akin to locating a needle within a haystack [19]. The inherent complexity of the traditional drug discovery process renders it highly intricate, costly, and susceptible to potential setbacks [18]. As a result, a pronounced and inevitable trend has emerged whereby the initial identification of potential chemical components is accomplished through simulation before actual experimentation. In recent years, in response to this imperative, artificial intelligence (AI) technologies, with neural networks serving as prominent exemplars, have been developed [5]. Through the utilization of such techniques, substantial reductions in both temporal and monetary expenditures associated with drug discovery have been achieved.

Deep generative models have demonstrated significant efficacy in natural language processing (NLP) and forecasting applications [14,15,29]. Recently, their potential has been recognized in drug discovery as well. A variety of generative models have been developed specifically for this end, including variational auto-encoders (VAEs) [3,10,13], generative adversarial networks (GANs) [4,8] and diffusion models [9]. GANs learn the training data distribution through a two-player mini-max game between a generator and a discriminator. The generator aims to produce samples that closely resemble real ones to deceive the discriminator, which, in turn, aims to distinguish between generated and real samples. This competitive cycle drives the learning process for a molecular generation. GANs outperform VAEs in generating realistic samples due to their explicit training objective. Although diffusion models can also generate realistic samples, they typically require multiple steps for a gradual generation. Despite advancements in sampling methods, the process remains slower compared to GANs. Therefore, GANs are our preferred choice for molecular generation.

Molecular generation can be approached from two directions: molecular graph-based generation and simplified molecular-input line-entry system (SMILES) string-based generation. A SMILES string is derived from a molecular graph, representing the molecule as a simple string-based sequence [26]. This simplification reduces the model's capacity requirements and makes it commonly used to describe molecules. We prefer SMILES strings because SMILES strings are sequences; thus, the SMILES generation problem can be modeled as sequence modeling. However, GANs cannot be directly applied to train discrete value sequences because they rely on differentiable value functions. Gradients necessary for training the generator of discrete variables are close to zero across most data space, rendering direct training using the value function ineffective. Backpropagation using only the discriminator result becomes challenging. To overcome these challenges in sequence generation with GANs, researchers have explored alternative reward functions to guide the training process.

Transformer-based objective-reinforced GAN (TransORGAN) [16] utilizes a Monte Carlo tree search that combines the transformer-based GAN and the policy gradient reinforcement learning (RL) for SMILES generation, particularly suited for discrete value generation. However, TransORGAN is time-consuming when handling many samples and only applies to small SMILES datasets, excluding larger ones. ScratchGAN [17] is a promising GAN model that, combined with REINFORCE gradient estimator [27] designed initially for text generation, can also be employed for SMILES generation. However, the actor-critic RL-based cumulative reward is unsuitable for SMILES string generation since the impact of rewards from future time step tokens on the current time step token decreases as future tokens become further from the current token, unlike the behavior observed in SMILES strings. This discrepancy makes scratchGAN less compatible with actual SMILES generation. To address this issue, we adopt a modified version of scratchGAN, called moment-actor-critic RL-based GAN (MacGAN), to generate molecules from SMILES strings. This modification reduces the time required for sample generation and enables its application to large chemical datasets. We conduct experiments to assess the effectiveness of MacGAN in terms of validity, uniqueness, novelty, and diversity.

Furthermore, we compared the desirable chemical properties, i.e., the drug-likeness (QED), solubility (logp), and synthesizability (SA) for the generated molecules. The obtained results confirm the efficacy of our approach. Our primary contributions can be summarized as follows:

- **RL-based discrete GAN:** Unlike Monte Carlo-based GANs, MacGAN combines a discrete GAN and REINFORCE gradient estimator to handle large chemical datasets.
- **Moment rewards for generator update:** MacGAN employs moment rewards instead of cumulative rewards, resulting in reduced computation time and a more accurate portrayal of the inherent properties of SMILES strings.
- **Performance improvement:** The molecules generated using the proposed MacGAN model exhibit a high degree of validity, uniqueness, and diversity and have similar distributions of chemical properties to the original training dataset.

2   Related Work

Deep generative models have proven effective and exhibited remarkable performance in image processing and NLP domains. Extending these models to molecular generation to improve drug discovery efficiency is a natural progression. Noteworthy examples of molecular generation using variational autoencoders (VAEs) include GrammarVAE [13], which employs parse trees to capture the grammars of SMILES strings and generates syntactically valid molecules, although lacking semantic representation. The syntax-direct VAE (SDVAE) model [3] addresses this limitation by incorporating both grammar and semantics. Junction tree-based VAE (JT-VAE) [10] employs a graph message-passing network to generate molecular graphs. However, VAE models often produce less realistic molecules due to their inherent training objectives.

Similarly, Graph-AF [22] utilizes a flow-based regression model for the molecular graph generation. Nonetheless, compared to SMILES string-based methods, molecular graph-based approaches demand more substantial computational resources due to their inherent complexity, and the flow-based model requires a specific architecture, increasing its complexity. Equivariant Diffusion Model (EDM) [9] combines diffusion models and equivariant networks to generate three-dimensional (3D) molecules, effectively handling both continuous and discrete features. However, this method also incurs notable computational resource requirements due to the nature of diffusion models and 3D molecular generation.

Sequence GAN (SeqGAN) [28] incorporates RL methods into GAN training to handle discrete value sequences with a Monte Carlo-based sample generation approach. Objective-reinforced GAN (ORGAN) [8] extends SeqGAN [28] by incorporating domain-specific properties as rewards. TransORGAN [16] further extends ORGAN [8] by utilizing a transformer architecture to capture semantic information of molecules and Variant SMILES for better learning syntax rules through data augmentation. In addition, TransORGAN employs the variant SMILES technique to enhance the transformer's semantic and syntactic learning capabilities. MolGAN [4] generates small molecular structures using an implicit and likelihood-free graph-structure-based GAN.

Although GAN-based methods have shown effectiveness, their reliance on time-consuming Monte Carlo-based sample generation or molecular graph approach makes them unsuitable for large molecule datasets. In order to address this limitation, this study adopts the MacGAN, which utilizes moment reward and the RL techniques in conjunction with GAN. By incorporating these advancements, MacGAN is suitable for generating large molecule datasets.

3   MacGAN Overview

Fig. 1 overviews the architecture of the proposed MacGAN model, a framework that combines GAN and moment reward for generating SMILES sequences. The figure outlines the general GAN architecture, including the training process for the discriminator and the generator. Additionally, it illustrates the specific components of MacGAN, emphasizing the utilization of moment reward.

3.1   GAN

A GAN learns the training data distribution through a two-player mini-max game between a generator and a discriminator. The generator aims to produce samples that closely resemble real ones to deceive the discriminator, which, in turn, aims to distinguish between generated and real samples. This competitive cycle drives the learning process for SMILES generation. The mini-max game can be expressed as follows:

$$\min_{\theta} \max_{\phi} V(G_\theta, D_\phi) = \mathbb{E}_{x \sim p_{data}(x)}[\log D_\phi(x)] + \tag{1}$$

$$\mathbb{E}_{z \sim p_z(z)}[\log(1 - D_\phi(G_\theta(z)))],$$

where $p_{data}(x)$ represents the underlying distribution of real data, while $p_z(z)$ refers to the input noise distribution used by the generator to generate SMILES strings.

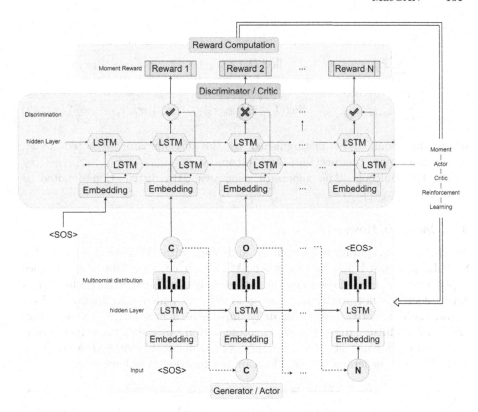

Fig. 1. Overview of the MacGAN model. The generator utilizes prior generated atoms to predict future ones, while the discriminator assigns rewards based on its predicted probabilities for each atom generated by the generator.

3.2 Autoregressive GAN for SMILES Strings

The generator of a GAN aims to generate complete SMILES strings. However, as described in scratchGAN, the discriminator can distinguish between the generated sequences and real SMILES strings, posing a challenge for the generator. During early training, the generator lacks a learning signal, as it only receives a reward when it produces a complete sequence. This is because the discriminator can easily differentiate the generated SMILES. To address this issue, MacGAN also employs a strategy of assigning rewards to each token in an autoregressive generation process, as in scratchGAN. This approach can be expressed as follows:

$$\max_{\phi} \sum_{t=1}^{T} \mathbb{E}_{p_{data}(x_t|x_1,\cdots,x_{t-1})}[\log D_\phi(x_t|x_1,\cdots,x_{t-1},x_{t+1},\cdots,x_T)],$$

$$+ \sum_{t=1}^{T} \mathbb{E}_{p_\theta(x_t|x_1,\cdots,x_{t-1})}[\log(1 - D_\phi(x_t|x_1,\cdots,x_{t-1},x_{t+1},\cdots,x_T)], \qquad (2)$$

where the atom to be distinguished by the discriminator, denoted as x_t, is dependent on the previous and future atoms $x_1, \cdots, x_{t-1}, x_{t+1}, \cdots, x_T$ within the molecule. The generator generates data distribution denoted as $p_\theta(x_t|x_1, \cdots, x_{t-1})$.

3.3 Moment Reward

To optimize computational resources and align with the characteristics of a SMILES string, it is not necessary for SMILES strings to exhibit long-distance dependencies akin to those found in natural language. Furthermore, the dependency does not necessarily decrease gradually with increasing distance. As a result, the cumulative reward strategy implemented in scratchGAN to gradually diminish future atom rewards for present atoms as the distance increases is not suitable. In contrast, MacGAN employs the moment reward for assigning a reward to each generated atom. This can be expressed as follows:

$$R_t = 2D_\phi(\hat{x}_t|\hat{x}_1, \cdots, \hat{x}_{t-1}, \hat{x}_{t+1}, \cdots, \hat{x}_T) - 1, \qquad (3)$$

where $\hat{x}_t$ represents the current generated atom, while $\hat{x}_1, \cdots, \hat{x}_{t-1}$ denote the previously generated atoms, $\hat{x}_{t+1}, \cdots, \hat{x}_T$ denote the future generated atoms. This is an autoregressive process as discussed earlier.

Similar to standard RL methods, MacGAN also incorporates the reduction of a baseline to mitigate variance and emphasize the relative reward of each atom. The baseline utilized in MacGAN is computed as the exponential moving average (EMA) of the mean reward across the batch data. The loss function for the generator can be expressed as follows:

$$Loss_G = -\sum_{n=1}^{N} \sum_{t=1}^{T_n} (R_t^n - b) \log p_\theta(\hat{x}_t|\hat{x}_1, \cdots, \hat{x}_{t-1})/(\sum_{n=1}^{N} T_n), \qquad (4)$$

$$b_i = \alpha b_{i-1} + (1 - \alpha)\hat{R}_i, \qquad (5)$$

where N is the batch size, b_i represents the current baseline, b_{i-1} denotes the previous baseline, and $\hat{R}_i$ represents the current mean reward of atoms in the batch. The value of α is set to 0.9. The baseline is updated per batch.

4 Experiment

4.1 Dataset

We utilize the complete QM9 dataset [20], which comprises 133,885 organic compounds containing up to 9 heavy atoms. These heavy atoms include carbon

(C), oxygen (O), nitrogen (N), and fluorine (F) and are sourced from the GDB-17 chemical database [21].

4.2 Evaluation Measures

We employ four criteria to evaluate the performance of MacGAN: validity, uniqueness, novelty, and diversity.

- **Validity** represents the ratio of chemically valid SMILES strings among all the generated SMILES strings.
- **Uniqueness** measures the ratio of unique SMILES strings among all the valid SMILES strings.
- **Novelty** indicates the ratio of SMILES strings absent from the training dataset among the unique SMILES strings.

To assess diversity, we calculated the similarity of chemical structures between molecules using the Tanimoto coefficient [25] based on the MorganFingerprint [2]. All scores range from 0 to 1, with higher values indicating better results.

4.3 Desired Chemical Properties

To assess the desired chemical properties, we utilized three criteria to evaluate the valid SMILES strings as molecules.

- **Drug-likeness (QED)** [1] determines the similarity of the compound to a drug.
- **Solubility (LogP)** measures the hydrophilicity of a molecule, represented by the logarithm of the octanol-water partition coefficient (logP), which expresses the concentration ratio of a substance in a mixture of octanol and water.
- **Synthesizability (SA)** utilizes the synthetic accessibility (SA) score [6] to quantify the ease of synthesizing a molecule.

4.4 Model Setup

In MacGAN, both the generator and discriminator adopt LSTM as the central architecture, similar to scratchGAN. The generator employs input transformation layers after the embedding layer, one LSTM cell with a hidden dimension size of 64, and linear output layers, and these input and output layers have a dropout probability [24] of 0.1. We set the maximum SMILES string length as 19, and the gradient of the iterable parameters is clipped from -0.1 to 0.1. The discriminator employs a bidirectional LSTM layer with a hidden dimension size of 64 after the embedding layer and linear output layers with a dropout rate of 0.1. The gradient of the iterable parameters is also clipped from -0.1 to 0.1, and an L2 regularization decay of 1×10^{-6} is applied.

The MacGAN model is trained for 20,000 steps, and its performance is evaluated every 100 steps using 10k samples on NVIDIA GeForce RTX 3090. We select the model with the highest validity score based on the selection criteria for the best model, including a validity score of at least 85.0% and a uniqueness score of at least 80.0% as the best model. This training process was repeated for five rounds, and the results were averaged. Both the generator and discriminator in MacGAN utilize the Adam optimizer [11] with a learning rate of 2×10^{-4}. The batch size is 1024.

4.5 Experimental Results

On average, the training time for MacGAN is 178.20 min. We compared Mac-GAN with MolGAN [4] and likelihood-based methods utilizing VAEs: CharacterVAE [7], GrammarVAE [13], GraphVAE [23], these results are also training on the complete QM9 dataset and evaluation on 10k samples. Table 1 compares results, with the best results highlighted in bold. MacGAN surpasses VAE-based methods across all evaluated aspects except for novelty. Its validity score of 94.41% and uniqueness measure of 83.27% exemplify its strong performance. Although MacGAN falls short of MolGAN in terms of validity (98.1%) and novelty (94.2%), it still demonstrates a commendable level of validity while boasting the highest uniqueness among all the examined methods. MolGAN suffers from a severe overfitting problem, which leads to the generation of molecules with a very limited uniqueness (i.e., 10.4%). The consequences of this overfitting issue are evident in generating carbon chains that exhibit comparable structural characteristics. While these carbon chains generally adhere to validity criteria, it is imperative to acknowledge the futility of exclusively relying on validity as a metric for comparison without due consideration for their uniqueness.

These findings highlight MacGAN's remarkable proficiency in generating SMILES strings, positioning it as a formidable contender in a molecular generation. With its validity score ranking second highest among the methods and its uniqueness metric outperforming all others, MacGAN demonstrates a balanced performance across validity, uniqueness, and novelty. Such outcomes reinforce that MacGAN holds considerable potential for generating structurally valid and diverse molecular structures. While its novelty score may be lower than MolGAN's, MacGAN compensates by demonstrating superior validity and uniqueness, which are critical in drug discovery.

Therefore, the comprehensive evaluation shows that MacGAN is a promising approach for molecule generation tasks. The balanced performance of MacGAN strengthens its position as an effective tool for generating high-quality molecules.

In addition to the aforementioned comparisons, we compared the valid molecules generated by MacGAN with those present in the training dataset, focusing on three crucial chemical properties: Drug-likeness (QED), Solubility (LogP), and Synthesizability (SA). Table 2 shows the results of comparing the chemical properties between the generated and training datasets. Specifically, MacGAN generates SMILES strings with better QED, yielding a mean value of 0.49. In contrast, the mean QED value of the molecules in the training dataset is 0.47. This improvement suggests that MacGAN can generate molecules with higher drug-likeness, which is critical in drug discovery.

Table 1. Comparison results with different baselines on complete QM9 dataset.

Baselines	Validity ↑	Uniqueness ↑	Novelty ↑
CharacterVAE [7]	10.3%	67.5%	90.0%
GrammarVAE [13]	60.2%	9.3%	80.9%
GraphVAE [23]	55.7%	76.0%	61.6%
GraphVAE [23]/imp	56.2%	42.0%	75.8%
GraphVAE [23] NoGM	81.0%	24.1%	61.0%
MolGAN [4]	**98.1%**	10.4%	**94.2%**
Avg. MacGAN	94.41%	**83.27%**	65.55%

⋆ The abbreviation "imp" indicates implicit node probabilities, while "NoGM" indicates the absence of graph matching. The term "Avg." represents the average results. The values highlighted in bold are the maximum values attained within the respective columns.

Similarly, MacGAN demonstrates an improved mean value for LogP at 0.37 compared to the training dataset. This indicates that the generated molecules possess a higher solubility potential, essential for drug formulation and bioavailability. Furthermore, MacGAN exhibits an enhanced mean value for SA at 0.32, surpassing that of the molecules in the training dataset. This signifies that the generated molecules are more synthetically accessible, indicating their feasibility for chemical synthesis.

These findings highlight the ability of MacGAN to generate SMILES strings with favorable chemical properties, which is of great significance in drug discovery and design. By consistently generating molecules with higher QED, LogP, and SA scores, MacGAN demonstrates its potential to accelerate the identification and development of novel therapeutic candidates. In summary, the comparison between the generated valid molecules and the molecules from the training dataset supports the efficacy of MacGAN in generating SMILES strings with improved chemical properties.

To further analyze the generated molecules, we focus on Fig. 2, which present Violin plots illustrating the distribution of QED, LogP, and SA for molecules generated by MacGAN compared to molecules from the training dataset. These graphical representations provide visual evidence that aligns with the numerical findings presented in Table 2, corroborating the notion that the molecules generated by MacGAN consistently exhibit higher values for QED, LogP, and SA compared to the molecules in the training dataset.

Notably, we observe a distinct pattern when focusing on the most densely populated region. The highest density distribution of QED for the generated molecules is approximately 0.52, surpassing the corresponding density of approximately 0.48 observed for the molecules in the training set. This indicates that MacGAN has a propensity for generating molecules with a greater propensity for drug-likeness, further highlighting its potential for generating novel compounds with desirable pharmacological properties.

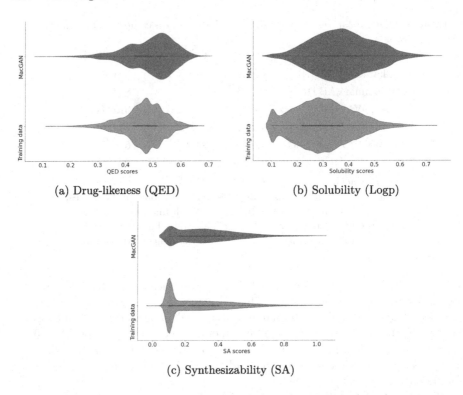

(a) Drug-likeness (QED)   (b) Solubility (Logp)

(c) Synthesizability (SA)

Fig. 2. Violin plots comparing the Drug-likeness (QED), Solubility (Logp) and Synthesizability of molecules generated by MacGAN and molecules from the training dataset.

Similarly, the density of LogP in the generated molecules is approximately 0.37, notably higher than the density of approximately 0.27 observed for the training molecules. This suggests that MacGAN has learned to generate molecules with improved solubility characteristics, a key aspect in drug formulation and bioavailability. Furthermore, examining the density of SA, we observe that the generated molecules exhibit a density of approximately 0.15, slightly higher than the density of approximately 0.14 observed for the training molecules. This indicates that MacGAN has been successful in generating molecules that are synthetically accessible, implying their feasibility for chemical synthesis in the laboratory. MacGAN also exhibits a diversity score of 0.92, indicating its strong performance in generating diverse molecular structures.

In summary, the violin plots provide a visual representation that reinforces the numerical findings. The molecules generated by MacGAN consistently exhibit higher numerical densities for QED, LogP, and SA scores, signifying their superiority over the molecules in the training dataset. These results further accentuate MacGAN's capability to generate molecules with improved chemical properties, thus strengthening its potential for drug discovery and design application.

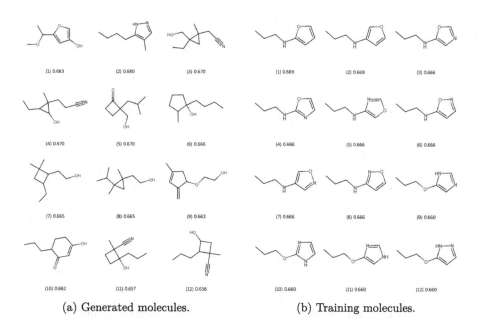

(a) Generated molecules.   (b) Training molecules.

Fig. 3. Top 12 Drug-likeness (QED) molecules.

Additionally, Figs. 3, 4 and 5 depicts the top 12 samples showcasing the values of QED, Logp, and SA scores for molecules generated by MacGAN, juxtaposed with samples from the training dataset. Notably, a distinct structural dissimilarity is observed between the generated and training samples, indicating the emergence of novel molecular configurations within the top value range for MacGAN-generated molecules.

These positive results may partly stem from the simplicity of the QM9 dataset, which primarily consists of small organic molecules. Therefore, for a more comprehensive evaluation of MacGAN's performance, further experiments on more complex and larger datasets are warranted. Future work will explore the application of MacGAN to diverse and challenging datasets, such as those containing macrocycles or incorporating specific functional groups, to assess its effectiveness in generating molecules with desired properties in a broader chemical space. These investigations will provide a more thorough understanding of MacGAN's capabilities and its potential impact on computational chemistry and drug discovery.

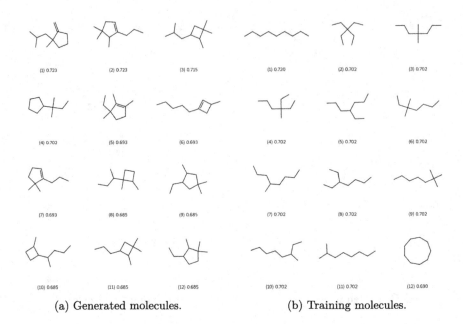

(a) Generated molecules.   (b) Training molecules.

Fig. 4. Top 12 Solubility (Logp) molecules.

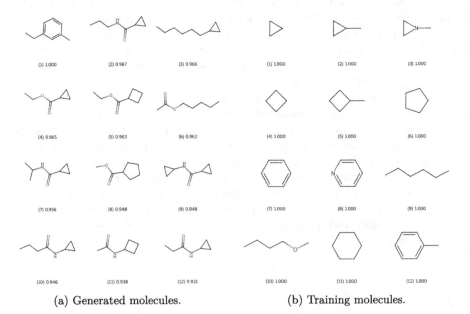

(a) Generated molecules.   (b) Training molecules.

Fig. 5. Top 12 Synthesizability molecules.

Table 2. Chemistry property results on complete QM9 dataset.

	Max Len	Min Len	Mean Len	Mean QED ↑	Mean Logp ↑	Mean SA ↑
Generation	19	9	15.41	0.49	0.37	0.32
Training	29	1	15.22	0.47	0.30	0.27

5   Conclusion

We employ the moment-actor-critic-based GAN (MacGAN) to generate SMILES strings. MacGAN leverages the robust architecture of the scratch GAN while incorporating a more straightforward reward mechanism, making it suitable for larger datasets compared to computationally intensive Monte Carlo-based methods.

MacGAN demonstrates remarkable proficiency in generating SMILES strings, striking a balance between validity, uniqueness, and novelty while exhibiting desirable chemical properties. These findings highlight its potential applicability in drug discovery and design. However, the performance of Mac-GAN on more complex and larger datasets remains to be discovered and warrants further investigation. In future work, we will evaluate MacGAN's performance on diverse and challenging datasets to gain insights into its effectiveness in generating molecules with desired properties in a broader chemical space.

Acknowledgements. This research has been supported by KAKENHI (20K11830) Japan.

References

1. Bickerton, G.R., Paolini, G.V., Besnard, J., Muresan, S., Hopkins, A.L.: Quantifying the chemical beauty of drugs. Nat. Chem. 4(2), 90–98 (2012)
2. Cereto-Massagué, A., Ojeda, M.J., Valls, C., Mulero, M., Garcia-Vallvé, S., Pujadas, G.: Molecular fingerprint similarity search in virtual screening. Methods 71, 58–63 (2015)
3. Dai, H., Tian, Y., Dai, B., Skiena, S., Song, L.: Syntax-directed variational autoencoder for molecule generation. In: Proceedings of the International Conference on Learning Representations (2018)
4. De Cao, N., Kipf, T.: MolGAN: an implicit generative model for small molecular graphs. arXiv preprint arXiv:1805.11973 (2018)
5. Deng, J., Yang, Z., Ojima, I., Samaras, D., Wang, F.: Artificial intelligence in drug discovery: applications and techniques. Briefings Bioinf. 23(1), bbab430 (2022)
6. Ertl, P., Schuffenhauer, A.: Estimation of synthetic accessibility score of drug-like molecules based on molecular complexity and fragment contributions. J. Cheminformatics 1, 1–11 (2009)
7. Gómez-Bombarelli, R., et al.: Automatic chemical design using a data-driven continuous representation of molecules. ACS Cent. Sci. 4(2), 268–276 (2018)
8. Guimaraes, G.L., Sanchez-Lengeling, B., Outeiral, C., Farias, P.L.C., Aspuru-Guzik, A.: Objective-reinforced generative adversarial networks (organ) for sequence generation models. arXiv preprint arXiv:1705.10843 (2017)

9. Hoogeboom, E., Satorras, V.G., Vignac, C., Welling, M.: Equivariant diffusion for molecule generation in 3D. In: International Conference on Machine Learning, pp. 8867–8887. PMLR (2022)

10. Jin, W., Barzilay, R., Jaakkola, T.: Junction tree variational autoencoder for molecular graph generation. In: International Conference on Machine Learning, pp. 2323–2332. PMLR (2018)

11. Kingma, D.P., Ba, J.: Adam: a method for stochastic optimization. In: International Conference on Learning Representations (2015)

12. Kumar, S.A., et al.: Machine learning and deep learning in data-driven decision making of drug discovery and challenges in high-quality data acquisition in the pharmaceutical industry. Future Med. Chem. 14(4), 245–270 (2022)

13. Kusner, M.J., Paige, B., Hernández-Lobato, J.M.: Grammar variational autoencoder. In: International Conference on Machine Learning, pp. 1945–1954. PMLR (2017)

14. Li, C., Chen, Z., Zheng, J.: An efficient transformer encoder-based classification of malware using API calls. In: 2022 IEEE 24th International Conference on High Performance Computing & Communications; 8th International Conference on Data Science & Systems; 20th International Conference on Smart City; 8th International Conference on Dependability in Sensor, Cloud & Big Data Systems & Application (HPCC/DSS/SmartCity/DependSys), pp. 839–846. IEEE (2022)

15. Li, C., He, M., Qaosar, M., Ahmed, S., Morimoto, Y.: Capturing temporal dynamics of users' preferences from purchase history big data for recommendation system. In: 2018 IEEE International Conference on Big Data (Big Data), pp. 5372–5374. IEEE (2018)

16. Li, C., Yamanaka, C., Kaitoh, K., Yamanishi, Y.: Transformer-based objective-reinforced generative adversarial network to generate desired molecules. In: Proceedings of the Thirty-First International Joint Conference on Artificial Intelligence, IJCAI-22, pp. 3884–3890 (2022)

17. de Masson d'Autume, C., Mohamed, S., Rosca, M., Rae, J.: Training language GANs from scratch. In: Advances in Neural Information Processing Systems, vol. 32 (2019)

18. Mullard, A.: New drugs cost us $2.6 billion to develop. Nat. Rev. Drug Discovery 13(12), 877 (2014)

19. Müller, T.D., Blüher, M., Tschöp, M.H., DiMarchi, R.D.: Anti-obesity drug discovery: advances and challenges. Nat. Rev. Drug Discovery 21(3), 201–223 (2022)

20. Ramakrishnan, R., Dral, P.O., Rupp, M., Von Lilienfeld, O.A.: Quantum chemistry structures and properties of 134 kilo molecules. Sci. Data 1(1), 1–7 (2014)

21. Ruddigkeit, L., Van Deursen, R., Blum, L.C., Reymond, J.L.: Enumeration of 166 billion organic small molecules in the chemical universe database GDB-17. J. Chem. Inf. Model. 52(11), 2864–2875 (2012)

22. Shi, C., Xu, M., Zhu, Z., Zhang, W., Zhang, M., Tang, J.: GraphAF: a flow-based autoregressive model for molecular graph generation. In: International Conference on Learning Representations (2020). https://openreview.net/forum?id=S1esMkHYPr

23. Simonovsky, M., Komodakis, N.: GraphVAE: towards generation of small graphs using variational autoencoders. In: Kůrková, V., Manolopoulos, Y., Hammer, B., Iliadis, L., Maglogiannis, I. (eds.) ICANN 2018. LNCS, vol. 11139, pp. 412–422. Springer, Cham (2018). https://doi.org/10.1007/978-3-030-01418-6_41

24. Srivastava, N., Hinton, G., Krizhevsky, A., Sutskever, I., Salakhutdinov, R.: Dropout: a simple way to prevent neural networks from overfitting. J. Mach. Learn. Res. 15(1), 1929–1958 (2014)

25. Tanimoto, T.T.: An elementary mathematical theory of classification and prediction, IBM report (november, 1958), cited in: G. salton, automatic information organization and retrieval (1968)

26. Weininger, D.: Smiles, a chemical language and information system. 1. introduction to methodology and encoding rules. J. Chem. Inf. Comp. Sci. **28**(1), 31–36 (1988)

27. Williams, R.J.: Simple statistical gradient-following algorithms for connectionist reinforcement learning. In: Reinforcement Learning, pp. 5–32 (1992)

28. Yu, L., Zhang, W., Wang, J., Yu, Y.: SeqGAN: sequence generative adversarial nets with policy gradient. In: Proceedings of the AAAI Conference on Artificial Intelligence, vol. 31 (2017)

29. Zhang, X., Li, C., Morimoto, Y.: A multi-factor approach for stock price prediction by using recurrent neural networks. Bull. Netw. Comput. Syst. Softw. **8**(1), 9–13 (2019)

Multi-modal Graph Convolutional Network for Knowledge Graph Entity Alignment

Yinghui You, Yuyang Wei, Yanlong Zhang, Wei Chen[✉], and Lei Zhao

School of Compute Science and Technology, Soochow University, Suzhou, China
{yhyou,yyweisuda,ylzhang828}@stu.suda.edu.cn,
{robertchen,zhaol}@suda.edu.cn

Abstract. Entity Alignment (EA) plays a crucial role in the integration of multiple knowledge graphs (KGs). With the blooming of KGs, the auxiliary multi-modal data, such as attributions and images, are widely used to enhance alignment performance. However, most existing techniques for multi-modal knowledge exploitation separately pre-train uni-modal features and heuristically merge these features, failing to adequately consider the interplay between different modalities. To tackle this problem, we propose a novel model entitled MGCEA (**M**ulti-modal **G**raph **C**onvolutional network for knowledge graph **E**ntity **A**lignment), which considers the guidance of neighborhood structure in cross-modal embedding enhancement. Specifically, MGCEA pre-trains multiple modal features to initialize their corresponding embeddings. Then a multi-modal embedding enhancement mechanism, which consists of a multi-modal graph convolution network and an attention network, is developed to achieve cross-modal enhancement guided by the neighborhood structure and learn an effective joint embedding. Moreover, a joint loss based on contrast learning is introduced to optimize model parameters by considering intra-modal relationships and cross-modal interactions. The extensive experiments conducted on two benchmarks demonstrate that MGCEA significantly outperforms the state-of-the-art multi-modal knowledge graph entity alignment baselines.

Keywords: Entity Alignment · Knowledge Graph · Graph Convolutional Network · Multi-modal Representation Learning

1 Introduction

Knowledge graphs (KGs), which store facts in a structured manner, have been widely used to promote knowledge-driven downstream tasks, such as recommendation systems [2], question answering [28], information extraction [10], and text generation [12]. In recent years, a growing number of KGs have been constructed based on different data sources for diverse purposes, where the knowledge may

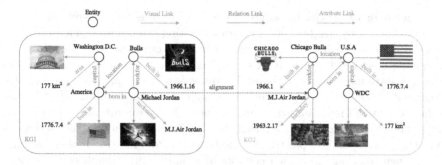

Fig. 1. An example of multi-modal knowledge graph entity alignment.

be complementary to downstream tasks. To integrate this complementary knowledge, entity alignment (EA) [4], which aims to discover entities equivalent to the same real-world object across different KGs, has attracted increasing attention.

Over the past decade, extensive embedding-based methods have been developed to align entities across KGs, which focus on exploiting semantics or concept to facilitate knowledge analysis. To improve the EA performance, some studies [11,22,23] have exploited entity attributes as auxiliary data to enhance entity representation. These approaches have proven effective because entity attributes contain essential information about the nature of entities, and equivalent entities in different KGs often share identical attributes. To integrate more auxiliary data, such as images, multi-modal knowledge graphs (MMKGs) have been proposed to further enrich entity information. To leverage these auxiliary data, some multi-modal EA studies [5,16] have been proposed to directly merge different modal features to obtain fusion representations. EVA [15] suggests that each modality contributes unequally to the joint embedding, and it preliminarily explores multi-modal fusion with an attention-based mechanism that can identify valuable information from different modalities. In addition to assigning different attentive weights to various modalities, MCLEA [14] also models the intra-modal relationships to distinguish the embeddings of equivalent entities from those of other entities for each modality, and the cross-modal interactions to reduce the gaps between modalities for each entity. MSNEA [5] delves into the inter-modal effect and utilizes visual features to guide relational feature learning.

Although these studies have explored the inter-modal and cross-modal interactions, they ignore the guidance of neighborhood structure in cross-modal embedding enhancement. By way of illustration, in Fig. 1, the entities "Michael Jordan" in KG1 and "M.J.Air Jordan" in KG2 refer to the same object in real-world, and we assume each entity has two modalities including neighborhood structure and image. Intuitively, an effective embedding is necessary for high-performance entity alignment, and the neighbors can provide significant information for a specific entity in a graph. In view of this, the entity "Bulls" is much more important than "Washington D.C." when enhancing the representation of "Michael Jordan" from the perspective of structure, since the former is the neighbor of "Michael Jordan" while the latter is not. Despite the great benefits brought by neighborhood entities, existing studies have not taken this essential information into account while enhancing cross-modal embedding.

Having observed the drawback of existing work, we propose a novel model namely Multi-modal Graph Convolutional network for knowledge graph Entity Alignment (MGCEA), by considering the guidance of neighborhood structure in the cross-modal representation enhancement. First, we initialize multi-modal features including neighborhood structure, image, relation, and attribute embeddings with corresponding pre-trained methods. Then, a multi-modal embedding enhancement mechanism, which consists of a multi-modal graph convolutional network and an attention network, is designed to enhance cross-modal embeddings guided by the neighborhood structure. Specifically, an adjacency matrix is constructed to represent the neighborhood structure, which guides the graph convolutional network performed on other modal features. With the multi-modal embeddings enhanced, an attention network is utilized to fuse these embeddings by capturing valuable information from various modalities, and learning an effective joint embedding. Moreover, a joint loss based on contrast learning is introduced to optimize model parameters by modeling intra-modal relationships and cross-modal interaction. In summary, the main contributions of this paper are as follows:

- To the best of our knowledge, we are the first to study the guidance of neighborhood information in the cross-modal representation enhancement to align equivalent entities across different multi-modal knowledge graphs.
- A multi-modal graph convolutional network is designed to guide the cross-modal embedding enhancement by neighborhood structures, and an attention network is introduced to fuse enhanced embedding by capturing valuable information from various modalities and learning a joint multi-modal embedding.
- Extensive experiments are conducted on two multi-modal benchmarks, and the experimental results demonstrate that the proposed model significantly outperforms the state-of-the-art baselines.

The rest of the paper is organized as follows. Section 2 presents the related work, while Sect. 3 formulates the problem and provides a detailed description of MGCEA. The experimental results are reported in Sect. 4, which is followed by the conclusion and the future work in Sect. 5.

2   Related Work

2.1   Entity Alignment

To make full use of entity information from different data sources, many entity alignment methods have been proposed in recent years, which can be divided into two categories, i.e., translation-based and GNNs-based methods.

Translation-based approaches primarily focus on the semantic or conceptual aspects to enhance knowledge analysis, e.g., TransE [1] interprets relationships as translations operating on low-dimensional embeddings of entities. MTransE [6] learns transitions to map the embedding vectors to their corresponding vectors. Instead of training map embeddings, IPTransE [27] uses an iterative and parameter-sharing strategy. BootEA [20] also utilizes the iterative strategy by marking possible alignments and correcting errors through alignment editing methods. Considering the degree of difference between alignment entity pairs, SEA [17] uses awareness of the degree difference in adversarial training and incorporates the unaligned entities to enhance the performance. In addition, AttrE [22] utilizes extra attributes to generate attribute character embeddings, and IMUSE [11] employs a bivariate regression to merge the alignment results of relations and attributes better. Despite the great contributions made by translation-based methods, they struggle to achieve satisfactory results when dealing with complex relational graphs. GNNs-based methods recursively capture feature representations of adjacent entities, enabling them to learn structured embeddings of KG, e.g., GCN-Align [23] utilizes graph convolutional networks to encode the structured information of entities and combines both relation and attribute embeddings for EA. HMAN [26] incorporates multi-aspect features into entity embeddings and exploits highway networks to improve convergence speed. AliNet [21] controls the aggregation of direct and distant neighborhood information within multiple hops through a gating mechanism. MCEA [18] utilizes a multi-scale graph convolutional network structure that allows the corresponding nodes of long-tailed entities to accumulate richer information for EA. Although these methods have achieved promising results, they cannot deal with multi-modal knowledge from real scenes on entity alignment tasks.

2.2   Multi-modal Knowledge Graph

In recent years, EA approaches aiming at the effective utilization of multi-modal representation and fusion representation have emerged, e.g., PoE [16] link prediction by considering the alignment links between entities as SameAs relationships. MMEA [4] fuses multi-modal knowledge embeddings from a separate space to a common space. Considering that vector representations in Euclidean space may lead to negative embedding effects, HMEA [9] uses the hyperbolic graph convolutional networks [3] to learn entity structured representation and visual representation. MSNEA [5] incorporates visual features to guide relational and attribute learning, and assign weights to important attributes for EA. Notably, most of these methods cannot identify valuable information from different multi-modal embeddings for modeling joint representations. In view of this, EVA [15] utilizes visual knowledge to create an initial seed dictionary and extends the training set through iterative learning. MultiJAF [7] dynamically learns the significance of multiple modalities (e.g., structures, attributes, and images) through an attention mechanism in its fusion network. Although these methods can adaptively learn the influence of different modalities on the joint representation, they tend to ignore the inter-modal influence, resulting in a situation where a single modality cannot fully exploit the information of other modalities. MCLEA [14], based on EVA, employs the output distribution between the joint embeddings and uni-modal embeddings to predict aligned entities. However, it fails to sufficiently consider the influence brought by the neighborhood of the graph structure, thus limiting the utilization of multi-modal knowledge.

3   Methodology

In this section, we first give the problem definition and a preliminary overview of MGCEA, then introduce the multi-modal pre-trained embedding, multi-modal embedding enhancement mechanism, and the objective.

3.1   Definition and Model Overview

Definition 1. *Multi-modal Knowledge Graph. Given a multi-modal knowledge graph, it can be formalized as $\mathcal{G} = (\mathcal{E}, \mathcal{R}, \mathcal{A}, \mathcal{P}, \mathcal{T})$ where $\mathcal{E}, \mathcal{R}, \mathcal{A}, \mathcal{P}$ denote the sets of entities, relations, attributes, and images, respectively. Here, the set of relations between entities is represented by the triple $\mathcal{T} \in \mathcal{E} \times \mathcal{R} \times \mathcal{E}$.*

Definition 2. *Multi-modal Entity Alignment Task. Given two multi-modal knowledge graphs $\mathcal{G}_1 = (\mathcal{E}_1, \mathcal{R}_1, \mathcal{A}_1, \mathcal{P}_1, \mathcal{T}_1)$ and $\mathcal{G}_2 = (\mathcal{E}_2, \mathcal{R}_2, \mathcal{A}_2, \mathcal{P}_2, \mathcal{T}_2)$, the task of multi-modal entity alignment aims to find the set of entity pairs $\mathcal{S} = \{(e_1, e_2) \mid e_1 \equiv e_2, e_1 \in \mathcal{E}_1, e_2 \in \mathcal{E}_2\}$, where the symbol $\equiv$ denotes the equivalence of two entities.*

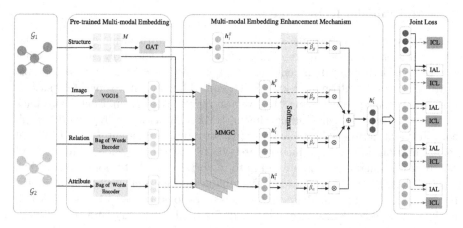

Fig. 2. The overview of the proposed model MGCEA.

Model Overview. The architecture of the proposed model is presented in Fig. 2. Specifically, we first initial the multi-modal representation of each modality by a multi-modal pre-trained embedding approach, then a multi-modal embedding enhancement mechanism is designed to learn multi-modal enhancement features, and generate effective joint representation. Lastly, the objective is given by a joint loss based on contrast learning.

3.2   Multi-modal Pre-trained Embedding

We use $\{g, p, r, a\}$ to denote the set of available modalities, where g, p, r, and a represent the modality of structure, visual, relation, and attribute, separately. Note that, we expand the knowledge coverage by integrating the target KGs ($\mathcal{G}_1$ and $\mathcal{G}_2$) into a unified KG ($\mathcal{G}$) to facilitate the refinement of the KGs by discovering potential flaws [4]. In this subsection, the multi-modal features are initialized with corresponding pre-trained methods.

Structure Embedding. Equivalent entities are often connected to each other through their neighbors in MMKGs. To capture the structured information of $\mathcal{G}$ and discover potential aligned entities, we utilize the graph attention network (GAT) to generate embeddings that are aware of the entity's neighborhood. The hidden state $\boldsymbol{h}_i \in \mathbb{R}^{1 \times d}$ of entity e_i is obtained by aggregating the one-hop neighbors Φ_i^n (including entity e_i), which can be formally represented as follows:

$$\boldsymbol{h}_i = ReLU \left(\sum_{j \in \Phi_i^n} \alpha_{ij} \boldsymbol{h}_j \right) \tag{1}$$

α_{ij} is an attention weight between (e_i, e_j), defined as:

$$\alpha_{ij} = \frac{\exp \left(LeakyReLU \left(\boldsymbol{\theta}^\top [\boldsymbol{W}\boldsymbol{h}_i \oplus \boldsymbol{W}\boldsymbol{h}_j] \right) \right)}{\sum_{u \in \Phi_i^n} \exp \left(LeakyReLU \left(\boldsymbol{\theta}^\top [\boldsymbol{W}\boldsymbol{h}_i \oplus \boldsymbol{W}\boldsymbol{h}_u] \right) \right)} \tag{2}$$

where $\boldsymbol{\theta} \in \mathbb{R}^{2 \times d}$, $\boldsymbol{W} \in \mathbb{R}^{d \times d}$ are the learnable weight matrix and the diagonal weight matrix, respectively. The symbol $\oplus$ represents the concatenation operation. To enhance the stability of the self-attentive learning process, we adopt a multi-head strategy, generating K ($K = 2$) independent representations $\boldsymbol{h}_i$ for each layer of the GAT. These representations are then concatenated to form the embedding of entity e_i for each layer. Finally, we employ a two-layer GAT model to aggregate neighborhood information and obtain the output of the last GAT layer as the graph structure embedding $\boldsymbol{h}_i^g$ of entity e_i.

Visual Embedding. In order to extract the visual features, we utilize the model VGG16 [19] pre-trained on dataset ILSVRC 2012 that is derived from ImageNet [8]. To adapt the pre-trained model for MGCEA, we introduce a new learnable fully-connected layer to replace the last fully-connected layer of VGG16. This change enables us to obtain more generalized visual representations of entities and to map a high-dimensional visual feature to a low-dimensional vector. Concretely, for an entity e_i and its corresponding image p_i, the initial visual feature embedding of entity e_i is expressed as:

$$\overline{h}_i^p = W_p \cdot \text{VGG16}(p_i) + b_p \tag{3}$$

where W_p is a learnable parameter, and b_p is a bias of the fully-connected layer.

Relation and Attribute Embedding. Due to the presence of noise from neighborhood entities polluting the representation of relations and attributes from KG [26] in the GNNs-based network, we regard relations and attributes of entities as a bag-of-words feature to capture the information. Concretely, we build the count-based *N-hot* vectors o_i^r and o_i^a for the features of relations and attributes, respectively, and only consider the top-1000 most frequent relations and attributes to avoid data sparsity issues. After this, a parameterized separate fully-connected layer is applied to transform these features as follows:

$$\overline{h}_i^m = W_m \cdot o_i^m + b_m \tag{4}$$

where $W_m \in \mathbb{R}^{d_m \times d}$ is learnable parameter and $m \in \{r, a\}$. For entity e_i, the dimensions of its relation and attribute features are d_r and d_a, respectively.

3.3 Multi-modal Enhancement Embedding Mechanism

After obtaining the initial embeddings, a multi-modal convolutional network is designed to enhance the representations of multiple modalities, and an attention network is employed to fuse these multi-modal embeddings by capturing valuable information from each modality.

Multi-modal Convolutional Network (MMGC). Inspired by [24] and its multi-modal representation in the field of multi-modal recommendation, we design MMGC that leverages the neighborhood structure between entities to guide the learning of multiple modalities. Similar to GCN [13], feature transformation and information propagation are the core of MMGC, but we reduce the complexity of the model by iteratively removing the nonlinearity between graph convolution layers and collapsing the resulting function into a single linear transformation, which has proven successful in capturing the information of the graph structure [25].

   In our model, MMGC is applied to guide the enhancement of cross-modal features. Specifically, we utilize the adjacency matrix of entities to convolve the embedding of each modality separately, thus making each embedding contains richer neighborhood information. The adjacency matrix is first given as follows:

$$M = D^{-\frac{1}{2}} \tilde{A} D^{-\frac{1}{2}} \tag{5}$$

where $\tilde{A} = A + I$ is the adjacency matrix of graph $\mathcal{G}$ with added self-connections. Here, I is an identity matrix and D denotes the diagonal matrix of $\tilde{A}$. The normalized adjacency matrix M, which represents the graph structure between entities for multi-modal information aggregation, is then utilized to guide the learning of the multi-modal embeddings $\overline{h}^m$ for achieving the enhanced representation h^m of modality m,

$$h^m = \overline{h}^m + \left\| H^{(L)} \right\| \tag{6}$$

$$H^{(l)} = M \cdot H^{(l-1)} \tag{7}$$

$$H^{(0)} = \overline{h}^m \tag{8}$$

where $\|\cdot\|$ denotes L_2 normalization, and the multi-modal initial embedding matrix $\overline{h}^m$ of all entities is regarded as the input of MMGC, $m \in \{p, r, a\}$. Notably, the i-th row of h^m (i.e., h_i^m) is the representation of modality m of the entity e_i. L denotes the number of MMGC layers.

Attention-Based Modality Fusion. Previous studies [5,9] lack effective exploration in multi-modal fusion, neglecting the diverse importance of different modalities for entity embedding. To tackle the problem, we adopt an attention-based network to fuse multi-modal data effectively. By automatically learning the correlation and importance of modalities, we can better integrate the feature representations of different modalities and improve the performance of the multi-modal EA task. Specifically, we first learn the dynamic weight of each modality and then generate the joint embedding by weighted concatenation. The final joint embedding h_i^c of entity e_i is designed as follows:

$$h_i^c = \bigoplus_{m \in \{g,p,r,a\}} \beta_m h_i^m \tag{9}$$

$$\beta_m = \mathrm{softmax}\left(h_i^m\right) = \frac{\exp\left(h_i^m\right)}{\sum_{k \in \{g,p,r,a\}} \exp\left(h_i^k\right)} \tag{10}$$

where β_m represents the trainable weight of a specific modality m.

3.4   Objective

To optimize our objective, the multi-modal enhanced embeddings and effective joint embedding are applied to model the intra-modal contrast loss (ICL) and inter-modal alignment loss (IAL) [14]. In addition, iterative learning is used to extend the training set to fully exploit the information contained in each modality. Specifically, we generate a negative entity set by destroying aligned entity pairs according to the 1-to-1 alignment constraint [20]. For each entity pair $\left(e_i^1, e_i^2\right)$ in the set $\mathcal{S}$, we define its negative set $N_i = \left\{e_j^1 \mid \forall e_j^1 \in \mathcal{E}_1, j \neq i\right\} \cup$

$\{e_j^2 \mid \forall e_j^2 \in \mathcal{E}_2, j \neq i\}$, where N_i comes from Ω, which is a minibatch of $\mathcal{S}$ specified to guide the training. The aligned probability distribution of modality m is defined as:

$$\varsigma_m \left(e_i^1, e_i^2\right) = \frac{\psi_m \left(e_i^1, e_i^2\right)}{\psi_m \left(e_i^1, e_i^2\right) + \sum_{e_j \in N_i} \psi_m \left(e_i^1, e_j\right)} \tag{11}$$

where $\psi_m \left(e_u, e_v\right) = \exp\left(\left(\boldsymbol{h}_u^m\right)^\top \boldsymbol{h}_v^m / \tau_1\right)$ and τ_1 is the temperature parameter. As each input of $\varsigma_m \left(e_i^1, e_i^2\right)$ is directed and asymmetric, we align each modality in both directions. Therefore, the ICL of modality m is defined as follows:

$$\mathcal{L}_m^{ICL} = -\log \left(\varsigma_m \left(e_i^1, e_i^2\right) + \varsigma_m \left(e_i^2, e_i^1\right)\right) / 2 \tag{12}$$

In addition to training different modal embeddings with independent ICLs, we constrain the distribution between uni-modal features and joint features by minimizing the bidirectional KL divergence, which encourages each modality to effectively utilize the complementary information from other modalities. The IAL is defined as:

$$\mathcal{L}_m^{IAL} = \frac{1}{2} \left[\delta \left(\varsigma_c \left(e_i^1, e_i^2\right) \| \varsigma_m \left(e_i^1, e_i^2\right)\right) + \delta \left(\varsigma_c \left(e_i^2, e_i^1\right) \| \varsigma_m \left(e_i^2, e_i^1\right)\right)\right] \tag{13}$$

where δ denotes the KL divergence, $\varsigma_c \left(e_i^1, e_i^2\right)$ and $\varsigma_m \left(e_i^1, e_i^2\right)$ represent the joint embedding distribution and the uni-modal embedding distribution of modality m, respectively. The probability distributions are calculated similarly to Eq. 11, but with a different temperature parameter τ_2.

Finally, the overall loss of MGCEA is defined as follows:

$$\mathcal{L} = \mathcal{L}_c^{ICL} + \sum_{m \in \{g,p,r,a\}} \left(\frac{1}{\mu_m^2} \mathcal{L}_m^{ICL} + \frac{1}{\lambda_m^2} \mathcal{L}_m^{IAL} + \log \left(\mu_m \lambda_m\right)\right) \tag{14}$$

where μ_m and λ_m are hyperparameters used to balance the losses in ICL and IAL for modality m, and $\mathcal{L}_c^{ICL}$ denotes the joint representation loss of ICL.

4 Experiments

In this section, we present the experiments for the empirical evaluation of the proposed model MGCEA. We first introduce the datasets, then give the details of the evaluation metrics and parameter settings. Next, we discuss the experimental results, and give the parametric analysis finally.

Table 1. Statistics of Datasets.

Datasets	#Ent.	#Rel.	#Attr.	#Rel tr.	#Attr tr.	#Image	#Ref.
DB15K	12,842	279	225	89,197	48,080	12,837	12,846
YAGO15K	15,404	32	7	122,886	23,532	11,194	11,199
FB15K	14,951	1,345	116	592,213	29,395	13,444	–

4.1   Datasets

The proposed model is evaluated on two real-world datasets, i.e., FB15K-DB15K and FB15K-YAGO15K, which are constructed in [16]. The detailed statistics of them are presented in Table 1, including the numbers of entities (#Ent.), relations (#Rel.), attributes (#Attr.), relation triples (#Rel tr.), attribute triples (#Attr tr.), and images (#Image). In addition, it includes the number of alignment seeds (#Ref.) between FB15K and the corresponding aligned datasets, i.e., DB15K and YAGO15K, respectively. Notably, not all entities have associated images or equivalent counterparts in another knowledge graph.

4.2   Experimental Settings

We evaluate the developed model on the task of multi-modal entity alignment. In our experiments, the number of hidden units is set to 300 for each GAT layer. The dimensions of visual, relational, and attribute embeddings are set to 100. The epoch number is set to 1000 and an optional iterative training strategy is applied after 500 epochs. The batch size and learning rate are set to 512 and 0.001, respectively. The temperature parameters τ_1 and τ_2 are set to 0.1 and 4, respectively. The dimension of the visual encoder is set to 4096 for VGG16, the dimension of the bag-of-words feature of relation and attribute is set to 1000, and the number of layers in MMGC is set to 4. AdamW is employed to optimize the parameters of MGCEA. Hits@n and Mean Reciprocal Rank (MRR) are used as evaluation metrics. Hits@n represents the percentage of correct relations ranked at top-n, while MRR denotes the average of the reciprocal ranks of correct relations. Higher values of Hits@n and MRR indicate better performance.

4.3   Baselines

We compare MGCEA with two categories based on whether visual knowledge is incorporated: uni-modal and multi-modal entity alignment methods.

1) Uni-modal methods. MTransE [6] weighs the loss of entity alignment by using hyperparameters in the loss function. GCN-Align [23] adopts GCN to encode structured information, while AliNet [21] introduces distant neighbors to expand the overlap between their neighborhood structures. SEA [17] implements adversarial training by perceiving degree differences and utilizes labeled entities with rich information about unlabeled entities for alignment. BootEA [20] iteratively marks possible alignments as training data and reduces the accumulation of errors during iterations with alignment editing methods.

2) Multi-modal methods. PoE [16] combines the multi-modal features and measures the credibility of facts by matching the underlying semantics of the entities and mining the relations contained in the embedding space. MMEA [4] first obtains multi-modal embeddings which are then fused from different

spaces to a unified space. EVA [15] fuses multi-modal information into joint embeddings with an attention-based mechanism. MSNEA [5] uses inter-modal effects to obtain an enhanced multi-modal entity representation and designs a multi-modal contrast learning module to avoid the overwhelming influence of weak modalities. MCLEA [14] designs two contrast representation learning to learn intra-modal and inter-modal interactions.

4.4   Main Results

To demonstrate the effectiveness of MGCEA, we compare it with both uni-modal and multi-modal methods. Table 2 shows the Hits@1, Hits@10 and MRR of MGCEA and baselines on both datasets split into training/testing data of 2:7 alignment seeds, the remains are treated as valid sets on the task of multi-modal entity alignment [5]. Our model utilizes a structured neighborhood matrix to guide the enhanced representation of visual and relational modalities. From the results, we can observe the following phenomena:

1) MGCEA outperforms all baselines in terms of all evaluation metrics. Concretely, Hits@1, Hits@10 and MRR exhibit relative improvements of 17.43%, 8.03%, 13.58% and 6.24%, 7.73%, 6.59% respectively on datasets FB15K-DB15K and FB15K-YAGO15K compared with the best baseline MCLEA. This reason for improvements is that our model leverages the guidance of neighborhood structure to enhance cross-modal embeddings, which has been ignored by MCLEA. The excellent performance of MGCEA proves the advantages of graph neighborhood information for learning multi-modal representations.

2) BootEA has the best performance in all uni-modal methods but is still worse than some multi-modal methods (MCLEA and MGCEA). This is because BootEA fails to take advantage of the multi-modal knowledge, which contains rich semantic information about the entities. In addition, the multi-modal approaches perform better than uni-modal approaches in most cases, which demonstrates the usefulness and effectiveness of multi-modal knowledge for EA, and that adequately leveraging multi-modal knowledge can improve performance to a certain extent.

3) Note that multi-modal methods (e.g., PoE and MMEA) perform worse than some uni-modal methods such as AliNet and BootEA. This can be attributed to the fact that these multi-modal methods do not adequately account for modality heterogeneity and thus incorporate noise.

Table 2. Main experiments with 20% alignment seeds on FB15K-DB15K and FB15K-YAGO15K. The best results are highlighted in bold and the underlined values are the second best result.

Model	FB15K-DB15K			FB15K-YAGO15K		
	Hits@1	Hits@10	MRR	Hits@1	Hits@10	MRR
MTransE	0.28%	2.08%	0.012	0.13%	1.35%	0.008
GCN-Align	3.61%	14.40%	0.074	2.76%	10.96%	0.058
SEA	11.10%	34.57%	0.188	11.82%	33.74%	0.193
AliNet	28.97%	50.58%	0.363	28.10%	46.73%	0.346
BootEA	37.86%	63.89%	0.467	28.04%	51.17%	0.357
PoE	12.00%	25.60%	0.167	10.90%	24.1%	0.154
MMEA	26.12%	54.91%	0.358	20.52%	45.46%	0.289
EVA	12.56%	31.95%	0.191	7.97%	23.04%	0.130
MSNEA	27.64%	55.62%	0.372	22.92%	48.39%	0.315
MCLEA	45.10%	72.89%	<u>0.545</u>	<u>41.70%</u>	<u>65.87%</u>	<u>0.501</u>
MGCEA	**52.96%**	**78.74%**	**0.619**	**44.30%**	**70.96%**	**0.534**

4.5   Ablation Study

To verify the effectiveness of introducing the multi-modal convolution network to enhance the multi-modal representations, we compare MGCEA with three groups of variants. 1) Modality-free MGCEAs, which ignore specific modalities, including $w/o\,P$, $w/o\,R$ and $w/o\,A$ denote modalities without visual, relation and attribute, respectively. 2) MGCEA with/without neighborhood structure guidance, including a variant with neighborhood structure guiding attribute($w\,GA$), the variant without neighborhood structure guiding relation ($w/o\,GR$) or visual ($w/o\,GP$). 3) MCLM-free MGCEAs, which lack intra-modal or cross-modal modeling, i.e., $w/o\,ICL$ and $w/o\,IAL$. Table 3 presents the performance of these variants and the full MGCEA model, from which we have the following observations.

Table 3. Ablation study.

Model	FB15K-DB15K			FB15K-YAGO15K		
	Hits@1	Hits@10	MRR	Hits@1	Hits@10	MRR
MGCEA	**52.96%**	<u>78.74%</u>	**0.620**	44.30%	**70.96%**	**0.534**
$w/o\,P$	52.92%	78.53%	0.617	44.25%	70.23%	0.532
$w/o\,R$	47.01%	74.40%	0.564	36.14%	62.55%	0.451
$w/o\,A$	48.39%	74.87%	0.576	41.53%	67.39%	0.503
$w/o\,GP$	46.79%	70.85%	0.550	36.79%	63.57%	0.459
$w/o\,GR$	52.89%	78.67%	0.620	<u>44.34%</u>	70.82%	0.534
$w\,GA$	52.15%	77.45%	0.610	**44.39%**	70.00%	0.531
$w/o\,ICL$	52.06%	76.97%	0.609	42.81%	67.81%	0.515
$w/o\,IAL$	52.90%	**78.76%**	0.619	44.20%	70.89%	0.534

1) The results of the first group of variants (Modality-free MGCEAs) demonstrate that each modality contributes to the multi-modal entity alignment task. Among these modalities, the visual has the least impact on the model. This suggests that visual knowledge does not dominate the final results of MGCEA, and the potential reason is that both datasets contain limited valuable visual information.

2) For the second group of variants (MGCEAs with/without neighborhood structure guidance), both *w/o GP* and *w/o GR* perform worse than the full MGCEA model, which indicates the effectiveness of visual and relational learning mechanisms guided by graph neighborhood structure. The variant *w GA* uses the graph neighborhood structure to guide the enhanced representation learning of attribute modality, resulting in the introduction of noise into the model. Because attributes are entity-specific information, introducing neighbor information may disrupt the interaction between the entity and its attributes. Furthermore, by observing the first and second groups of variants in combination, *w/o P* is superior to *w/o GP*, illustrating the potential negative effects of employing inappropriate strategies to exploit visual knowledge, again confirming the effectiveness of visual modality enhancement embedding.

3) From the last group of variants (MCLM-free MGCEAs), the experimental results of *w/o ICL* and *w/o IAL* are nearly lower than MGCEA, indicating the importance of ICL in learning the intra-modal proximity and the interdependence between different modalities.

Overall, these ablation experiments provide insights into the individual contributions of different modalities and the importance of graph neighborhood structure guidance in MGCEA. The validity of our model is proved once again.

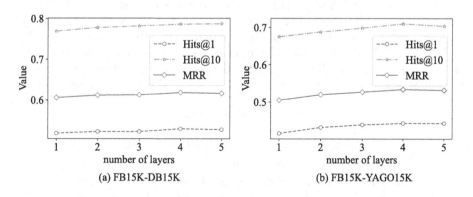

Fig. 3. Impact of MMGC Layers.

4.6 Parameter Analysis

We investigate the influence of several important parameters on the model performance, including the number of layers in the multi-modal convolution neural network and the different proportions of alignment seeds.

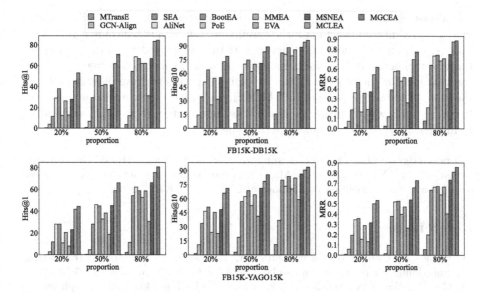

Fig. 4. Comparison results with different proportions of alignment seeds on FB15K-DB15K and FB15K-YAGO15K.

Impact of Layers in MMGC. To explore the impact of the number of layers in the multi-modal convolution neural network, we vary it from 1 to 5. The experimental results on two datasets are reported in Fig. 3, from which we can observe that with the increase of the number of layers, the model performance first increases and then decreases slightly. When the number is set to 4, the model achieves the highest Hits@1, Hits@10, and MRR. The model performance is first improved because more neighborhood information is aggregated to enhance its feature learning capability. However, adding too many layers will lead to a decrease in performance because of the introduction of distant neighbors, resulting in noise and feature duplication.

Seed Sensitivity. To evaluate the sensitivity of our model, we conduct experiments by using different proportions of alignment seeds as training sets (e.g., 20%, 50% and 80%), and the results are presented in Fig. 4. From this, we can observe that when 20% and 50% of alignment seeds are used for training, MGCEA shows a clear advantage over state-of-the-art methods by efficiently utilizing limited data. This observation further emphasizes the robustness and effectiveness of MGCEA in dealing with limited pre-aligned data, allowing it to achieve competitive performance.

5   Conclusion and Future Work

In this paper, a novel model called MGCEA is proposed to solve the multi-modal entity alignment problem with full consideration of feature modeling across modalities. Primarily, MGCEA utilizes the adjacency matrix to guide multi-modal representation learning for obtaining cross-modal enhancement representations, and an attention-based network is then employed to learn the importance of various modalities for joint representations, which is followed by a joint contrast loss for the objective. The extensive experimental evaluations validate the state-of-the-art performance of MGCEA on two publicly available datasets. In future work, we can utilize more auxiliary information (e.g., video and audio) to further improve the alignment performance.

Acknowledgements. This work is supported by the National Natural Science Foundation of China No. 62272332, the Major Program of the Natural Science Foundation of Jiangsu Higher Education Institutions of China No. 22KJA520006.

References

1. Bordes, A., Usunier, N., García-Durán, A., Weston, J., Yakhnenko, O.: Translating embeddings for modeling multi-relational data. In: NIPS, pp. 2787–2795 (2013)
2. Cao, Y., Wang, X., He, X., Hu, Z., Chua, T.: Unifying knowledge graph learning and recommendation: towards a better understanding of user preferences. In: WWW 2019: The World Wide Web Conference, pp. 151–161 (2019)
3. Chami, I., Ying, Z., Ré, C., Leskovec, J.: Hyperbolic graph convolutional neural networks. In: NIPS, pp. 4869–4880 (2019)
4. Chen, L., Li, Z., Wang, Y., Xu, T., Wang, Z., Chen, E.: MMEA: entity alignment for multi-modal knowledge graph. In: KSEM, vol. 12274, pp. 134–147 (2020)
5. Chen, L., et al.: Multi-modal siamese network for entity alignment. In: SIGKDD, pp. 118–126 (2022)
6. Chen, M., Tian, Y., Yang, M., Zaniolo, C.: Multilingual knowledge graph embeddings for cross-lingual knowledge alignment. In: IJCAI, pp. 1511–1517 (2017)
7. Cheng, B., Zhu, J., Guo, M.: MultiJAF: multi-modal joint entity alignment framework for multi-modal knowledge graph. Neurocomputing **500**, 581–591 (2022)
8. Ferrada, S., Bustos, B., Hogan, A.: IMGpedia: a linked dataset with content-based analysis of wikimedia images. In: ISWC, vol. 10588, pp. 84–93 (2017)
9. Guo, H., Tang, J., Zeng, W., Zhao, X., Liu, L.: Multi-modal entity alignment in hyperbolic space. Neurocomputing **461**, 598–607 (2021)
10. Han, X., Liu, Z., Sun, M.: Neural knowledge acquisition via mutual attention between knowledge graph and text. In: AAAI, pp. 4832–4839 (2018)
11. He, F., et al.: Unsupervised entity alignment using attribute triples and relation triples. In: DASFAA, vol. 11446, pp. 367–382 (2019)
12. Ji, H., Ke, P., Huang, S., Wei, F., Zhu, X., Huang, M.: Language generation with multi-hop reasoning on commonsense knowledge graph. In: EMNLP, pp. 725–736 (2020)
13. Kipf, T.N., Welling, M.: Semi-supervised classification with graph convolutional networks. CoRR arXiv:1609.02907 (2016)
14. Lin, Z., Zhang, Z., Wang, M., Shi, Y., Wu, X., Zheng, Y.: Multi-modal contrastive representation learning for entity alignment. In: COLING, pp. 2572–2584 (2022)
15. Liu, F., Chen, M., Roth, D., Collier, N.: Visual pivoting for (unsupervised) entity alignment. In: AAAI, pp. 4257–4266 (2021)
16. Liu, Y., Li, H., García-Durán, A., Niepert, M., Oñoro-Rubio, D., Rosenblum, D.S.: MMKG: multi-modal knowledge graphs. In: ESWC, vol. 11503, pp. 459–474 (2019)

17. Pei, S., Yu, L., Hoehndorf, R., Zhang, X.: Semi-supervised entity alignment via knowledge graph embedding with awareness of degree difference. In: WWW 2019: The World Wide Web Conference, pp. 3130–3136 (2019)

18. Qi, D., Chen, S., Sun, X., Luan, R., Tong, D.: A multiscale convolutional gragh network using only structural information for entity alignment. Appl. Intell. **53**(7), 7455–7465 (2023)

19. Simonyan, K., Zisserman, A.: Very deep convolutional networks for large-scale image recognition. In: ICLR (2015)

20. Sun, Z., Hu, W., Zhang, Q., Qu, Y.: Bootstrapping entity alignment with knowledge graph embedding. In: IJCAI, pp. 4396–4402 (2018)

21. Sun, Z., et al.: Knowledge graph alignment network with gated multi-hop neighborhood aggregation. In: AAAI, pp. 222–229 (2020)

22. Trisedya, B.D., Qi, J., Zhang, R.: Entity alignment between knowledge graphs using attribute embeddings. In: AAAI, pp. 297–304 (2019)

23. Wang, Z., Lv, Q., Lan, X., Zhang, Y.: Cross-lingual knowledge graph alignment via graph convolutional networks. In: EMNLP, pp. 349–357 (2018)

24. Wei, Y., Wang, X., Nie, L., He, X., Hong, R., Chua, T.: MMGCN: multi-modal graph convolution network for personalized recommendation of micro-video. In: ACM MM, pp. 1437–1445 (2019)

25. Wu, F., Jr., A.H.S., Zhang, T., Fifty, C., Yu, T., Weinberger, K.Q.: Simplifying graph convolutional networks. In: ICML, vol. 97, pp. 6861–6871 (2019)

26. Yang, H., Zou, Y., Shi, P., Lu, W., Lin, J., Sun, X.: Aligning cross-lingual entities with multi-aspect information. In: EMNLP, pp. 4430–4440 (2019)

27. Zhu, H., Xie, R., Liu, Z., Sun, M.: Iterative entity alignment via joint knowledge embeddings. In: IJCAI, pp. 4258–4264 (2017)

28. Zhu, Y., Zhang, C., Ré, C., Fei-Fei, L.: Building a large-scale multimodal knowledge base for visual question answering. CoRR arXiv:1507.05670 (2015)

Subgraph Federated Learning with Global Graph Reconstruction

Zhi Liu[1], Hanlin Zhou[1], Feng Xia[2], Guojiang Shen[1], Vidya Saikrishna[3], Xiaohua He[1], Jiaxin Du[1(✉)], and Xiangjie Kong[1]

[1] Zhejiang University of Technology, Hangzhou 310014, China
jiaxin.joyce.du@gmail.com
[2] RMIT University, Melbourne, VIC 3000, Australia
[3] Federation University, Ballarat, VIC 3353, Australia

Abstract. Missing cross-subgraph information is a unique problem in Subgraph Federated Learning (SFL) and severely affects the performance of the learned model. Existing cutting-edge methods typically allow clients to exchange data with all other clients to predict missing neighbor nodes. However, such client-to-client data exchanges are highly complex and lead to expensive communication overhead. In this paper, we propose FedGGR: subgraph federated learning with global graph reconstruction. FedGGR is a practical and effective framework. Specifically, the core idea behind it is to directly learn a global graph on the server by a graph structure learning module instead of predicting the missing neighbors on each client. Compared to existing methods, FedGGR does not require any data exchange among clients and achieves remarkable enhancements in model performance. The experimental results on four benchmark datasets show that the proposed method excels with other state-of-the-art methods. We release our source code at https://github.com/poipoipoi233/FedGGR.

Keywords: Federated learning · Subgraph federated learning · Graph neural networks

1 Introduction

Graphs are prevalent to represent the interactions among entities in many real-world scenarios, such as social networks [1], traffic networks [2], and molecules [3]. To mine graph data, modern approaches generally resort to Graph Neural Networks (GNNs) [4]. Following a message-passing mechanism to aggregate each node's neighbor information and create node embedding for all nodes, GNNs can achieve outstanding performance in various graph mining tasks.

Most existing studies centrally train GNNs on a single complete graph, where node features and all connection information are gathered on a central server [5]. However, in certain real-world scenarios, each subgraph of such a complete graph may be stored at different data owners and only accessible locally. Consider an

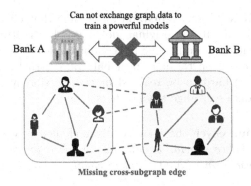

Fig. 1. A motivation scenario of subgraph federated learning. Two banks keep respective customer relation graphs for an identical graph mining task. However, they cannot share graph data to train a model collaboratively due to privacy concerns. The dashed red lines represent missing cross-subgraph edges. (Color figure online)

example in the financial domain, as shown in Figure 1. Two banks maintain their respective customer relation graphs, both aiming to address an identical task, such as fraud detection. However, due to privacy concerns or interest conflicts, each bank cannot share its graph data with others and merge them into a large graph for collaborative training. Therefore, such graph islands problem severely hinders the deployment of GNNs in practice.

Federated Learning (FL) [6] has experienced substantial progress in the past few years. It allows data owners, also known as clients, to train a global model collaboratively without sharing local data [7]. Employing FL to train GNNs in distributed subgraph scenarios, so-called Subgraph Federated learning (SFL), seems the simplest way to achieve collaborative training across different subgraph owners.

Nevertheless, training GNNs within federated subgraph scenarios introduces a unique challenge: *dealing with missing cross-subgraph neighbors and edges*. Specifically, nodes in each subgraph can potentially connect with nodes in other subgraphs, but no clients record such important cross-subgraph links. This issue is not trivial since these missing edges and potential neighbor nodes in other subgraphs may carry valuable information. Several recent studies have demonstrated that simply ignoring the missing cross-subgraph information will significantly impair the model accuracy of SFL [8,9].

Prior Work. Many cutting-edge works have made efforts to deal with the cross-subgraph information. For instance, FedSage+ [10] and FedNI [11] enable each client to predict cross-subgraph neighbors for local subgraph augmentation. Despite their proven effectiveness, these methods suffer from the drawback that clients must exchange large amounts of additional data (e.g., all node embeddings, model weights, and gradients) with all other clients. It incurs expensive communication overheads that grow exponentially with the number of clients.

Additionally, FedGCN [9] and FedGraph [8] assume that clients or server has detailed information about the original global graph structure. Therefore, in these methods, clients can acquire the node embeddings of their cross-subgraph neighbors directly based on the known global graph structure. Unfortunately, such assumptions are hard to be realized in reality. Moreover, these methods still require direct data exchange among clients.

Research Problem. Given the above realistic challenge, we propose the following question: *Can we provide a practical approach to learning the missing cross-subgraph information without client-to-client data exchange?*

Present Work. In this paper, we propose a subgraph federated learning framework named FedGGR. We designed it based on a simple and practical idea: directly learning the underlying global graph on the server side instead of predicting missing neighbor nodes on each client. Specifically, to realize this idea, we design and develop a global graph structure learning (GSL) module on the server to learn an optimal global graph structure and global node embeddings for all nodes. Each learned global node embedding would contain cross-subgraph knowledge and be sent to the corresponding client for local subgraph augmentation. Compared to existing works, this novel design avoids client-to-client data exchange and can fully explore potential cross-subgraph information to provide performance gain. However, the attendant challenge is that the standard FL framework does not support server-side model training. To further overcome this challenge, we draw inspiration from split learning [12] to design a training procedure. Concretely, FedGGR aggregates the gradient of the global node embeddings to update the server-side model and uses FedAvg to update the model on the client side.

We summarise our contributions in mainly three aspects:

- We propose a subgraph federated learning framework FedGGR, which can effectively reconstruct the missing cross-subgraph information. Moreover, it overcomes the limitations of relevant works that require client-to-client data exchange.
- We novelly propose a global GSL module on a server to learn the global graph directly. Besides, we further design a split learning-based training procedure to optimize the module. To the best of our knowledge, it is the first systematic attempt to address missing cross-subgraph information via graph structure learning.
- We conduct extensive experiments on four benchmark graph datasets. The results show that our FedGGR outperforms the state-of-the-art baselines.

2   Related Work

2.1   Subgraph Federated Learning (SFL)

An increasing number of studies on SFL have been published in recent years. For instance, FedSage+ [10] is a cornerstone study. In FedSage+, each client first requests additional data from other clients for subgraph augmentation. Then clients take the augmented subgraph to train a GNN model in the FL manner. Besides, FEDPUB [13] and FedEgo [14] focus on designing personalized FL strategies to address subgraph data heterogeneity among clients. Despite their effectiveness, these methods do not involve handling missing edges explicitly. As a result, performance degradation may inevitably occur. Another research direction is optimizing model convergence and communication overhead in the SFL setting. The representative works include FedGCN [9] and FedGraph [8]. However, these methods typically assume that clients or servers have information about all cross-subgraph edges and neighbors. Although the above works have provided valuable insights into SFL scenarios, reconstructing missing cross-subgraph information without exchanging data among clients remains a research gap. In this work, we aim to use our FedGGR to address this non-trivial challenge.

2.2   Graph Structure Learning (GSL)

The message-passing mechanism enables GNNs to take full advantage of the given graph structure information to learn the node embeddings. However, it also leads to GNNs being highly sensitive to the quality of the given graph structures [15]. This challenge motivates the development of graph structure learning [16], which aims to simultaneously learn an optimal graph structure and GNN parameters to obtain better performance on graph mining tasks. Certain studies, such as [17,18], have also incorporated GSL technologies into the federated setup. Nevertheless, no prior study has utilized GSL to address the issue of missing cross-subgraph information.

2.3   Split Learning

Split learning [12] is a distributed machine learning paradigm. Different from FL, the critical idea of split learning is to split a large neural network into multiple sub-networks, and each sub-network is trained on different parties [19]. In detail, the training of each sub-network is based on the exchange of the intermediate results (e.g., embedding) among participants. Several works have made progress toward combining split learning into FL. The application case includes federated spatiotemporal modeling [20] and federated news recommendation [21]. In this paper, we use an approach based on split learning to optimize the model parameter of our server-side global graph structure learning model.

3   Problem Setting

Consider $\mathcal{G} = (\mathcal{V}, \boldsymbol{X}, \boldsymbol{A})$ as a global graph, where $\mathcal{V}$ is a set of $N = |\mathbf{V}|$ nodes, $\boldsymbol{X} \in \mathbb{R}^{N \times f}$ is the node feature matrix, and $\boldsymbol{A} \in \mathbb{R}^{N \times N}$ is the adjacency matrix that describes the pairwise relationship among nodes.

In SFL scenario involve m clients, each client $i \in [1, m]$ holds a small subgraph $\mathcal{G}_i = (\mathcal{V}_i, \boldsymbol{X}_i, \boldsymbol{A}_i)$ of the global graph $\mathcal{G}$ and all cross-subgraph edges present in the global graph are missing. A central server with sufficient computing power will coordinate clients for collaborative training. In this work, our goal is to design a practical SFL framework to deal with the performance degradation caused by missing cross-subgraph information. We assume that nodes of different subgraphs are disjoint and choose semi-supervised node classification as our downstream task, which is widely used in the performance evaluation of SFL frameworks.

4   Methodology

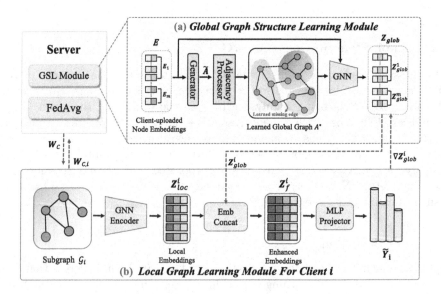

Fig. 2. Proposed FedGGR. Before federated training, clients upload compressed node embeddings to the server (described in Sect. 4.2). We present a detailed training pipeline in Sect. 4.5.

4.1   Framework Overview

We begin by giving a general overview of the framework. Figure 2 shows two core modules of our framework: the *global GSL module* on the server side and the *local graph learning module* on the client side.

More specifically, the global GSL module aims to directly learn the global graph structure and generate a global node embedding matrix Z_{glob}, which incorporates cross-subgraph neighbor information. Then, the server will send each sub-matrix Z_{glob}^i to the corresponding client i as supplementary knowledge. In comparison, the local graph learning module aims to use a GNN model to learn the local node embeddings on their subgraph and then fuse the local node embeddings with the global node embeddings to perform downstream tasks.

Overall, our design has the following advantages. Firstly, directly learning the global graph on the server is an effective and easy-to-implement way to reconstruct the missing cross-subgraph information, which does not require clients to exchange any additional data with other clients. Secondly, fusing the global and local node embeddings will enhance each node's representation ability, which can significantly boost the performance of downstream tasks (demonstrated in Sect. 5.3).

4.2 Local Pre-training

Here, we first introduce a simple pre-training step before federated training. As previously noted, we aim to learn the global graph on the server directly. However, existing GSL technologies need the feature vector of each node as the starting point for training.

To address the above challenge, we propose to send each node's compressed embedding to the server. Technically, each client first trains a simple Graph Auto Encoder (GAE) [22] variant based on its local subgraph data. Subsequently, the trained GAE can compress each node's feature vector into a low-dimensional node embedding for uploading.

Take the client i as an example. Given a graph feature matrix $X_i \in \mathbb{R}^{|\mathcal{V}_i| \times F}$, the forward process of our GAE is as follows:

$$E_i = \text{Encoder}(X_i, A_i), X_i^{rec} = \text{Decoder}(E_i, A_i), \tag{1}$$

where $E_i \in \mathbb{R}^{|\mathcal{V}_i| \times p}$ denotes the compressed node embedding matrix. $X_i^{rec} \in \mathbb{R}^{|\mathcal{V}_i| \times F}$ is the reconstructed node feature matrix. For simplicity, both Encoder and Decoder are two-layer GCN [23]. We optimize them by minimizing the mean squared error loss between X_i and X_i^{rec}:

$$\mathcal{L}_i^{rec} = \|X_i, X_i^{rec}\|_F \tag{2}$$

After pre-training, each client i will use the encoder of trained GAE to generate E_i and upload it to the server.

4.3 The Local Graph Learning Module

Given a client i with subgraph $\mathcal{G}_i$, the local graph learning module first uses a GNN-based encoder f_θ to extract the local node embeddings:

$$Z^i_{loc} = f_\theta(\mathcal{G}_i) = f_\theta(\boldsymbol{X}_i, \boldsymbol{A}_i), \tag{3}$$

where θ is the parameters of f_θ, $Z^i_{loc} \in \mathbb{R}^{|\mathcal{V}_i| \times F_l}$ is the local node embedding matrix, and F_l is the dimension number. In FedGGR, we choose a two-layer GCN [23] as the encoder for a fair comparison with baseline methods.

To provide each node with knowledge about its missing cross-subgraph neighbor, we concatenate the local node embedding matrix Z^i_{loc} with the global node embedding matrix $Z^i_{glob} \in \mathbb{R}^{|\mathcal{V}_i| \times F_g}$ generated from the global GSL module (described in Sect. 4.4). We denote the enhanced node embedding matrix as $Z^i_f \in \mathbb{R}^{|\mathcal{V}_i| \times q}$, where $q = F_l + Fg$.

Finally, we use a one-layer MLP projector with the softmax function to obtain the final classification result:

$$\widetilde{\boldsymbol{Y}}_i = \text{SoftMax}(\text{MLP}(Z^i_f)), \tag{4}$$

where the final output $\widetilde{\boldsymbol{Y}}_i \in \mathbb{R}^{|V_i| \times c}$ denotes the label prediction, and c denotes the number of classes. We use the cross-entropy loss function as the local loss function for each client i:

$$\mathcal{L}_i = \text{CrossEntropyLoss}(\widetilde{\boldsymbol{Y}}_i, \boldsymbol{Y}_i), \tag{5}$$

where Y_i denotes the set of labeled nodes on the subgraph $\mathcal{G}_i$.

4.4 The Global Graph Structure Learning Module

The global GSL module, as shown in Fig. 2 (a), consists of three components: a graph generator, an adjacency processor, and a GNN model.

Graph Generator. It takes the embedding matrix $\boldsymbol{E} = \text{CONCAT}(\boldsymbol{E}_1, \cdots, \boldsymbol{E}_m)$ as input, where $\boldsymbol{E}_i$ is generated by the local pre-training step. The output of the graph generator is an initial global adjacency matrix $\widetilde{\boldsymbol{A}}$. We choose the MLP-D generator proposed by [16] as the graph generator in our FedGGR. Formally, the MLP-D generator is defined as:

$$\widetilde{\boldsymbol{A}} = \text{KNN}(\text{MLP}(\boldsymbol{E})), \tag{6}$$

where KNN is the K nearest neighbor algorithm, MLP is a two-layer MLP projector. The weight of this MLP is zero, except for the primary diagonal.

Adjacency Processor. Following [16], we use the adjacency processor to make the initial global adjacency matrix normalized and symmetric. Specifically, the adjacency processor performs the following transformation on the adjacency matrix $\widetilde{\boldsymbol{A}}$ to obtain a refined global adjacency matrix $\boldsymbol{A}^*$:

$$\boldsymbol{A}^* = \frac{1}{2}\boldsymbol{D}^{-\frac{1}{2}}(P(\widetilde{\boldsymbol{A}}) + P(\widetilde{\boldsymbol{A}})^T)\boldsymbol{D}^{-\frac{1}{2}}, \tag{7}$$

where $\boldsymbol{D}$ is the degree matrix of $\widetilde{\boldsymbol{A}}$, and P is the ELU [24] activation function.

Global GNN Model. The inputs of the global GNN model are the refined global adjacency matrix $\boldsymbol{A}^*$ and the node embedding matrix $\boldsymbol{E}$. It outputs the corresponding global embedding for each node in the global graph. The forward propagation process of the global GNN model is as follows:

$$\boldsymbol{Z}_{glob} = f_\varphi(\boldsymbol{E}, \mathbf{A}^*), \tag{8}$$

where $\boldsymbol{Z}_{glob} \in \mathbb{R}^{|\mathcal{V}| \times F_g}$ is the global node embedding matrix, F_g is its dimension number. $f_\varphi(\cdot)$ is a GNN encoder with the parameter φ. We also choose a two-layer GCN as the GNN encoder here. In $\boldsymbol{Z}_{glob}$, the submatrix $\boldsymbol{Z}^i_{glob}$ contains the global node embeddings corresponding to the nodes of the client i. Each global node embedding has already integrated cross-subgraph information thanks to the learned global adjacency matrix and the message-passing mechanism of GNNs. The server will send each $\boldsymbol{Z}^i_{glob}$ to the corresponding client i.

4.5 Objective and Training Procedure

In this subsection, we state our training objective: We look for a set of optimal parameters (W_C^*, W_S^*) that can minimize the classification losses of all clients.

$$(W_C^*, W_S^*) = \underset{(W_C, W_S)}{\arg\min} \frac{1}{M} \sum_{i=1}^{M} \frac{|\mathcal{V}_i|}{|\mathcal{V}|} \mathcal{L}_i(\mathcal{G}_i; W_C; W_S), \tag{9}$$

where $\mathcal{L}_i$ represents the local loss of client i. W_C and W_S denote the learnable parameters in the local graph learning and global GSL modules, respectively.

The training procedure consists of two parts: (1) clients pre-train a GAE model. Subsequently, they use the trained GAE to generate compressed node embeddings, which they then upload to the server. (2) In each communication round, FedGGR employs the FedAvg [6] algorithm to update the W_C collaboratively, and the server updates W_S by aggregating the gradients $\nabla \boldsymbol{Z}_{glob}$ of the global node embeddings $\boldsymbol{Z}_{glob}$. We present a detailed training workflow in Algorithm 1.

Algorithm 1. The training algorithm of FedGGR

Input: $\mathcal{G}_i$, $i = 1, \cdots, m$, learning rate η, number of nearest neighbors k
Output: Trained W_S and W_C
Client Pre-training: each client i pre-trains a GAE model and then uploads the compressed node embedding E_i to the server.
Server Executes:
  1: Initialize:$W_S^{(0)}$ and $W_C^{(0)}$;
  2: $E \leftarrow \text{CONCAT}(E_1, \cdots, E_M)$;
  3: **for** each round $t = 1, 2, \cdots$ **do**
  4:    $Z_{glob} \leftarrow \text{GlobalGSL}(E, k)$ // Eq. (6), (7), and (8);
  5:    **for** each client i **in parallel do**
  6:       $W_{C,i}^{(t)}, \nabla Z_{glob}^i \leftarrow \text{ClientTraining}(i, W_C^{(t)}, Z_{glob}^i)$
  7:    **end for**
  8:    $\nabla Z_{glob} \leftarrow \text{CONCAT}(\nabla Z_{glob}^1, \cdots, \nabla Z_{glob}^m)$;
  9:    $\nabla W_S^{(t)} \leftarrow Z_{glob}.backward(\nabla Z_{glob})$;
 10:    $W_C^{(t+1)} \leftarrow \sum_i \frac{|V_i|}{|V|} W_{C,i}^{(t)}$;
 11:    $W_S^{(t+1)} \leftarrow W_S^{(t)} - \eta \nabla W_S^{(t)}$;
 12: **end for**
ClientTraining$(i, W_C^{(t)}, Z_{glob}^i)$:
  1: $W_{C,i}^{(t)} \leftarrow W_C^{(t)}$;
  2: **for** each local epoch $j = 1, \cdots, Q$ **do**
  3:    $\mathcal{L}_i \leftarrow$ Eq. (5) with Z_{glob}^i
  4:    $\nabla W_{C,i}^{(t)}, \nabla Z_{glob}^i \leftarrow \mathcal{L}_i.backward()$
  5:    $W_{C,i}^{(t)} \leftarrow W_{C,i}^{(t)} - \eta \nabla W_{C,i}^{(t)}$
  6: **end for**
  7: return $W_{C,i}^{(t)}$ and ∇Z_{glob}^i to server

---

5   Experiment

In this section, we conduct experiments to verify our proposed FedGGR. We run all experiments on FederatedScope-GNN (FS-G) [25], a comprehensive Python package for federated graph learning. We aim to answer three questions:

1. **RQ1:** How does FedGGR perform compared with the state-of-the-art SFL baseline methods?
2. **RQ2:** Are the learned global graph and the global node embeddings really useful?
3. **RQ3:** How do critical hyper-parameters impact the performance of FedGGR?

5.1   Experimental Setups

Datasets and Graph Partitioning. We test FedGGR on four benchmark datasets, i.e., Cora [26], CiteSeer [26], PubMed [27], and Amazon-Computers [28]. The dataset statistics are shown in Table 1. We split the complete global graph into 3, 5, and 10 subgraphs for all datasets using the *random_splitter*, a graph partitioning algorithm provided by FS-G [25]. Specifically, the *random_splitter* will randomly split the set of nodes in the original graph into m subsets and then deduce subgraphs based on the node subsets. It should be noted that the *random_splitter* will produce more missing cross-subgraph edges and decrease the homophilic degree in each subgraph compared to the community detection-based graph partitioning algorithm (e.g., Louvain [29]). We also specify that the nodes of each subgraph are non-overlapping since it will produce the highest number of missing edges.

Baselines. We make comparisons with various baselines to demonstrate the performance of our FedGGR. The chosen baselines are as follows:

1. **Typical GNNs**: We use the FedAvg [6] to train several representative GNNs, i.e., GCN [30], GAT [31], and GarphSAGE [32] in the SFL setting.
2. **FedSage+** [10]: This method is the most representative SFL baseline. The basic idea of this method is to generate the missing neighbors on each client with the help of exchanging additional data among clients.
3. **GCLF+** [33]: This baseline is a well-known personalized FL method designed for the graph-level FL setting. In GCFL+, the server clusters the clients according to the gradient sequence of each client's local model, and the server will generate multiple global models for each client cluster. We slightly adapt this method to make it applicable to our SFL setting.
4. **FED-PUB** [13]: It is a state-of-the-art SFL baseline. In this method, each client's local GNN takes a random graph as input to compute the functional embeddings. Then, each client uploads the functional embeddings to the server. The server then calculates the subgraph similarity matrix according to the function embeddings and generates customized models for each client based on the similarity matrix.

Table 1. Graph Data Statistics

Dataset	Node	Edge	Feature	Classes
Cora	2,708	10,556	1,433	7
CiteSeer	3,327	9,104	3,703	6
PubMed	19,717	88,648	500	3
Computers	13,752	491,722	767	10

Implementation Details. Following [25,34], we divide the set of nodes into train/valid/test sets, with a ratio of 60:20:20. We search the optimal number of hidden units for GCN within $\{64, 128, 256\}$. The F_l and F_g are selected in the range of $\{64, 128, 256\}$. For the graph generator in the global GSL module, we vary the number of neighbors k over $\{5, 10, ..., 40\}$. For training, we use the Adam optimizer to optimize all model parameters. We vary the learning rate

$\eta \in \{1e^{-3}, 1e^{-2}\}$ and the number of local epochs $Q \in \{1, 2, 3\}$. The initial hyper-parameter settings of baseline methods are based on the original papers and official public code. We further perform an optimal hyper-parameter search for all baseline methods to make a fair comparison.

Table 2. Comparison results on representative node classification datasets with *random splitter*. The reported results are mean accuracy with standard deviation on the test sets over five different runs.

Method	Cora			CiteSeer		
	3 Clients	5 Clients	10 Clients	3 Clients	5 Clients	10 Clients
GCN	82.3 0.5	80.5 0.8	78.7 0.6	76.6 0.5	75.4 0.6	76.3 0.8
GAT	82.6 0.5	80.1 0.3	78.5 0.6	76.8 0.5	75.7 0.5	76.0 0.8
GraphSAGE	83.2 0.5	80.2 0.4	78.8 0.7	76.1 0.2	74.9 0.3	73.8 0.2
GCFL+	83.4 0.6	80.1 0.4	78.4 1.3	77.4 0.6	76.1 0.1	73.9 0.6
FedSage+	83.7 0.6	81.1 0.2	77.2 0.2	77.3 0.4	75.6 0.5	75.4 0.8
FED-PUB	83.9 0.6	81.1 1.1	76.2 0.8	76.9 0.7	76.7 0.4	76.7 0.4
FedGGR(ours)	84.1 0.6	81.8 1.1	79.5 0.6	79.0 0.9	77.6 0.9	77.0 0.7
Method	PubMed			Amazon-Computers		
	3 Clients	5 Clients	10 Clients	3 Clients	5 Clients	10 Clients
GCN	85.6 0.6	84.8 1.0	85.4 1.2	88.4 0.9	87.6 0.8	83.8 0.2
GAT	84.6 0.6	85.0 1.1	84.4 0.1	87.9 0.5	86.6 0.5	85.3 0.7
GraphSAGE	85.8 1.6	85.7 0.6	88.6 0.1	85.1 0.3	84.0 0.3	83.7 0.8
GCFL+	85.5 0.3	85.9 0.5	86.3 0.1	85.8 0.4	85.2 0.1	84.9 0.1
FedSage+	87.4 0.4	86.0 1.5	86.2 1.1	87.2 0.5	86.7 0.3	85.1 0.1
FED-PUB	89.7 0.3	89.4 0.1	88.8 0.3	89.1 0.6	88.9 0.8	86.2 0.8
FedGGR(ours)	90.0 0.1	89.5 0.2	89.1 0.4	89.8 0.1	89.0 0.2	86.8 0.2

5.2 Comparison with State-of-the-art Methods (RQ1)

Table 2 presents the node classification performance of our FedGGR and baseline methods. Overall, we have the following observations.

Firstly, our FedGGR consistently outperforms all baselines on four benchmark datasets. Notably, FedGGR does not employ additional strategies to address the data heterogeneity for simplicity. However, it can be easily integrated with existing personalized federated learning methods to improve performance further.

Secondly, the performance of all methods decreases with the increase in the number of clients on most datasets. This phenomenon is reasonable since the number of missing edges and the heterogeneity between subgraphs increase with the number of clients.

Finally, we observe that the performance of some methods does not decrease but fluctuates as the number of clients increases on the PubMed dataset. This unusual phenomenon may be because some unnecessary noisy edges are removed during the subgraph partitioning process.

5.3   Ablation Study (RQ2)

We do an ablation study on the global GSL module in this subsection. In fact, the GCN trained by FedAvg can be seen as a variant of FedGGR without the global GSL module. As shown in Table 2, FedGGR performs better than the GCN by an average of 2.7% on the four datasets. This observation demonstrates that the FedGGR significantly benefits from the global GSL module.

For further demonstration, we build variants of FedGGR by replacing the GNN encoders of the local graph learning module with other representative GNNs, i.e., GAT [31] and GPR-GNN [34]. Figure 3 presents the test accuracy of these variants on the CiteSeer dataset with and without the global GSL module. It shows that the global GSL module can provide considerable performance gains regardless of the backbone model.

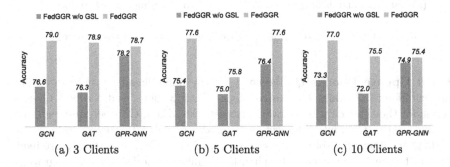

Fig. 3. Node classification accuracy of FedGGR and its variant on the CiteSeer dataset.

5.4   Sensitivity Analysis (RQ3)

In this subsection, we analyze the sensitivity of critical hyper-parameters in our FedGGR. Specifically, the analyzed hyper-parameters include the number of neighbors k in KNN post-processing of the global GSL module, the dimension number of the global and local embeddings F_l and F_g, and the number of local update epochs Q.

Number of Neighbors k. This hyperparameter determines the number of neighbors for each node when generating the global graph structure. We change the value of k in the $\{5, 10, ..., 40\}$ under different datasets and varied numbers of clients. We have two main observations. Firstly, as shown in Fig. 4(a), the best choice of k is different under different numbers of clients. Secondly, Fig. 4(b) shows that the change of k has an insignificant effect on the final test accuracy. This observation differs from the typical results observed in centralized graph structure learning works, where an unsuitable k will significantly reduce the accuracy. For instance, an inappropriate k value causes a drop in accuracy of more than 5% for the Cora dataset in work [16]. A explanation for this observation is that the local node embeddings in FedGGR mitigate the harmful effects of bad global embeddings generated by setting an inappropriate k.

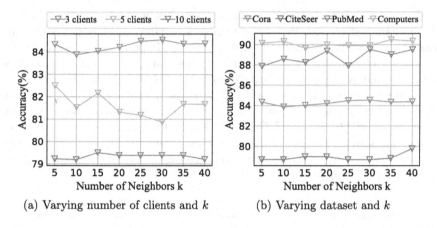

(a) Varying number of clients and k (b) Varying dataset and k

Fig. 4. The node classification accuray about k under different settings.

Local Update Epochs Q. We compare the node classification performance under different Q values with three clients. Table 3 shows that the best Q value is one for the three citation network datasets, i.e., Cora, CiteSeer, and PubMed. While for the Computer dataset, setting the Q value to three produces the highest accuracy. Notably, increasing the number of local updates is more likely to lead to overfitting, especially when the number of nodes in each subgraph is small. We recommend that readers tune this hyperparameter according to the dataset size for better performance.

Table 3. Comparison results by varying the number of local update epochs.

Datasets	Cora	CiteSeer	PubMed	Computers
$Q=1$	**0.841**	**0.790**	**0.900**	0.863
$Q=2$	0.827	0.788	0.893	0.890
$Q=3$	0.826	0.789	0.887	**0.898**

The Output Dimension F_l and F_g. We investigate the impact of different combinations of F_l and F_g on accuracy. Figure 5 shows the results on the PubMed dataset. We see that the accuracy is relatively stable when changing these two hyper-parameters, demonstrating the robustness of our FedGGR. Compared to F_l, FedGGR is less sensitive to the selection of F_g. In addition, we observe that the performance tends to be higher when the F_l is equal to F_g.

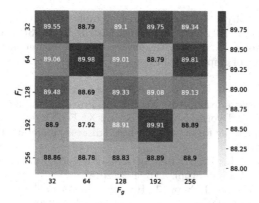

Fig. 5. Accuracy w.r.t. combinations of F_l and F_g on the PubMed dataset with three clients.

6 Conclusion

In this paper, we focus on dealing with missing cross-subgraph information in subgraph federated learning. We propose FedGGR, a quite practical subgraph federated learning framework. It enables the server to directly learn the global graph using a graph structure learning module and a split learning-based training procedure. Compared to the relevant methods, FedGGR not only eliminates the need for direct data exchange among clients but also offers remarkable performance gains in model accuracy. Extensive experiments on the four benchmark datasets validate the effectiveness of our proposed method.

References

1. Kong, X., Gao, H., Shen, G., Duan, G., Das, S.K.: FedVCP: a federated-learning-based cooperative positioning scheme for social internet of vehicles. IEEE Trans. Comput. Soc. Syst. **9**(1), 197–206 (2021)
2. Hou, M., Xia, F., Gao, H., Chen, X., Chen, H.: Urban region profiling with spatio-temporal graph neural networks. IEEE Trans. Comput. Soc. Syst. **9**(6), 1736–1747 (2022)
3. Xia, J., et al.: Mole-BERT: rethinking pre-training graph neural networks for molecules. In: The Eleventh International Conference on Learning Representations, pp. 1–18 (2023)
4. Zhou, J., et al.: Graph neural networks: a review of methods and applications. AI Open **1**, 57–81 (2020)
5. Xia, F., et al.: CenGCN: centralized convolutional networks with vertex imbalance for scale-free graphs. IEEE Trans. Knowl. Data Eng. **35**(5), 4555–4569 (2022)
6. McMahan, B., Moore, E., Ramage, D., Hampson, S., Arcas, B.A.: Communication-efficient learning of deep networks from decentralized data. In: Artificial Intelligence and Statistics, pp. 1273–1282. PMLR (2017)
7. Kong, X., et al.: A federated learning-based license plate recognition scheme for 5G-enabled internet of vehicles. IEEE Trans. Ind. Inf. **17**(12), 8523–8530 (2021)

8. Chen, F., Li, P., Miyazaki, T., Wu, C.: Fedgraph: federated graph learning with intelligent sampling. IEEE Trans. Parallel Distrib. Syst. **33**(8), 1775–1786 (2021)

9. Yao, Y., Joe-Wong, C.: FedGCN: convergence and communication tradeoffs in federated training of graph convolutional networks, pp. 1–31 (2022). arXiv preprint arXiv:2201.12433

10. Zhang, K., Yang, C., Li, X., Sun, L., Yiu, S.M.: Subgraph federated learning with missing neighbor generation. Adv. Neural. Inf. Process. Syst. **34**, 6671–6682 (2021)

11. Peng, L., Wang, N., Dvornek, N., Zhu, X., Li, X.: FedNI: federated graph learning with network inpainting for population-based disease prediction. IEEE Trans. Med. Imaging 1–12 (2022)

12. Gupta, O., Raskar, R.: Distributed learning of deep neural network over multiple agents. J. Netw. Comput. Appl. **116**, 1–8 (2018)

13. Baek, J., Jeong, W., Jin, J., Yoon, J., Hwang, S.J.: Personalized subgraph federated learning, pp. 1–20 (2022). arXiv preprint arXiv:2206.10206

14. Zhang, T., Chen, C., Chang, Y., Shu, L., Zheng, Z.: FedEgo: privacy-preserving personalized federated graph learning with ego-graphs, pp. 1–25 (2022). arXiv preprint arXiv:2208.13685

15. Liu, Y., Zheng, Y., Zhang, D., Chen, H., Peng, H., Pan, S.: Towards unsupervised deep graph structure learning. In: Proceedings of the ACM Web Conference 2022, pp. 1392–1403 (2022)

16. Fatemi, B., El Asri, L., Kazemi, S.M.: SLAPS: self-supervision improves structure learning for graph neural networks. Adv. Neural. Inf. Process. Syst. **34**, 22667–22681 (2021)

17. Chen, F., Long, G., Wu, Z., Zhou, T., Jiang, J.: Personalized federated learning with a graph. In: Proceedings of the Thirty-First International Joint Conference on Artificial Intelligence, IJCAI-22, pp. 2575–2582 (2022)

18. Zhao, G., Huang, Y., Tsai, C.H.: FedGSL: federated graph structure learning for local subgraph augmentation. In: 2022 IEEE International Conference on Big Data (Big Data), pp. 818–824. IEEE (2022)

19. Thapa, C., Arachchige, P.C.M., Camtepe, S., Sun, L.: SplitFed: when federated learning meets split learning. In: Proceedings of the AAAI Conference on Artificial Intelligence, vol. 36, pp. 8485–8493 (2022)

20. Meng, C., Rambhatla, S., Liu, Y.: Cross-node federated graph neural network for spatio-temporal data modeling. In: Proceedings of the 27th ACM SIGKDD Conference on Knowledge Discovery & Data Mining, pp. 1202–1211 (2021)

21. Yi, J., Wu, F., Wu, C., Liu, R., Sun, G., Xie, X.: Efficient-FedRec: efficient federated learning framework for privacy-preserving news recommendation. In: Proceedings of the 2021 Conference on Empirical Methods in Natural Language Processing, pp. 2814–2824 (2021)

22. Kipf, T.N., Welling, M.: Variational graph auto-encoders, pp. 1–3 (2016). arXiv preprint arXiv:1611.07308

23. Kipf, T.N., Welling, M.: Semi-supervised classification with graph convolutional networks. In: 5th International Conference on Learning Representations, pp. 1–14 (2017)

24. Clevert, D., Unterthiner, T., Hochreiter, S.: Fast and accurate deep network learning by exponential linear units (ELUs). In: 4th International Conference on Learning Representations, pp. 1–14 (2016)

25. Xie, Y., et al.: FederatedScope: a flexible federated learning platform for heterogeneity. Proc. VLDB Endow. **16**(5), 1059–1072 (2023)

26. Sen, P., Namata, G., Bilgic, M., Getoor, L., Gallagher, B., Eliassi-Rad, T.: Collective classification in network data. AI Mag. **29**(3), 93–106 (2008)

27. Namata, G., London, B., Getoor, L., Huang, B., Edu, U.: Query-driven active surveying for collective classification. In: 10th International Workshop on Mining and Learning with Graphs, vol. 8, pp. 1–8 (2012)
28. Shchur, O., Mumme, M., Bojchevski, A., Günnemann, S.: Pitfalls of graph neural network evaluation, pp. 1–11 (2018). arXiv preprint arXiv:1811.05868
29. Blondel, V.D., Guillaume, J.L., Lambiotte, R., Lefebvre, E.: Fast unfolding of communities in large networks. J. Stat. Mech: Theory Exp. **2008**(10), P10008 (2008)
30. Wang, X., Zhu, M., Bo, D., Cui, P., Shi, C., Pei, J.: AM-GCN: adaptive multi-channel graph convolutional networks. In: Proceedings of the 26th ACM SIGKDD International Conference on Knowledge Discovery & Data Mining, pp. 1243–1253 (2020)
31. Velickovic, P., Cucurull, G., Casanova, A., Romero, A., Liò, P., Bengio, Y.: Graph attention networks. In: 6th International Conference on Learning Representations, pp. 1–12 (2018)
32. Hamilton, W., Ying, Z., Leskovec, J.: Inductive representation learning on large graphs. Adv. Neural. Inf. Process. Syst. **30**, 1–11 (2017)
33. Xie, H., Ma, J., Xiong, L., Yang, C.: Federated graph classification over MNon-IID graphs. Adv. Neural. Inf. Process. Syst. **34**, 18839–18852 (2021)
34. Chien, E., Peng, J., Li, P., Milenkovic, O.: Adaptive universal generalized pagerank graph neural network. In: 9th International Conference on Learning Representations, pp. 1–24 (2021)

SEGCN: Structural Enhancement Graph Clustering Network

Yuwen Chen[1], Xuefeng Yan[1,2]([✉]), Peng Cui[3], and Lina Gong[1]

[1] College of Computer Science and Technology, Nanjing University of Aeronautics
and Astronautics, Nanjing 211106, Jiangsu, China
yxf@nuaa.edu.cn
[2] Collaborative Innovation Center of Novel Software Technology and
Industrialization, Nanjing 210023, Jiangsu, China
[3] Institute of Combat Software and Simulation, Dalian Naval Academy, Dalian
116018, Liaoning, China

Abstract. Deep graph clustering, which reveals the intrinsic structure
and underlying relationship of graph node data, has become a highly concerning and challenging research task for graph-structured data. In recent
years, existing graph clustering methods have gotten better performance
without human guidance by combining auto-encoder and graph convolution networks. However, the existing methods exist problems: 1) the
aggregation of noise information in the graph structure information leads
to poor model learning effect; 2) the methods focus on learning the local
structure information of the graph, and fail to learn the global structure
information of the graph, resulting in incomplete learning of the graph.
To overcome the shortcomings, we propose a **Structural Enhancement
Graph Clustering Network(SEGCN)** to learn the graph topology information hidden in the dependencies of nodes attribute and the global
structure information. Specifically, SEGCN adopts the method of calculating correlation to extract the graph topology information hidden in
the dependencies of node attributes and design a dynamic fusion strategy
to enrich the original graph structure information. Meanwhile, SEGCN
enriches the embedding information extracted from the local structure by
fusing information on the global structure based on the global structure
dynamic fusion method. Moreover, to cluster the subsequent, SEGCN
applies the pseudo-cluster labels learned by a dual self-supervised strategy to guide the learning of the node attribute representation and optimize the weight matrix of networks. Extensive experimental results on
four benchmark datasets indicate that SEGCN achieves excellent performance on all datasets and accomplishes an accuracy of 71.56% and NMI
of 44.50% on the CITE dataset.

Keywords: graph clustering · structural enhancement · global
structure · convolutional networks · self-supervised

This work is supported by the Basic Research for National Defense under Grant
Nos.JCKY2020605C003.

X. Song et al. (Eds.): APWeb-WAIM 2023, LNCS 14331, pp. 174–190, 2024.
https://doi.org/10.1007/978-981-97-2303-4_12

1   Introduction

Graph data, such as the citation network data representing the citation relationship of papers and the social network data composed of mutual communication between people, are ubiquitous in the Internet era. Analyzing the implicit relations existing in graph structure data has become very popular in the field of artificial intelligence [10,13,24,27–29] such as graph clustering and link prediction. Among them, graph clustering is an important and challenging task for data analysis, whose purpose is to cluster the graph nodes into different clusters to maximize the similarity within the class and minimize the similarity outside the class. At present, the graph clustering methods introduce deep learning to learn feature representations with discriminative information from graph structure and attribute data to obtain advanced performance. Meanwhile, due to the limited manual annotation of graph structure data, unsupervised learning in deep graph clustering plays a crucial role, which has been one of the most representative clustering techniques.

Among existing graph clustering methods, it is extremely important and challenging to design a good objective function for optimizing models and to build an effective architecture so that neural networks can collect comprehensive and discriminative information and reveal the intrinsic structure. According to the different input data, they are mainly divided into three categories: (1) taking node attributes as input; (2) taking graph structure as input; (3) both node attributes and graph structure are taken as input. In the first two methods [3,5,9,12,23,25], a part of the graph data information (node attributes or graph structure) is taken as input and unique network architectures are designed for the graph clustering task. Although having made excellent progress, the methods only focus on learning feature information from graph node attributes or graph structure. Thus, various studies [2,8,14,15,20] simultaneously learn graph node attributes and graph structure by introducing graph convolution networks (GCNs).

However, the methods with GCN samples only the direct neighbors(one-hop nodes) for each layer of the network. Although the k-layer GCN can sample k-hop neighbor nodes, the value of k is limited because of the over-smoothing problem. In addition, the nodes in some graphs are not strongly related but are still directly connected. For example, although two people are friends on a social network, their preferences may be completely different. Therefore, the existing GCN model has some problems: 1) the aggregation of noise information in the graph structure information leads to poor model learning effect; 2) the methods focus on learning the local structure information of the graph, and fail to learn the global structure information of the graph, resulting in incomplete learning of the graph.

Thus, we propose a structural enhancement graph clustering network (SEGCN), which explores the information of the graph node attribute dependence and global structure information through the topology enhancement module and the global structure dynamic fusion module. The topology enhancement module introduces a dynamic learning parameter to merge the normalized correlation matrix of nodes with the original adjacency matrix to enrich the graph structure information. The global structure dynamic fusion module uses a message-passing operation to aggregate the feature representation information globally. Finally, we designed a unified model framework based on the dual self-supervised strategy to learn the feature representation suitable for the graph clustering task. The main contributions of this paper are listed as follows:

(1) We propose a topology enhancement module, which exploits the structural relations between nodes from the correlation of node attributes to enhance the original graph structure. The model reduces the noise effect caused by weakly correlated neighbor nodes of the graph.
(2) We improve the attention-driven graph clustering network with a global structure dynamic fusion module, which can aggregate graph feature information from global aspects.
(3) To verify the validity of our SEGCN and the effectiveness of each module, we conduct extensive experiments to compare with the state-of-the-art methods on four benchmark datasets. Besides, we create a GitHub repository[1] to distribute the code of SEGCN.

2   Related Work

As described in Sect. 1, graph clustering learning methods can be divided into three categories: taking node attributes as input, taking graph structure as input, and both node attributes and graph structure as input. Next, we described each of the three categories in detail.

Taking the Node Attribute as Input. Auto-encoder(AE) [5] adopts the double-layer structure of encoder and decoder, the difference between the original data and the reconstructed data is used as a loss function to guide the neural network to learn a mapping relationship. Based on the AE framework, deep embedded cluster (DEC) [26] introduces Kullback-Leibler (KL) divergence minimization to learn the depth representation of graph data. Deep transfer clustering [4] combines data representation learning and clustering of unlabeled new visual categories as a unified framework.

Taking the Graph Structure as Input. VIB-GSL [18] proposes the principle of Information bottleneck (IB) for graph structure learning. By learning the compressed graph structure, VIB-GSL extracts more information from the data and designs a unified and general architecture for mining the underlying task correlation.

---

[1] https://github.com/hacker-honker/SEGCN

Taking Both the Node Attribute and Graph Structure as Input . To simultaneously learn the node and structural information of graph, some clustering methods based on graph convolutional networks (GCNs) are proposed [7,8,14–16,19–22,24,30]. IDEC [3] adopt an incomplete auto-encoder, and the loss function used the sum of relative entropy and reconstruction loss to ensure the representativeness of the embedded space features. Kipf and Welling [8] propose to use graph auto-encoders (GAE) and variational graph auto-encoders (VGAE) to learn the node representation in the neighborhood of each node through multi-layer neural networks, and implicitly aggregate the graph structure into the node attribute representation. A deep attention-embedded graph clustering network (DAEGC) [21] adopts the attention mechanism to learn the feature representation of graph data and the reconstruction loss and clustering loss are combined to simultaneously learn the node embedding representation and clustering. Reverse regularization graph auto-encoder (ARGA) [14] introduces the policy learning graph representation of adversarial training based on the graph encoder to improve the clustering performance. Based on the DEC framework, the structural deep clustering network (SDCN) [2] introduces GCN to learn the graph structure information and aggregate the information in the neighborhood of nodes to improve the identifiability between node feature representations. The attention-driven graph clustering network (AGCN) [16] fuses multi-scale graph embedding from each GCN and AE layer, preserving more graph information.

3   Method

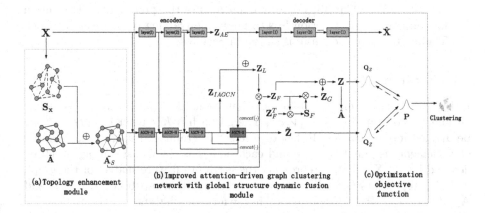

Fig. 1. The details of the architecture of SEGCN.

In this section, we present our proposed Structural Enhancement Graph Clustering Network (SEGCN), which includes three parts: topology enhancement module, improved attention-driven graph convolutional network module with global structure dynamic fusion module (IAGCN), and optimization objective function (i.e., self-supervised learning strategy and combination loss and optimization strategy). According to Fig. 1, we detailedly describe the architecture of the SEGCN.

(1) **Topology enhancement module.** Firstly, the module digs out the similarity relationship between nodes and then carries out dynamic fusion with the original graph structure. The enhanced graph topology more strongly strengthens the edge relationship between the strongly correlated nodes than the weakly correlated nodes, thus weakening the noise effect caused by the weakly correlated nodes. We cover the module in more detail in Sect. 3.2.

(2) **Improved attention-driven graph clustering network with global structure dynamic fusion module.** Attention-driven graph clustering network(AGCN [16]) has learned the multi-level feature information of node feature and graph structure well. However, AGCN [16] still only aggregates the local structure information and lacks the learning of non-local structure information. To overcome the shortcoming, we adopt the enhanced topology structure as the input and introduce the global structure dynamic fusion method. The methods can learn the global structure of nodes and the information of the distant neighbor nodes and strongly correlated nodes of non-local neighbor nodes and aggregate the global feature embedding information, mining more graph structure information. The module in more detail is shown in Sect. 3.3.

(3) **Optimization objective function.** We design a training model that integrates the clustering task and graph representation learning under a unified framework and adopts a dual self-supervision strategy to perform unsupervised learning. The objective function adopts the probability distribution loss and the reconstruction loss to guide the training of the overall model. The module in more detail is shown in Sect. 3.4.

3.1 Notations

In this part, we present the notations that we used in our SEGCN. We define an undirected graph as $\mathcal{G} = \{\mathcal{V}, \mathcal{E}\}$. The graph has K cluster centers and can be divided into K classes. $\mathcal{V}$ denotes the set of graph nodes, the total number of $\mathcal{V}$ elements is denoted by N and each node has d-dimensional eigenvectors, denoted by x_i. $\mathbf{X} \in \mathbb{R}^{N \times d}$ represents the attribute feature matrix of the graph nodes. $\mathcal{E}$ represents the set of the connection relationship between nodes, and $\mathbf{A} = (a_{ij})_{N \times N} \in \mathbb{R}^{N \times N}$ represents $\mathcal{E}$ in the form of adjacency matrix to describe the graph structure information. In adjacency matrix $\mathbf{A}$, $a_{ij} = 1$ if $(v_i, v_j) \in \mathcal{E}$, otherwise $a_{ij} = 0$, where v_i and v_j denote the i-th and j-th elements of $\mathcal{V}$. The normalized adjacency $\widetilde{\mathbf{A}} \in \mathbb{R}^{N \times N}$ is obtained by calculating $\mathbf{D}^{-\frac{1}{2}} (\mathbf{A} + \mathbf{I}) \mathbf{D}^{-\frac{1}{2}}$, where $\mathbf{D} = diag(d_i) \in \mathbb{R}^{N \times N}$ is the corresponding degree matrix with $d_i = \sum_{v_j \in V} a_{ij}$ and $\mathbf{I} \in \mathbb{R}^{N \times N}$ is the identity matrix which represents a self-loop structure between nodes in $\mathcal{V}$.

3.2 Topology Enhancement Module

By learning graph structure information, the graph clustering network adopts the message-passing strategy to aggregate node information, aiming at mining complex higher-order neighborhood information. In our proposed topology enhancement module, we first extract the topology structure from the attribute information by computing the correlation. The correlation matrix $\left(\mathbf{X} \in \mathbb{R}^{N \times d} \right)$ can be expressed by many algorithms (e.g. cosine similarity and Euclidean distance). In our proposed method, the sample correlation matrix is calculated by

$$\mathbf{S}_X = softmax(\mathbf{XX}^T) \tag{1}$$

where softmax($\cdot$) is the function that maps the range of elements values of the correlation matrix to $[0, 1]$, and the sum of elements is constrained to be 1.

Then, we dynamically fuse the extracted topology structure with the original graph structure. Specifically, we adopt the learnable parameter α to combine the normalized adjacency matrix and the correlation matrix of attributes with a linear combination operation:

$$\tilde{\mathbf{A}}_S = \tilde{\mathbf{A}}_{ij} + \alpha \mathbf{S}_X \tag{2}$$

where $\tilde{\mathbf{A}}_{ij} = \mathbf{D}^{-\frac{1}{2}} \left(\mathbf{A} + \mathbf{I} \right) \mathbf{D}^{-\frac{1}{2}}$, $\tilde{\mathbf{A}}_{ij} \in \mathbb{R}^{N \times N}$ denotes the normalized adjacency matrix of graph.

The uniqueness of the proposed module is that enriches the original graph structure and reduces the noise impact of weak correlation nodes.

3.3 Improved Attention-Driven Graph Clustering Network with Global Structure Dynamic Fusion Module

In this section, we design the improved attention-driven graph clustering network with a global structure dynamic fusion module, including a symmetric encoder-decoder module and an improved attention-driven graph convolutional network (IAGCN) and global structure dynamic fusion module. The global structure dynamic fusion module combines the feature outputs of two modules linearly and aggregates the global structure information at the feature level through message passing operation. Next, we describe the implementation of the three modules in detail.

Symmetric Encoder-Decoder Module. We use AE [5] to obtain the embedding X_i of the i-th encoder layer and the reconstructed data $\hat{X}$. The latent embedding of the final encoder layer denotes $\mathbf{Z}_{AE}$. The reconstruction loss function between the original data $X \in \mathbb{R}^{N \times d}$ and the reconstructed data $\hat{X} \in \mathbb{R}^{N \times \hat{d}}$ is formulated as:

$$\mathcal{L}_{AE} = \left\| X - \hat{X} \right\|_F^2 \tag{3}$$

Improved Attention-Driven Graph Convolutional Network (IAGCN).
In the section, we first extract the latent embedding Z'_i from the AGCN-H module of AGCN [16]. The method dynamically merges the hidden embeddings of GCN and auto-encoder through the concatenation operation and the attention mechanism.

Then, the learned embedding matrix $Z'_i \in \mathbb{R}^{N \times d'_i}$ of i-th GCN layer is used as the input to calculate the embedding Z_{i+1} of $(i+1)$-th GCN layer. The specific formula is as follows:

$$Z_{i+1} = LeakyReLU(\mathrm{D}^{-\frac{1}{2}}((\tilde{A_S})_i)\mathrm{D}^{-\frac{1}{2}}Z'_i\mathrm{W}_i) \tag{4}$$

where $(\tilde{A_S})_i$ being the topology enhanced normalized adjacency matrix. Especially, the adjacency matrix input of the first GCN layer is $\tilde{\mathbf{A}}$. W_i denotes the weight matrix of the neural network layer. $LeakyReLU(\cdot)$ denotes activation function. Z_{i+1} denotes the embedding of the $(i+1)$-th layer. And we use $\mathbf{Z}_{IAGCN}$ to denote the embedding of the last layer of the network. The multi-scale embedding is obtained from the embedding of each layer network by using the concatenation method. Finally, we obtain the probability distribution from multi-scale embedding. The specific formula is as follows:

$$\mathbf{Q}_{\tilde{Z}} = \mathbf{AGCN\text{-}S}(Z_1 \cdots Z_i \cdots Z_l, \tilde{A}_S, \mathrm{X}_l) \tag{5}$$

where $\mathbf{Q}_{\tilde{Z}}$ denotes the probability distribution of multi-scale fusion embedding. $Z_i \in \mathbb{R}^{N \times d_i}$ denotes the embedding generated at the i_{th} GCN layer, d_i is the dimension of the embedding, l denotes the last layer of the neural network, $\mathrm{X}_l \in \mathbb{R}^{N \times d_l}$ is output of the l_{th} encoder layer of symmetric encoder-decoder network. The AGCN-S module is based on attention mechanism and GCN, more detailed descriptions of the module can be found in the original paper introducing AGCN [16].

Global Structure Dynamic Fusion Module. Inspired by the paper of DFCN [19], we improve IAGCN by introducing the global structure dynamic fusion module. The module fuses the embeddings learned from the local structure and global structure to obtain a better graph feature representation.

First, we combine the latent embedding of symmetric encoder-decoder module $\left(\mathbf{Z}_{AE} \in \mathbb{R}^{N \times d}\right)$ and IAGCN $\left(\mathbf{Z}_{IAGCN} \in \mathbb{R}^{N \times d'}\right)$ with a linear combination operation:

$$\mathbf{Z}_L = \alpha_1 \mathbf{Z}_{AE} + \alpha_2 \mathbf{Z}_{IAGCN} \tag{6}$$

where α_1 and α_2 are learnable coefficients that dynamically balance the importance of the two embeddings. In this paper, α_1 and α_2 are initially set to 0.5, and the gradient descent method is utilized to optimize the two coefficients after continuous iteration of the model.

Then, we enhance the relationship between features of the embedding $\mathbf{Z}_L \in \mathbb{R}^{N \times d'}$ by using the topology enhanced normalized adjacency matrix:

$$\mathbf{Z}_F = \tilde{\mathbf{A}}_s \mathbf{Z}_L \tag{7}$$

Based on enhancing the initial fusion embedding, we introduce a correlation learning mechanism to explore the correlation between nodes and other nodes in the embedding space and learn the global structure of the graph. Firstly, we calculate $\mathbf{S}_F \in \mathbb{R}^{N \times N}$ from the learned embedding $\mathbf{Z}_F$ through Eq.(1). Taking $\mathbf{S}_F$ as the correlation coefficient between nodes. Then, the global correlation information is aggregated into the embedded representation by Eq.(8):

$$\mathbf{Z}_G = \mathbf{S}_F \mathbf{Z}_F \tag{8}$$

Finally, in the process of learning the global embedding $\mathbf{Z} \in \mathbb{R}^{N \times d'}$ through the dynamic fusion mechanism, the global feature information is transmitted by skip connection:

$$\mathbf{Z} = \beta \mathbf{Z}_G + \mathbf{Z}_F \tag{9}$$

where β is initialized to 0 as a learnable weight parameter, and the weight is updated during the training of the neural network. To minimize the reconstruction loss between the raw adjacency and the reconstructed adjacency, the loss function is formulated as:

$$\mathcal{L}_Z = \left\| \tilde{\mathbf{A}} - \hat{\mathbf{A}} \right\|_F^2 \tag{10}$$

where $\hat{\mathbf{A}} = softmax(\mathbf{Z}\mathbf{Z}^T) \in \mathbb{R}^{N \times N}$ indicates the reconstructed normalized adjacency matrix of the global embedding $\mathbf{Z}$.

3.4   Optimization Objective Function

Self-supervised Learning Strategy. The global embedding $\mathbf{Z}$ is updated by a dual self-supervised learning strategy, which makes feature learning more in-depth and accurate, and enhances the learning ability of the clustering model. Firstly, we use $\mathbf{Z}$ to generate the soft assignment distribution $\mathbf{Q_Z}$ and the target distribution $\mathbf{P}$. The generation steps are shown in Eq. (11) and Eq. (12):

$$q_{ij} = \frac{\left(1 + \|z_i - u_i\|^2 / v\right)^{-\frac{v+1}{2}}}{\sum_{j'} \left(1 + \left\|z_i - u_{j'}\right\|^2 / v\right)^{-\frac{v+1}{2}}} \tag{11}$$

$$p_{ij} = \frac{q_{ij}^2 / \sum_i q_{ij}}{\sum_{j'} (q_{ij'}^2 / \sum_i q_{ij'})} \tag{12}$$

The first step in the target distribution generation process is to calculate the similarity as a soft assignment. Specifically, the Student's t-distribution is taken as the kernel of the target distribution, and the similarity between the i-th feature representation (z_i) and the j-th cluster center (u_j) obtained by K-means algorithm in advance in the embedding space is measured via Eq. (11). In Eq. (11), v indicates that the degree of freedom of the Student's t-distribution, q_{ij} represents the probability that the i-th graph node falls into the range of the j-cluster center. $\mathbf{Q_Z} \in \mathbb{R}^{N \times K}$ denotes the probability distribution that all nodes are in a certain cluster.

The second step is to increase confidence in clustering assignment results. Thus, we adopt Eq. (12) to make the gap between all sample nodes and the cluster center smaller. In Eq. (12), p_{ij} is an element of the target distribution matrix $\mathbf{P} \in \mathbb{R}^{N \times K}$ which indicates the probability that the i-th node falls within the range of the j-th cluster center, with values ranging from 0 and 1.

Then, we adopt the KL divergence as the loss function to bring the target distribution $\mathbf{P}$ and the Q-distributions close to each other. The Q-distributions include the soft assignment distribution $\mathbf{Q_Z}$ and the probability distribution $\mathbf{Q_{\tilde{Z}}}$ from Eq. (5). The specific formula is as follows:

$$
\begin{aligned}
\mathcal{L}_{KL} &= \lambda_1 * KL\left(\mathbf{P}, \mathbf{Q}_Z\right) + \lambda_2 * KL\left(\mathbf{P}, \mathbf{Q}_{\tilde{Z}}\right) \\
&= \lambda_1 \sum_i \sum_j p_{ij} log \frac{p_{ij}}{z_{ij}} + \lambda_2 \sum_i \sum_j p_{ij} log \frac{p_{ij}}{\tilde{z}_{ij}}
\end{aligned}
\tag{13}
$$

where λ_1 and λ_2 are predefined hyper-parameters that balance the importance of two distributions.

Finally, we perform clustering over target distribution $\mathbf{P}$ by $\mathbf{Y} = argmax(\mathbf{P})$ where argmax($\cdot$) function takes the maximum value over the dim = 1 of target distribution $\mathbf{P}$. $\mathbf{Y}$ is the set of predicted labels.

Combination Loss and Optimization Strategy. The combined loss of the overall model mainly includes two main parts: (1) the reconstruction losses of the attribute matrix and adjacency matrix; (2) the KL divergence loss of the target distribution. The loss function is shown as follows:

$$
\mathcal{L} = \mathcal{L}_{AE} + \mathcal{L}_Z + \mathcal{L}_{KL}
\tag{14}
$$

The overall training and learning steps of the SEGCN framework are described in detail as shown in Algorithm 1.

Algorithm 1. Training Procedure of Structural Enhancement Graph Clustering Network

---

Input: Input node attribute feature X, Original adjacency matrix A, Graph Cluster number $\mathbf{K}$; Trade-off parameters λ_1, λ_2; Iteration number $iter_{Max}$;
Output: Cluster results $\mathbf{Y}$

1: Initialize $iter_{Max} = 200$; Set $Z_0 = X$; Set $X_0 = X$;
2: Initialize and pre-training the parameters of auto-encoder;
3: **while** $iter < iter_{Max}$ **do**
4:      Calculate the normalized similarity matrix $\mathbf{S}_X$ by Eq. (1);
5:      Obtain self-learnable normalized adjacency matrix $\mathbf{A}_S$ by Eq. (2)
6:      Obtain $\mathbf{X}$ and $\hat{\mathbf{X}}$ by Auto-encoder and calculate the loss by Eq. (3);
7:      Learning from IAGCN module to fusion embedding and obtain the probability distribution via Eq.(4) and Eq. (5);
8:      Obtain the combined embedding $\mathbf{Z}_L$ via Eq. (6);
9:      Calculate the global embedding $\mathbf{Z}$ by the self-correlation mechanism via Eq. (7)–Eq. (10);
10:      Calculate the cluster center embedding μ of the latent embedding $\mathbf{Z}$ by K-means algorithm;
11:      Calculate the soft assignment distribution matrix $\mathbf{Q}$ between the embedding $\mathbf{Z}$ and the clustering center embedding μ via Eq. (11);
12:      Calculate the target distribution matrix $\mathbf{P}$ on the soft assignment distribution matrix $\mathbf{Q}$ via Eq. (12);
13:      Calculate the loss between the two $\mathbf{Q}$ distributions and the target distribution $\mathbf{P}$ by KL-divergence via Eq. (13);
14:      Obtain the predicted labels $\mathbf{Y}$ from target distribution $\mathbf{P}$;
15:      Calculate the combination loss via Eq. (14);
16:      Continuous backpropagation and parameter updating through iteration in the SEGCN model;
17:      $iter = iter + 1$;
18: **return Y**

---

4 Experiment

4.1 Benchmark Datasets

The proposed SEGCN is evaluated on four public datasets, including two graph datasets (ACM[2] and CITE[3]) and two non-graph datasets (USPS [6] and HHAR [17]). The information of four datasets is briefly described in Table 1. The original data of non-graph datasets is regarded as the node data of the graph and generated undirected k-nearest neighbor (KNN [1]) graph, the adjacency matrix $\mathbf{A}$ is obtained in this graph, the details of generation process are following AGCN [16].

---

[2] http://dl.acm.org/.
[3] http://citeseerx.ist.psu.edu/.

Table 1. Description of the datasets

Dataset	Type	nodes	Classes	Dimension
USPS	Image	9298	10	256
HHAR	Record	10299	6	561
ACM	Graph	3025	3	1870
CITE	Graph	3327	6	3703

4.2   Experimental Setup and Evaluation

Detailed descriptions of these methods in Table 2 can be found in the Sect. 2 or in the original paper [2,3,5,8,14,16,21,26].

To ensure the experimental correctness of other methods, we establish the network structure and set the parameters according to papers [2,3,26]. We set the dimension of the auto-encoder and GCN layers to 500,500,2000 and 10, respectively. The training for the proposed SEGCN is conducted in two phases. In the first phase, we set the learning rate to 0.001 and iteration to 30 for pre-training the AE network. In the second phase, we set iterations to 200 for training the whole network of SEGCN (i.e., $iter_{Max} = 200$). The learning rate is set at 0.001 for USPS, HHAR, and ACM datasets and 0.0001 for the CITE dataset. The range of parameter λ_1 and λ_2 initialization is $[0.01, 0.1, 1, 10, 100]$. After the experiments, We set the best parameters for SEGCN. λ_1 and λ_2 are set to $\{10, 100\}$ and $\{1, 0.1\}$ for USPS and HHAR datasets, respectively. λ_1 and λ_2 are set to the same $\{0.1, 0.01\}$ for ACM and CITE datasets. We set the batch size of the network to 256. We set the parameters of the ARGA method according to the original paper [14]. For a fair and correct comparison with other methods, we directly cite the experimental results in the AGCN [16]. We perform 10 repeated experiments on the proposed SEGCN to evaluate the performance of the model, and finally, report the results in the form of mean ± std. The whole experiment is conducted in PyTorch and a GPU (GeForce RTX 2080 Ti).

Similar to AGCN [16], Accuracy (ACC), Normalized Mutual Information (NMI), Average Rand Index (ARI), and macro F1-score (F1) are to evaluate the clustering performance of all methods, The larger the result of each indicator, the better the clustering performance of the method.

4.3   Clustering Results

In this part, the proposed SEGCN is compared with nine state-of-the-art clustering methods to illustrate its effectiveness. We summarize the clustering performance of SEGCN and nine methods on four datasets in Table 2. The clustering results on datasets are displayed in the form of mean ± std, where the best results are marked in red font and the second-best results are marked in blue font. From the results in Table 2, we can summarize and analyze the advantages and disadvantages of the SEGCN.

Table 2. The clustering results on datasets

Dataset	Metric	AE [5]	DEC [26]	IDEC [3]	GAE [8]	VGAE [8]	DAEGC [21]	ARGA [14]	SDCN [2]	AGCN [16]	Our
USPS	ACC	71.04±0.03	73.31±0.17	76.22±0.12	63.10±0.33	56.19±0.72	73.55±0.40	66.80±0.70	78.08±0.19	80.98±0.28	84.40±2.21
	NMI	67.53±0.03	70.58±0.25	75.56±0.06	60.69±0.58	51.08±0.37	71.12±0.24	61.60±0.30	79.51±0.27	79.64±0.32	82.61±1.43
	ARI	58.83±0.05	63.70±0.27	67.86±0.12	50.30±0.55	40.96±0.59	63.33±0.34	51.10±0.60	71.84±0.24	73.61±0.43	79.44±2.46
	F1	69.74±0.03	71.82±0.21	74.63±0.10	61.84±0.43	53.63±1.05	72.45±0.49	66.10±1.20	76.98±0.18	77.61±0.38	79.33±2.47
HHAR	ACC	68.69±0.31	69.39±0.25	71.05±0.36	62.33±1.01	71.30±0.36	76.51±2.19	63.30±0.80	84.26±0.17	88.11±0.43	84.43±2.82
	NMI	71.42±0.97	72.91±0.39	74.19±0.39	55.06±1.39	62.95±0.36	69.10±2.28	57.10±1.40	79.90±0.09	82.44±0.62	82.77±1.80
	ARI	60.36±0.88	61.25±0.51	62.83±0.45	42.63±1.63	51.47±0.73	60.38±2.15	44.70±1.00	72.84±0.09	77.07±0.66	75.08±1.61
	F1	66.36±0.34	67.29±0.29	68.63±0.33	62.64±0.97	71.55±0.29	76.89±2.18	61.10±0.90	82.58±0.08	88.00±0.53	83.32±4.52
ACM	ACC	81.83±0.08	84.33±0.76	85.12±0.52	84.52±1.44	84.13±0.22	86.94±2.83	86.10±1.20	90.45±0.18	90.59±0.15	91.00±0.10
	NMI	49.30±0.16	54.54±1.51	56.61±1.16	55.38±1.92	53.20±0.52	56.18±4.15	55.70±1.40	68.31±0.25	68.38±0.45	69.99±0.20
	ARI	54.64±0.16	60.64±1.87	62.16±1.50	59.46±3.10	57.72±0.67	59.35±3.89	62.90±2.10	73.91±0.40	74.20±0.38	75.31±0.24
	F1	82.01±0.08	84.51±0.74	85.11±0.48	84.65±1.33	84.17±0.23	87.07±2.79	86.10±1.20	90.42±0.19	90.58±0.17	90.98±0.11
CITE	ACC	57.08±0.13	55.89±0.20	60.49±1.42	61.35±0.80	60.97±0.36	64.54±1.39	56.90±0.70	65.96±0.31	68.79±0.23	71.56±0.40
	NMI	27.64±0.08	28.34±0.30	27.17±2.40	34.63±0.65	32.69±0.27	36.41±0.86	34.50±0.80	38.71±0.32	41.54±0.30	44.50±0.50
	ARI	29.31±0.14	28.12±0.36	25.70±2.65	33.55±1.18	33.13±0.53	37.78±1.24	34.40±1.50	40.17±0.43	43.79±0.31	47.46±0.46
	F1	53.80±0.11	52.62±0.17	61.62±1.39	57.36±0.82	57.70±0.49	62.20±1.32	54.80±0.80	63.62±0.24	62.37±0.21	62.27±0.26

(1) SEGCN consistently outperforms other methods on most metrics across the three datasets. According to the results on CITE, our SEGCN outperforms AGCN [16] by 4.02%, 7.12%, and 8.38% increments for ACC, NMI, and ARI. This result shows that by learning the node correlation and global structure, SEGCN explores more graph structure information and reduces the noise effects of weakly correlated nodes, resulting in better clustering performance.

(2) For non-graph HHAR [17], the adjacency matrix obtained by the KNN [1] is different from the formal graph structure, which disturbs the learning of SEGCN, but SEGCN also achieves good clustering performance.

(3) From the results, we can observe that the performances of the clustering methods based on GCN (e.g. GAE/VGAE [8], ARGA [14] and DAEGC [21]) is much lower than that of our methods. It is because SEGCN, compared with these methods, extracts the correlation information from the node attributes and pays attention to the global structure information.

(4) The AE-based methods of AE [5], DEC [26], and IDEC [3]) have strong learning ability for non-graph data, but can not achieve good performance because of the lack of structure information, SEGCN is better than these methods in graph structure learning ability.

4.4 Ablation Studies

We analyze the influence of the topology enhancement module and the global structure dynamic fusion module in the SEGCN through the ablation experiments. The results of experiments are shown in Table 3. 'Baseline' indicates the results of AGCN [16]; 'TE' indicates only adding the topology enhancement module; 'GE' indicates only adding the global structure dynamic fusion module; 'SE' indicates adding the topology enhancement module and the global structure dynamic fusion module without reconstruction of the adjacency matrix. 'SEGCN' indicates the whole model. The best results are highlighted in bold.

Table 3. Results of ablation studies on four datasets.

Dataset	Model	ACC	NMI	ARI	F1
USPS	Baseline	80.98 ± 0.28	79.64 ± 0.32	73.61 ± 0.43	77.61 ± 0.38
	+TE	81.70 ± 0.65	78.71 ± 0.95	74.58 ± 0.88	76.92 ± 1.06
	+GE	83.97 ± 1.95	82.26 ± 1.01	78.25 ± 2.61	$\mathbf{79.51 \pm 1.09}$
	+SE	83.51 ± 1.94	81.73 ± 1.33	77.43 ± 2.75	79.18 ± 1.07
	SEGCN	$\mathbf{84.40 \pm 2.21}$	$\mathbf{82.61 \pm 1.43}$	$\mathbf{79.44 \pm 2.46}$	79.33 ± 2.47
HHAR	Baseline	$\mathbf{88.11 \pm 0.43}$	82.44 ± 0.62	$\mathbf{77.07 \pm 0.66}$	$\mathbf{88.00 \pm 0.53}$
	+TE	85.78 ± 2.48	82.51 ± 1.65	75.49 ± 1.90	85.87 ± 2.32
	+GE	82.96 ± 2.93	$\mathbf{83.31 \pm 2.11}$	74.70 ± 1.38	81.01 ± 5.27
	+SE	85.25 ± 2.87	82.16 ± 1.65	74.99 ± 2.65	84.97 ± 3.76
	SEGCN	84.43 ± 2.82	82.77 ± 1.80	75.08 ± 1.61	83.32 ± 4.52
ACM	Baseline	90.59 ± 0.15	68.38 ± 0.45	74.20 ± 0.38	90.58 ± 0.17
	+TE	$\mathbf{91.89 \pm 0.23}$	$\mathbf{71.85 \pm 0.56}$	$\mathbf{77.49 \pm 0.58}$	$\mathbf{91.88 \pm 0.23}$
	+GE	90.84 ± 0.14	69.57 ± 0.44	74.91 ± 0.38	90.81 ± 0.15
	+SE	90.96 ± 0.10	69.89 ± 0.17	75.22 ± 0.23	90.94 ± 0.10
	SEGCN	91.00 ± 0.10	69.99 ± 0.20	75.31 ± 0.24	90.98 ± 0.11
CITE	Baseline	68.79 ± 0.23	41.54 ± 0.30	43.79 ± 0.31	62.37 ± 0.21
	+TE	69.26 ± 0.36	42.01 ± 0.45	44.35 ± 0.50	$\mathbf{62.61 \pm 0.30}$
	+GE	71.11 ± 0.23	43.95 ± 0.32	46.87 ± 0.32	62.22 ± 0.50
	+SE	71.22 ± 0.21	44.03 ± 0.26	46.99 ± 0.29	62.07 ± 0.23
	SEGCN	$\mathbf{71.56 \pm 0.40}$	$\mathbf{44.50 \pm 0.50}$	$\mathbf{47.46 \pm 0.46}$	62.27 ± 0.26

Effectiveness of Topology Enhancement Module. The topology enhancement module denotes the method by which the original graph structure is enhanced by the correlation matrix calculated by attribute features. We can observe from Table 3 that the 'TE' model including the topology enhancement module is superior to the original model in most of the three datasets, especially in the two graph datasets. Compared with the baseline, all indicators ascend to a certain extent in the ACM dataset, and the ACC and NMI improve by 1.43% and 5.07%, respectively. It is active to use the topology enhancement module. This module enhances the correlation between nodes and reduces the intra-class gap of the same kind of nodes.

Analysis of Global Structure Dynamic Fusion Module. The module can strengthen the connection between the feature of non-adjacent nodes in the embedding space and discover deeper graph information. From Table 3, we can see that the performance of the 'GE' model with the global structure dynamic fusion module in the three datasets is improved by about 0.27% to 5.8% than baseline, so we can conclude that learning the global information of the embedding is helpful to improve the clustering performance.

Analysis of Hyper-parameter λ. The self-supervised strategy is realized based on the interaction between the embedding space and the target distribution. The KL divergence loss function between the soft assignment distribution and probability distribution and the target distribution is an important part that guides the model learning, so the settings of λ_1 and λ_2 are critical. Figure 2 illustrates the clustering performance variation of SEGCN when λ_1 and λ_2 are selected and combined in $[0.01, 0.1, 1, 10, 100]$. It can be concluded from the figure that 1) the setting of hyper-parameters has a great impact on the clustering performance of the SEGCN model, and the clustering performance varies greatly with different parameter sizes. 2) the best clustering performance can be obtained by setting different hyper-parameters on different datasets.

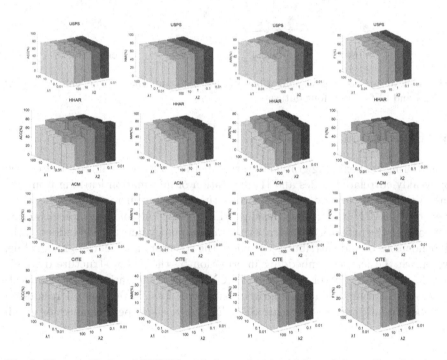

Fig. 2. The sensitivity of SEGCN with the variations of λ_1 and λ_2 on four datasets.

4.5   Visualization Results

We adopt the t-distributed stochastic neighbor embedding (t-SNE) algorithm [11] to visualize the clustering results of the SEGCN in Fig. 3. From Fig. 3, we can observe that there is strong aggregation among nodes within the same class and strong separation across different classes.

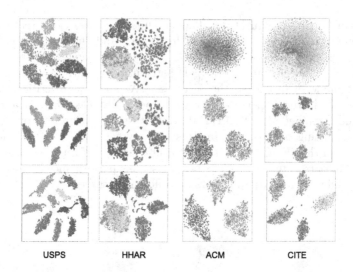

Fig. 3. Two-dimensional visualizations of raw data and the learned embeddings of AGCN and SEGCN on four datasets

5   Conclusion

Considering that existing graph clustering methods ignore the noise effect caused by weakly correlated nodes and the limitation of information learning from distant neighbor nodes, we propose a Structural Enhancement Graph Clustering Network(SEGCN) to improve the clustering performance of the graph clustering network. In the network, we extract the graph topology information hidden in the dependencies of node attributes to extend the structure information. Meanwhile, we enrich the learned embedding information by the global structure dynamic fusion method. Moreover, we train SEGCN by the combination of the dual self-supervised learning strategy and reconstruction loss. Experimental results on four benchmark datasets indicate SEGCN has excellent performance for graph clustering.

References

1. Altman, N.S.: An introduction to kernel and nearest-neighbor nonparametric regression. Am. Stat. **46**(3), 175–185 (1992)
2. Bo, D., Wang, X., Shi, C., Zhu, M., Lu, E., Cui, P.: Structural deep clustering network. In: Proceedings of the Web Conference 2020, pp. 1400–1410 (2020)
3. Guo, X., Gao, L., Liu, X., Yin, J.: Improved deep embedded clustering with local structure preservation. In: IJCAI, pp. 1753–1759 (2017)
4. Han, K., Vedaldi, A., Zisserman, A.: Learning to discover novel visual categories via deep transfer clustering. In: Proceedings of the IEEE/CVF International Conference on Computer Vision, pp. 8401–8409 (2019)
5. Hinton, G.E., Salakhutdinov, R.R.: Reducing the dimensionality of data with neural networks. Science **313**(5786), 504–507 (2006)

6. Hull, J.J.: A database for handwritten text recognition research. IEEE Trans. Pattern Anal. Mach. Intell. **16**(5), 550–554 (1994)
7. Kim, D., Oh, A.: How to find your friendly neighborhood: graph attention design with self-supervision. arXiv preprint arXiv:2204.04879 (2022)
8. Kipf, T.N., Welling, M.: Variational graph auto-encoders. arXiv preprint arXiv:1611.07308 (2016)
9. Kumar, A., Rai, P., Daume, H.: Co-regularized multi-view spectral clustering. Adv. Neural Inf. Process. Syst. **24**, 1–9 (2011)
10. Li, X., Wang, S., Li, B.: Editorial for application-driven knowledge acquisition. World Wide Web **23**, 2649–2651 (2020)
11. Van der Maaten, L., Hinton, G.: Visualizing data using t-sne. J. Mach. Learn. Res. **9**(11) (2008)
12. Nie, F., Cai, G., Li, J., Li, X.: Auto-weighted multi-view learning for image clustering and semi-supervised classification. IEEE Trans. Image Process. **27**(3), 1501–1511 (2017)
13. Nikolentzos, G., Dasoulas, G., Vazirgiannis, M.: k-hop graph neural networks. Neural Netw. **130**, 195–205 (2020)
14. Pan, S., Hu, R., Fung, S.F., Long, G., Jiang, J., Zhang, C.: Learning graph embedding with adversarial training methods. IEEE Trans. Cybern. **50**(6), 2475–2487 (2019)
15. Park, J., Lee, M., Chang, H.J., Lee, K., Choi, J.Y.: Symmetric graph convolutional autoencoder for unsupervised graph representation learning. In: Proceedings of the IEEE/CVF International Conference on Computer Vision, pp. 6519–6528 (2019)
16. Peng, Z., Liu, H., Jia, Y., Hou, J.: Attention-driven graph clustering network. In: Proceedings of the 29th ACM International Conference on Multimedia, pp. 935–943 (2021)
17. Stisen, A., et al.: Smart devices are different: assessing and mitigatingmobile sensing heterogeneities for activity recognition. In: Proceedings of the 13th ACM Conference on Embedded Networked Sensor Systems, pp. 127–140 (2015)
18. Sun, Q., et al.: Graph structure learning with variational information bottleneck. In: Proceedings of the AAAI Conference on Artificial Intelligence, vol. 36, pp. 4165–4174 (2022)
19. Tu, W., et al.: Deep fusion clustering network. In: Proceedings of the AAAI Conference on Artificial Intelligence, vol. 35, pp. 9978–9987 (2021)
20. Veličković, P., Cucurull, G., Casanova, A., Romero, A., Lio, P., Bengio, Y.: Graph attention networks. arXiv preprint arXiv:1710.10903 (2017)
21. Wang, C., Pan, S., Hu, R., Long, G., Jiang, J., Zhang, C.: Attributed graph clustering: a deep attentional embedding approach. arXiv preprint arXiv:1906.06532 (2019)
22. Wang, X., Zhu, M., Bo, D., Cui, P., Shi, C., Pei, J.: AM-GCN: adaptive multi-channel graph convolutional networks. In: Proceedings of the 26th ACM SIGKDD International Conference on Knowledge Discovery & Data Mining, pp. 1243–1253 (2020)
23. Wu, J., Lin, Z., Zha, H.: Essential tensor learning for multi-view spectral clustering. IEEE Trans. Image Process. **28**(12), 5910–5922 (2019)
24. Wu, Z., Pan, S., Chen, F., Long, G., Zhang, C., Philip, S.Y.: A comprehensive survey on graph neural networks. IEEE Trans. Neural Netw. Learn. Syst. **32**(1), 4–24 (2020)
25. Xie, D., Gao, Q., Deng, S., Yang, X., Gao, X.: Multiple graphs learning with a new weighted tensor nuclear norm. Neural Netw. **133**, 57–68 (2021)

26. Xie, J., Girshick, R., Farhadi, A.: Unsupervised deep embedding for clustering analysis. In: International Conference on Machine Learning, pp. 478–487. PMLR (2016)
27. Zang, Y., et al.: GISDCN: a graph-based interpolation sequential recommender with deformable convolutional network. In: Bhattacharya, A., et al. (eds.) DASFAA 2022, vol. 13426, pp. 289–297. Springer, Heidelberg (2022). https://doi.org/10.1007/978-3-031-00126-0_21
28. Zhang, J., et al.: Self-supervised convolutional subspace clustering network. In: Proceedings of the IEEE/CVF Conference on Computer Vision and Pattern Recognition, pp. 5473–5482 (2019)
29. Zhang, Y., Li, B., Gao, H., Ji, Y., Yang, H., Wang, M.: Fine-grained evaluation of knowledge graph embedding models in downstream tasks. In: Wang, X., Zhang, R., Lee, Y.K., Sun, L., Moon, Y.S. (eds.) APWeb-WAIM 2020, vol. 12317, pp. 242–256. Springer, Heidelberg (2020). https://doi.org/10.1007/978-3-030-60259-8_19
30. Zhu, Y., Xu, W., Zhang, J., Liu, Q., Wu, S., Wang, L.: Deep graph structure learning for robust representations: a survey. arXiv preprint arXiv:2103.03036 (2021)

Designing a Knowledge Graph System for Digital Twin to Assess Urban Flood Risk

Feng Ye[1]([✉]), Yu Wang[1], Dong Xu[1], Xuejie Zhang[1], and Gaoyang Jin[2]

[1] Hohai University, Nanjing 211100, People's Republic of China
{yefeng1022,wy1227}@hhu.edu.cn
[2] China Water Resources Pearl River Planning, Surveying and Designing Co., Ltd.,
Guangzhou 510610, People's Republic of China

Abstract. The digital twin is widely used to create virtual counterparts of physical entities, and it is also as a promising approach for assessing urban flood risk. A key challenge of supporting such DTs is to formalize objects related to urban flood, and integrate multi-modal data and distilled expert knowledge. However, no research has applied or implemented the DT with KG to assess unban flood risk. To ad-dress the problem, we propose the UrbanFloodKG system, a knowledge graph system that models and integrates data in a digital twin formalism for urban flood risk assessment. The system provides a complete solution for constructing a knowledge graph of urban flooding events, with a data layer, graph layer, algorithm layer, and digital twin layer. It implements functions for knowledge extraction, as well as integrating knowledge representation learning models and graph neural network models to support link prediction and node classification tasks. We conduct model comparison experiments on data related to flood events in Guangzhou, which demonstrates the effectiveness of our proposed solution.

Keywords: Urban Flood · Knowledge Graph · Digital Twin and Knowledge Reasoning

1 Introduction

The digital twin (DT) is widely used to create virtual counterparts of physical entities, enabling smart services such as tracking, monitoring, and optimization. This technology has been successfully pervading different application domains, including industry [28], healthcare [21], and smart cities [26]. Urban flood is one of the most devastating natural disasters in the world, which has caused casualties and property damage [2]. Thus, constructing a DT for assessing urban flood risk is very critical and valuable. However, a key challenge of supporting such DTs is how to formalize objects related to urban flood, and integrate multi-modal data and distilled expert knowledge. For example, for different urban regions, experts may adopt distinct data and methods for assessing flood risk. Knowledge graph (KG) [12,14], interlinking domain knowledge and physical asset data

X. Song et al. (Eds.): APWeb-WAIM 2023, LNCS 14331, pp. 191–205, 2024.
https://doi.org/10.1007/978-981-97-2303-4_13

in a uniform graph representation, becomes the reasonable enabling element for DT. As far as we know, no research has applied or implemented the DT with KG to assess unban flood risk, although some research works also provide good insights, such as [10].

To unleash the knowledge potential of urban flood data for DTs, we propose an Urban Flood Knowledge Graph (UrbanFloodKG) system that enables various tasks of the DT to assess urban flood risk. The system has multiple layers of data, graph, algorithm and DT to meet various needs in urban flood risk assessment research. Then, we conduct comparative experiments of multiple models to demonstrate the effectiveness of the proposed UrbanFloodKG system.

The main contributions of our work are as follows:

- We design a UrbanFloodKG system which indicates the potential of using KG to support DT and presents a way on to extract and infer knowledge from urban flood data, directly benefit various tasks of the DT.
- We integrate various KG representation algorithms and graph neural network (GNN) models. Through comparative experiments of multiple models, we compare their performance and demonstrate the effectiveness of the algorithms to support DT.

2   Related Work

Here we summarize the related work into two aspects of KG-based DT systems and urban flood risk assessment.

In [15], the authors construct the KG of flood control operations, which implements the correlation and response between the natural elements of the basin and the basin management business, and supports the DT demonstration system of the Yangtze River. However, the details of the KG's support for the DT demonstration systems are not mentioned, and whether the KG has reasoning ability is not specifically intro-duced. In [17], the researchers propose a data-model-knowledge integrated representation data model for a DT railway, which explicitly describes the spatiotemporal, and interaction relationships among railway features through a conceptual KG. The results of adopting KG bring more clear understanding of the complex interactive relationships in landslides scenario of DT railway. In [19], the authors suggest the DT architecture to support the complete AI lifecycle, based on a KG that centralizes all information. However, how to drill knowledge and reason from the KG is not covered in the above two papers. In [20], Liu combines urban computing with KG system, exploring new modes of urban data management and analysis. A production logistics resource allocation approach based on the dynamic spatial-temporal KG [37] is proposed. The dynamic spatial-temporal KG model is established for representing the DT replica with spatial-temporal consistency, followed by reasoning and completion of relationships based on production logistics task information. According to [37], their solutions provide a reasonable and feasible way for reference to adopt KG enabled DT for assessing urban flood risk.

For urban flood risk assessment, physics-based and data-driven approaches are the mainstream, such as [5,8,32]. In [8], a 2D-surface and a 1D-sewer integrated hydrodynamic model are proposed to accurately simulate the whole urban flooding processes and assess the flood risk to people. A novel approach to simulate dynamic flow interactions between storm sewers and overland surface for different land covers in urban areas is developed and introduced in [5]. Wang et al. [32] explore the impact of different methods for handling terrain datasets on flood modeling results using multiple information sources, while Lyu et al. [23] evaluate the flood risk of the Shanghai subway system in a subsidence environment by considering factors such as regional and subway longitudinal subsidence using the FAHP-FCA method. Through the above works are very effective, the structures of the models are complex, and they are difficult to generalize due to the limitation of the study area and climatic conditions, which leads to a large amount of key information related to domain knowledge that is not used effectively. Formalizing objects related to urban flood, and integrating complex data and distilled expert knowledge to construct KG for enabling DT is still room for research.

To sum up, no research has applied or implemented the DT with KG to assess uban flood risk. Therefore, we intend to focus on adopting KG for data fusion and knowledge representation of urban flood information, providing an effective system for enabling DT to assess urban flood risk.

3   Preliminary

Here, we formally define some concepts according to [27,38].

Definition 3.1 (Knowledge Graph Construction). Knowledge graph construction f is a procedure that maps a data source into a knowledge graph: $f : D \times f_k(D) \to \mathcal{G}$, where D is the set of data sources, and $f_k(D)$ is background knowledge of the data target, which can be domain knowledge. Notably, knowledge graph construction is usually unable to continue without background knowledge that is provided by pre-designed rules or a language model of representations.

Definition 3.2 (Digital Twin for Complex Objects). The proposed DT for complex objects can be depicted as: $DT = (PE, VE, SS, DD, CN)$, where PE refers to the physical objects; VE is the virtual objects; SS stands for services for PE and VE; DD refers to DT data, and CN is the connection among PE, VE, Ss, and DD. For DD, it can further be formalized as: $DD = (Dp, Dv, Ds, Dk, Df)$, where Dp is the data from the PE, Dv is the data from the VE, Ds is the data from the Ss, Dk represents the domain knowledge, and Df denotes the fused data of Dp, Dv, Ds, and Dk. DD includes data from both physical and virtual aspects as well as their fusion, which enriches the data greatly.

Therefore, according to the Definition 3.1, 3.2, we build the UrbanFloodKG system, which is introduced in the following.

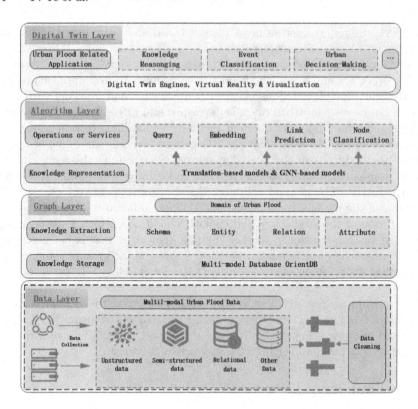

Fig. 1. The architecture of UrbanFloodKG system.

4   The Proposed UrbanFloodKG System

4.1   System Overview

The high-level system architecture of the UrbanFloodKG system is shown in Fig. 1. There are different layers described as follows:

- **Data Layer.** The data layer is responsible for collecting urban flood data from multiple sources and cleaning the collected data.
- **Graph Layer.** In graph layer, we define schema, extract entities and relationships from urban flood data, and then enrich them with additional attributes. The constructed triplets are then transformed into graph data structures and stored into the multi-model database OrientDB [36].
- **Algorithm Layer.** Translation-based models and GNN-based models are employed to convert triplets in the KG into vector embeddings. On the basis of these algorithms, various specific operations or service functions can be implemented, including link prediction, node classification, embedding access, query via Cypher and so on, which provide the logical basis for enabling higher-layer applications.

- **Digital Twin Layer.** The DT layer provides the user interface and presents the results of analysis and pre-diction in the form of virtual reality and simulation, including knowledge reasoning, event classification, and decision-making tasks for urban flood risk assessment. Built upon the operations in algorithm layer, the DT layer calls various operations or opera-tion combinations to support specific applications.

4.2 Data Layer

The data layer provides the functions of data collection, and data cleaning. Urban flood data can be divided into three types: structured, semi-structured, and unstructured data. Structured data is recorded in tabular form and contains detailed information about urban flood events. Semi-structured data is presented in textual form, while unstructured data comes from sensor-captured images.

4.3 Graph Layer

According to urban flood data, before constructing the overall scheme, some tasks are introduced. Named entity recognition [18] and relation extraction, which are classic tasks in the field of natural language processing, are useful for extracting knowledge from unstructured text. Here, we adopt the UIE [22] framework to extract knowledge from text and identifies key attributes as event attribute information, and uses the Word2vec [6] model to extract feature vectors from texts as event attribute. Similarly, object detection technology in the field of computer vision is used to extract knowledge from unstructured images. Specifically, the ResNet [13] model is adopted to extract feature vectors from images as event attribute information.

Schema Definition. The schema or ontology describes the high-level structure of a KG, including the type of entities and relations therein. Figure 2 depicts the overall schema of UrbanFloodKG, where the nodes represent types of entities and the edges describe their relationships in UrbanFloodKG.

Entity Identification. Based on urban flood data, the basic entities include the following types:

- **Event.** Event represents each specific urban flood event.
- **Influence.** Influence represents the impact caused by urban flood events.
- **Reason.** Reason represents the causes of urban flood events summarized by professionals.
- **Leader.** Leader represents the personnel responsible for handling urban flood events.
- **Department.** Department represents the department to which personnel responsible for handling urban flooding events belong.
- **Region.** Department represents the department to which personnel responsible for handling urban flood events belongs.

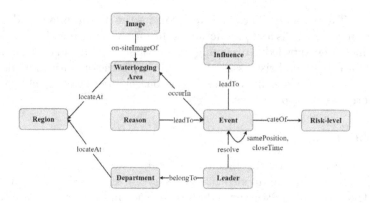

Fig. 2. The schema of UrbanFloodKG. Each rectangle represents a node

- **Water-logging Area.** Region represents the administrative region of the city to which the urban flood event belongs.
- **Image.** Image records some on-site pictures of the urban flood events.
- **Risk-level.** Risk-level represents the evaluation of the risk level of urban flood events by experts in the later stage.

Relation Extraction. Based on the identified entity types in the Urban-FloodKG, we extract typical relations to describe semantic connections between entities, which are classified as follows:

- **Spatial Relations.** The spatial relationships model the spatial relationship between entities. The locatAt and occurIn relationships simulate the spatial relationship between event entities and location entity as well as region entities, respectively.
- **Personal Relations.** Personal relationships focus on individual knowledge, such as the belongTo relation-ship between the department entities and the leader's workplace, and the resolve relationship between the leader entities and the event entities they are responsible for handling.
- **Causal Relations.** Causal relationships help us understand and predict connections and outcomes. The leadTo relationship connects event entities to reason and influence entities, forming a complete chain of events.
- **Affiliated Relations.** Affiliated relationships represent connections between entities. The on-siteImageOf corresponds to the on-site image entities to the water-logging area entities, and the cateOf connects the event entity to the risk-level entities. The samePosition and clos-eTime connect event entities with the same location or close occurrence time, effec-tively considering the association between entities.

Attribute Enrichment. To fuse more urban data into the UrbanFloodKG, the system further enriches the entities with attribute provided, which are listed as follows:

- **Event Attributes.** The event attributes are listed in Table 1. These pieces of information can be supple-mented with available data.

Table 1. Event Attributes.

Attribute	Classified Group
StartTime (ms)	timestamp
Depth (m)	0–0.3, 0.3–0.5, 0.5–0.8, >0.8
Duration (h)	0–1, 1-3, 3–5, >5
Lanes (number)	0–3, 3-6, 6–9, >9

- **Influence Attributes.** The impact of the event is expressed in text form, and its attribute is the relevant textual description of the impact caused by the event.
- **Reason Attributes.** The cause of the event is expressed in text form, and their attributes are the relevant textual description of the causes that triggered the event.
- **Leader Attributes.** The leader attributes include the name and phone number.
- **Department Attributes.** The department attributes include the name and location information.
- **Region Attributes.** The region attributes include the boundary, area, and number of buildings information of the region.
- **Water-logging Area Attributes.** The water-logging area attributes include the specific location information and the area information of the water-logging area.
- **Image Attributes.** The on-site image attributes of the event are extracted as feature maps using the ResNet model.
- **Risk-level Attributes.** The risk-level attributes of the event include three types: mild, moderate, and severe.

Knowledge Storage. UrbanFloodKG system uses object-oriented approach to store triples as node and edge classes. Specifically, Table 2 shows three types of node and two types of edge in OrientDB. Each node or edge has a unique Rid identifier and name. In particular, each entity node is considered as an instance of a schema node, so there is a schema edge named hasInstance between them. The name of the entity node is a combination of the name of the schema to which it belongs and its own Rid. Fig 3 shows a portion of the KG.

In particular, each Entity Node is considered as an instance of a Schema Node, so there is a Schema Edge named hasInstance between them. The name of the Entity Node is a combination of the name of the schema to which it belongs and its own Rid. Figure 3 shows a portion of the KG.

Table 2. The classes in OrientDB. There are three types of nodes: Schema Node, Entity Node, and Attribute Node, two types of edges: Schema Edge and Entity Edge.

Class	Type	Name
Schema	Schema Node	Schema name
Entity	Entity Node	SchemaName_Rid
Attribute	Attribute Node	Attribute value
SchemaEdge	Schema Relation Edge	hasInstance, Relation name
EntityEdge	Entity Relation Edge	Attribute name, Relation name

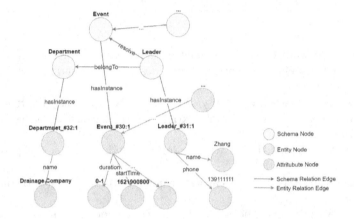

Fig. 3. The portion of UrbanFloodKG. Each Entity Node is connected to multiple Attribute Nodes through an Entity Relation Edge, and the edge name corresponds to the name of the attribute.

4.4   Algorithm Layer

Translation-Based Models. The urban flood events information is stored in the form of nodes and edges. It is very difficult to directly analyze them, and existing research has proposed KG representation learning [31]. It learns low-dimensional continuous representation vectors of entities and relationships while preserving the inherent structure and semantics of the KG.

Given a knowledge graph $\mathcal{G} = (\mathcal{E}, \mathcal{R}, \mathcal{F})$, which contains embedding vectors $\mathcal{E}$ and $\mathcal{R}$ of entities and relationships, as well as a set of triples $\mathcal{F}$. The knowledge graph representation learning algorithm designs various scoring functions ϕ to calculate scores for entity and relationship embedding, so as to compute higher scores for valid triples than for invalid ones. Based on predefined loss function $\mathcal{L}$ such as cross-entropy loss and hinge loss [25], the KG representation learning algorithm updates embedding parameters until convergence.

Representative scoring functions used for KG representation include translation-based models [3] and tensor decomposition-based models [1], among others. For example, given a triple (h, r, t), the TransE model based on the distance assumption designs a scoring function that assumes the head entity and tail entity are close to each other through a specific relationship operation. This system integrates two Translation-based models: TransE [3] and HolE [24]. In addition, it also supports two decomposition-based models: ComplEx [30] and DisMult [35].

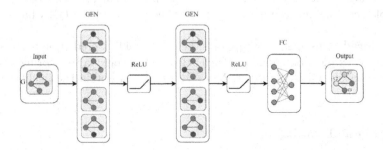

Fig. 4. The GEN model for urban flooding event classification outputs the corresponding classification (C1, C2, C3) for each node.

GNN-Based Models. Some GNN-based models can be integrated into the system. We have integrated a GEN-based model [16] and set it as the default model. In the forward propagation process of the GEN network, the feature vector update formula for each urban flooding event node is:

$$\mathbf{x}'_i = MLP(\mathbf{x}_i + AGG(ReLU(\mathbf{x}_j + \mathbf{e_{ji}}) + \epsilon : j \in \mathcal{N}(i))) \tag{1}$$

This formula describes an information aggregation and updating process in the GEN network. In this process, the feature vector of a node is updated through a multi-layer perceptron (MLP) and an aggregation function (AGG). Specifically, $\mathbf{x}_i$ represents the feature vector of node i, $\mathbf{x}_j$ represents the feature vector of node j, $\mathbf{e_{ji}}$ represents the edge feature vector, which represents the attribute information of the edge from node j to node i, and ϵ is a learnable parameter vector used to adjust the scaling and offset of the node feature vector. The AGG function of the model uses a summation aggregation function.

The model consists of two layers of GEN networks and a fully connected layer, as shown in Fig. 5. The input is a graph G, which includes the feature vectors of each node, the edge indices of adjacent nodes, and the feature vectors of each edge.

Operations. The UrbanFloodKG system supports typical functions in traditional graph systems, such as Cypher and embedding access. To adapt the system to urban flood risk assessment, we have developed four types of operations. Ampligraph [7] and PyG [11] are adopted to implement the latter three proposed functions.

- **query.** This operation accepts the Cypher query from the user, which returns the corresponding results on the UrbanFloodKG.
- **get_emb.** This operation provides the direct interface to access the embedding of the entity or relation in the UrbanFloodKG. Especially, the embeddings of various Translation-based models can be provided based on the input augment model.
- **node_cla.** This operation achieves the node classification tasks in KG. By providing the feature of vector of entity the system calls this operation to implement classification based on the embedding learnt by GNN based models.
- **link_pred.** This operation supports the relational link prediction between two entities of src_ent and tar_ent, which calculates a score via the scoring function via Translation-based models. Moreover, user can obtain the likelihood of a triple fact.

4.5   Digital Twin Layer

Fig. 5. The display interface of DT layer. UrbanFloodKG system has been integrated for support the application in DT layer.

The DT layer is developed using virtual reality and visualization technology. For a specific application in DT layer, we can call various operations as well as operation combinations in algorithm layer for the task demands. Figure 5 shows the display interface of DT layer, and some services has been integrated for support the application in DT layer.

Representative applications are introduced in three categories:

- **Knowledge Reasoning.** The DT layer is used to show the results of knowledge reasoning. For example, users can input (event, cause, ?), and the DT layer will show causes for urban flooding for that event.

- **Event Classification.** Users can input information about urban flood events like rainfall, water level, duration, etc. The DT layer will analyze the event's characteristics, evaluate the risk level, and provide a risk score.
- **Urban Decision Making.** The DT layer helps related officials better understand and predict the risk and impacts of urban floods. It provides data and knowledge to support decision making for urban flood.

5 Experiment and Discussion

5.1 Dataset and Environment

We collect a total of 10,000 urban flood event messages from 2010 to 2018 in Guangzhou. In the node classification task, we extract the node feature matrix, edge feature matrix, and edge index matrix required for GEN training based on the dataset. The experiments are implemented with an AMD Ryzen7 5800 CPU, a 32 GB RAM and one NVIDIA GeForce RTX 4080 GPU.

5.2 Link Prediction Analysis

We formulate the traditional causal prediction problem into the link prediction problem on the UrbanFloodKG, which is stated as follows:

UrbanFloodKG-based Reason Prediction Problem. Given the Urban-Flood-KG $\mathcal{G} = \{\mathcal{E}, \mathcal{R}, \mathcal{F}\}$, a recorded information of urban flood event like reason e_r led to the event e_e can be expressed as (e_r, r_{leadTo}, e_e) with e_r, and e_e as entities and r_{leadTo} as relation therein.

Hence, the Reason Prediction problem of potential reasons of an urban flood event e_e, can be formulated as the link prediction problem of $(?, r_{leadTo}, e_e)$ in UrbanFloodKG. Especially, the DT layer calls the operation link_pred to predict if there exist leadTo links between reason entities and event entities.

To evaluate the proposed framework, we sample some data from our own constructed UrbanFlood dataset. We split the dataset by 7:1:2 as the train-/valid/test datasets. The first 70% records of each event's id form the training set, and the middle 10% records are the valid set, while the left records are used for testing.

By using four knowledge graph representation learning models: TransE, ComplEx, DistMult, and HolE, we evaluate the effectiveness of our system for reason prediction using the MRR (Mean Reciprocal Rank), MR (Mean Rank), and Hits@n metrics. The experimental results, shown in Table 3, demonstrate the effectiveness of our system for reason prediction.

Table 3. The comparison result of reason prediction task.

Model	MRR	MR	Hits@10	Hits@3	Hits@1
ComplEx	0.36	1166.96	0.69	0.54	0.16
DistMult	0.39	1010.31	0.72	0.56	0.19
HolE	0.49	884.47	0.85	0.71	0.27
TransE	0.38	457.92	0.66	0.49	0.13

5.3 Node Classification Analysis

In this part, we summary the issues about risk assessment in urbanFlood research that can be considered as node classification problems.

Risk-level classification. For the risk assessment of urban flooding events, in addition to triggering from the events' own attributes, we believe that two hypotheses can be proposed:

- The more urban flooding events associated with time and location, the higher the risk of individual events.
- The higher the degree of the event node, the greater the risk.

Therefore, assessing event risk should account for both temporal and spatial elements, beyond just the events themselves. In our urban flood dataset, the expert-evaluated risk levels of each event are represented by the risk-level entity. Especially, the DT layer calls the operation node_cla to classify each event node.

To evaluate the proposed framework, we also sample some data from our own constructed UrbanFlood dataset. We split the dataset into train, validation, and test sets with a ratio of 7:1:2. The first 70% of records for each event ID form the training set, the next 10% of records are used for validation, and the remaining records are used for testing.

We compare different models and evaluate them using accuracy, precision, recall, and F1-score metrics. The experimental results are shown in Table 4.

Table 4. The comparison result of risk-level classification prediction task.

Model	Accuracy	Precision	Recall	F1-score
GATv2 [4]	0.3564	0.5883	0.3564	0.2669
SG [33]	0.3832	0.7636	0.3832	0.2123
AGNN [29]	0.6741	0.6640	0.6741	0.6558
TAG [9]	0.3815	0.7640	0.3815	0.2107
GIN [34]	0.3914	0.7618	0.3914	0.2202
GEN	0.9136	0.9129	0.9136	0.9126

The GEN model-based classification network achieves 91% accuracy, outperforming other graph neural network models. It considers both node features and properties, representing nodes of various urban flood events as vectors for knowledge representation and supporting node classification.

6   Conclusion

In this paper, we propose the UrbanFloodKG system, which is capable of enabling DT for urban flood risk assessment. Based on the collected real dataset, we classify the practical application problems of flood risk assessment into link prediction and node classification task, and evaluate the effectiveness of the models using multiple approaches, thereby providing a new perspective for urban flood risk assessment. In the future, we will further improve the system and enhance reasoning ability to better empower the urban DT to help human reduce the actual economic losses.

Acknowledgements. The paper is supported by the Research on Key Technologies for Improving Flood Control Safety System of Nansha District, Guangzhou (823005916); the Jiangsu Province Water Conservancy Science and Technology Project (2022003); the Major Science and Technology Project of the Ministry of Water Resources (SKS-2022139).

References

1. Balažević, I., Allen, C., Hospedales, T.M.: Tucker: tensor factorization for knowledge graph completion. In: Proceedings of the 2019 Conference on Empirical Methods in Natural Language Processing and the 9th International Joint Conference on Natural Language Processing, Proceedings of the Conference, pp. 5185–5194. Association for Computational Linguistics (2019)
2. Bin, L., Xu, K., Pan, H., Zhang, Y., Shen, R.: Urban flood risk assessment characterizing the relationship among hazard, exposure, and vulnerability. Environ. Sci. Pollut. Res. **30**, 86463–86477 (2023)
3. Bordes, A., Usunier, N., Garcia-Duran, A., Weston, J., Yakhnenko, O.: Translating embeddings for modeling multi-relational data. In: Proceedings of the 27th Annual Conference on Neural Information Processing Systems 26. Neural Information Processing Systems Foundation (2013)
4. Brody, S., Alon, U., Yahav, E.: How attentive are graph attention networks? In: Proceedings of 10th International Conference on Learning Representations (2021)
5. Chang, T., Wang, C., Chen, A.: A novel approach to model dynamic flow interactions between storm sewer system and overland surface for different land covers in urban areas. J. Hydrol. **524**, 662–679 (2015)
6. Church, K.: Emerging trends word2vec. Nat. Lang. Eng. **23**(1), 155–162 (2017)
7. Costabello, L., Pai, S., Van, C., McGrath, R., McCarthy, N., Tabacof, P.: Ampligraph: a library for representation learning on knowledge graphs. https://doi.org/10.5281/zenodo.2595043 (2023), (Accessed 12 Aug 2023)
8. Dong, B., Xia, J., Zhou, M., Li, Q., Ahmadian, R., Falconer, R.: Integrated modeling of 2d urban surface and 1d sewer hydrodynamic processes and flood risk assessment of people and vehicles. Sci. Total Environ. **827**, 154098 (2022)

9. Du, J., Zhang, S., Wu, G., Moura, J., Kar, S.: Topology adaptive graph convolutional networks. arXiv preprint arXiv:1710.10370 (2017)
10. Feng, J., Zhu, Y.L., Hang, T.T., Lu, J.M., Wu, Y.R., Wang, W.P.: Knowledge Graph Research and Field Practice. Posts & Telecom Press, Beijing (2022)
11. Fey, M., Lenssen, J.: Fast graph representation learning with pytorch geometric. arXiv preprint arXiv:1903.02428 (2019)
12. Gutierrez, C., Sequeda, J.F.: Knowledge graphs. Commun. ACM **64**(3), 96–104 (2021)
13. He, K., Zhang, X., Ren, S., Sun, J.: Deep residual learning for image recognition. In: Proceedings of the 29th IEEE Conference on Computer Vision and Pattern Recognition, pp. 770–778. IEEE Computer Society (2016)
14. Hogan, A., et al.: Knowledge graphs. ACM Comput. Surv. **54**(4), Article No.: 71, 1–37 (2021)
15. Huang, Y., Yu, S., Luo, B., Li, R., Li, C., Huang, W.: Development of the digital twin changjiang river with the pilot system of joint and intelligent regulation of water projects for flood management. Shuili Xuebao **53**(3), 253–269 (2022)
16. Li, G., Xiong, C., Thabet, A., Ghanem, B.: Deepergcn: all you need to train deeper gcns. arXiv preprint arXiv:2006.07739 (2020)
17. Li, H., et al.: Integrated representation of geospatial data, model, and knowledge for digital twin railway. Inter. J. Digital Earth **15**(1), 1657–1675 (2022)
18. Li, J., Sun, A., Han, J., Li, C.: A survey on deep learning for named entity recognition. IEEE Trans. Knowl. Data Eng. **34**(1), 50–70 (2022)
19. Lietaert, P., Meyers, B., Van Noten, J., Sips, J., Gadeyne, K.: Knowledge graphs in digital twins for ai in production. In: Dolgui, A., Bernard, A., Lemoine, D., von Cieminski, G., Romero, D. (eds.) Advances in Production Management Systems. Artificial Intelligence for Sustainable and Resilient Production Systems, vol. 630, pp. 249–257. Springer, Cham (2021). https://doi.org/10.1007/978-3-030-85874-2_26
20. Liu, Y., Ding, J., Fu, Y., Li, Y.: Urbankg: an urban knowledge graph system. ACM Trans. Intell. Syst. Technol. **14**(4), 1–25 (2023)
21. Liu, Y., et al.: A novel cloud-based framework for the elderly healthcare services using digital twin. IEEE Access **7**, 49088–49101 (2019)
22. Lu, Y., et al.: Unified structure generation for universal information extraction. In: Proceedings of the 60th Annual Meeting of the Association for Computational Linguistics, pp. 5755–5772. Association for Computational Linguistics (ACL), Dublin, Ireland (2022)
23. Lyu, H., Shen, S., Zhou, A., Zhou, W.: Flood risk assessment of metro systems in a subsiding environment using the interval fahp-fca approach. Sustain. Urban Areas **50**, 101682 (2019)
24. Nickel, M., Rosasco, L., Poggio, T.: Holographic embeddings of knowledge graphs. In: Proceedings of the AAAI Conference on Artificial Intelligence, pp. 1955–1961. AAAI Press (2016)
25. Ruffinelli, D., Broscheit, S., Gemulla, R.: You can teach an old dog new tricks! on training knowledge graph embeddings. In: Proceedings of the 8th International Conference on Learning Representations(2020)
26. Shahat, E., Hyun, C.T., Yeom, C.: City digital twin potentials: a review and research agenda. Sustainability **13**(6), 3386 (2021)
27. Tao, F., Zhang, M., Liu, Y., Nee, A.: Digital twin driven prognostics and health management for complex equipment. CIRP Ann. Manuf. Technol. **67**, 169–172 (2018)

28. Tao, F., Zhang, H., Liu, A., Nee, A.Y.C.: Digital twin in industry: State-of-the-art. IEEE Trans. Industr. Inf. **15**(4), 2405–2415 (2019)
29. Thekumparampil, K., Wang, C., Oh, S., Li, L.: Attention-based graph neural network for semi-supervised learning. arXiv preprint arXiv:1803.03735 (2018)
30. Trouillon, T., Welbl, J., Riedel, S., Ciaussier, E., Bouchard, G.: Complex embeddings for simple link prediction. In: Proceedings of the 33rd International Conference on Machine Learning, pp. 3021–3032. International Machine Learning Society (IMLS) (2016)
31. Wang, Q., Mao, Z., Wang, B., Guo, L.: Knowledge graph embedding: a survey of approaches and applications. IEEE Trans. Knowl. Data Eng. **29**(12), 2724–2743 (2017)
32. Wang, Y., Chen, A., Fu, G., Djordjević, S., Zhang, C., Savić, D.: An integrated framework for high-resolution urban flood modelling considering multiple information sources and urban features. Environ. Model. Softw. **107**, 85–95 (2018)
33. Wu, F., Zhang, T., de Souza, A., Fifty, C., Yu, T., Weinberger, K.: Simplifying graph convolutional networks. In: Proceedings of 36th International Conference on Machine Learning, pp. 11884–11894. International Machine Learning Society (IMLS) (2019)
34. Xu, K., Jegelka, S., Hu, W., Leskovec, J.: How powerful are graph neural networks? In: Proceedings of 7th International Conference on Learning Representations (2019)
35. Yang, B., Yih, W., He, X., Gao, J., Deng, L.: Embedding entities and relations for learning and inference in knowledge bases. In: Conference Track Proceedings of 3rd International Conference on Learning Representations (2015)
36. Ye, F., Sheng, X., Nedjah, N., Sun, J., Zhang, P.: A benchmark for performance evaluation of multi-model database vs polyglot persistence. J. Database Manag. **34**(3), 1–20 (2023)
37. Zhao, Z., Zhang, M., Chen, J., Qu, T., Huang, G.: Digital twin-enabled dynamic spatial-temporal knowledge graph for production logistics resource allocation. Comput. Indus. Eng. **171**, 108454 (2022)
38. Zhong, L., Wu, J., Li, Q., Peng, H., Wu, X.: A comprehensive survey on automatic knowledge graph construction. arXiv preprint arXiv:2302.05019v1 (2023)

TASML: Two-Stage Adaptive Semi-supervised Meta-learning for Few-Shot Learning

Zixin. Ren[1], Ze. Tao[1], Jian. Zhang[1]([✉]), Guilin. Jiang[2], and Liang. Xu[2]

[1] Central South University, Changsha 410083, China
{renzixin,taoze.tz,jianzhang}@csu.edu.cn
[2] Hunan Chasing Financial Holdings Co., Ltd., Changsha 410035, China
{jiangguilin,xuliang}@hnchasing.com

Abstract. Human vision system(HVM) is a remarkable computational system that excels at analyzing complex visual scenes and recognizing objects with high accuracy and efficiency, and interrelated hierarchies for integrated processing of visual inputs is a signature of human vision intelligence. Similar to the process of human recognition of unknown objects, meta-learning enables models to "learn to learn" by providing a small number of "support" samples to solve the few-shot or one-shot problem in real scenarios. Inspired by this, we propose a two-stage adaptive semi-supervised meta-learning(TASML) framework with a hierarchical structure to solve the few-shot learning task. Specifically, parallels to the transition from primary to advanced visual cortex procedure in HVM, the invariant information of the target are obtained and features fusion is performed through an unsupervised representation sensing phase. Subsequently, a gradient-based meta-learning mechanism is used to simulate the execution of object targeting and recognition processes in HVM. In addition, a global context-aware(GCA) module is proposed in conjunction with our framework to follow the HVM's access to context and semantics of visual object. Extensive experiments conducted on Mini-ImageNet and CIFAR-100 datasets validate the effectiveness of our framework and module on few-shot classification tasks, which achieves 76.66%, 77.8% accuracy on $5way\text{-}5shot$ classification and 63.96%, 62.8% accuracy on $5way\text{-}1shot$ classification for two datasets, respectively.

Keywords: Meta-learning · Semi-Supervised learning · Few-shot classification

1 Introduction

Cognitive neuroscience researchers have found that human vision system(HVM) processes images in a hierarchical manner [1–3], from low-level features to high-level objects and concepts. In HVM, visual input from the retina is first sent to the lateral geniculate nucleus (LGN) of the thalamus and then relays information to the primary visual cortex (V1). V1 focus on detecting relatively simple

X. Song et al. (Eds.): APWeb-WAIM 2023, LNCS 14331, pp. 206–221, 2024.
https://doi.org/10.1007/978-981-97-2303-4_14

features like edges, bars, and blobs in the field of view. After extracting perceptual features of object, the category is locked and confirmed through visual cognitive mechanism. The secondary visual areas V2 and V3 receive inputs from V1 and process basic shapes, contours and depth cues in scene. Visual information is then sent to posterior inferotemporal cortex (PIT) to detect intricate two- or three-dimensional shapes and components of objects. The anterior temporal lobe combines content from PIT with contextual associations to recognize whole objects and categories.

With large-scale datasets and huge computational resources, deep learning methods have achieved much in the field of computer vision, yet its heavy reliance on labeled data limits its application in real-world scenarios. Meta-learning [4–7] has shown promise by learning to learn from few examples and adapting to new tasks quickly, to date, most existing meta-learning methods only utilize labeled data during meta-training and meta-testing, and the lack of prior knowledge and labeled data also leads to the fact that achieving robustness and generalization of meta models remains a major challenge [8,9].

As human partiality may represents the most efficient way to solve vision tasks, we propose a two-stage semi-supervised meta-learning approach that aims to acquire a magnificent ability to generalize and distinguish various objects by modeling simulated hierarchical structures. Similar to the transition from simple feature detection in view to basic shape and surface perception in high-level visual region of HVM, we introduce an unsupervised contrastive learning strategy to extract object features from the shallower to the deeper. In this stage, using the invariant information present in sample pair, the model achieves a gradual migration from low-level features to the perception of high-level semantics. And for the feature combination process in HVM, we forming a comprehensive representation of the inputs to guide downstream tasks. After feature perception, HVM attempts to match the perceptual representation of the object to stored visual memories or patterns of previously seen objects. Meta-learning follow the "learn to learn" paradigm which provides assignable prior knowledge to support object category discrimination. Using the representation obtained in previous phase, the later meta-learning stage performs pattern matching function as it in HVM. Meanwhile, in HVM, when a object is targeted, its meaning and semantic category is accessed, which in turn affects the recognition. Based on this, we propose a global context-aware module to capture long-range background information for object perception and cognition.

In this paper, we introduce a two-stage framework TASML for few-shot learning, which combines of an unsupervised invariant representation learning stage and a task adaptive meta-learner. In addition, we use a global context-aware(GCA) module to further improve few-shot classification performance.

The contributions of this paper can be summarized as follows:

- We propose a two-stage semi-supervised meta-learning framework for few-shot learning inspired by human visual object recognition mechanism, which organized by an object perception procedure and a feature cognition procedure. Gradient-based meta-learning methods usually require significant resources for model learning, the flexibility of our framework allows adjustment of feature extraction backbone scale without increasing the training difficulty of the following meta-learning stage.

- A feature aggregation module is introduced for global context-aware perception and cognition which further improve the long-range contextual information capture and usage.
- Experiments conducted on two frequently used datasets Mini-ImageNet and CIFAR-100, the results and visualization demonstrate that the proposed framework and the model constructed on it can simulate the hierarchical structure of object recognition in human vision system and achieve considerable few-shot classification accuracy.

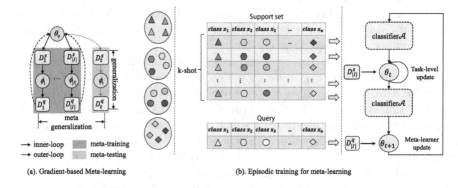

Fig. 1. Illustration of MAML learning process. (a).Training procedure of gradient-based meta-learning, which consists of an inner-loop and an outer-loop; (b).Task sampling process and model update sequence of gradient optimization based meta-learning algorithm.

2   Related Work

2.1   Brain-Inspired Model for Visual Object Recognition

In order to build a plausible brain-inspired model, a biological network of visual object representation is brought out to simulate the recognition procedure of human brain in terms of both visual functional mechanisms and neural activity prediction [2,3].

Benefit from the simulation of the cellular hierarchy of brain from simple to complex, CNN-based methods have achieved many successes. Dapello et al. [10] use increased depth but narrow receptive fields to simulate the hierarchical and increasingly spatially invariant processing in human visual cortex. Park et al. [11] uses a shallow hierarchical with fewer layers to reduce computational requirements, and maintain strong performance with long-range horizontal connections inspired by biological visual systems.

2.2   Meta-learning for Few-Shot Learning

Meta-learning algorithms enable models to rapidly and efficiently adapt to new tests and generalize to unseen classes, and can be divided into three main categories according to its optimization objectives or paradigms: **1) Metric-based methods** aim to learn a proper feature embedding or a valid metric space, and solve the classification problem by measuring sample-related or sample-to-class distance. Matching Networks [4] use an attention module to compare the cosine distance among examples. Relation Networks [5] extend [4] by proposing relation modules to model second-order relationships between features. Afrasiyabi et al. [12] holds an insight that bipartite representations should be produced for both intra-class matching and inter-class discrimination. Snell et al. [6] propose Prototypical Networks, which assign query example to its closest prototype according to l2-distance. Since then, Oreshkin et al. [13] try to learn a linear transformation on prototype representations for each task, which leads a task-adaptive metrics. **2) Memory-augmented methods** use external memory modules to read, write and modify object representations, which allow models to store and retrieve information to facilitate rapid learning on new tasks. R2-D2 [14] uses a large memory module to store embeddings and demonstrates how a memory component can dramatically improve data-efficiency. **3) Model-agnostic methods** aim to learn an optimization algorithm over tasks rather than a specific model. MAML [7] and follow-ups [15,16] have explored this field, resulting in fast adaptation to new tasks, however, second-order derivatives lead to instability of the training process and model optimization during test stage can be extremely expensive. Subsequent studies work on optimize or alleviate the above questions [17–20]. Yao et al. [21] propose task interpolation to reduce the number of training tasks by interpolating between training tasks to generate dense sampling of related tasks. In our TASML, we also follow the inner-outer-loop training process, but we leverage a two-stage framework to reduce the costs while keeping a powerful feature extraction capacity through freezing the representation mining network at meta-learning stage.

Moreover, benefited from the emergence and development of weakly supervised learning paradigms, **semi-supervised meta-learning** methods have been heavily studied and proposed. Ren et al. [22] use soft k-means to improve [6] with additional unlabeled data. SelfNet [23] is based on local fisher discriminant analysis, considering geometry of both labeled and unlabeled examples to learn a more discriminative representation.

3   Methodology

3.1   Preliminary

Few-shot Learning and Model-Agnostic Meta-Learning. Meta-learning aims to improve learning efficiency of model by leveraging prior experience from a variety of tasks under few-shot scene. In few-shot learning, the episodic training paradigm is used, which consists of two phases called meta-training and meta-testing, and extracts knowledge from task T_i. Every task T_i is defined as $N\text{-}way$

K-shot, which means classifying N classes drawn from $D_{labeled} = \{(x_i, y_i)\}_i$ and $y_i \in Y = \{C_1, C_2, ..., C_N\}$ with K samples. When it comes to meta-testing phase, model is finetuned with the help of support set $S_{support}$ and finally evaluated its performance on query set Q_{query}. MAML is a learning framework consists of an internal and external optimization loop, inner-loop to achieve learning of support set and the outer-loop increases the potential of the model for future tasks by exploiting the query set samples. The training procedure is illustrated in Fig. 1.

Contrastive Learning. Contrastive learning is always an unsupervised or self-supervised manner, it refers to learning a mapping function $F(x)$ from input x to representation r. And r is a stable and effective compress of object x, which can be able to guide downstream tasks like image classification, object detection and image retrieval efficiently. By using a projection head $Proj(r)$ to enable similarity measuring between representation pairs (r_i, r_j), contrastive learning maximize the similarity between semantically similar samples. We call a sample pair $(\tilde{x}_i, \tilde{x}_j)$ consisted of two transform versions of sample x created by data augmentation module $Aug(x)$ *positive pair*. Each of the different augmentations obtains a subset of the feature area from raw sample, which leads to discover and capture the most discriminative high-level semantic information of x.

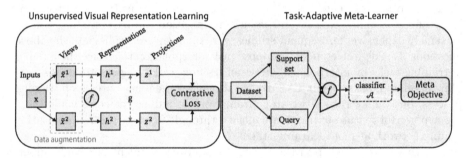

Fig. 2. Illustration of TASML training pipeline. TASML mainly consists of two stages:(1).Contrastive loss function guides visual features learning. Through abundant data augmentation for samples, invariant representations will be gained without any additional label information.(2).Supervised meta-learning. Episodic training paradigm enriches task diversity through massive random sampling.

3.2 The Two-Stage Semi-supervised Meta-learning Framework

Meta-learning constructs solutions for few-shot scenarios by using prior knowledge from a large amount of randomly sampled tasks. We propose a two-stage hierarchical network inspired by HVM, the main components are a robust feature exploiter followed by a gradient-based meta-learner and we also propose a global context-aware module to get better perception and cognition of the object.

While the first stage training a visual representation extractor $E_\theta(x)$ to better denotes object x through self-supervised contrastive learning paradigm on unlabeled dataset $D_{unlabeled}$. The network is optimized to minimize the contrastive loss of sample pairs conducted by data augmentations. After that, we freeze E_θ with learned parameters θ and replace the projection head by a novel meta-learner M_ϕ. M_ϕ is trained in episodes with given few samples on labeled dataset $D_{labeled}$. We sample every supervised meta training task T_i according to typical few-shot learning inputs setting type (N, K), which means the task is combined with N categories of objects and each class contains K samples.

Meanwhile, our framework allows the meta-learning stage to freeze the powerful and complex feature extraction network, overcoming the resource dependence of gradient-based meta-learning methods due to the second-order derivative computing and the two-layer optimization structure. The pipeline of our TASML is illustrated in Fig. 2.

3.3 Unsupervised Visual Representation Learning

Visual representation learning is formulated as a margin-based ranking objective, where the model is trained to differentiating between positive and negative pairs by a margin. Randomly sample a batch $\{x_k\}_{k=1}^{N}$ of N examples and for each sample we create two random augmentations, which results in $2N$ positive sample pairs $(\tilde{x}_{2k}, \tilde{x}_{2k-1})_{k=1...N}$. The training task is aimed to minimize the loss function between all positive pairs and it is defined as follow:

$$Loss = \sum_{i \in I} Loss_i = \sum_{i \in I} \log \frac{\exp(dis(z_i, z_{j(i)}/\tau)}{\sum_{k=1}^{2N} \mathbb{I}_{[k \neq i]} \exp(dis(z_i, z_k/\tau)}, \qquad (3.1)$$

where index $i \in I \equiv \{1...2N\}$ and $j(i)$ means the positive, the others are called negatives. $dis(u, v) = u^T v / \parallel u \parallel \parallel v \parallel$ is the dot product between vector u and v normalized by l_2 norm (i.e., *cosine distance*) to calculate similarity between a sample pair. And $z_l = Proj(r_l)$ is a vector produced by projection head on its representation r_l. τ is a temperature indicating scalar. $\mathbb{I}_{[k \neq i]}$ is an indicator function to exclude the sample itself, which means the denominator holds a total of $(2N - 1)$ terms consisted of one positive pair and $(2N - 2)$ negatives.

3.4 Gradient-Based Meta-learning for Few-Shot Learning

We define a meat-learner M_ϕ which is trained by MAML to further solve the downstream tasks followed by visual representation learning.

Under the basic meta-training assuming, the extensive tasks T_i are all drawn from a distribution $p(T)$, each task consists of a support set $S_i = \{(x_{i,j}^s, y_{i,j}^s)\}_{j=1}^{N_s}$ and a query set $Q_i = \{(x_{i,j}^q, y_{i,j}^q)\}_{j=1}^{N_q}$, where N_s and N_q represents the number of support and query samples, respectively.

During the meta-training stage, M_ϕ is adapted to the support set of i-th task T_i with several gradients steps which called inner-update, then evaluated and update model parameters ϕ on its query set. This bi-level optimization problem can be formulated as follow:

$$\phi_0^* := \min_\phi \mathbb{E}_{\mathcal{T}_i \sim p(\mathcal{T})}[\mathcal{L}(M_{\varphi_i}(X_i^q), Y_i^q)], \qquad (3.2)$$

$$s.t. \quad \varphi_i = \phi_0 - \alpha \nabla_{\phi_0} \mathcal{L}(M_{\varphi_i}(X_i^s), Y_i^s), \qquad (3.3)$$

where $\mathcal{L}$ is loss function for specific downstream mission(i.e., cross-entropy for classification and mean square error for regression problem) and α is the inner-update learning rate. $X_i^{s/q}$ is sample set and $Y_i^{s/q}$ are labels related. And model parameters ϕ is updated across tasks through:

$$\phi \leftarrow \phi - \beta \nabla_\phi \sum_{\mathcal{T}_i \sim p(\mathcal{T})} \mathcal{L}(M_{\varphi_i}(X_i^s), Y_i^s), \qquad (3.4)$$

where β is outer-update step size. When it comes to meta-testing and for a given task $\mathcal{T}_t$, the model M_ϕ is finetuned on support set to resulting the corresponding task adaptive parameter ϕ_t.

3.5   Global Context-Aware Module

In human vision system, both the context and background information in which the object is located support the final judgment of target recognition, we take advantage of this characteristic and propose a feature aggregation module for global context-aware perception and cognition, which parallel capture long-range contextual information in both channel dimension and spatial dimension.

Channel-wise Attention Module. As shown in the upper part of Fig. 3, given the feature map $X \in \mathbb{R}^{C \times H \times W}$ from visual representation capture component, where $N = H \times W$. Do matrix multiplication as the scoring function on A and its transpose version A^{trans} to calculate their response similarity:

$$s(x_i, q) = F(A, a_j) $$
$$= x_i^T a_j, \qquad (3.5)$$

where q is query and $F(*)$ donates scoring function. Then use softmax function to obtain channel-wise attention map $M_{ch} \in \mathbb{R}^{C \times C}$:

$$m_{ij}^{ch} = \frac{\exp(s_i)}{\sum_{j=1}^C \exp(s_j)}$$
$$= \frac{\exp(A_j \cdot A_i)}{\sum_{j=1}^C \exp(A_j \cdot A_i)}, \qquad (3.6)$$

where m_{ij}^{ch} represents how much the j-th channel affects the i-th channel. Then another matrix multiplication is done between X and M_{ch}, the result is reshaped to $\mathbb{R}^{C \times H \times W}$. The final channel-wise output $O_{ch} \in \mathbb{R}^{C \times H \times W}$ comes from a scaled element-wise sum operation with original feature map X, which models long-range semantic dependencies between different channels and various feature maps:

$$O_i^{ch} = \alpha \sum_{j=1}^{C} (m_{ij}^{ch} X_j) + X_i, \tag{3.7}$$

where α is a learnable scale parameter, and it gradually learns from 0.

Position-wise Attention Module. The feature map X $\in \mathbb{R}^{C \times H \times W}$ is fed into a convolution block to gain two different feature maps $\{B, C\}$ with the same size as $\mathbb{R}^{C \times H \times W}$ and reshaped to $\mathbb{R}^{C \times N, N = H \times W}$. Pixel-wise spatial attention map $M_{sp} \in \mathbb{R}^{N \times N}$ is calculated on similarity score between C and transposed B obtained through dot product function $s(x_i, q) = x_i^T q$, M_{sp} is defined as follow:

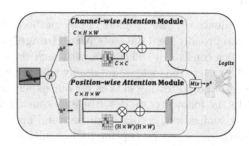

Fig. 3. Global context-aware module. The upper flow is a channel-wise attention compute procedure and the bottom is for position-wise attention map capture. An aggregation operation is followed to obtain the final feature map.

$$m_{ji}^{sp} = \frac{exp(B_i \cdot C_j)}{\sum_{j=1}^{N} exp(B_i \cdot C_j)}, \tag{3.8}$$

where m_{ji}^{sp} metrics the i-th position's influence on j-th position.

A higher correlation response will generate when two positions share more similarity with each other. And we parallel generate map D $\in \mathbb{R}^{C \times H \times W}$ and reshape it to $\mathbb{R}^{C \times N, N = H \times W}$. The spatial-wise response map $O_{sp} \in \mathbb{R}^{C \times H \times W}$ is calculated on pristine feature map X with a matrix multiplication result between D and spatial attention map M_{sp}:

$$O_i^{sp} = \beta \sum_{j=1}^{N} (m_{ij}^{sp} D_j) + X_i, \tag{3.9}$$

where β scared the terms with an initiation 0 and learns to getting higher weight for it. Still, every pixel in the output is a combined effect of all positions and original representation, which means intra-class distance is compacted and semantic consistency is guaranteed, the processing stream is shown in the Fig. 3.

Feature Aggregation. To further leverage the global contextual info, we use pixel-wise summation on channel attention map $O_{ch} \in \mathbb{R}^{C \times H \times W}$ and spatial attention map $O_{sp} \in \mathbb{R}^{C \times H \times W}$ to obtain a symphysis feature map S $\in \mathbb{R}^{C \times H \times W}$:

$$s_{ij}^c = Mix(O^{ch}, O^{sp})$$
$$= \sum_{i=1}^{H} \sum_{j=1}^{W} (O_{ij}^{ch} + O_{ij}^{sp}), \tag{3.10}$$

where $Mix(*)$ is a feature fusion strategy and we define it as pixel-wise summation. And c donates the c-th channel of both O_{ch} and O_{sp}. Coordinate pair(i, j) indicates the element on a specific feature map $O \in \mathbb{R}^{H \times W}$. With this strategy, we get a better visual representation interpretation ability, and make the subsequent reasoning more justifiable.

4   Experiments

To evaluate the effectiveness of our method, we carry out extensive experiments on two broadly used datasets **Mini-ImageNet** [4] and **CIFAR-100** [24]. Both of which consists of 100 classes with each class includes 600 natural color images. And for every dataset, we randomly split it into 64 training classes, 16 validation classes, and 20 testing classes for our experiments and evaluation.

In the following, we describe experiments details and results, ablation study and visualization, trying to understand and restore the process of HVM when performing object perception and cognition.

4.1   Few-Shot Image Classification

We carry out image classification tasks on two commonly used few-shot learning dataset to evaluate the performance of the proposed model. All the experiments follow typical few-shot learning inputs setting type (N, K). Additionally, we resize the inputs to 84×84 and 32×32 respectively, corresponding to [4] and [24]. We choose mean accuracy with 95% corresponding confidence interval as indicator. Backbones are trained from scratch without pretrain, and at testing stage, the average prediction accuracy is conducted over 600 episodes.

Table 1. Average accuracy (%)with different feature extractor complexity.

Dataset	ResNet-12		ResNet-18		ResNet-34		ResNet-50	
	1shot	5shot	1shot	5shot	1shot	5shot	1shot	5shot
Mini-ImageNet	54.83	63.8	57.96	68.9	58.54	66.36	**62.8**	**76.66**
CIFAR-100	59.33	71.44	63.67	74.46	63.33	74.17	**63.96**	**77.8**

Results. Since MAML [7] needs to save and compute second-order derivatives during the training process, the demand on computational resources is huge, and generally only a smaller backbone can be selected to support the task with source limited facilities. Benefit from our proposed two-stage learning framework, we can select an arbitrarily network as the backbone without increasing the model training complexity in later stage, similar to the metric-based approach, we reduce the computation and memory usage by freezing the feature extraction network during the meta-training stage. As shown in Table 1, we note that the experimental results show the backbone depth applied by the network has direct influence on classification accuracy. As the model grows deeper and more complex, the enhanced feature extraction capability of the model can directly benefits the final determination, with a steady performance improvement especially on the 5*shot* few-shot task. Meanwhile, the classification accuracy on 1*shot* setting on CIFAR-100 dataset is not affected much by the model size and shows some fluctuations, indicating that the learning strategy in meta-learning stage has more influence on the classification results under extremely few-shot contexts, because in the case of only a single sample as the support set, the extracted features are very limited, and how to make better use of these features is the key to the correct judgment of the model.

Table 2. Comparison to prior works. And * denotes weakly-supervised meta-learning methods. † indicates the backbone is pretrained using all training datapoints. The results are sorted in ascending order by 5*shot* classification accuracy under same settings, respectively.

Method	Backbone	Mini-ImageNet	
		1-shot	5-shot
Prototypical Net [6]	Conv-4	42.38	62.39
Soft k-Means+ Cluster* [22]	Conv-4	49.03	63.08
Masked Soft k-Means* [22]	Conv-4	50.41	64.39
MetaGAN* [25]	Conv-4	50.35	64.43
Soft k-Means* [22]	Conv-4	50.09	64.59
TPN* [26]	Conv-4	52.78	66.42
SALA* [27]	Conv-4	54.14	67.23
Semi DSN* [28]	Conv-4	52.18	68.48
FSWL* [29]	Conv-4	52.98	72.77
LSTALF [30]	ResNet-12	56.86	72.92
Prototypical Net [6]	ResNet-12	53.49	72.97
TPN* [26]	ResNet-12	60.9	73.4
MTL [31]	ResNet-12†	59.4	74.6
TASML*(ours)	ResNet-34	**58.54**	66.36
UDS* [32]	ResNet-34	54.23	74.48
FSL [33]	ResNet-50	49.33	69.17
TASML*(ours)	ResNet-50	**62.8**	**76.66**
Soft k-Means+Cluster* [22]	WRN-28	49.61	63.49
Soft k-Means* [22]	WRN-28	50.95	66.61
Masked Soft k-Means* [22]	WRN-28	51.46	67.02
LEO [34]	WRN-28†	61.68	**77.47**
Adv.ResNet [35]	WRN-40†	55.2	69.6

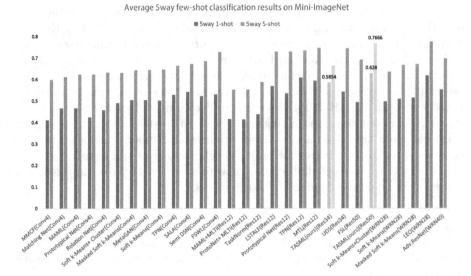

Fig. 4. Comparison to prior works. The abscissa indicates method name and its backbone.

Results on Mini-ImageNet. We try to compare the performance of our proposed method with other classical meta-learning algorithms [4–7], latest researches on gradient-based or metric-based methods [30,31,33,36,37] and weakly supervised meta-learning algorithms [22,25–29,32]. From Tab. 2 and Fig. 4, we can observe that the overall classification accuracy of the model tends to increase as the complexity of the network increasing. Due to the flexibility of the two-stage framework, we can change and increase the complexity of the feature extraction network to improve the feature perception of the model. Compared to other weakly-supervised meta-learning algorithms, we obtained classification accuracy higher than 4.74%, 5.8%, and 4.42% under 1*shot* setting for each of *Soft k-Means* [22], *Soft k-Means + Cluster* [22], and *Masked Soft k-Means* [22] with Conv-4 as backbone respectively, by enhancing the model's capability to extract image features with ResNet-12. Moreover, pretraining the model with all datapoints can improve the classification performance of the model as the results shown for MTL [31], Adv.ResNet [35] and LEO [34]. We achieved the second highest accuracy 76.66% on 5*way*-5*shot* classification tasks using ResNet50 without any pretrain, and outperformed all other methods shown on 5*way*-1*shot* tasks, shows that our proposed model can still remain competitive in the case of extremely few labeled samples.

Table 3. Average 5way few-shot classification results on CIFAR-100.

Method	Backbone	CIFAR-100	
		1-shot	5-shot
LSTALF [30]	ResNet-12	40.71	54.75
TADAM [13]	ResNet-12	40.4	56.54
Cosine classifier [38]	ResNet-12	39.03	58.06
TASML(ours)	ResNet-12	**59.33**	**71.44**
TASML(ours)	ResNet-18	**63.67**	74.46
Dual TriNet [39]	ResNet-18	62.77	**77.81**

Results on CIFAR-100. Tab. 3 gives the experimental results on CIFAR-100. Under the same network structure, our proposed method is able to obtain comparable or better performance than the latest and state-of-art algorithms, especially in 1*shot* classification task. The results show that the model proposed can fully utilize the sample information to support the final decision making process even if there is only one sample. Equally, as shown in Tab. 1, the scale of the model also affects its performance to some but not all extent.

4.2   Ablation Study

To gain more insight into our approach, we demonstrate the effectiveness and contribution of the proposed global context-aware module through ablation experiments in this section. We conduct experiments on two different datasets using various backbones, the results are shown in Tab. 4. Results suggest that GCA module can significantly improve the classification efficiency in few-shot scenarios by fusing the perception and cognition of both channel-wise and pixel-wise features. When using the GCA module, the model has a more significant and stable improvement with 1*shot* setting on Mini-ImageNet, which means only one sample per class can also provide the corresponding support.

Table 4. Ablation experiments for GCA module. Baseline donates the model follows our framework without GCA and CGA indicates the module is activated.

Dataset		ResNet-12		ResNet-18		ResNet-34		ResNet-50	
		1-shot	5shot	1-shot	5shot	1-shot	5-shot	1-shot	5-shot
Mini-ImageNet	GCA	55.47	67.87	57.96	68.9	58.54	66.36	62.8	76.66
	Baseline	54.83	63.8	55.27	69.3	56.2	68.46	58.4	71.44
CIFAR-100	GCA	59.33	71.44	63.67	74.46	63.33	74.17	63.96	77.8
	Baseline	52.34	66.55	53.56	67.53	57.47	68.85	58.0	73.0

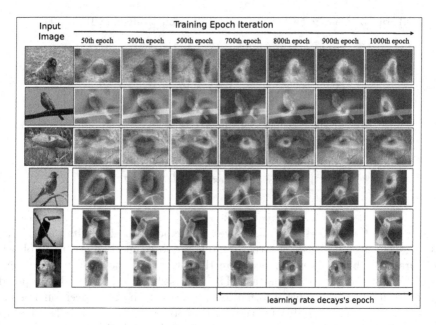

Fig. 5. Illustration of class activation mapping. Training iterations increasing and learning rate decay.

4.3   Visualization

In this section, we visualize changes in class activation mappings with training iterations and learning rate reduction, trying to simulate and reveal the extraction and utilization of information within the visual area by human vision system.

We reconstruct the contribution distribution of the predicted outputs as the number of iterations increasing with ResNet50 as backbone. As shown in Fig. 5, from left to right, the map highlight the different discriminative pixels or regions used for image classification. When visual signals are input, rods and cones in the retina detect basic edges, motion and color features. Accordingly, in the early stage of model training as shown in the left, the model decision is mainly supported by the large range perception of images, discrimination is usually based on the entire image area without a specific focus. With iterations of training procedure, the model gradually learns to focus on the features of the object itself, which is similar to visual cortex detects simple features and feature combination to represents parts of object in HVM. When learning rate changes, usually the learning ability of the model also changes abruptly, model's attention is more focused on the target to be classified itself, and the determination of the category is more dependents on its discriminative features, such as the beak of a bird, the umbrella of a mushroom.

The visualization results show the two-stage model we proposed well simulates the perceptual and cognitive processes of HVM in performing object recognition and classification. In addition to simple features extraction and fusion, the meta-learner simulates the matching of familiar patterns in HVM as well as memory-based learning and inference to obtain the final classification result of the image.

5 Conclusion

In this paper, we introduce a few-shot learning framework TASML, which combines of an unsupervised visual representation learning stage and a multi-task adaptive meta-learner learning stage. We also use GCA module to further simulate contextual integration in human brain. And we conducted extensive experiments and the results demonstrate the feasibility of our framework and effectiveness of the methodology and module. In the future, we will continue to explore and optimize the adaptability of our framework to achieve supporting for a variety of downstream visual tasks under few-shot scenario, such as image segmentation, target localization, etc.

References

1. Cichy, R.M., Khosla, A., Pantazis, D., Torralba, A., Oliva, A.: Comparison of deep neural networks to spatio-temporal cortical dynamics of human visual object recognition reveals hierarchical correspondence. Sci. Rep. **6**(1), 27755 (2016)
2. Federer, C., Xu, H., Fyshe, A., Zylberberg, J.: Improved object recognition using neural networks trained to mimic the brain's statistical properties. Neural Netw. **131**, 103–114 (2020)
3. Zhuang, C., Yan, S., Nayebi, A., Schrimpf, M., Frank, M.C., DiCarlo, J.J., Yamins, D.L.: Unsupervised neural network models of the ventral visual stream. Proc. Natl. Acad. Sci. **118**(3), e2014196118 (2021)
4. Vinyals, O., Blundell, C., Lillicrap, T., Wierstra, D., et al.: Matching networks for one shot learning. Adv. Neural Inf. Process. Syst. **29**, 1–9 (2016)
5. Hu, H., Gu, J., Zhang, Z., Dai, J., Wei, Y.: Relation networks for object detection. In: Proceedings of the IEEE Conference on Computer Vision and Pattern Recognition, pp. 3588–3597 (2018)
6. Snell, J., Swersky, K., Zemel, R.: Prototypical networks for few-shot learning. Adv. Neural Inf. Process. Syst. **30** (2017)
7. Finn, C., Abbeel, P., Levine, S.: Model-agnostic meta-learning for fast adaptation of deep networks. In: International Conference on Machine Learning, pp. 1126–1135. PMLR (2017)
8. Munkhdalai, T., Yu, H.: Meta networks. In: International Conference on Machine Learning, pp. 2554–2563. PMLR (2017)
9. Yin, M., Tucker, G., Zhou, M., Levine, S., Finn, C.: Meta-learning without memorization. arXiv preprint arXiv:1912.03820 (2019)
10. Dapello, J., Marques, T., Schrimpf, M., Geiger, F., Cox, D., DiCarlo, J.J.: Simulating a primary visual cortex at the front of cnns improves robustness to image perturbations. Adv. Neural. Inf. Process. Syst. **33**, 13073–13087 (2020)
11. Park, Y., Baek, S., Paik, S.B.: A brain-inspired network architecture for cost-efficient object recognition in shallow hierarchical neural networks. Neural Netw. **134**, 76–85 (2021)

12. Afrasiyabi, A., Larochelle, H., Lalonde, J.F., Gagné, C.: Matching feature sets for few-shot image classification. In: Proceedings of the IEEE/CVF Conference on Computer Vision and Pattern Recognition, pp. 9014–9024 (2022)
13. Oreshkin, B., Rodríguez López, P., Lacoste, A.: Tadam: task dependent adaptive metric for improved few-shot learning. Adv. Neural Inf. Process. Syst. **31** (2018)
14. Bertinetto, L., Henriques, J.F., Torr, P.H., Vedaldi, A.: Meta-learning with differentiable closed-form solvers. arXiv preprint arXiv:1805.08136 (2018)
15. Nichol, A., Schulman, J.: Reptile: a scalable metalearning algorithm, vol. 2, no. 3, p. 4. arXiv preprint arXiv:1803.02999 (2018)
16. Rajeswaran, A., Finn, C., Kakade, S.M., Levine, S.: Meta-learning with implicit gradients. Adv. Neural Inf. Process. Syst. **32** (2019)
17. Ye, H.J., Chao, W.L.: How to train your maml to excel in few-shot classification (2022)
18. Baik, S., Choi, M., Choi, J., Kim, H., Lee, K.M.: Meta-learning with adaptive hyperparameters. Adv. Neural. Inf. Process. Syst. **33**, 20755–20765 (2020)
19. Baik, S., Choi, J., Kim, H., Cho, D., Min, J., Lee, K.M.: Meta-learning with task-adaptive loss function for few-shot learning. In: Proceedings of the IEEE/CVF International Conference on Computer Vision, pp. 9465–9474 (2021)
20. Baik, S., Hong, S., Lee, K.M.: Learning to forget for meta-learning. In: Proceedings of the IEEE/CVF Conference on Computer Vision and Pattern Recognition, pp. 2379–2387 (2020)
21. Yao, H., Zhang, L., Finn, C.: Meta-learning with fewer tasks through task interpolation. arXiv preprint arXiv:2106.02695 (2021)
22. Ren, M., et al.: Meta-learning for semi-supervised few-shot classification. arXiv preprint arXiv:1803.00676 (2018)
23. Feng, R., Ji, H., Zhu, Z., Wang, L.: Selfnet: a semi-supervised local fisher discriminant network for few-shot learning. Neurocomputing **512**, 352–362 (2022)
24. Krizhevsky, A.: Learning multiple layers of features from tiny images. Master's thesis, University of Tront (2009)
25. Zhang, R., Che, T., Ghahramani, Z., Bengio, Y., Song, Y.: Metagan: an adversarial approach to few-shot learning. Adv. Neural Inf. Process. Syst. **31** (2018)
26. Li, X., et al.: Learning to self-train for semi-supervised few-shot classification. Adv. Neural Inf. Process. Syst. **32** (2019)
27. Wang, X., Cai, J., Ji, S., Li, H., Wu, F., Wang, J.: Self-adaptive label augmentation for semi-supervised few-shot classification. arXiv preprint arXiv:2206.08150 (2022)
28. Simon, C., Koniusz, P., Nock, R., Harandi, M.: Adaptive subspaces for few-shot learning. In: Proceedings of the IEEE/CVF Conference on Computer Vision and Pattern Recognition, pp. 4136–4145 (2020)
29. He, X., Lin, J., Shen, J.: Weakly-supervised object localization for few-shot learning and fine-grained few-shot learning. arXiv preprint arXiv:2003.00874 (2020)
30. Gao, F., Luo, X., Yang, Z., Zhang, Q.: Label smoothing and task-adaptive loss function based on prototype network for few-shot learning. Neural Netw. **156**, 39–48 (2022)
31. Sun, Q., Liu, Y., Chua, T.S., Schiele, B.: Meta-transfer learning for few-shot learning. In: Proceedings of the IEEE/CVF Conference on Computer Vision and Pattern Recognition, pp. 403–412 (2019)
32. Hu, Z., Li, Z., Wang, X., Zheng, S.: Unsupervised descriptor selection based meta-learning networks for few-shot classification. Pattern Recogn. **122**, 108304 (2022)
33. Wertheimer, D., Hariharan, B.: Few-shot learning with localization in realistic settings. In: Proceedings of the IEEE/CVF Conference on Computer Vision and Pattern Recognition, pp. 6558–6567 (2019)

34. Rusu, A.A., et al.: Meta-learning with latent embedding optimization. arXiv preprint arXiv:1807.05960 (2018)
35. Mehrotra, A., Dukkipati, A.: Generative adversarial residual pairwise networks for one shot learning. arXiv preprint arXiv:1703.08033 (2017)
36. Yao, H., et al.: Improving generalization in meta-learning via task augmentation. In: International Conference on Machine Learning, pp. 11887–11897. PMLR (2021)
37. Bronskill, J., Gordon, J., Requeima, J., Nowozin, S., Turner, R.: Tasknorm: rethinking batch normalization for meta-learning. In: International Conference on Machine Learning, pp. 1153–1164. PMLR (2020)
38. Chen, W.Y., Liu, Y.C., Kira, Z., Wang, Y.C.F., Huang, J.B.: A closer look at few-shot classification. arXiv preprint arXiv:1904.04232 (2019)
39. Chen, Z., Fu, Y., Zhang, Y., Jiang, Y.G., Xue, X., Sigal, L.: Multi-level semantic feature augmentation for one-shot learning. IEEE Trans. Image Process. **28**(9), 4594–4605 (2019)

An Empirical Study of Attention Networks for Semantic Segmentation

Hao Guo[1], Hongbiao Si[2], Guilin Jiang[2], Wei Zhang[3], Zhiyan Liu[4],
Xuanyi Zhu[2], Xulong Zhang[5], and Yang Liu[1(✉)]

[1] Hunan Chasing Securities Co., Ltd., Changsha, China
liuyang87@hnchasing.com
[2] Hunan Chasing Financial Holdings Co., Ltd., Changsha, China
[3] Hunan Chasing Digital Technology Co., Ltd., Changsha, China
[4] Hunan Chasing Trust Co., Ltd., Changsha, China
[5] Ping An Technology (Shenzhen) Co., Ltd., Shenzhen, China

Abstract. Semantic segmentation is a vital problem in computer vision. Recently, a common solution to semantic segmentation is the end-to-end convolution neural network, which is much more accurate than traditional methods. Recently, the decoders based on attention achieve state-of-the-art (SOTA) performance on various datasets. But these networks always are compared with the mIoU of previous SOTA networks to prove their superiority and ignore their characteristics without considering the computation complexity and precision in various categories, which is essential for engineering applications. Besides, the methods to analyze the FLOPs and memory are not consistent between different networks, which makes the comparison hard to be utilized. What's more, various methods utilize attention in semantic segmentation, but the conclusion of these methods is lacking. This paper first conducts experiments to analyze their computation complexity and compare their performance. Then it summarizes suitable scenes for these networks and concludes key points that should be concerned when constructing an attention network. Last it points out some future directions of the attention network.

Keywords: Machine Learning · Deep Learnig · Semantic Segmentation · Attention

1 Introduction

Recently, the human brain inspires the study of the attention mechanism. That is, the human brain can quickly select an area from vision signal to focus [10]. Specifically, when observing an image, humans will learn the position where attention should be concentrated in the future by previous observation. Besides, humans always pay low attention to the surrounding area of an image, rather than reading all pixels of the whole image at one time and adjusting the focus over time [6]. This mechanism includes hard attention and soft attention [5]. Hard attention uses a fixed matrix to focus on a certain part of the picture.

However, it can not be updated and can not work for different samples [23]. Therefore, almost all methods focus on soft attention, which utilizes a learnable weight matrix in the convolution layer.

Initially, attention is used for machine translation to capture long-range features. It helps the deep network to significantly improve its performance. The attention function projects query,key and value to an output which is obtained by a weighted sum of the values, where the weight is computed by the corresponding query and key [20]. The input of an attention function includes the query, keys, and values, which are all vectors, and the output is also vectors. The attention function calculates the query's attention to the key, and self-attention is its attention to itself, meaning query and key are both obtained from the same input. And then it was utilized in encoder-decoder network in semantic segmentation.

However, these networks always are compared with the mIoU of previous SOTA methods to prove their superiority and ignore that different attention networks are suitable for different scenes, considering the computation complexity and precision in various categories, which is important for engineering applications. What's more, these methods do not follow a critical standard to compare their FLOPs. Therefore, the comparisons are hard to be utilized. Besides, the conclusion of these attention networks is lacking. Therefore, this paper conducts an empirical study to analyze and summarize typical attention networks to guide future research.

2   Related Work

Segmentation. Semantic segmentation is a task which assigns a category to all pixels in a picture [25]. Many methods have been developed to tackle problems ranging from automatic driving [21], virtual reality [19], human-machine interaction [18], scene understanding [27], medical image analysis [15], etc. Recently, the common strategy to solve semantic segmentation problems is the deep structure network, which is much more accurate than traditional methods.

A common architecture of a semantic segmentation network is widely regarded as an encoder network, followed by an end-to-end structure decoder network. The encoder is a pretrained backbone, such as Resnet [7]. The function of the decoder is to map the feature semantic calculated by the encoder to the pixel space to make pixel-level classification.

Attention Networks. Non-Local Net bridges self-attention in machine translation and the Non-Local operations in computer vision. The self-attention networks introduced in the following part can be regarded as typical examples of this network. The proposed Non-Local method calculates the interaction between every two points to obtain global dependencies without the limitation of convolution kernel size. Non-Local method is similar to utilize a convolution kernel with size of the feature map, which can maintain more contextual information. The structure of Non-Local block is shown in Fig. 1.

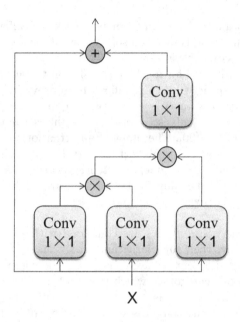

Fig. 1. An overview of the Non-Local block.

In the figure, "⊕" means element-wise sum, and "⊗" means matrix multi-plication. Each operation includes softmax function. The blue boxes mean 1×1 convolutions.

Based on Non-Local Net, the following works can be divided into two types, as shown in Fig. 2. One is enriching the obtained contextual information, such as DNLNet, DANet, RecoNet, and FLANet, and the other is reducing the computation cost, such as EMANet, ISSANet, CCNet, and AttaNet.

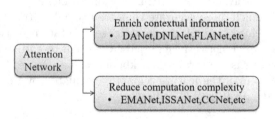

Fig. 2. The classification of attention network.

2.1 Enrich Contextual Information Based Methods

Non-Local Net captures global information in the spatial dimension. How-ever,the channel dimension is overlooked. Compared with Non-Local Net, Dual

Attention Network (DANet) [4] proposes two attention module to capture dependencies along the both dimension, respectively. Disentangled Non-Local Neural Networks (DNLNet) [24] decouples Non-Local operation to a whitened pairwise term to obtain the relationship between each pixel and a unary term to obtain the saliency of every pixel. However, these methods aim to construct a 2D similarity matrix to describe 3D context information. Therefore, a specific dimension is eliminated during the multiplication, which might damage the information representation. Tensor Low-Rank Reconstruction (RecoNet) [1] splits feature map to tensors in height, width,and channel directions, and then reconstruct the context feature. Consequently, compared with the 2D similarity matrix, the reconstructed 3D context information can capture long-range features without eliminating any dimension. But the attention missing issue caused by matrix multiplication still exists. To solve this problem, Fully Attentional Network (FLANet) [16] capture channel and spatial attention in a single similarity map by proposed attention block.

2.2   Reduce Computation Complexity Based Methods

Non-Local network enables each pixel to capture the global information. However, the self-attention mechanism produces a large attention map, which requires high spatial and temporal complexity. The bottleneck is that the attention map of each pixel needs to calculate the whole feature map. EMANet [12] utilizes the expectation maximization (EM) algorithm iteratively to obtain a set of compact ba ses and input them to the attention mechanism, which greatly reduces the complexity. Interlaced Sparse Self-Attention (ISSA) proposed an efficient scheme to factorize the computation of the dense matrix as the product of two sparse matrices. CCNet [9] introduces a recurrent criss-cross attention (RCCA) operation based on the criss-cross attention to solve this problem. The RCCA module can execute the criss-cross attention twice to capture the full-image dependencies.

3   Experiment

3.1   Datasets

Cityscapes [3] is a city scene dataset that can be used for semantic segmentation. It includes 20000 images with coarse annotations and 5000 images with fine pixel-level annotations and has 19 classes for semantic segmentation, and each image has a 1024 × 2048 resolution. The 5000 fine annotated images are divided into 2975, 500, and 1525 images for training, validation, and testing.

ADE20K [29] is a challenging scene parsing benchmark. This dataset contains 20K images for training and 2K images for validation. Images are densely labeled as 150 stuff/object categories and collected from different scenes with more scale variations.

3.2   Implementation Details

Parameter Settings. In the experiment, the learning rate is $1e^{-2}$, and models are trained for 80000 iterations with batch size of 8. The encoder network is ResNet-101 which is a widely used backbone network. The crop size is 769×769 for Cityscapes, and 520×520 for ADE20K. The poly learning rate policy is utilized which is $(1 - iter/maxiter)^{0.9}$. And the weight decay coefficients and momentum are 0.0005 and 0.9, which are common settings. And the synchronized batch normalization is utilized. FLOPs, memory, and mIoU are used to evaluate runtime, memory, and precision, respectively.

FLOPs Calculation. However, there are some inconveniences by comparing FLOPs. The first problem is that some papers do not provide the analysis of FLOPs, such as DANet and Non-Local Net. Besides, most papers do not provide the method for computing the FLOPs. Therefore their approaches to comparing the FLOPs might be different. Some papers use the open-source tools which follow the method in [14] to compute the FLOPs in the whole network. However, this method is used to compute the FLOPs in the backbone, which does not involve matrix multiplication like $A \cdot B$. Therefore, the computation of matrix multiplication is overlooked, and the comparison is unfair.

What's more, some papers analyze the whole network's FLOPs, and some compare the FLOPs in the attention module. And Some networks, such as EMANet, calculate their FLOPs with input size of 513×513, and some calculate their FLOPs with input size of 769×769. These results are difficult to utilize for a fair comparison.

In this paper, following method [14] which is also used by CCNet, we conduct experiments to calculate FLOPs in the attention module with input size of 769×769. The FLOPs of convolutional kernels are defined as follows

$$FLOPs = 2HW(C_{in}K^2 + 1)C_{out} \tag{1}$$

where H, W and C_{in} are height, weight and input channel numbers, K is the size of kernel, and C_{out} is the size of output channel.

4   Analysis

This section shows experiment results on attention networks and analyses methods by which they achieve their performance. Table 1 shows FLOPs of attention networks, and Table 2 shows quantitative results on Cityscapes dataset.

Table 1. Flops and Memory usage of typical attention networks with input size 769×769.

Method	FLOPs(G)	Memory(M)
Denoised NL	8	428
EMANet	14	100
CCNet	16	127
FLANet	19	436
ISANet	41	141
Non-Local Net	108	1411
DNLNet	108	1450
DANet	113	1462

Non-local Net. Non-Local Net achieves scores of 78.5% mIoU on Cityscapes and 43.2% mIoU on ADE20K. Compared with the baseline, which predicts results by the backbone, Non-Local Net gets 4% mIoU increment on Cityscapes validation set [9]. It demonstrates the proposed Non-Local operations directly capture global dependencies and significantly improve the segmentation performance. However, it also leads to high computation complexity. The FLOPs of Non-Local Net is much higher than its variations that are designed to reduce the computation cost.

DANet. DANet achieves scores of 80.6% mIoU on Cityscapes and 43.8% mIoU on ADE20K. Compared with Non-Local Net, the increment of computation complexity comes from the channel attention module. Compared with the attention map obtained by Non-Local Net, which size is $HW \times HW$, the attention map generated by the channel attention module is $C \times C$, which is not a big number. Therefore, DANet does not increase high computation on Non-Local Net. Table 1 shows the FLOPs is slightly over Non-Local Net, but the accuracy is much higher. It demonstrates the proposed Dual Attention block successfully enriches the contextual information without cost too much computation.

DNLNet. DNLNet achieves scores of 79.4% mIoU on Cityscapes and 43.7% mIoU on ADE20K. Compared with Non-Local Net, DNLNet splits the computation of Non-Local block into two terms, a whitened pairwise term and a unary term. The unary term tends to model salient boundaries and the whitened pairwise tends to obtains within-region relationships. Decoupling these two terms can enhance their learning. The unary term is achieved by 1×1 convolution, and whitened pairwise term is consist of the spatial attention module with the whitened operation. Therefore, the increment of computation complexity only comes from the unary term, which can be overlooked, and the computation complexity is similar to Non-Local Net.

FLANet. FLANet achieves scores of 82.1% mIoU on Cityscapes and 46.68% mIoU on ADE20K, which are the best results among these attention networks. Unlike DANet, which considers channel and spatial dimensions separately, FLANet proposes Fully Attentional Block to encode channel and spatial attention in a single similarity map. And the improved accuracy shows it is important to integrate both dimensions. Besides, compared with DANet, FLANet utilizes global average pooling to obtain a global view in channel Non-Local mechanism. Therefore it takes less computation.

Denoised Non-local. Denoised NL achieves scores of 81.1% mIoU on Cityscapes and 44.3% mIoU on ADE20K. Besides, it has the least computation cost among these attention networks. Denoted NL utilizes convolution layers to obtain a smaller input feature map, which dramatically reduces the computation cost. What's more, different from DANet and FLANet, which enrich the contextual information by leading in the channel dimension, Denoised NL enriches the neighbor information to reduce noises in the attention map. The experiment results demonstrate that the contextual information can be more precise without inter-class and intra-class noises.

EMANet. EMANet achieves scores of 78.6% mIoU on Cityscapes and 43.3% mIoU on ADE20K. Compared with Non-Local Net, EMANet simplifies the input to a set of compact bases by EM algorithm and can converge in only three times iterations. Therefore, EMANet greatly reduces the complexity, the computation is much more efficient than Non-Local Net. What's more, the performance is slightly over Non-Local Net on Cityscapes dataset.

ISANet. ISANet achieves scores of 81.9% mIoU on Cityscapes and 43.5% mIoU on ADE20K. Since ISANet factorizes the computation of the dense affinity matrix to the product of two sparse affinity matrices, the FLOPs of ISANet is much less than Non-Local Net. Besides, the ISANet can adjust the group number of the sparse affinity matrices to achieve the best computation efficiency. In the experiment, we set the group number to 8 to minimize the computation cost.

CCNet. CCNet achieves scores of 78.8% mIoU on Cityscapes and 44.0% mIoU on ADE20K. Compared to Non-Local Net, which links each pixel with all pixels, CCNet links each element to the row and column where the element is located, which is called criss-cross attention. It can obtain the contextual information of all the pixels on its criss-cross path. CCNet can adjust the recurrent times of RCCA module. In the experiment, we set the recurrent number to two, which achieves a good trade-off between speed and accuracy. Although indirectly obtained global dependencies cannot generate better results than direct calculation, CCNet utilizes the prior knowledge that the similarity in criss-cross path is more important. The RCCA module utilizes the similarity in criss-cross path twice to enhance the attention map. Therefore, the accuracy is better than Non-Local Net.

Table 1 shows the comparison of FLOPs and memory among different attention modules. The table shows that Denoised NL, FLANet, EMANet, and CCNet achieve the highest computation efficiency. They take much fewer GFLOPS than other networks. Besides, Fig. 3 shows that FLANet achieves a good trade-off between accuracy and speed.

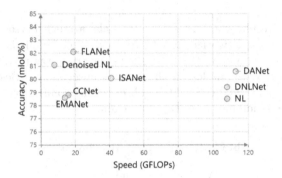

Fig. 3. The comparison of accuracy and speed of attention networks.

This section also analyses the per-class mIoU of attention networks for the Cityscape dataset to find their characteristics. The quantitive results of experiments can be found in Table 2. From the table, all networks can get high scores for classes that are big objects and account for a large percentage of the dataset, such as roads and buildings. However, the results are not well for other classes that are hard to distinguish and recognize, such as walls, fences, and poles. Most classes such as Non-Local Net, EMANet, CCNet, and DNLNet cannot predict trains correctly. However, attention networks that enrich the contextual information, such as DANet and FLANet, can achieve high accuracy on trains with non-trivial increments. Besides, Denoised NL and FLANet achieve the best accuracy on per-class results, and the overall accuracy is much higher than other networks. For road, wall, build, wall, sidewalk, pole, sign, traffic light, terrain, person, sky, train, car, motorbike, and bike, FLANet achieves the highest accuracy. Besides, for vegetation, rider, truck, and bus, Denoised NL achieves the highest accuracy. This section conducts experiments and visualizes the prediction on the class bus and train, and the qualitative results can check Fig. 4 and Fig. 5.

In Fig. 4, the yellow box indicates Denoised NL achieves the highest performance on the class bus. Denoised NL captures clear edge information, and the prediction inside the bus is consistent. Compared with Denoised NL, EMANet and CCNet can capture clear edge information, but the prediction inside the bus shows an inconsistency. This observation shows why the mIoU is much lower than other networks.

Table 2. Results on the Cityscapes val set.

Method	road	swalk	build	wall	fence	pole	tlight	sign	veg.	terrain	sky	person	rider	car	truck	bus	train	mbike	bike	mIoU (%)
Non-Local Net	98.1	84.8	93.2	62.3	62.2	67.3	74.1	81.0	92.5	63.1	95.1	83.6	66.6	95.7	82.9	87.9	73.9	63.9	79.1	78.5
EMANet	98.2	85.9	93.3	62.3	63.0	67.2	73.9	81.2	92.7	64.1	95.2	84.1	66.9	95.8	77.5	84.0	58.3	65.2	79.4	78.6
CCNet	98.2	85.4	92.9	57.5	62.5	67.0	73.7	81.1	92.7	63.0	95.0	84.0	67.7	95.7	83.4	86.7	69.0	62.2	78.8	78.8
DNLNet	98.1	85.6	93.1	60.6	64.0	67.4	73.8	81.0	92.6	65.7	95.0	83.8	66.5	95.0	78.6	86.9	72.1	67.3	79.3	79.4
ISANet	98.2	85.5	93.0	58.8	63.8	66.8	74.4	80.2	92.6	64.0	95.2	83.5	66.0	95.8	84.5	91.3	79.0	68.9	79.3	80.1
DANet	98.2	85.8	93.4	61.1	65.9	67.4	74.0	81.6	92.8	64.9	95.2	83.9	67.1	95.7	84.5	88.5	84.0	71.4	79.1	80.6
Denoised NL	98.4	87.1	93.2	59.4	64.5	67.6	73.0	80.3	93.1	67.4	95.4	84.5	69.1	95.6	84.6	**92.3**	84.3	68.9	79.3	81.1
FLANet	98.5	87.2	94.3	64.1	67.5	68.6	78.9	82.6	93.0	73.5	96.2	84.7	68.7	96.6	83.0	91.7	**86.2**	74.3	80.0	**82.1**

In Fig. 5, the results show FLANet can make a much better prediction on the class train. Other networks cannot capture correct details in the train. For example, the inconsistency in the train shows the train was predicted as a wall or bus. And the tree on the left of the train is misclassified. Besides, the wire on the top cannot be predicted. These observations confirm why FLANet can achieve non-trivial increments on the class train.

Table 3. Results on the ADE20K val set

Method	mIoU(%)
Non-Local Net	43.2
EMANet	43.3
ISANet	43.5
DNLNet	43.7
DANet	43.8
CCNet	43.9
Denoised NL	44.1
FLANet	**44.2**

This section also conducts experiments on ADE20K datasets. Following previous works, it does not analyse the per-class mIoU. Since the ADE20K dataset contains 150 classes, it is pretty hard to analyse the characteristic of each class for all networks. Table 3 shows the overall performance by mIoU. The results show that FLANet and Denoised NL can also achieve the highest accuracy.

Besides, Table 4 shows the comparison of these attention networks with SOTA methods on Cityscapes test set. Denoised NL and FLANet also outperform most SOTA methods, except HSSA. HSSA is the latest work which get the best prediction by MaskFormer [2] which is transformer structure. Therefore, the computation cost of HSSA is much higher than FLANet. When HSSA utilizes ResNet-101 as backbone, the mIoU is 83.02%, which is outperformed by FLANet. In conclusion, for most segmentation tasks, FLANet is a good choice. It achieves the highest accuracy with high computation efficiency.

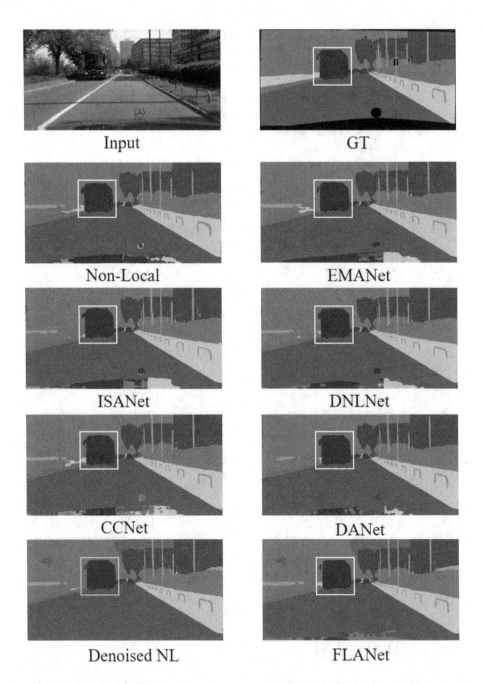

Fig. 4. The qualitative results on Cityscapes. This figure shows the prediction of the class bus of attention networks. The white box is the class that should be focused on, and the yellow box is the prediction generated by the network which achieves the best accuracy in this class. (Color figure online)

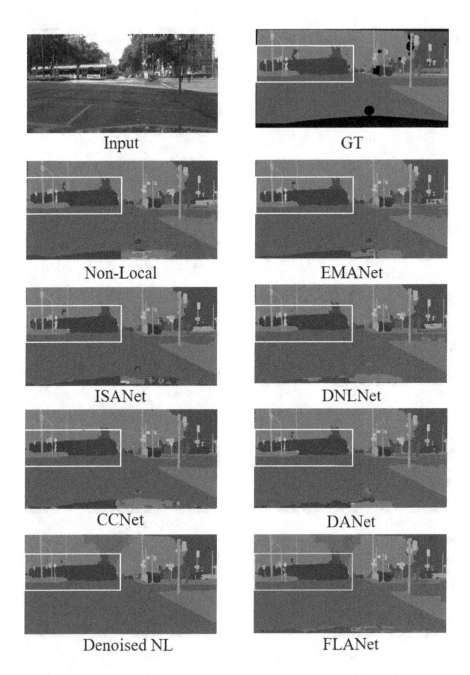

Fig. 5. The qualitative results on Cityscapes. This figure shows the prediction of the class train of attention networks. The white box is the class that should be focused on, and the yellow box is the prediction generated by the network which achieves the best accuracy in this class. (Color figure online)

Table 4. The Comparison on Cityscapes test set

Method	mIoU(%)
Non-Local Net	78.3
EMANet	78.4
CCNet	78.6
DNLNet	79.3
ISANet	79.9
DANet	80.4
Segmenter [17]	81.3
ACNet [8]	82.3
GFF [13]	82.3
SETR [28]	82.2
OCR [26]	82.4
DRANet [8]	82.9
SegFormer [22]	83.1
Denoised NL	83.5
FLANet	83.6
HSSN [11]	**83.7**

5   Conclusions and Future Works

This paper first introduces the background and development of semantic segmentation and attention. Then it presents an overview of several attention networks. Besides, this paper conducts experiments to evaluate their performance by accuracy and speed. Last, it analyses these attention networks through qualitative results and quantitative results.

In conclusion, for most segmentation tasks, FLANet is a good choice. It achieves the highest accuracy with high computation efficiency. And Denoised NL, which has the fastest speed, can be utilized for the task that requires high computation efficiency. Besides, to construct an attention network that can achieve high accuracy, we can enrich the contextual information in the attention mechanism from four aspects. First, globally shared and unshared attention can be disentangled. Second, both channel-wise and spatial-wise attention are important. Third, when utilizing channel-wise and spatial-wise attention, the attention missing issues should be avoided. Forth, attention noises should be eliminated. To reduce the computational cost, we can consider two aspects. First, the input can be simplified. Besides, the process of obtaining the dense affinity matrix can be simplified.

This paper summarizes that FLANet and Denoised NL are proposed to enrich contextual information. However, FLANet proposes to encode channel attention and spatial attention in one attention map. Denoised NL proposes to encode neighbor information in the attention map. Both networks still lack contextual information from others. We may merge these two networks into one that can solve these two problems in future work.

Besides, the transformer-based methods achieve non-trivial increment on various tasks, including semantic segmentation. Since transformer structure is based on multi-head attention, similar to introduced attention networks, these networks also modify structure based on attention module. Therefore, if these attention networks can be transferred to a transformer structure, it is also valuable to be studied.

References

1. Chen, W., et al.: Tensor low-rank reconstruction for semantic segmentation. In: Vedaldi, A., Bischof, H., Brox, T., Frahm, J.M. (eds.) ECCV 202, vol. 12362, pp. 52–69. Springer, Heidelberg (2020). https://doi.org/10.1007/978-3-030-58520-4_4
2. Cheng, B., Schwing, A., Kirillov, A.: Per-pixel classification is not all you need for semantic segmentation. Adv. Neural. Inf. Process. Syst. **34**, 17864–17875 (2021)
3. Cordts, M., et al.: The cityscapes dataset for semantic urban scene understanding. In: Proceedings of the IEEE Conference on Computer Vision and Pattern Recognition, pp. 3213–3223 (2016)
4. Fu, J., et al.: Dual attention network for scene segmentation. In: Proceedings of the IEEE Conference on Computer Vision and Pattern Recognition, pp. 3146–3154 (2019)
5. Guo, M.H., et al.: Attention mechanisms in computer vision: a survey. In: Computational Visual Media, pp. 1–38 (2022)
6. Hayhoe, M., Ballard, D.: Eye movements in natural behavior. Trends Cogn. Sci. **9**(4), 188–194 (2005)
7. He, K., Zhang, X., Ren, S., Sun, J.: Deep residual learning for image recognition. In: Proceedings of the IEEE Conference on Computer Vision and Pattern Recognition, pp. 770–778 (2016)
8. Hu, X., Yang, K., Fei, L., Wang, K.: ACNET: attention based network to exploit complementary features for rgbd semantic segmentation. In: 2019 IEEE International Conference on Image Processing (ICIP), pp. 1440–1444. IEEE (2019)
9. Huang, Z., Wang, X., Huang, L., Huang, C., Wei, Y., Liu, W.: Ccnet: criss-cross attention for semantic segmentation. In: Proceedings of the IEEE International Conference on Computer Vision (ICCV), pp. 603–612 (2019)
10. Itti, L., Koch, C., Niebur, E.: A model of saliency-based visual attention for rapid scene analysis. IEEE Trans. Pattern Anal. Mach. Intell. **20**(11), 1254–1259 (1998)
11. Li, L., Zhou, T., Wang, W., Li, J., Yang, Y.: Deep hierarchical semantic segmentation. In: Proceedings of the IEEE/CVF Conference on Computer Vision and Pattern Recognition, pp. 1246–1257 (2022)
12. Li, X., Zhong, Z., Wu, J., Yang, Y., Lin, Z., Liu, H.: Expectation-maximization attention networks for semantic segmentation. In: 2019 IEEE/CVF International Conference on Computer Vision (ICCV), pp. 9167–9176 (2019)
13. Li, X., Zhao, H., Han, L., Tong, Y., Tan, S., Yang, K.: Gated fully fusion for semantic segmentation. In: Proceedings of the AAAI Conference on Artificial Intelligence, vol. 34, pp. 11418–11425 (2020)

14. Molchanov, P., Tyree, S., Karras, T., Aila, T., Kautz, J.: Pruning convolutional neural networks for resource efficient inference. arXiv preprint arXiv:1611.06440 (2016)
15. Ravanbakhsh, M., et al.: Human-machine collaboration for medical image segmentation. In: ICASSP 2020-2020 IEEE International Conference on Acoustics, Speech and Signal Processing (ICASSP), pp. 1040–1044. IEEE (2020)
16. Song, Q., Li, J., Li, C., Guo, H., Huang, R.: Fully attentional network for semantic segmentation. arXiv preprint arXiv:2112.04108 (2021)
17. Strudel, R., Garcia, R., Laptev, I., Schmid, C.: Segmenter: transformer for semantic segmentation. In: Proceedings of the IEEE/CVF International Conference on Computer Vision, pp. 7262–7272 (2021)
18. Sun, A., Zhang, X., Ling, T., Wang, J., Cheng, N., Xiao, J.: Pre-avatar: an automatic presentation generation framework leveraging talking avatar. In: 2022 IEEE 34th International Conference on Tools with Artificial Intelligence (ICTAI), pp. 1002–1006 (2022). https://doi.org/10.1109/ICTAI56018.2022.00153
19. Valenzuela, A., Arellano, C., Tapia, J.: An efficient dense network for semantic segmentation of eyes images captured with virtual reality lens. In: 2019 15th International Conference on Signal-Image Technology & Internet-Based Systems (SITIS), pp. 28–34. IEEE (2019)
20. Vaswani, A., et al.: Attention is all you need. In: Advances in Neural Information Processing Systems (NeurIPS), pp. 5998–6008 (2017)
21. Wang, P., et al.: Understanding convolution for semantic segmentation. In: 2018 IEEE Winter Conference on Applications of Computer Vision (WACV), pp. 1451–1460. IEEE (2018)
22. Xie, E., Wang, W., Yu, Z., Anandkumar, A., Alvarez, J.M., Luo, P.: SegFormer: simple and efficient design for semantic segmentation with transformers. Adv. Neural. Inf. Process. Syst. **34**, 12077–12090 (2021)
23. Xu, K., et al.: Show, attend and tell: neural image caption generation with visual attention. In: International Conference on Machine Learning, pp. 2048–2057. PMLR (2015)
24. Yin, M., et al.: Disentangled non-local neural networks. In: Vedaldi, A., Bischof, H., Brox, T., Frahm, J.M. (eds.) ECCV 2020. LNCS, vol. 12360, pp. 191–207. Springer, Heidelberg (2020)
25. Yuan, J., Deng, Z., Wang, S., Luo, Z.: Multi receptive field network for semantic segmentation. In: 2020 IEEE Winter Conference on Applications of Computer Vision (WACV), pp. 1883–1892. IEEE (2020)
26. Yuan, Y., Chen, X., Wang, J.: Object-contextual representations for semantic segmentation. In: Vedaldi, A., Bischof, H., Brox, T., Frahm, J.M. (eds.) ECCV 2020. LNCS, vol. 12351, pp. 173–190. Springer, Heidelberg (2020). https://doi.org/10.1007/978-3-030-58539-6_11
27. Zhao, H., et al.: Psanet: point-wise spatial attention network for scene parsing. In: Proceedings of the European Conference on Computer Vision (ECCV), pp. 267–283 (2018)
28. Zheng, S., et al.: Rethinking semantic segmentation from a sequence-to-sequence perspective with transformers. In: Proceedings of the IEEE/CVF Conference on Computer Vision and Pattern Recognition, pp. 6881–6890 (2021)
29. Zhou, B., Zhao, H., Puig, X., Fidler, S., Barriuso, A., Torralba, A.: Scene parsing through ade20k dataset. In: Proceedings of the IEEE Conference on Computer Vision and Pattern Recognition, pp. 633–641 (2017)

Epidemic Source Identification Based on Infection Graph Learning

Xingyun Hong[1], Ting Yu[1], Zhao Li[2], and Ji Zhang[3(✉)]

[1] Zhejiang Laboratory, Hangzhou, China
{xyhong,yuting}@zhejianglab.com
[2] Hangzhou Yugu Technology Co., Ltd., Hangzhou, China
[3] The University of Southern Queensland, Toowoomba, Australia
Ji.Zhang@usq.edu.au

Abstract. Source identification plays a critical role in various domains, including patient zero tracing, social network rumor source detection, and more, as it enables tracing propagation processes and implementing effective measures to block transmission. However, most algorithms employed to identify propagation sources in networks heavily rely on prior knowledge of the underlying propagation models and associated parameters, as different propagation patterns yield diverse outcomes. To address this limitation, we present an approach called Infection Graph Learning for Sourcing (IGLS), which utilizes multiple snapshots to learn the propagation pattern. In this method, the source identification problem is redefined as a novel graph node classification scenario solved using Graph Convolutional Networks (GCN). The IGLS model takes both the state of nodes and the network's structure into consideration while introducing a new loss function specifically designed for the task. Furthermore, to tackle the multiple source identification problem, multi-task learning is incorporated into the IGLS model structure, identifying the number of sources and their locations simultaneously. This represents the first application of a deep learning model to address problems involving an unknown number of sources. We conducted experiments on both synthetic and real-world networks, and the results demonstrate the effectiveness and superiority of our proposed method.

Keywords: Source identification · Infection graph learning · Graph Convolutional Network

1 Introduction

The spread of infectious diseases poses a significant and ongoing threat to human health and well-being. Large-scale epidemics, such as the recent COVID-19 pandemic, have demonstrated the devastating impact that diseases can have on individuals, communities, and entire countries. In the face of such outbreaks, it is critical to identify the source of infection quickly and accurately to break

X. Song et al. (Eds.): APWeb-WAIM 2023, LNCS 14331, pp. 236–251, 2024.
https://doi.org/10.1007/978-981-97-2303-4_16

the transmission chain and implement appropriate measures to protect public health and maintain social stability [1-3].

The source identification problem is a critical task in controlling the spread of diseases. This problem is of utmost importance in situations where prompt and targeted interventions are required to prevent the propagation of the disease. The problem is typically defined as determining the origin of the disease based on the propagation model and human contact network, given observations of either all or some patients at a specific time point (snapshot).

Identifying the source of propagation can be abstracted as finding the source within an undirected network, with the predicted sources expected to be consistent with the actual ones. However, this problem poses several challenges due to multiple network node states and incomplete observations. The state transition chain can be complex, and missing information about the nodes further complicates source identification. Additionally, the dynamics of disease spread can be influenced by various factors, such as the patient's location and the time of the day, making the initial infection source hard to identify.

The spread of disease or information is governed by two key factors, which are interrelated and can significantly impact the outcome of the process. The first factor, the contact network, in which each node represents an individual, and an edge exists between two nodes if the corresponding individuals engage in close contact, whether physically or virtually. The contact network can be formalized as a graph, with nodes and edges representing individuals and their connections, respectively. The second factor is the propagation dynamic model [4-6], which describes the progression of the process over time. For contagious diseases, individual states consist of susceptible (S), exposed (E), infectious (I), and recovered (R). Corresponding models include SI [7], SIR [8], SIS, SEIR, and others. Similarly, in social networks, the Independent Cascade Model (IC) [9] and Linear Threshold Model (LT) [9] are more prevalent, each of which provides a framework for understanding the spread of information.

Source identification methods can be broadly categorized into two primary groups: methods based on Bayesian theory and those employing statistical approaches. Methods in the first group typically involve inferring or estimating the maximum posteriori probability of certain observations, given a propagation source. This group includes various techniques such as Dynamic Message-Passing (DMP) [10], Monte-Carlo Simulation [11], and the approach based on minimum description length (NetSleuth) [12]. Although these methods can be effective, their main drawback is that they require prior knowledge of the propagation time or parameters of the underlying dynamic model. On the other hand, there are centrality-based methods that focus on the network's structure and identify sources based on their importance within the network. These methods include Rumor Centrality (RC) [13], Jordan Centrality (JC) [14], Dynamical Age (DA) [15], and Unbiased Betweenness Centrality (UBC) [16]. However, one limitation of some of these approaches is that their accuracy has only been proven in the context of specific propagation models, potentially limiting their generalizability to other scenarios.

As deep learning continues to advance, graph-based neural networks, such as Graph Convolutional Network (GCN) [17] and Graph Attention Network (GAT) [18], have become popular in supervised learning applications. These approaches are well-suited for problems involving graph-structured data [19], such as social networks or biological networks.

To address the limitations of existing methods for source identification, we propose a novel model-based approach called Infection Graph Learning for Sourcing (IGLS). Our model leverages the Graph Convolutional Network (GCN) layer as its backbone, which is effective in learning the topological structure of a network and the attributes of its nodes. Unlike traditional node classification problems, source identification aims to infer the source of a network given known node information at a snapshot. Therefore, it is crucial to learn the relationship between the infection graph and source nodes. To achieve this, our proposed model considers both node states and network structures in an end-to-end learning process. Furthermore, an attention mechanism is adopted to learn the interactions between different nodes while considering node states during message passing. Additionally, our model is capable of not only solving single source identification problems, but also addressing cases where the number of sources is unknown through multi-task learning. To our knowledge, this is the first time a deep learning model has been used to handle the problem of unknown number of sources.

The main contributions of this paper are as follows:

1) We propose the IGLS model to solve the source identification problem. This model utilizes GCN layers and takes into consideration both node states and network structure to better address the task. It also uses attention mechanisms to learn interactions between different nodes during message passing.
2) The IGLS model can be extended to handle the problem of multiple source detection through multi-task learning. It also has a novel loss function designed to align with the characteristics of the source identification task, which further improves the model's performance.
3) We conduct experiments on both synthetic and real-world networks to demonstrate the superiority and effectiveness of the IGLS model compared to existing methods. The results show that our proposed model achieves state-of-the-art performance in both single and multiple source identification scenarios.

The rest of this paper is organized as follows: In Sect. 2, we introduce preliminaries, including the formulation of the source identification problem and the related propagation models. Section 3 provides a survey of related work in source identification. In Sect. 4, we present the proposed Infection Graph Learning for Sourcing (IGLS) model in detail, including the model architecture and the learning process. Section 5 reports the experimental results and analyses the performance of the proposed model in comparison to existing methods. Finally, in Sect. 6, we give conclusion and discuss potential future work.

2 Preliminaries

2.1 Problem Description

Given a contact network $G = (V, E)$, disease spread from propagation sources s. After time t, we observe the states of each node in the network, and the snapshot is denoted as $O(t)$. Therefore, the task of source identification is to determine the sources given the snapshot. Based on Bayesian theory, the problem can be formulated as follows [13]:

$$s^* = \arg\max_{s \in V} P(s|O) \tag{1}$$

2.2 Propagation Model

The dynamic propagation process is typically invisible, making it difficult to infer the origin of the disease or information. However, understanding the underlying propagation mechanism is critical for identifying the source. Propagation models are thus used to simulate the process and aid in source identification.

For instance, the SIR model is commonly used in infectious disease propagation. In this model, each node in the network can be in one of three possible states: susceptible (S), infectious (I), or recovered (R). At the onset of the propagation, only one or a few nodes are infectious, while the rest are susceptible. During the propagation, every infectious node has the potential to infect surrounding susceptible nodes with the same infection rate. Infectious nodes may also recover and become immune to future infections.

$$\frac{dS(t)}{dt} = -\frac{\beta I(t)}{N} S(t) \tag{2}$$

$$\frac{dI(t)}{dt} = \frac{\beta I(t)}{N} S(t) - \gamma I(t) \tag{3}$$

$$\frac{dR(t)}{dt} = \gamma I(t) \tag{4}$$

where β and γ represent the infection rate and recovery rate, respectively. N is the total number of nodes in the network, which is equal to $S(t) + I(t) + R(t)$ at time t, where $S(t)$, $I(t)$, and $R(t)$ denote the number of susceptible, infectious, and recovered nodes, respectively.

3 Related Work

There are several methods for source identification, including statistical and model-based approaches.

In the statistical approach, Shah proposed the Rumor Centrality (RC) algorithm [13], which assumes the probability of a node being the source is proportional to the number of all possible infection sequences from that node to all other nodes. Another method is the Unbiased Betweenness Centrality (UBC) [16], which takes into account both Betweenness Centrality and the degree of the nodes. However, statistical approaches rely solely on centrality metrics to identify the source and do not consider dynamic propagation process. As a result, some of these methods are only accurate with specific propagation models.

Dynamic Message-Passing (DMP) [10] is based on Bayesian theory. It enables us to calculate the probability of a given snapshot by using a recurrence equation that assumes a node to be the propagation source. The source node is identified as the one that can maximize the probability. However, in the presence of loops in the network, the recurrence equation becomes invalid. Additionally, DMP requires knowledge of the propagation time or parameters of the underlying dynamic model, which can be challenging to estimate in practice.

NetSleuth [12] is a method based on the minimum description length principle. This approach encodes the predicted sources' length and the most likely transmission path, thus transforming the source identification problem into an optimization problem. This iterative algorithm can also be applied to the unknown number of sources problem, where the number of sources is determined by whether the description length decreases. However, it is important to note that NetSleuth requires prior knowledge of the infection rate, which may not always be available in practice.

LPSI [20] is a method for identifying the sources of information propagation in a network based on label propagation. In this approach, the infection status is iteratively propagated in the network as a label, and the local peaks of the label distribution are identified as the source nodes.

In recent years, researchers have explored the use of graph neural networks to solve source identification problem. One popular approach is the Graph Convolutional Network (GCN). The definition of GCN $l+1^{th}$ layer can be written as follows:

$$H^{l+1} = \sigma(\tilde{D}^{-\frac{1}{2}}\tilde{A}\tilde{D}^{-\frac{1}{2}}H^lW^l) \tag{5}$$

where H^l is the output of lth GCN layer. σ is activate function Elu. A is the adjacent matrix of graph. $\tilde{A} = A + I$, I is the identity matrix. $\tilde{D} = d(A)$ is the degree of each node. W^l is a trainable weight matrix. By leveraging the spatial features of the graph and the interconnections between nodes, GCN can effectively learn to distinguish between infected and uninfected nodes, enabling it to accurately identify the sources of information propagation.

GCNSI [21] is the first attempt that applies GCN on multiple rumor source detection problem. It uses LPSI results as input features. Shah [22] also uses GCN to find patient zero, which introduces residual connections between GCN layers to improve its performance(referred to as GCN-P0 in the experiment). Guo [23] proposes IGCN with attention mechanism to ensure robustness and effectiveness of the algorithm. SIGN [24] considers neighborhood loss to reduce average error distance. Chen [25] formulates a backward contact tracing problem as ML estimation integrated with GNN to search for superspreaders.

4   Our Model

Our proposed model, IGLS (Infection Graph Learning for Sourcing), is designed to address the challenging problem of identifying sources of information propagation in epidemic spreading scenarios. This problem can be framed as a novel node classification problem, where the objective is to infer the infection sources in the entire graph given the node status information. To address this problem, the IGLS model characterizes the entire graph using node embeddings, leveraging the graph structure information to learn the infectious pattern. The attention mechanism is applied to learn the coefficient of interaction, which enables the model to capture the varying degrees of interaction between nodes that arise due to differences in their infection status.

IGLS is a versatile model that can handle both single-source and multiple-source identification problems. While single-source identification is a relatively straightforward problem, multiple-source identification is more challenging. This is because the number of propagation sources is often unknown in real-world scenarios. To address this challenge, our proposed method takes a novel approach by incorporating the source number as a model target, making it a multi-task learning model. Specifically, the difference in the estimated source numbers acts as a penalty term in the loss function, encouraging the model to learn both the source locations and the number of sources accurately. By jointly optimizing these tasks, our proposed model can effectively identify the sources of information propagation in complex and dynamic networks, even when the number of sources is unknown or uncertain.

4.1   Architecture

Our proposed model can be divided into several parts, including the input layer, GCN layers, graph embedding layer, dense layer, and output layer. The input layer is responsible for generating the initial node status embeddings. The GCN layers update the node embeddings with neighborhood information, enabling the model to capture the complex interaction patterns between nodes in the graph. The graph embedding layer is designed to capture the graph-level structural features, providing additional context for the model to make accurate predictions. The dense layer concatenates the node and graph embeddings and then feeds them into the final activation function to generate the output. The overall structure of our model is illustrated in Fig. 1.

4.2   Input Generation

To build our model, we begin by generating the input data based on a snapshot of the contact network $G = (V, E)$. This involves selecting a dynamic propagation model and its associated parameters, such as the propagation time and the random sources of infection. By simulating a forward diffusion process using different parameter settings, we can generate a diverse set of propagation results, which can be used to train and test our neural network model. This diversity is an advantage of using neural networks, as it allows the model to learn from a broad range of scenarios and improve its performance.

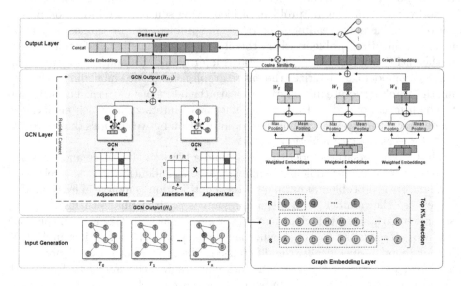

Fig. 1. Architecture of the model.

4.3   GCN Layer

In conventional GCN, every node feature is updated by its neighbors with the weight of the normalized degree. However, in the source identification task, nodes can have different statuses, such as infected or recovered, and these statuses can have different impacts on their neighboring nodes. In addition, the intensity of interactions may also vary depending on the type of target node. Therefore, we introduce an attention mechanism to learn the interaction coefficients between different statuses. The attention coefficients $a_{i,j}$ between nodes i and j are learned through the following formula:

$$a_{i,j} = g(V^T(W^s h_i + W^t h_j + b^s) + b^v) \tag{6}$$

where $h_i \in R^D$ and $h_j \in R^D$ are the initial vectors of nodes i and j, and nodes with the same status share the same embedding. W^s and W^t are weight matrices, V^T and b^s are bias vectors, and b^v is a scalar bias. The activation function $g(\cdot)$ is applied to the inner product of the learned embeddings and weights to generate the attention coefficients $a_{i,j}$. The attention matrix is expressed as $a_m \in R^{N \times N}$.

The symmetrically normalized Laplacian matrix L' is then updated as $L = \tilde{D}^{-\frac{1}{2}} \tilde{A} \tilde{D}^{-\frac{1}{2}}$, and $L' = norm(a_m) \odot L$. The forward propagation formula of GCN is the combination of the joint action of two adjacency matrices:

$$H^{l+1} = \sigma(LH^l W^l + L'H^l W^l) + H^l \tag{7}$$

where $W^l \in R^{D \times D}$, and $\sigma(\cdot)$ is the sigmoid activation function. To preserve information from the previous layer, residual connections are applied in each layer.

4.4   Graph Embedding Layer

To further improve the nodes' representational ability, we combine the global structural information extracted by the graph embedding layer with the importance score of each node, which is calculated using the GCN output.

Specifically, we first calculate the importance score of each node using the following formula:

$$z = \sigma(H^L Q) \tag{8}$$

where $Q \in R^D$ is the weight matrix, and $\sigma(\cdot)$ is the sigmoid activation function. Instead of selecting the top K nodes with the highest scores directly, we also consider the nodes' status. Nodes are categorized according to their status, and the top K nodes are selected in each category, resulting in a mask matrix M. Two pooling layers are then applied to reduce the dimension.

Finally, the readout of the graph can be stated as follows:

$$r = \sum_{state}^{\{S,I,R\}} W_a(max(H^L M Z) + mean(H^L M Z)) \tag{9}$$

where $state$ represents different node statuses, and $W_a \in R^D$ is the weight vector. The max and mean functions are applied to the masked node representations Z, and their outputs are concatenated and weighted by W_a to obtain the final graph representation r.

4.5   Output Layer

Output layer employs both node and graph embedding to predict whether node is source or not:

$$\hat{y} = \sigma([H^L, r]P + cos(H^L, r)) \tag{10}$$

where $P \in R^{2*D}$ is the weight vector. $cos(\cdot)$ is to calculate similarity, which takes the correlation between node and the whole graph embedding into consideration.

4.6   Loss Function

To ensure optimal performance, both accuracy and ranking metrics are taken into account when designing the loss function. The loss function incorporates both binary cross-entropy and KL divergence. Binary cross-entropy measures the dissimilarity between the predicted probability distribution and the true binary label of the nodes. In addition, the model also uses KL divergence, which measures the difference between the predicted probability distribution and a target probability distribution. Specifically, the loss function is defined as:

$$l_b = -\frac{1}{N}\sum_{i=1}^{N}y_i log(\hat{y}_i) + (1 - y_i)log(1 - \hat{y}_i) \tag{11}$$

$$l_k = -\alpha\sum_{i=1}^{N}p_i log(\frac{p_i}{softmax(\hat{y}_i)}) \tag{12}$$

$$l = l_p + l_k + \lambda||w||_2 \tag{13}$$

where p_i represents the softmax value of the real label. For instance, in a SIR model, where the status of infected and recovered nodes is set to 1 and the status of susceptible nodes is set to 0, the softmax function is applied to the predicted probabilities to obtain the real label probabilities. The less the difference in the distribution, the better the model ranking performance. In addition, L2 regularization term $||w||_2$ is also included. Here, w represents all weight parameters in the model, and λ is a weight coefficient.

For multi-source problems, an additional penalty term is added to the loss function to account for the difference between the predicted and actual number of sources. The penalty term is given by $|s^* - s|$, where s^* is the actual number of sources and s is the predicted number of sources. The overall loss function for the multi-source problem is given by:

$$l = |s^* - s| \cdot (l_b + l_k) + \lambda||w||_2 \tag{14}$$

the penalty term encourages the model to accurately predict the number of sources in the graph. This is particularly important for applications where the number of sources has a significant impact on the performance of the model, such as in disease outbreak prediction.

4.7 Model Complexity

In the classical GCN, the computational complexity of the convolution layer is $O(N^2D + ND^2)$, where N is the number of nodes, D is the dimension of the node embeddings.

In the IGLS model, to accelerate computation, the attention module is rewritten in the form of matrix calculations. The overall time complexity is $O(N^2 + ND^2)$. As for the graph embedding layer, the time complexity of computing the importance score, ranking and top K selection and weighted pooling add up to $O(2ND + NlogN)$. The output layer is a fully connected layer with a similarity function. The time complexity for this layer is $O(3ND)$.

In summary, when the hidden unit size is much smaller than the node number in the network, the overall time complexity of the IGLS model is $O(N^2)$, which is the same as the classical GCN model and more efficient than the LPSI model ($O(N^3)$). When the hidden unit size is close to the network size, the overall time complexity is $O(ND^2)$.

5   Experiment

5.1   Datasets and Baselines

In this section, we describe the datasets used to evaluate the performance of the proposed model, as well as the baselines used for comparison.

Four datasets are selected to evaluate the performance of the model, which are listed in Table 1.

Table 1. Dataset.

Network	Nodes	Edges	Avg(degree)
BA	1000	1996	3.99
Karate	34	78	4.59
Dolphin	62	159	5.13
US Power Grid	4941	6594	2.67

- BA [26] is a scale-free network. It is a commonly used complex networks. For our experiments, we set the number of nodes to be 1000 and the number of edges to attach from a new node to existing nodes to be 2, and then generated a BA network based on the network growth strategy.
- Karate [27] is a social network of friendships between 34 members of a karate club at a US university in the 1970s.
- Dolphin [28] is an undirected social network of frequent associations between 62 dolphins in a community living off Doubtful Sound, New Zealand.
- US Power Grid [29] is an undirected, unweighted network representing the topology of the Western States Power Grid of the United States. It is a real-world example of a large-scale infrastructure network.

5.2   Evaluation Metrics

In the problem of epidemic source identification, it can be difficult to accurately locate sources. Therefore, it is important to consider not only accuracy, but also ranking and distance-related metrics when evaluating the performance of our proposed method. Several metrics can be used for this evaluation.

Accuracy: It measures the proportion of correctly identified sources among all identified sources for a given evaluation dataset D:

$$acc = \frac{\sum_D |s \cap s^*|}{\sum_D |s|} \tag{15}$$

Rank: In the epidemic source identification problem, it can be difficult to predict the exact location of the sources accurately. Therefore, we also consider the ranking of the true sources in the predicted results, which is expressed as R. A higher value indicates better ranking performance in identifying true sources.

$$R = 1 - \frac{1}{N \sum_D |s|} \sum_{s \in D} r_s \qquad (16)$$

Hop: Hop is a distance-related metric that measures the minimum number of hops between the true sources and the identified sources. In the multi-source problem, the predicted and true number of sources may be different, so we need to first match the identified sources with the true sources before calculating Hop.

$$hop = \frac{1}{min(|s^*|, |s|)} \left(\min_{p \in permutation(s)} \sum_{i=1}^{K} \frac{d(s_i^*, s_i)}{K} \right) \qquad (17)$$

In the multi-source problem, it is important to accurately estimate the number of sources in addition to identifying their locations. Therefore, we also consider the mean absolute error (MAE) of the estimated and real number of sources as a metric for evaluating the performance of our proposed method.

Table 2. Single source identification performance on different networks.

Network	Karate			Dolphin			BA			US Power Grid		
Algorithm	R	ACC	HOP	R	ACC	HOP	R	ACC	HOP	R	ACC	HOP
RC	0.767	0.241	2.141	0.840	0.235	2.348	0.900	0.169	2.843	0.996	0.304	2.702
UBC	0.763	0.255	2.235	0.830	0.241	2.711	0.908	0.206	3.227	0.996	0.317	2.910
Net Sleuth	0.742	0.258	2.514	0.805	0.229	2.787	0.882	0.175	3.568	0.994	0.301	3.663
LPSI	0.778	0.311	**1.948**	0.846	0.259	2.262	0.922	0.217	2.702	0.996	0.329	2.852
GCN	0.616	0.136	2.776	0.779	0.156	2.580	0.835	0.173	2.877	0.997	0.282	2.614
GCN-P0	0.820	0.348	2.236	0.903	0.325	2.191	0.958	0.309	2.549	**0.998**	0.355	2.490
IGLS	**0.843**	**0.368**	2.083	**0.914**	**0.344**	**2.129**	**0.970**	**0.310**	**2.450**	**0.998**	**0.365**	**2.476**

5.3   Experimental Setting

In our experiments, we used the SIR model as the underlying propagation model. The infection probability p was randomly sampled from a uniform distribution $U(0, 1)$, while the recover probability r was sampled from a uniform distribution $U(0, p)$.

For each network, we performed simulations by randomly selecting source nodes, resulting in a total of 10,000 infection graphs for each network. The dataset was divided into training, validation, and test sets in a 6:2:2 proportion, with the test set used for performance comparison among the methods.

Our proposed model was built using the PYG framework. The number of GCN layers was set to 3, and the embedding size of each node was set to 3, which is consistent with the hidden unit size in the GCN layer. To accelerate the training process, the model was designed to adapt to multiple batches, with a batch size of 8. The Adam optimizer was used, with an initial learning rate of 0.01, and a regularization term of 5e-4. For each graph neural network model, we ran the model 10 times, each time for 100 epochs and with early stopping strategy, then calculated the average to avoid random error.

As for baseline models, the parameter r in UBC was set to 2, and the α in LPSI was chosen to be 0.5. For multiple source detection, the infection rate in Net Sleuth was set to 0.5.

5.4   Source Identification Performance

Table 2 presents the single source identification performance of different algorithms on different networks. In almost all the networks, IGLS outperforms other algorithms. Among unsupervised methods, LPSI achieves the best performance, which indicates that aggregating neighbor node information contributes significantly to the source identification. In terms of Graph Convolution Network, the metrics of the original GCN model are even lower than some traditional algorithms. The result of GCN-P0 demonstrates that residue connection can improve model performance. Compared with GCN-P0, IGLS shows a significant improvement in all evaluating metrics.

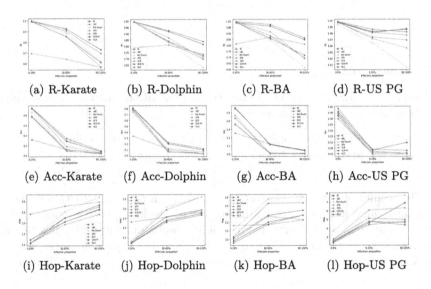

(a) R-Karate	(b) R-Dolphin	(c) R-BA	(d) R-US PG
(e) Acc-Karate	(f) Acc-Dolphin	(g) Acc-BA	(h) Acc-US PG
(i) Hop-Karate	(j) Hop-Dolphin	(k) Hop-BA	(l) Hop-US PG

Fig. 2. Results under different infection proportion.

Results under different infection proportions on each network are demonstrated in Fig. 2. According to the infection proportion, the infection graphs are categorized into three scales. As the infection scale increases, there is a significant decline in the performance of each method, which is aligned with the common sense that it's easier to locate the source when the disease is still at its early stage. In most cases, IGLS still remains the best even when the infection proportion is high. Other methods may achieve the highest performance in one metric, but far from the best on other indicators.

In the multiple source identification problem, Table 3 shows the results. Though Net Sleuth can address the problem of unknown number of infection sources, the results are relatively inaccurate. In Karate and Dolphin network, basic GCN and GCN-P0 give more precise outcomes on the infection source numbers. However, the performance on other two metrics is inferior to the optimal. In most circumstances, IGLS has low MAE on infection source numbers and achieves better accuracy and error distance compared to GCN-P0.

Table 3. Multiple source identification performance on different networks.

Network	Karate			Dolphin			BA			US Power Grid		
Algorithm	MAE	ACC	HOP	MAE	ACC	HOP	MAE	ACC	HOP	MAE	ACC	HOP
Net Sleuth	2.222	0.063	3.330	1.775	0.033	3.558	1.952	0.033	4.452	1.948	0.058	12.487
GCN	1.539	0.175	2.481	**1.289**	0.159	3.036	**1.389**	0.169	3.523	**1.193**	0.166	10.238
GCN-P0	**1.296**	0.300	2.514	1.527	0.282	**2.782**	1.395	0.226	3.304	1.228	0.172	9.470
IGLS	1.403	**0.312**	**2.439**	1.462	**0.293**	2.802	**1.389**	**0.231**	**3.183**	**1.193**	**0.199**	**8.654**

5.5   Ablation Study

In order to analyze the effect of each component of the model, we conducted ablation tests with the following settings:

- **IGLS w/ strc**: Only consider graph structure of the IGLS model.
- **IGLS w/ attn**: Only consider attention mechanism in the IGLS.
- **IGLS w/o kl-loss**: Remove distribution difference IGLS's loss function.

The results of the ablation tests (in Table 4) show that both the attention mechanism and graph structure learning contribute to improving model performance compared to GCN-P0. Furthermore, combining these modules further enhances the identification effect. In the Karate and Dolphin networks, adding the KL divergence loss item leads to an improvement in the Rank indicator.

5.6   Impact of Parameters

In the IGLS model, there are several parameters that can have an impact on model performance. To evaluate the impact of these parameters, we conducted experiments on the BA network using different values of the batch size, number

Table 4. Ablation experiment results on different networks.

Network	Karate			Dolphin			BA			US Power Grid		
Algorithm	R	ACC	HOP	R	ACC	HOP	R	ACC	HOP	R	ACC	HOP
IGLS	**0.843**	**0.368**	2.083	**0.914**	0.344	**2.129**	0.970	0.310	**2.450**	**0.998**	**0.366**	2.470
IGLS w/ strc	0.837	0.360	2.179	0.912	**0.347**	2.154	**0.974**	0.312	2.451	**0.998**	0.362	2.475
IGLS w/ attn	0.841	0.367	2.098	0.910	0.337	2.158	0.967	**0.315**	2.488	**0.998**	0.363	**2.457**
IGLS w/o kl-loss	0.841	0.367	**2.071**	0.913	0.343	**2.129**	0.971	0.314	2.457	**0.998**	0.365	2.476
GCN-P0	0.820	0.348	2.236	0.903	0.325	2.191	0.958	0.309	2.549	**0.998**	0.355	2.490

of GCN layers, hidden size, and learning rate. The comparison metric used was $R(rank)$ and results are presented in Fig. 3.

As the number of GCN layers in the IGLS model increases, the ranking metric first improves and reaches its peak at 4 layers, and then declines. This suggests that a higher model complexity may lead to overfitting. The best learning rate for the IGLS model is 0.1. Model performance is also strongly related to the batch size of the training data, with smaller batch sizes leading to better performance. However, smaller batch sizes also result in longer training times. Finally, we found that the best hidden unit size is 4, indicating that the IGLS model doesn't necessarily need to be very complex to achieve the best results.

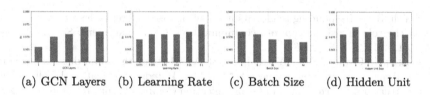

(a) GCN Layers   (b) Learning Rate   (c) Batch Size   (d) Hidden Unit

Fig. 3. Impact of Parameters.

5.7 Model Efficiency

The experiment also demonstrates the time complexity of the IGLS model. It takes 17 s to complete an training epoch on the BA network with a batch size of 8. In the prediction process, when the batch size is set to 64, the time required to process 2,000 graphs is approximately 1.7 s, which is efficient for practical applications.

6 Conclusion and Future Work

In this paper, we have proposed a novel framework, Infection Graph Learning for Sourcing (IGLS), to address the propagation source identification problem. Our proposed model integrates an attention mechanism to learn the interactions between different node states and considers the network structure information. Additionally, we have addressed the issue of multiple source detection by modifying the loss function in our model. We have conducted experiments on both synthetic and real-world datasets, which demonstrate the effectiveness of our proposed model in accurately identifying sources.

However, our work still has some limitations, so there are a plethora of avenues that we can explore in our future work. First, the current experiment only focuses on the SIR model as the underlying propagation model. Future research can explore the effectiveness of IGLS in identifying sources in other propagation models, such as the SEIR model or the SEIS model. Moreover, in reality, some states may not be observable in the network, future work can explore the effectiveness of IGLS in the situation of missing node status. Finally, the contact network between people is constantly changing. Thus, dynamic network information should also be incorporated in the future work.

Acknowldgemnts. This research is supported by Natural Science Foundation of China (No. 62172372), Zhejiang Provincial Natural Science Foundation (No. LZ21F030001, No. LQ22F020033), and the Exploratory Research Project of Zhejiang Lab (No. 2022KG0AN01).

References

1. Smith, R.D.: Responding to global infectious disease outbreaks: lessons from SARS on the role of risk perception, communication and management. Soc. Sci. Med. **63**(12), 3113–3123 (2006)
2. Kostka, J., Oswald, Y.A., Wattenhofer, R.: Word of mouth: rumor dissemination in social networks. In: Shvartsman, A.A., Felber, P. (eds.) International Colloquium on Structural Information and Communication Complexity, pp. 185–196. Springer, Heidelberg (2008). https://doi.org/10.1007/978-3-540-69355-0_16
3. Lappas, T., Terzi, E., Gunopulos, D., et al.: Finding effectors in social networks. In: Proceedings of the 16th ACM SIGKDD International Conference on Knowledge Discovery and Data Mining, pp. 1059–1068 (2010)
4. Pastor-Satorras, R., Castellano, C., Van, M.P., et al.: Epidemic processes in complex networks. Rev. Mod. Phys. **87**(3), 925 (2015)
5. Hethcote, H.W.: The mathematics of infectious diseases. SIAM Rev. **42**(4), 599–653 (2000)
6. Duan, W., Fan, Z., Zhang, P., et al.: Mathematical and computational approaches to epidemic modeling: a comprehensive review. Front. Comput. Sci. **9**(5), 806–826 (2015)
7. Anderson, R.M., May, R.M.: Infectious Diseases of Humans: Dynamics and Control. Oxford University Press, Oxford (1992)
8. Allen, L.J.S.: Some discrete-time SI, SIR, and SIS epidemic models. Math. Biosci. **124**(1), 83–105 (1994)
9. Kempe, D., Kleinberg, J., Tardos, É.: Maximizing the spread of influence through a social network. In: Proceedings of the Ninth ACM SIGKDD International Conference on Knowledge Discovery and Data Mining, pp. 137–146 (2003)
10. Lokhov, A.Y., Mézard, M., Ohta, H., et al.: Inferring the origin of an epidemic with a dynamic message-passing algorithm. Phys. Rev. E **90**(1), 012801 (2014)

11. Antulov, F.N., et al.: Identification of patient zero in static and temporal networks. Robustness and limitations. Phys. Rev. Lett. **114**(24), 248701 (2015)
12. Prakash, B.A., Vreeken, J., Faloutsos, C.: Efficiently spotting the starting points of an epidemic in a large graph. Knowl. Inf. Syst. **38**(1), 35–59 (2014)
13. Shah, D., Zaman, T.: Detecting sources of computer viruses in networks: theory and experiment. In: Proceedings of the ACM SIGMETRICS International Conference on Measurement and Modeling of Computer Systems, pp. 203–214 (2010)
14. Zhu, K., Ying, L.: Information source detection in the SIR model: a sample-path-based approach. IEEE/ACM Trans. Networking **24**(1), 408–421 (2016)
15. Fioriti, V., Chinnici, M.: Predicting the sources of an outbreak with a spectral technique. arXiv preprint arXiv: 1211.2333 (2012)
16. Comin, C.H., Costa, L.: Identifying the starting point of a spreading process in complex networks. Phys. Rev. E **84**(5), 056105 (2011)
17. Kipf, T.N., Welling, M.: Semi-supervised classification with graph convolutional networks. In: Proceedings of ICLR (2017)
18. Velikovi, P., Cucurull, G., Casanova, A., et al.: Graph attention networks. In: Proceedings of ICLR (2018)
19. Liu, Y., Cao, J., Wu, J., et al.: Modeling the social influence of COVID-19 via personalized propagation with deep learning. World Wide Web **26**, 2075–2097 (2023)
20. Zheng, W., Chaokun, W., Jisheng, P., Xiaojun, Y.: Multiple source detection without knowing the underlying propagation model. In: AAAI, pp. 217–223 (2017)
21. Dong, M., Zheng, B., Quoc, V.H.N., et al.: Multiple rumor source detection with graph convolutional networks. In: Proceedings of the 28th ACM International Conference on Information and Knowledge Management, pp. 569-578 (2019)
22. Shah, C., Dehmamy, N., Perra, N., et al.: Finding patient zero: learning contagion source with graph neural networks. arXiv preprint arXiv:2006.11913 (2020)
23. Guo, Q., Zhang, C., Zhang, H., et al.: IGCN: infected graph convolutional network based source identification. In: 2021 IEEE Global Communications Conference (GLOBECOM), pp. 1–6. IEEE (2021)
24. Li, L., Zhou, J., Jiang, Y., et al.: Propagation source identification of infectious diseases with graph convolutional networks. J. Biomed. Inform. **116**, 103720 (2021)
25. Chen, S., Yu, P.D., Tan, C.W., et al.: Identifying the superspreader in proactive backward contact tracing by deep learning. In: 2022 56th Annual Conference on Information Sciences and Systems (CISS), pp. 43–48. IEEE (2022)
26. Réka, A., Albert-László, B.: Statistical mechanics of complex networks. Rev. Mod. Phys. **74**(1), 47 (2002)
27. Wayne, W.Z.: An information flow model for conflict and fission in small groups. J. Anthropol. Res. **33**(4), 452–473 (1977)
28. David, L., Karsten, S., Oliver, J.B., et al.: The bottlenose dolphin community of doubtful sound features a large proportion of long-lasting associations. Behav. Ecol. Sociobiol. **54**(4), 396–405 (2003)
29. Duncan, J.W., Steven, H.S.: Collective dynamics of "small-world" networks. Nature **393**(6684), 440 (1998)

Joint Training Graph Neural Network for the Bidding Project Title Short Text Classification

Shengnan Li[1,2], Xiaoming Wu[1,2]([✉]), Xiangzhi Liu[1,2], Xuqiang Xue[1,2], and Yang Yu[1,2]

[1] Key Laboratory of Computing Power Network and Information Security, Ministry of Education, Shandong Computer Science Center (National Supercomputer Center in Jinan), Qilu University of Technology (Shandong Academy of Sciences), Jinan, China
wuxm@sdas.org
[2] Shandong Provincial Key Laboratory of Computer Networks, Shandong Fundamental Research Center for Computer Science, Jinan, China

Abstract. With the advent of the information era, the demand for data processing speed and scale is far beyond the capability of past manual methods. Therefore, in the face of complex and complicated bidding information, how to select the bidding projects that meet their needs and process them in a shorter time, the classification of project titles has become an urgent problem to be solved. Considering the characteristics of the short text of bidding titles, We propose BESGN by using the large-scale pre-training model BERT to obtain text contextual information, which facilitates learning to generate good representations of the target text. We also construct a bipartite graph structure using graph neural networks for inter-neighborhood node messaging to capture information between different granularities in the text and overcome its heterogeneity. The two are fused to maximize the preservation of contextual information, compensate for the limitations of short texts, and enable automatic labeling and classification of bidding projects. We conduct experiments on bidding and benchmark datasets and compare BESGN with other classification methods. The results show that BESGN outperforms other models in terms of classification accuracy, especially in handling short texts.

Keywords: bidding information · Graph Neural Network · Simplifying Graph Convolutional Network · BERT

1 Introduction

As the information age progresses, the country will actively guide the construction and operation of electronic bidding and tendering trading platforms with social capital. The electronic bidding and procurement system and the standard technical system will be gradually established and improved, achieving electronic

X. Song et al. (Eds.): APWeb-WAIM 2023, LNCS 14331, pp. 252–267, 2024.
https://doi.org/10.1007/978-981-97-2303-4_17

bidding and procurement for projects that must be tendered by law throughout the process. This will form a nationwide, precise classification, transparent and standardized, interoperable electronic bidding and procurement system, realizing the transformation of the bidding industry to information technology and intelligence. However, due to the lack of a unified portal for publishing bidding and procurement information, bidders often do not know what to do when faced with huge and scattered bidding information. At the same time, the irregularity and ambiguity of bidding information classification also reduce the efficiency of resource allocation. Therefore, how to enable bidders to access and process bidding projects quickly and conduct market trend analysis. The classification of project title becomes an urgent issue.

Automatic classification of bidding project titles is conducive to significantly reducing manual processing costs, reducing data errors due to human subjectivity, and improving classification accuracy, which can ensure fair, open, and just bidding principles [1]. For bidders, it solves the problem of poor utilization of their bidding resources; for proposers, it pinpoints the bidding documents, avoids spending too much time and cost on targets unrelated to them, and improves the efficiency of collecting target bidding information.

The title of bidding documents belongs to a short text, which has the characteristics of the small number of words, sparseness, and lack of context [2], while Chinese has more words and multiple meanings of words compared with English, so its semantic ambiguity is strong. This leads to a highly sparse text representation of traditional machine learning methods based on word frequency statistics, which usually difficult to achieve satisfactory classification results [3]. And they ignore contextual semantics and order of words, the features that are crucial for the classification of short texts. Therefore, [4] obtained semantic information from hierarchical lexical databases and statistical information contained in the involved corpus to model short texts. However, due to the instability of external knowledge bases and the development of feature extraction techniques, the approach of using external information to expand the short text was gradually abandoned and turned to mine the information contained in the short text itself using feature extraction. [5] extracted potential topics and entities to construct heterogeneous graphs and used a two-layer attention mechanism of node-level and type-level attention to assist in short text classification.

Although the above methods have excellent performance in text classification tasks, they still face the following problems: In text classification, the goal of machine learning is to learn the mapping of input features to labels based on a given label and apply it to predict the class of new data, so label information plays an important role in classification results. However, current graph neural networks do not consider label information in the process of building the graph structure, ignore the semantic relationship between input documents and labels, and most graph neural network models are rarely combined with large-scale pretrained models; no explicit implicit semantic structure of documents is assumed, making the learned representations ineffective and difficult to interpret.

To address the above issues, this paper proposes BESGN, which takes full advantage of the pre-trained model BERT and represents the potential contextual meaning of bidding project titles as embeddings. A simplified graph convolutional network (SGC) is used to construct a bipartite graph structure for the same nodes and different nodes and propagate it among neighbors to combine the advantages of both aspects for classifying bidding short texts. The main contributions of this paper can be summarized as follows:

(i) We propose a heterogeneous graph construction strategy that fuses labels, enabling the label information of a document node to be passed to its neighbor nodes, thus capturing more label information contained in the document features and the relationship information between words, documents, and labels, compensating for the sparsity of short texts.

(ii) To address the problem that graph convolutional neural networks ignore the differences between nodes, a bipartite graph structure is introduced, and intra- and inter-graph fusion operations are performed to reduce the interference of noisy information and to obtain the important information associated with the nodes.

(iii) Comparison experiments are done on our model with five short text classification datasets, and the experimental results show that our method outperforms the state-of-the-art methods on five benchmark datasets and our own bidding project title dataset, especially on short texts.

2   Related Work

2.1   Text Classification

Text classification is assigning predefined labels to text, which is also an essential and vital task in many natural language processing (NLP) [6]. It has been widely applied to sentiment analysis [7], spam detection [8], etc. Early text classification usually requires the extensive and tedious manual acquisition of sample features of the text, which are then fed into traditional machine learning models. Among them, the main methods for manually acquiring sample features are bag-of-words, n-gram, and word frequency-inverse document frequency (TF-IDF). Traditional machine learning algorithms include Naive Bayes (NB), K-nearest neighbor (KNN), and Support Vector Machine (SVM).

Some recent deep learning approaches have been widely used in text classification tasks, and the most representative models are convolutional neural networks (CNN) and recurrent neural networks (RNN) [9,10], among which the CNN-based text classification model (TextCNN) [11] uses one-dimensional convolution and one-dimensional maximum pooling operations to obtain a localized representation of text, and the RNN-based text classification model including LSTM [12] obtains the sequential representation of text. TextRCNN [13] replaces the convolution operation with a bidirectional LSTM on the basis of TextCNN, thus capturing contextual information to the maximum extent and preserving a more extensive range of word order when learning text representation.

To improve the intuitiveness and interpretability of text classification, attention [14] methods have been proposed, where the most commonly used self-attention network can obtain the weight distribution of words in a sentence, which can effectively improve the long-term dependability of text. Google's team proposed a total attention mechanism dependent architecture Transformer [15] and achieved good performance with self-attention [16] and feed-forward neural networks. And the pre-trained language model BERT [17] is a two-way Transformer encoder, which has achieved advanced results in several NLP tasks and is able to obtain rich contextual semantic information. However, the above methods cannot obtain structural information about the text.

Currently, graph neural networks have been widely used in text classification tasks, It applies a convolutional operation to the node feature vectors in the graph which combines information from neighboring nodes to update the feature vector of each node [18]. TextGCN [19] uses the relationship between words and words, words and text to model the whole corpus to build heterogeneous graphs, and does graph convolution operations to obtain global structural information of the text. SGC [20] builds on TextGCN to reduce the complexity of GCN by removing the weight matrix between the nonlinear transform and the compressed convolutional layers. However, the single graph structure is insufficient to effectively portray the rich information, so joint learning of multiple graphs is crucial. TensorGCN [21] proposes a generalized joint learning framework for multiple graphs to effectively fuse information from multiple sources of heterogeneous graphs by simultaneous intra-graph and inter-graph message passing.

2.2   Short Text Classification

With the increasing development of the Internet, a large number of unstructured short texts are generated, while classifying short text data is a key prerequisite for developing applications in different domains [22], such as question and answer, sentiment analysis, news commentary, and other scenarios [23–25]. Sequential models have been applied to short text classification in the past. Since short texts often lack context and adhere to a less strict syntactic structure [2], these models are more suitable for long texts. With the development of graph neural networks (GNN) [26], HGAT was proposed, which takes a two-layer attention mechanism (node-level and type-level attention) for short text classification, and uses entity and topic embeddings to solve the semantic sparsity problem of short texts. STGCN [27] used different topic models for extracting short text of topic information and constructing short text graphs by word co-occurrence and document word relations.

Although graph neural networks have been applied to short text classification, most existing models neglect the semantic relationship between input documents and labels, make insufficient use of label information as well as text classification models based on graph convolutional neural networks ignore the differences between nodes and do not capture the important information associated with nodes, and the document node representations obtained based on heterogeneous graph convolutional neural networks are not sufficiently differentiated and do not make full use of document hierarchy and contextual information.

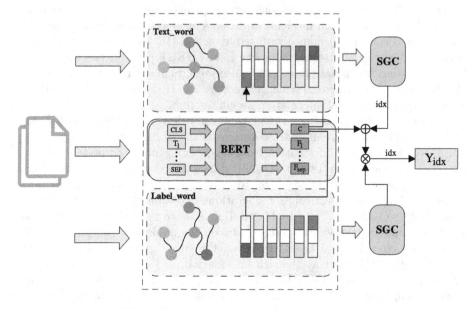

Fig. 1. The overall framework of BESGN.

3   Method

We propose a novel classification method, BESGN, for the short text of the bidding project title, which adds inductive BERT while using the transductive Simplifying Graph Convolutional Networks (SGC), and utilizing the association between labels and words, after two layers of convolution operation to stitch together the structural and semantic information learned from both. The dimensionality is set to the number of categories for classification, and thus the classification operation is performed. Therefore, we divide the modeling approach into the following steps: first, extract textual feature information by BERT; construct a bipartite graph structure of text-word graph and label-word graph based on the corpus of bidding project titles; track and cache BERT features with inductive features and perform residuals with a transductive SGC; make intra- and inter-graph nodes exchange information and aggregate neighbor nodes by doing simple graph convolutional operations on the bipartite graph structure; use text features and label features in the bipartite graph to classify and obtain classification labels. The structure of our overall model is shown in Fig. 1 we will describe it in detail in the following subsections.

3.1   Extracting Contextual Information

Since the content of the bidding project title is a short text with strong ambiguity, and the Chinese meaning is more confusing, so in order to enrich the semantic information of the short text of the bidding project title, we first compensate for its semantic sparsity by BERT training the short text words obtained after cleaning and splitting, and use its generated $[cls]$ specially marked features as the text feature vector $h_{[cls]} \in R^d$, and d is the hidden layer dimension of BERT.

3.2   Graph Structure Construction

Since previous text graph convolutional models usually only consider word nodes and document nodes in the process of graph construction, during this process ignores the semantic relationship between input documents and labels, resulting in partial loss of supervisory information. To effectively utilize the information contained in the text, we add label nodes to the text graph convolution model construction process in order to capture more information in the document features. Also, in order to overcome the problem that the document node representations obtained from the heterogeneous graph convolutional neural network are not sufficiently differentiated, we build a bipartite graph structure.

First, all words in the corpus are used as word nodes in the text-word graph A_{tw} and the label-word graph A_{lw}, and the word-to-word edge weights are calculated as PMI (Pointwise Mutual Information):

$$PMI(i,j) = \log \frac{p(i,j)}{p(i)p(j)} \tag{1}$$

$$p(i,j) = \frac{\#W(i,j)}{\#W} \tag{2}$$

$$p(i) = \frac{\#W(i)}{\#W} \tag{3}$$

Use sliding window on the pre-processed bidding project title words to obtain local co-occurrence sequence context information between them. $p(i,j)$ is the probability that word pair (i,j) appears simultaneously in the same sliding window, $\#W(i,j)$ is the number of sliding windows simultaneously containing both words i and j, and $\#W$ is the total number of sliding windows in a single text. When the PMI value is positive, representing a high semantic relevance between two words, an edge is added between these two word nodes; otherwise, no edge is added. Among them, due to the short length of the short text, using sliding windows of shorter length can collect the local co-occurrences of word pairs and mine their weights in the short text.

The edge weights between label nodes and word nodes and text nodes and word nodes are set to TF-ID(L)F (Term Frequency-Inverse Document (Label) Frequency), respectively.

$$TF_{i,j} = \frac{n_{i,j}}{\sum_k n_{k,j}} \tag{4}$$

$$ID(L)F_i = \log \frac{|D(L)|}{1 + |j : t_i \in d_j(l_j)|} \tag{5}$$

$$TF - IDF = TF \cdot ID(L)F \tag{6}$$

The adjacency matrix of the corresponding graph structure is generated from the above way of setting the edge weights between words and words, text and words, and labels and words. The structure of the graph is shown in Fig. 2.

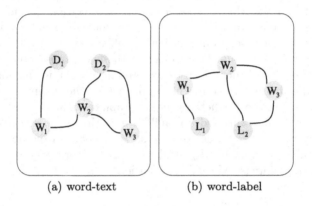

(a) word-text   (b) word-label

Fig. 2. Transductive graph with labels, texts, and words.

3.3   Feature Caching and Replacement

The adjacency matrix is generated in 3.2 while also initializing the feature matrices $H_{tw} \in R^{(n+m) \times d}$ and $H_{lw} \in R^{(c+m) \times d}$ for the corresponding graph structures, where n is the number of texts in the corpus, c is the number of labels, and m is the length of the word list. Before each round of training, the features of the text in H_{tw} and H_{lw} and the features of the labels are updated using BERT. SGC will do the whole batch operation for all texts, labels, and words in the corpus without considering the distinction between the training set and test set, i.e., transductive operation, but due to the memory limitation, BERT can only select some texts each time and cannot do the whole batch operation, so it is necessary to track and cache the text features and text ordinal numbers for BERT to do the batch operation so as to find the text features corresponding to the graph neural network, i.e., H_{tw} in the corresponding text node features, and the residuals are updated to H'_{tw}. This ensures that BERT and SGC can perform serial operations better and reduce the system memory usage, speeding up the training speed.

3.4   Graph Convolution Operation

SGC is used to reduce unnecessary complexity and redundant computations by successively removing the weight matrix between nonlinear and collapsed continuous layers in the GCN. Therefore, using SGC for both intra- and inter-graph propagation makes the obtained node representations more distinguishable and discriminative, bringing the documents closer to their corresponding labels and separating them from other labels, and also reducing resource consumption.

$$\hat{Y}_{SGC} = softmax\left(S^K X \Theta\right) \tag{7}$$

The mapped feature matrices H'_{tw} and H_{lw} and the adjacency matrices A_{tw} and A_{tw} are simultaneously input to the SGC for intra-graph propagation, i.e., the mapped feature matrices H'_{tw} and H_{lw} and the adjacency matrices A_{tw} and A_{lw} are subjected to a simplifying graph convolution network operation to obtain the intra-graph high-order neighbor information of each node, as follows.

$$\overline{H}_{tw} = \widetilde{A}_{tw}^k H'_{tw} W_{tw} + H_{[cls]} \tag{8}$$

$$\overline{H}_{lw} = \widetilde{A}_{lw}^k H_{lw} W_{lw} \tag{9}$$

where k is the number of SGC layers, W_{tw} and W_{lw} are trainable weight matrices, $\overline{H}_{tw}$ and $\overline{H}_{lw}$ are the final representations of text, word and label nodes after intra-graph propagation, and $\widetilde{A}$ is the adjacency matrix tensor of the bipartite graph structure adjacency matrix for the normalization operation $\widetilde{A} = (D^{-\frac{1}{2}}(A + I)D^{-\frac{1}{2}})$, D is the diagonal node degree matrix, and I is the identity matrix.

3.5   Classification

The text feature matrices of the text-word graph and label-word graph feature matrices are extracted respectively, where $\overline{H}_{tw}^t \in R^{n \times d}$ and the label feature matrix $\overline{H}_{lw}^l \in R^{c \times d}$, where $\overline{H}_{tw}^t$ is the text feature after aggregating words and $\overline{H}_{lw}^l$ is the label feature after aggregating words.

$Y = \overline{H}_{tw}^t \cdot (\overline{H}_{lw}^l)^T$, i.e., intergraph propagation, to overcome its heterogeneity, and also the operation of classification. n is the number of texts, c is the number of labels. The loss function is defined as the cross-entropy error of all labeled texts.

$$\mathcal{L} = -\sum_{d \in \mathcal{Y}_D} \sum_{f=1}^{F} Y_{df} \ln Z_{df} \tag{10}$$

where $\mathcal{Y}_D$ is the index set of documents with labels, F is the dimensionality of the output features, which is equal to the number of classes. Y is the matrix of label indicators.

4   Experiment

In this section, we compare the proposed BESGN with baseline models to demonstrate the performance and reliability of our model. Meanwhile, we conducted comparison experiments on our bidding dataset.

4.1   Datasets

Since there are few studies on public datasets for bidding projects, we crawl the bidding data ourselves from local bidding service platforms (both for-profit and non-profit organizations) and obtain a total of 4847 data from 26 different categories. The training set, validation set, and test set are partitioned according to the ratio of 8:2, We also use five public datasets to evaluate our model, which are MR, R8, R52, Snippets, and AG News, the details of which are described below, and the statistics of the datasets are shown in Table 1.

MR: A dataset consisting of movie reviews, classified as binary sentiment, with 10,662 samples, each corresponding to one sentence [6], including 5,331 positive and 5,331 negative reviews.

R8 and R52: Two subsets of Reuters, belonging to news classification, R8 has 8 categories with 7,674 samples, including 5,485 training texts and 2,189 test texts. R52 has 52 categories with 9,100 samples, including 6,532 training texts and 2,568 test texts [28].

Snippets: A dataset consisting of snippets returned by Google search [29], divided into 8 classes, with a total of 12,340 samples, including 10,060 training texts and 2,280 test texts.

AG News: The dataset is drawn from academic news [30]. From these, we selected four categories with a total of 16,000 samples, containing 11,200 training texts and 4,800 test texts.

Table 1. Summary statistics of all datasets.

Datasets	Avg Length	#Doc	#Train	#Test	#Classes
MR	20	10662	7108	3554	2
R8	66	7674	5485	2189	8
R52	70	9100	6532	2568	52
Snippets	17	12340	10060	2280	8
AG News	28	16000	11200	4800	4
Bidding(ours)	11	4847	3837	970	26

4.2   Data Processing

To ensure the authenticity and reliability of the data and the subsequent results, we must collect a large amount of tender document data from public bidding platforms around the area, but these raw data crawled from the website come with a large amount of non-text data, such as special symbols, code links, etc. Therefore, we need to clean these raw data, remove stop words and split words step by step.

We use the re regularization function to clean the raw data of non-normalized characters and punctuation marks and de-duplicate. The purpose is to make our text more concise and brief after removing redundant words, and the subsequent aspects of the study also make the extraction of keywords more accurate. Currently, the commonly used Chinese deactivation word lists are hit_stopwords[1] and baidu_stopwords[2], and we use the deactivation word list of HIT to process the text, mainly for the adverbs, prepositions, and conjunctions contained in the text, which usually have no clear meaning, but only have a role in a complete sentence, such as "of," "in," etc. Also, to avoid subjectivity, we remove provinces, cities, and counties. The last step is to sort the filtered text into words. The Chinese text is different from the English text in that the English word sorting is mainly based on spaces and punctuation marks. While Chinese is word-based, at the same time for Chinese, the ambiguity feature is more prominent; in the same sentence, there may be two or more ways to slice and dice, and new word recognition is also a problem. Currently, China's more mature word separation tools are jieba[3], HanLP[4] (Chinese language processing package). We use jieba splitting to slice the text.

4.3 Baseline Models

To verify the validity of our proposed BESGN, we use WideMLP [31], SWEM [32], fastText [33], CNN [9], LSTM [12], Bi-LSTM [34], TextGCN [19], BERT [17], SGC [20] in the benchmark dataset. On the bidding dataset, we compare it with the following six current state-of-the-art models.

TextCNN: Encoding of input text using multi-channel one-dimensional convolution to implement CNN for text classification tasks.

Bi-LSTM: bidirectional long-short-term memory (LSTM), which encodes the preorder and postorder of the text, respectively, an improvement of RNN that effectively alleviates the gradient disappearance problem.

TextRCNN: the convolutional layer is replaced with a bidirectional RNN.

TextGCN: the text and words in the corpus are considered as nodes in the graph, and GCN is implemented for text classification tasks.

SGC: a simplified version of GCN with the nonlinear transformations removed from GCN, which works slightly better than TextGCN in text classification tasks.

BERT: Transformer encoder, a large-scale pre-trained language model.

---

[1] https://github.com/HIT-SCIR/ltp.
[2] https://github.com/Pirate-Xing/stopwords/blob/master.
[3] https://github.com/fxsjy/jieba.
[4] https://github.com/hankcs/HanLP.

4.4   Experimental Settings

This experiment was implemented by PyTorch using an Nvidia Tesla A100 40GB GPU for training and test data. For the baseline models, we use the same parameter settings as in the original paper or provide the data reported in the original paper [2,19]. We set the BESGN sliding window to 5, the batch size to 64, and the optimization using the Adam optimizer with a learning rate of 1e-5 for BERT and 1e-3 for SGC. The number of SGC layers is set to 2, and the hidden layer dimension is the same as that of BERT, both 768. The number of training rounds is 30, using the early stopping strategy. The average of ten consecutive experimental results is taken as the final result.

4.5   Results

The results of the benchmark experiments are shown in Table 2. Overall, our model outperforms the baseline model on all five short text datasets, which proves the effectiveness of BESGN in short text classification. It can be clearly seen that the classification accuracy of BESGN improves significantly in MR and Snippets with shorter average text length, which indicates that we make full use of label information to construct short text graphs effectively. In contrast, for datasets with longer average text lengths like R8 and R52, the label information is not so useful for text classification. Therefore, for short text datasets, the shorter the length is, the more obvious the improvement of our model is.

Table 2. Test performance (%) measured on short text datasets.

Model	MR	R8	R52	Snippets	AG News
WideMLP	76.72	97.27	93.89	67.28	–
SWEM	76.65	95.32	92.94	86.98	87.41
fastText	75.14	96.13	92.81	88.56	87.98
CNN	77.75	95.71	87.59	–	–
LSTM	75.06	93.68	85.54	65.13	–
Bi-LSTM	77.68	96.31	90.54	84.81	87.68
TextGCN	75.08	97.07	93.56	83.49	87.55
BERT	85.97	97.38	96.17	88.20	–
SGC	75.9	97.2	94.0	–	–
BESGN	**86.63**	**98.04**	**96.34**	**91.10**	**90.46**

Meanwhile, in MR and Snippets, TextGCN does not show as good performance as the traditional model, which may be because the average text length of these two datasets is shorter and its default setting of the sliding window is larger, thus causing text redundancy and leading to the problem of semantic ambiguity. It is more important for sentiment classification to obtain its sequential contextual information, while graph structure is more about capturing full-text structural information. In R8 and R52, the advantage of graph structure is obvious, as it can capture long-range dependencies to improve classification accuracy. In addition, BESGN performs best for all datasets because we first obtain the original sequential information of the text by pre-training the model and then make full use of the labeling information of the text to assist in constructing the bipartite graph to obtain its global structural information. The fusion of both features gives the model an advantage in short text classification processing.

Table 3. Testing results of the bidding dataset.

Model	Accuracy(%)
TextCNN	57.32
Bi-LSTM	47.84
TextRCNN	58.14
TextGCN	60.21
SGC	60.62
BERT	68.87
BESGN	**70.82**

We also conducted the corresponding performance tests on our own bidding dataset, and the experimental results are shown in Table 3, and BESGN still performs well. It can be seen that for the Chinese dataset, the performance of our model is greatly improved after adding the pre-trained model BERT, which further indicates that fusing the pre-trained model BERT and the bipartite graph structure features is effective and improves for BESGN's accuracy on short text classification.

4.6 Parameter Analysis

Sliding Window Sizes. We tested the effect of sliding window size on the classification accuracy of short text MR and R8, and the results are shown in Fig. 3. Different datasets have different average lengths, and also different sliding window sizes lead to different classification accuracies. We found that as the sliding window increases, the neighbor node feature of text learning increases, and the classification performance is enhanced. At a certain level, the sliding window is too large, which tends to lead to overfitting, and the classification accuracy will trend down.

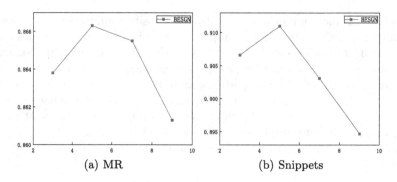

Fig. 3. Experimental results for different sliding window sizes and different datasets. Vertical axis denotes classification accuracy.

Graph Layers. As shown in Fig. 4, we also tested the performance effects of SGC with different number of layers on the classification performance of MR and Snippets, and we can see that the classification accuracy is not as good when $L = 1$, which may be because the model only learns the information of the neighboring first-order nodes and does not capture the higher-order information representation well. When $L = 2$, the optimal results are achieved for both datasets. However, when the number of layers keeps increasing, the classification results of both datasets show a significant decreasing trend, which may be because for short texts, which are short in length, accepting too much information about higher-order neighbors will make the representation learned by the model too smooth, and therefore, multi-layer SGC does not improve the performance.

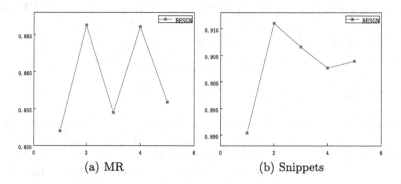

Fig. 4. Experimental results for different graph layers and different datasets. Vertical axis denotes classification accuracy.

5   Conclusion

In this paper, we propose BESGN, a classification model for the short text of bidding project titles. While constructing a bipartite graph structure for short text so that word information and label information are effectively propagated between graphs, we also incorporate the pre-training model BERT to obtain text order information and learn short text data from different levels to obtain both its contextual order information and its structural information, which undoubtedly greatly improves the accuracy of short text classification. The experiments prove that BESGN performs well on the short text dataset. In future work, we aim to reduce the complexity of the model while ensuring its classification level is stable.

Acknowledgements. This work is supported by "National Key Research and Development Project (No. 2021YFF0901300)", "Taishan Scholars Program (No. tsqn202211203)".

References

1. Liu, J., Jiao, Y., Wang, Y., Li, H., Zhang, X., Cui, G.: Research on the application of DNA cryptography in electronic bidding system. In: Pan, L., Liang, J., Qu, B. (eds.) Bio-inspired Computing: Theories and Applications, BIC-TA 2019. CCIS, vol. 1160, pp. 221–230. Springer, Singapore (2020). https://doi.org/10.1007/978-981-15-3415-7_18
2. Karl, F., Scherp, A.: Transformers are short text classifiers: a study of inductive short text classifiers on benchmarks and real-world datasets. arXiv preprint arXiv:2211.16878 (2022)
3. Song, G., Ye, Y., Du, X., Huang, X., Bie, S.: Short text classification: a survey. J. Multimedia 9(5) (2014)
4. Wenyin, L., Quan, X., Feng, M., Qiu, B.: A short text modeling method combining semantic and statistical information. Inf. Sci. 180(20), 4031–4041 (2010)
5. Linmei, H., Yang, T., Shi, C., Ji, H., Li, X.: Heterogeneous graph attention networks for semi-supervised short text classification. In: Proceedings of the 2019 Conference on Empirical Methods in Natural Language Processing and the 9th International Joint Conference on Natural Language Processing (EMNLP-IJCNLP), pp. 4821–4830 (2019)
6. Li, Q., et al.: A survey on text classification: from traditional to deep learning. ACM Trans. Intell. Syst. Technol. (TIST) 13(2), 1–41 (2022)
7. Yin, H., Song, X., Yang, S., Li, J.: Sentiment analysis and topic modeling for covid-19 vaccine discussions. World Wide Web 25(3), 1067–1083 (2022)
8. Jain, G., Sharma, M., Agarwal, B.: Spam detection in social media using convolutional and long short term memory neural network. Ann. Math. Artif. Intell. 85(1), 21–44 (2019)
9. Zha, W., et al.: Forecasting monthly gas field production based on the CNN-LSTM model. Energy 124889 (2022)
10. Gaafar, A.S., Dahr, J.M., Hamoud, A.K.: Comparative analysis of performance of deep learning classification approach based on LSTM-RNN for textual and image datasets. Informatica 46(5) (2022)

11. Kim, Y.: Convolutional neural networks for sentence classification. CoRR abs/1408.5882 (2014). http://arxiv.org/abs/1408.5882

12. Hochreiter, S., Schmidhuber, J.: Long short-term memory. Neural Comput. **9**(8), 1735–1780 (1997)

13. Lai, S., Xu, L., Liu, K., Zhao, J.: Recurrent convolutional neural networks for text classification. In: Proceedings of the AAAI Conference on Artificial Intelligence, vol. 29 (2015)

14. Bahdanau, D., Cho, K., Bengio, Y.: Neural machine translation by jointly learning to align and translate. arXiv preprint arXiv:1409.0473 (2014)

15. Vaswani, A., et al.: Attention is all you need. In: Advances in Neural Information Processing Systems, vol. 30 (2017)

16. Shaw, P., Uszkoreit, J., Vaswani, A.: Self-attention with relative position representations. arXiv preprint arXiv:1803.02155 (2018)

17. Devlin, J., Chang, M.W., Lee, K., Toutanova, K.: Bert: pre-training of deep bidirectional transformers for language understanding. arXiv preprint arXiv:1810.04805 (2018)

18. Xia, J., Li, M., Tang, Y., Yang, S.: Course map learning with graph convolutional network based on AuCM. World Wide Web 1–20 (2023)

19. Yao, L., Mao, C., Luo, Y.: Graph convolutional networks for text classification. In: Proceedings of the AAAI Conference on Artificial Intelligence, vol. 33, pp. 7370–7377 (2019)

20. Wu, F., Souza, A., Zhang, T., Fifty, C., Yu, T., Weinberger, K.: Simplifying graph convolutional networks. In: International Conference on Machine Learning, pp. 6861–6871. PMLR (2019)

21. Liu, X., You, X., Zhang, X., Wu, J., Lv, P.: Tensor graph convolutional networks for text classification. In: Proceedings of the AAAI Conference on Artificial Intelligence, vol. 34, pp. 8409–8416 (2020)

22. Zhao, K., Huang, L., Song, R., Shen, Q., Xu, H.: A sequential graph neural network for short text classification. Algorithms **14**(12), 352 (2021)

23. Peng, S., et al.: A survey on deep learning for textual emotion analysis in social networks. Digit. Commun. Netw. **8**(5), 745–762 (2022)

24. Chen, H., Wu, L., Chen, J., Lu, W., Ding, J.: A comparative study of automated legal text classification using random forests and deep learning. Inf. Process. Manag. **59**(2), 102798 (2022)

25. Hajibabaee, P., et al.: Offensive language detection on social media based on text classification. In: 2022 IEEE 12th Annual Computing and Communication Workshop and Conference (CCWC), pp. 0092–0098. IEEE (2022)

26. Scarselli, F., Gori, M., Tsoi, A.C., Hagenbuchner, M., Monfardini, G.: The graph neural network model. IEEE Trans. Neural Networks **20**(1), 61–80 (2008)

27. Ye, Z., Jiang, G., Liu, Y., Li, Z., Yuan, J.: Document and word representations generated by graph convolutional network and bert for short text classification. In: ECAI 2020, pp. 2275–2281. IOS Press (2020)

28. Huang, L., Ma, D., Li, S., Zhang, X., Wang, H.: Text level graph neural network for text classification. arXiv preprint arXiv:1910.02356 (2019)

29. Phan, X.H., Nguyen, L.M., Horiguchi, S.: Learning to classify short and sparse text & web with hidden topics from large-scale data collections. In: Proceedings of the 17th International Conference on World Wide Web, pp. 91–100 (2008)

30. Yang, Z., Dai, Z., Yang, Y., Carbonell, J., Salakhutdinov, R.R., Le, Q.V.: XLNet: generalized autoregressive pretraining for language understanding. In: Advances in Neural Information Processing Systems, vol. 32 (2019)

31. Galke, L., Scherp, A.: Bag-of-words vs. graph vs. sequence in text classification: questioning the necessity of text-graphs and the surprising strength of a wide MLP. arXiv preprint arXiv:2109.03777 (2021)
32. Shen, D., et al.: Baseline needs more love: on simple word-embedding-based models and associated pooling mechanisms. arXiv preprint arXiv:1805.09843 (2018)
33. Joulin, A., Grave, E., Bojanowski, P., Mikolov, T.: Bag of tricks for efficient text classification. arXiv preprint arXiv:1607.01759 (2016)
34. Liu, P., Qiu, X., Huang, X.: Recurrent neural network for text classification with multi-task learning. arXiv preprint arXiv:1605.05101 (2016)

Hierarchical Retrieval of Ancient Chinese Character Images Based on Region Saliency and Skeleton Matching

Ruijuan Cai[1,2,3] and Xuedong Tian[1,2,3(✉)] (iD)

[1] School of Cyber Security and Computer, Hebei University, Baoding 071002, China
xuedong_tian@126.com
[2] Hebei Machine Vision Engineering Research Center, Hebei University, Baoding 071002, China
[3] Institute of Intelligent Image and Document Information Processing, Hebei University, Baoding 071002, China

Abstract. Ancient Chinese characters exhibit characteristics of complex structure and diverse styles. To address the limitations of traditional Chinese character image retrieval techniques when applied to ancient Chinese characters, this paper proposes a hierarchical retrieval method for ancient Chinese characters images based on region saliency and skeleton matching (RSSM). The proposed method utilizes saliency joint weighting algorithm to effectively integrate the channel and spatial dimension information of deep convolutional features, enhancing the representation of key features. It focuses on capturing the detailed features of Chinese character contours and spatial structure, enabling coarse-grained retrieval of ancient Chinese characters. Furthermore, to further enhance retrieval accuracy, an improved shape descriptor, skeleton context, is introduced for fine-grained matching. The retrieval results are organized in ascending order of matching cost. The study constructs an ancient Chinese character image dataset named GJHZ. The Precision and mAP of RSSM achieved 90.71% and 90.59%, respectively. Experimental results demonstrate the superior performance of our method for ancient Chinese character image retrieval.

Keywords: Ancient Chinese character · Image retrieval · Deep feature · Skeleton Matching

1 Introduction

Ancient Chinese characters represent a vital medium for the cultural inheritance of the Chinese nation, holding significant importance in the continuity and dissemination of civilization. The advancement of historical document digitalization technology has opened up new opportunities for studying ancient Chinese characters. However, this advancement also brings forth a multitude of challenges, the processing and retrieval of ancient Chinese character images is the focus and difficulty of the research.

Due to the variability and complexity of Chinese character shapes and strokes in ancient texts, the difficulty of image processing is increased by problems such as paper aging and blurred handwriting. In addition, there are a large number of ancient Chinese characters in the image that are not included in the coding character set, making it difficult for researchers to effectively compare, retrieve, and identify ancient Chinese characters. Thus, this paper presents a hierarchical retrieval approach for ancient Chinese characters images based on region saliency and skeleton matching (RSSM).

There are three main aspects to RSSM: Firstly, we select the channels with special semantic and discriminative power in the deep convolutional feature map as the semantic detector of the image by optimizing the channel variance ranking and generate the corresponding filters to retain the key features of the image of ancient Chinese characters. Second, we combine the filters and saliency joint weighting (SJW) algorithm to enhance the saliency region of ancient Chinese characters, focusing on the spatial structure information of ancient Chinese characters to realize the coarse-grained retrieval of images. Third, to further improve the retrieval accuracy, the skeleton context descriptors of Chinese characters are extracted and the retrieval results are rearranged by fine-grained matching.

The rest of this paper is organized as follows: In Sect. 2, we review related works. We introduce our proposed RSSM method in Sect. 3. In Sect. 4, experimental investigations and performance comparisons are made. Section 5 provides a summary of the study.

2   Related Works

Feature extraction is the key to image retrieval of Chinese characters. Feature extraction methods mainly include two classes: structural features and statistical features.Structural features such as strokes [12], contours [26] etc. Xia et al. [19] utilized Elastic Histogram of Oriented Gradient and Derivative Dynamic Time Warping to achieve feature matching of similar calligraphic characters. To solve the problem of recognizing ancient Chinese characters with varying glyphs, Ma et al. [9] proposed a recognition method that integrates the component features of ancient Chinese characters with global and local point densities. Statistical features such as directional element features [7], wavelet transform [22], and elastic grid [15] etc. In order to identify comparable handwritten Chinese characters, Qu et al. [14] utilized an improved adaptive discriminative locality alignment method for their identification. In order to address the issue of the unstable structural relationship of Chinese characters, certain researchers have incorporated fuzzy set theory into handwritten image features to enhance the representativeness of handwritten Chinese character images [5,10]. Du et al. [5] combined double hesitation fuzzy features and elastic grid to extract the directional line element feature of ancient Chinese characters, and counted multiple indicators such as strokes and positions to achieve image retrieval of ancient Chinese characters.

Moreover, in the realm of image retrieval for calligraphic characters, certain researchers have employed the contour geometric structure [27], stroke density features [23], and skeleton information [4] of calligraphic characters as pivotal features to achieve effective retrieval. Gao et al. [13] introduced a novel descriptor called GIST-SC to represent calligraphic characters and integrated it with spectral hashing to achieve rapid retrieval of calligraphic character images.

While traditional feature extraction methods are meticulously designed based on the glyph characteristics of Chinese characters, they possess limitations in capturing high-level semantic information and handling the complexity of ancient Chinese characters. Consequently, feature extraction based on convolutional neural networks (CNN) has evolved into a mainstream approach for image retrieval [17,18]. Zhong et al. [25] enhanced model performance by merging domain-specific knowledge with CNN in the HCCR-Gabor-GoogLeNet model. Zhang et al. [24] proposed "DirectMap+ConvNet+Adaptation" combining the traditional normalized Gradient feature map in CNN to improve the performance of Chinese character recognition. Yang et al. [21] proposed an RNN(recurrent neural network) combined with attention for handwritten Chinese character recognition, which improves the recognition accuracy by continuously updating the feature attention Melnyk et al. [11] proposed a high-performance CNN for offline HCCR called Melnyk-Net, which improves the network interpretability by using improved pooling techniques. Tian et al. [16] incorporated deformable convolutional modules into CNN model to enhance its adaptability to geometric deformations, leading to improved retrieval accuracy of ancient Chinese character images.

3   Method

The proposed hierarchical retrieval system for ancient Chinese character images is based on regional saliency and skeleton matching. It operates in a hierarchical manner, progressing from local to whole and from coarse to fine comparisons. The system consists of two components: the coarse retrieval part, which combines regional channel screening algorithm and saliency joint weighting approach, and the fine matching part, which utilizes a skeleton matching method. The detailed algorithms are described in Algorithm 1 and Algorithm 2, and the specific steps are visualized in Fig. 1.

3.1   Visual Feature Extraction

The representation of image features plays a critical role in the accuracy of image retrieval. Prior research has utilized uni-dimensional features of fully connected layers to represent images for general image retrieval tasks. However, in the case of ancient Chinese character image retrieval, global semantic information is inadequate to distinguish their intricate details, necessitating a more accurate feature representation. The convolutional features that better represent spatial information and are more robust [6]. By aggregating discriminative local detail features in the main body of Chinese characters to improve image retrieval performance.

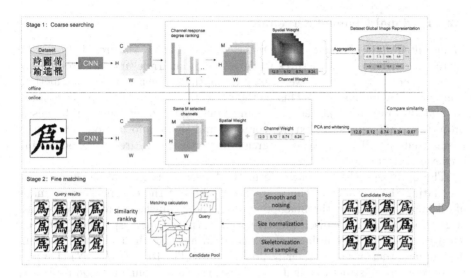

Fig. 1. Framework of the proposed RSSM image retrieval method for ancient Chinese characters.

Algorithm 1. The RSSM Framework for Coarse-Grained Retrieval

Input: deep convolutional features X, channel response degree ranking function f_v, spatial weighting function f_s, channel sensitivity weighting function f_c, the parameter W for whitening, the final number of dimension for aggregated feature D

Output: D-Dimensional global representation $G^{final} \in R^D$

1: $C \leftarrow f_v(X)$ ▷ channel response degree ranking
2: $S \leftarrow f_s(X, C)$ ▷ compute spatial weights
3: $X' \leftarrow X \otimes S$ ▷ spatial weighting
4: $R \leftarrow f_c(X')$ ▷ compute channel weights
5: $\Phi_k \leftarrow \sum_{i=1}^{W} \sum_{j=1}^{H} X'_k(i,j)$ ▷ intermediate vector
6: $G \leftarrow R \odot \Phi$ ▷ channel weighting
7: $G' \leftarrow norm(G)$ ▷ normalize
8: $G^{final} \leftarrow norm(PCA(G', W, D))$ ▷ PCA whitening and normalize again

---

We fine-tuned the ResNet50 network and extract the non-negative 3D response tensor $T \in R^{W \times H \times K}$ in the last convolutional layer of the network. Here, where W and H denote the spatial resolution (width and height), while $K = 2048$ represents the number of feature channels. We represent this 3D tensor of responses T as a set of 2D feature channel responses $X = \{X_k\}_{k=1}^{K}$, where $X_k \in R^{W \times H}(k = 1, 2, \ldots, K)$ denotes the 2D features responses of size $W \times H$ corresponding to the kth feature channel, and an element in the feature map is represented as $X_k(i,j) \in R(i = 1, 2, \ldots, W, j = 1, 2, \ldots, H)$.

3.2   Regional Channel Screening

In ancient Chinese character image retrieval, selecting key features that describe character outlines and structures while filtering out irrelevant background features is crucial. To this end, we employ the optimized channel variance method from the SBA method [20]. This method selects channels with special semantic and discriminative power as semantic detectors, generating "probabilistic proposals" that represent texture, contour, and shape information. These proposals facilitate effective feature aggregation across the entire image, enhancing retrieval performance.

The optimized channel variance method selects discriminative channels by calculating the variance of the aggregated values of the channel feature maps of all image samples in the image database, the channels of feature maps with large variances are more discriminative. Firstly, the channel responses of the deep features of the entire database are calculated, and the channels are ranked in descending order based on their response values. The top $M(M < N)$ channels are selected to construct a weight map. Next, the sum-pooling method is utilized to aggregate the deep feature X_k of the nth image, represented as $g_{k,n}$, and the K-dimensional vector of the nth images is represented as $g_n = \{g_{1,n}, g_{2,n}, \ldots, g_{K,n}\}$.

$$g_{k,n} = \sum_{i=1}^{W} \sum_{j=1}^{H} X_k(i,j) \tag{1}$$

The mean value of the kth channel is expressed as $\bar{g}$:

$$\bar{g} = \frac{1}{N} \sum_{n=1}^{N} g_{k,n} \tag{2}$$

where N is the number of database images. The variance of the kth channel is E_k:

$$E_k = \frac{1}{N} \sum_{n=1}^{N} (g_{k,n} - \bar{g})^2 \tag{3}$$

Sort the variances $E = \{E_1, E_2, \ldots, E_K\}$ of K channels, select the corresponding X_k with discriminative power based on the principle of maximum channel variance, and mark the indices of the corresponding channels sorted by the channel responses of E as $C = \{C_1, C_2, \ldots, C_K\}$.

3.3   Saliency Joint Weighting Method

Spatial Weighting. In this paper, we employ a spatial weighting matrix S' to weight the deep convolutional features of images, aiming to highlight salient regions in the feature maps and suppress background noise areas.

We denote the top M feature maps with the largest variances among all channels in C as $\hat{X} = \{X_{Cm}\}_{m=1}^{M}$. The eigenvalues of the corresponding positions in the M channels are squared and then summed by channel to generate the spatial weight matrix $S \in R^{W \times H}$.

$$S(i,j) = \sum_{m=1}^{M} X_{Cm}(i,j)^2 \tag{4}$$

S' represents the normalized weight of the response value in S at the corresponding position (i,j). The normalized weights S' are weighted by the original features $X = \{X_k\}_{k=1}^{K}$ to obtain the significance feature map $X' = \{X'_k\}_{k=1}^{K}$.

$$S'(i,j) = \left(\frac{S(i,j)}{\left(\sum_{i=1}^{W} \sum_{j=1}^{H} S(i,j)^\alpha \right)^{\frac{1}{\alpha}}} \right)^{\frac{1}{\beta}} \tag{5}$$

$$X'_k = X_k \otimes S' \tag{6}$$

where α and β are the experimental parameters set to 2.

Channel Weighting. When aggregating convolutional features, different channels exhibit varying responses to semantic content, reflected by the element values in feature maps. Frequently occurring features generate strong responses, affecting weaker channels. However, low-frequency features are crucial for distinguishing similar images. To optimize feature representation, we propose an element-sensitive channel response intensity weighting method that considers the sensitivity of each element in the feature.

Let Φ_k denote the total sum of response values of the kth channel. $X'_k(i,j)$ denote the response value at position (i,j) in the kth channel. The feature map X'_k is aggregated using sum-pooling to obtain the channel sparsity Z_k, as shown in Eq. 8.

$$\Phi_k = \sum_{i=1}^{W} \sum_{j=1}^{H} X'_k(i,j) \tag{7}$$

$$Z_k = \left(\frac{\Phi_k}{W \times H} \right)^2 \tag{8}$$

where W and H denote the width and height of the feature map, respectively. Z'_k represents the weight amplification of rare features in the overall response.

$$Z'_k = \log \left(\frac{\varepsilon + \sum_{k=1}^{K} Z_k}{\varepsilon + Z_k} \right) \tag{9}$$

To differentiate the differences between different channels, we need to construct channel weights R' using a combination of elemental values and sum-pooling sparsity after computing the salient feature map X'.

$$\Omega_k = \left(\sum_{i=1}^{W} \sum_{j=1}^{H} X'_k(i,j) \right)^2 \tag{10}$$

$$R_k = Z'_k \odot \Omega_k \tag{11}$$

where $\odot$ denotes element-wise product between vectors, based on the element-sensitive channel response intensity weights defined as R'_k:

$$R'_k = \log \left(\frac{\varepsilon + \sum_{k=1}^{K} R_k}{\varepsilon + R_k} \right) \tag{12}$$

where ε is a small constant for numerical stability.

After obtaining the spatial weights S' and channel weights R', the next step is to aggregate the deep features with these weights to obtain the saliency joint weighted descriptor $G \in R^K$, which is defined as Eq. 13.

$$G_k = R'_k \odot \Phi_k \tag{13}$$

Finally, the compact D-dimensional global feature representation is obtained after post-processing G with L2 parametric normalization, PCA dimensionality reduction and whitening.

3.4 Shape Fine Matching Based on Skeleton Context

The saliency joint weighting (SJW) algorithm enables the retrieval of candidate characters that share the spatial structure of the query sample. However, due to the inherent variation in stroke thickness seen in ancient Chinese characters, the algorithm may include similar but not identical characters in terms of stroke spatial distribution, while potentially overlooking characters with significant stroke distribution differences but identical to the query sample. This phenomenon influences the accuracy of retrieval. The first row of Fig. 5(a)–(c) illustrates the retrieval results of the SJW algorithm.

The ancient Chinese characters possess ideographic shapes, and their character skeleton has the capability to retain the structural and topological features between strokes of these characters, irrespective of the thickness of the strokes. Additionally, shape context [2] serves as a robust representation of the shape features in ancient Chinese characters. In this paper, we propose skeleton context-based shape matching algorithms by combining skeleton and shape contexts, enabling fine matching of candidate character skeleton images, the detailed algorithm for skeleton fine-grained matching is in Algorithm 2.

To calculate the skeleton context of ancient Chinese characters, essential preprocessing steps, such as denoising, size normalization and binarization, are applied to the character images. The image size is normalized to 128×128 pixels, with a minimum 10-pixel distance between character boundaries and image borders. Figure 2(a) and Fig. 2(b) depict the original and preprocessed

Algorithm 2. The RSSM Framework for Fine-Grained Matching

Input: Candidate image list C , query image q , normalized size M
Output: Reordered character image list $CostDistanceList$

1: $q \leftarrow Preprocess(q, M)$ ▷ image preprocessing operations
2: $S_q \leftarrow Skeletonization(q)$ ▷ obtain image skeleton map
3: $S'_q, centroid_q \leftarrow SkeletonProcess(S_q)$ ▷ deburring and uniform sampling of the skeleton
4: $SC_q \leftarrow SkeletonContext(S'_q, centroid_q)$
5: **for** each i in C **do**
6: $\quad i \leftarrow Preprocess(i, M)$ ▷ image preprocessing operations
7: $\quad S_i \leftarrow Skeletonization(i)$ ▷ obtain image skeleton map
8: $\quad S'_i, centroid_i \leftarrow SkeletonProcess(S_i)$ ▷ deburring and uniform sampling of the skeleton
9: $\quad SC_i \leftarrow SkeletonContext(S'_i, centroid_i)$
10: $\quad cost_i \leftarrow MatchDistance(SC_q, SC_i)$
11: $\quad cost_i$ added in $CostDistanceList$
12: **end for**
13: $sort\_in\_ascending\_order(CostDistanceList)$

---

character images. To mitigate stroke thickness influence, Lee's skeleton thinning algorithm [8] is employed to extract the skeleton from the preprocessed images (Fig. 2(c)). Further computational efficiency is achieved through burr removal [3] and a uniform sampling strategy, resulting in a sparse skeleton representation (Fig. 2(e)).

In the process of Chinese character skeleton matching and retrieval, it is crucial to take into account the shape deformation of the skeleton due to variations in Chinese character font slant and writing styles. To ensure geometric invariance, we improve the skeleton shape descriptor that employs the gray centroid method to guarantee Chinese character rotation invariance, and distance normalization to ensure scaling invariance.

The sparse skeleton is represented as a set of n points, denoted as $P = \{p_i | i = 1, 2, \ldots, n\}$. First, the centroid coordinates $p_0(x_0, y_0)$ of the sparse skeleton are calculated. For each point p_i , a log-polar coordinate system is constructed with p_i as the center and the line between p_i and the centroid p_0 as the X-axis. Next, the angle θ_j formed by the remaining skeleton point $p_j \in P, i \neq j$ on the X-axis and the Euclidean distance d_j between p_j and p_i are calculated. The Euclidean distance $D = \{d_1, d_2, \ldots, d_{n-1}\}$ between skeleton point p_i and the remaining skeleton points is normalized by the distance mean, and the normalized distance is defined as $D = \{\frac{d_1}{d}, \frac{d_2}{d}, \ldots, \frac{d_{n-1}}{d}\}$. Finally, we set angle $T = 12$ and the radius $R = 5$ and divide the coordinate system space into 60 bins. As shown in Fig. 2(f), we calculate the number of remaining skeleton points in each bin is computed based on two vectors corresponding to the skeleton point p_i: distance D and angle θ. This constitutes the skeleton context for the given skeleton point p_i, denoted as $h_i(k)$.

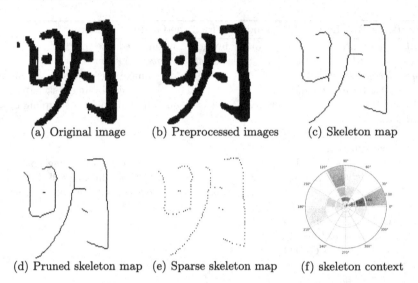

(a) Original image   (b) Preprocessed images   (c) Skeleton map

(d) Pruned skeleton map   (e) Sparse skeleton map   (f) skeleton context

Fig. 2. Output of the pre-processing algorithm: (f) shows the log-polar histogram distribution of the locations of the green dots in (e). (Color figure online)

$$h_i(k) = \#\{p_j \neq p_i : p_j \in bin(k)\}, k = 1, 2, \ldots, 60 \tag{14}$$

where $\#$ denotes the number of skeleton points in the kth bin. p_j is the skeleton point that falls into the kth bin different from the point p_i.

We utilizes the measures of Chi-square distance and cosine distance to calculate the skeleton similarity measures of query characters and candidate characters, which consider both the similarity measures of distance between skeleton points and adjacent angle to more accurately reflect the spatial distribution between image data points in high-dimensional space.

The approximate matching cost, denoted as $AMC(p_i, q_j)$, measures the similarity between skeleton point p_i of query character P and skeleton point q_j of candidate character Q. A smaller value indicates a closer approximation.

$$AMC(p_i, q_j) = \frac{1}{2} \sum_{k=1}^{60} \frac{[h_i(k) - h_j(k)]^2}{h_i(k) + h_j(k)} \cdot \left(1 - \frac{\sum_{k=1}^{60} p_{i,k} \times q_{j,k}}{\sqrt{\sum_{k=1}^{60} p_{i,k}^2}\sqrt{\sum_{k=1}^{60} q_{j,k}^2}}\right) \tag{15}$$

Here, $p_{i,k}$ represents the value of the kth dimension of the skeleton point p_i of the query character, and $q_{j,k}$ represents the value of the kth dimension of the skeleton point q_j of the candidate character.

The minimum matching cost of p_i and the corresponding point can be represented as $MinC_i$.

$$MinC_i = min\{AMC(p_i, q_j), j = 1, 2, \ldots, m\} \tag{16}$$

where, m denotes the number of skeleton points of the candidate characters.

The shape matching value between the query character P and candidate character Q is the summation of the approximate matching values of all their skeleton points.

$$Sim(P,Q) = \sum_{i=1}^{n}(MinC_i + \eta\|p_i - corresp(p_i)\|) \qquad (17)$$

In the above formula, $\|p_i - corresp(p_i)\|$ represents the Euclidean distance between the point p_i and its corresponding point $corresp(p_i)$. The penalty factor η is used to assign a greater penalty value for larger distances between two points. In the experiments, $\eta = 0.2$.

4   Experiments

4.1   Dataset and Evaluation

In this experiment, a substantial amount of real data samples were collected from the *Complete Library in Four Sections* [1], which is a typical document in the field of ancient literature research. Though correction, denoising, and character slicing, we acquired 92,295 images of ancient Chinese characters, constructing the GJHZ dataset. From the dataset, we selected more than 30,000 characters for categorization and labeling. Additionally, we applied data augmentation techniques to categories with limited samples, including scaling, rotation, character distortion, and noise addition, to enhance sample diversity. Consequently, the dataset comprises 845 categories and 42,250 labeled images, randomly divided into 33,800 for training and 8450 for testing in an 8:2 ratio. Table 1 illustrates the sample dataset. In this study, multiple metrics were used to evaluate the proposed method, including Precision, Recall, F1-score, mean average precision at topk (mAP@k) and average retrieval time(ART).mAP@k refers to the percentage of queries for which matching samples are ranked among the top k results.

4.2   Optimal Number of Channels

In the designed method, the spatial weights are constructed by selecting the top M channels with higher channel responses. These channels capture the special semantic information of ancient Chinese character images, thereby impacting the retrieval performance. To balance the preservation of character features and reduction of background noise, we explored various values of M, as shown in Table 2.

Table 2 presents the outcomes of downsampling the salience joint weighted descriptors to $D = 512$ using Principal Component Analysis (PCA). The mAP for image retrieval exhibits a general upward trend for $M \leq 25$, whereas it decreases for $M \geq 25$. Notably, $M = 25$ yields the best experimental result, indicating that the channel sorting algorithm effectively identifies discriminative semantic regions in ancient Chinese character images. Achieving better retrieval

Table 1. Sample Example of Dataset.

No.	Structure	Samples of Chinese characters	
1	Top-bottom Structure	東 GJHZ_0000010040193	藝 GJHZ_0000010070299
2	Left-right Structure	繕 GJHZ_0000010020057	儲 GJHZ_0000010020058
3	Semi-enclosed Structure	厲 GJHZ_0000010020183	風 GJHZ_0000010020076
4	Enclosed Structure	圓 GJHZ_00000100500915	困 GJHZ_0000110621215
5	Monomeric Structure	長 GJHZ_00000100602512	或 GJHZ_00000100410027

Table 2. mAP values for different number of channels on the dataset.

M	mAP@50	M	mAP@50
10	87.02	100	88.41
15	87.71	200	88.06
20	88.46	300	87.21
25	88.97	400	87.25
50	88.73	512	87.25

performance with fewer channels, we chose $M = 25$ as optimal while balancing computational efficiency and accuracy.

4.3   Ablation Study on Each Component

In this study, we assessed four channel selection and weighting approaches: Sum (sum pooling), Var (optimized channel variance), Var+SW (optimized channel variance with spatial weight matrix), SJW (saliency joint weighting with spatial weight matrix and channel weights), and Var+SJW (saliency joint weighting based on optimized channel variance). The results of the mAP comparisons for different components and dimensions under $M = 25$ are shown in Fig. 3.

Figure 3 illustrates the superior performance of Var+SJW, achieving a maximum mAP improvement of 5.13% and a minimum improvement of 3.78% compared to the Sum approach. Our method assigns higher weights to the structural regions of Chinese characters, highlighting their salient features. Moreover, the mAP value at $D = 512$ is comparable to, or slightly higher than, $D = 1024$ and $D = 2048$. Consequently, for enhanced retrieval speed, we selected $D = 512$ for our experiment.

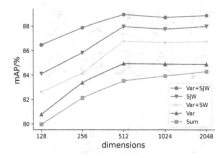

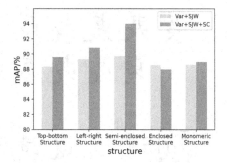

Fig. 3. Performance of different component combinations.

Fig. 4. coarse-grained retrieval and fine-grained matching of mAP values.

4.4 Fine-Grained Matching Experimental Results

To evaluate the retrieval performance of skeleton matching, we conducted comparison experiments of coarse-grained retrieval and fine-grained matching, denoted as Var+SJW and Var+SJW+SC, respectively. The results of these experiments are shown in Fig. 4. After employing skeleton context fine-grained matching, the mAP improved by 1.85% on average, except for the enclosed structure. Chinese characters with enclosed structures can affect the results of skeleton extraction and retrieval due to the adhesion of strokes to the enclosing borders.

In Fig. 5(a)–(c), the top 10 retrieval results for coarse-grained retrieval and fine-grained matching are shown in the first and second rows, respectively. Our observations reveal that skeleton matching can effectively avoid interference between different glyph shapes with the same spatial structure. Moreover, it can also relegate the unrelated Chinese character images to the lower part of the ranking, thus further enhancing the retrieval performance.

4.5 Performance Comparison

To demonstrate the performance of the proposed method, we constructed the methods of GIST-SC [13], AMR [23], MSRBC [21] and ACCINet [16] to conduct comparative experiments with the method of this paper, and the experimental results are shown in Table 3.

Table 3 presents the comparative results for precision, recall, F1-score, and mAP@50. It is evident that our method achieves the best precision and mAP with a precision of 90.71% and mAP@50 of 90.59%. However, concerning the average retrieval time (ART), the proposed method shows improvement compared to GIST-SC [13] and AMR [23], utilizing traditional features. Nonetheless, when compared to MSRBC [21] and ACCINet [16], which utilize deep features,

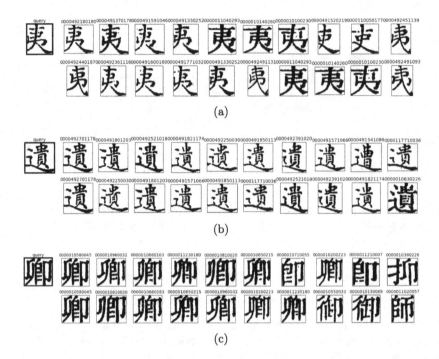

Fig. 5. Search results of coarse retrieval and fine matching for four Chinese characters.

Table 3. Comparing the mean Average Precision (mAP) values of different methods.

Method	Precision	Recall	F1-score	mAP@50	ART/s
GIST-SC [13]	68.57	47.79	53.87	63.33	41.01
AMR [23]	67.32	52.50	60.77	69.90	29.83
MSRBC [21]	74.64	55.43	68.17	78.12	2.98
ACCINet [16]	78.18	59.56	70.21	81.25	3.26
RSSM(Ours)	90.71	64.94	75.66	90.59	15.58

the method in this paper exhibits slightly higher computational time. This is due to the fact that skeleton matching requires a certain amount of computation.

5   Conclusion

In this paper, we present a hierarchical retrieval method for ancient Chinese character images based on region saliency and skeleton matching. The method employs a channel importance ranking algorithm to select discriminative key filters from deep convolutional features. In the feature aggregation stage, deep convolutional features are aggregated using a saliency joint weighting algorithm and the key filters to improve the discriminability between ancient Chinese characters. To achieve higher accuracy, a shape matching method based on skeleton context is designed for fine-grained matching of candidate characters. We employ sparse skeleton graphs as a means of depicting the shapes of Chinese characters, which helps to reduce the computational time and complexity while ensuring retrieval accuracy. Experimental results demonstrate that the proposed approach has certain advantages and can effectively improve the retrieval accuracy and speed of ancient Chinese character images. Future research will explore larger, diverse datasets to evaluate the method's performance and optimize the algorithm for efficient retrieval.

Acknowledgements. We would like to thank anonymous reviewers for their helpful comments and suggestions. This work was supported by the Natural Science Foundation of Hebei Province of China (grant number F2019201329).

References

1. Complete Library in Four Sections. http://skqs.guoxuedashi.net/
2. Belongie, S., Malik, J., Puzicha, J.: Shape matching and object recognition using shape contexts. IEEE Trans. Pattern Anal. Mach. Intell. **24**(4), 509–522 (2002)
3. Chang, Q., Wu, M., Luo, L.: Handwritten Chinese character skeleton extraction based on improved ZS thinning algorithm. Comput. Appl. Softw. **37**(7), 8 (2020)
4. Chen, J., Zhu, F.: Hierarchical matching for Chinese calligraphic retrieval based on skeleton similarity. J. Chin. Comput. Syst. (2010)
5. Du, S., Yang, F., Tian, X.: Ancient Chinese character image retrieval based on dual hesitant fuzzy sets. Sci. Program. **2021**, 1–9 (2021)
6. Kalantidis, Y., Mellina, C., Osindero, S.: Cross-dimensional weighting for aggregated deep convolutional features. In: Hua, G., Jégou, H. (eds.) ECCV 2016. LNCS, vol. 9913, pp. 685–701. Springer, Cham (2016). https://doi.org/10.1007/978-3-319-46604-0_48
7. Kato, N., Suzuki, M., Omachi, S., Aso, H., Nemoto, Y.: A handwritten character recognition system using directional element feature and asymmetric mahalanobis distance. IEEE Trans. Pattern Anal. Mach. Intell. **21**(3), 258–262 (1999)
8. Lee, T.C., Kashyap, R.L., Chu, C.N.: Building skeleton models via 3-D medial surface axis thinning algorithms. CVGIP Graph. Models Image Process. **56**(6), 462–478 (1994)
9. Ma, H., Zhonglin, Z.: A method of identification of ancient Chinese characters of multi technology fusion. J. Minzu Univ. China Nat. Sci. Ed. **27**(3), 4 (2018)
10. Mapari, S., Chaudhary, N., Naik, S., Metkewar, P.: Usage of fuzzy rule and SOM based model to identify a handwritten chemical symbol or structures. In: 2017 Second International Conference on Electrical, Computer and Communication Technologies (ICECCT), pp. 1–4. IEEE (2017)
11. Melnyk, P., You, Z., Li, K.: A high-performance CNN method for offline handwritten Chinese character recognition and visualization. Soft Comput. **24**(11), 7977–7987 (2020)
12. Oi, J., Long, H., Shao, Y., Du, Q.: Research on Chinese character similarity algorithm based on eigenvector and stroke coding. J. Chongqing Univ. Posts Telecommun. (Nat. Sci. Ed.) **31**(6), 7 (2019)

13. Pengcheng, G., Jiangqin, W., Yuan, L., Yang, X., Tianjiao, M.: Fast Chinese calligraphic character recognition with large-scale data. Multimedia Tools Appl. **74**, 7221–7238 (2015)
14. Qu, X., Xu, N., Wang, W., Lu, K.: Similar handwritten Chinese character recognition based on adaptive discriminative locality alignment. In: 2015 14th IAPR International Conference on Machine Vision Applications (MVA), pp. 130–133 (2015). https://doi.org/10.1109/MVA.2015.7153150
15. Ran, G., Huang, S., He, Z., Yang, J.: Standardized elastic dual-mesh Chinese character feature extraction based on overlap and fuzzy technology. Comput. Eng. Des. **37**(1), 5 (2016)
16. Tian, X., Wang, Z., Zuo, L.: Deformable convolutional network retrieval model for ancient Chinese character images. China Sciencepaper **15**(4), 8 (2020). (in Chinese)
17. Tzelepi, M., Tefas, A.: Deep convolutional image retrieval: a general framework. Signal Process. Image Commun. **63**, 30–43 (2018)
18. Wei, X.S., Luo, J.H., Wu, J., Zhou, Z.H.: Selective convolutional descriptor aggregation for fine-grained image retrieval. IEEE Trans. Image Process. **26**(6), 2868–2881 (2017)
19. Xia, Y., Yang, Z., Wang, K.: Chinese calligraphy word spotting using elastic hog features and derivatives dynamic time warping. J. Harbin Inst. Technol. 21–27 (2014)
20. Xu, J., Wang, C., Qi, C., Shi, C., Xiao, B.: Unsupervised semantic-based aggregation of deep convolutional features. IEEE Trans. Image Process. **28**(2), 601–611 (2018)
21. Yang, X., He, D., Zhou, Z., Kifer, D., Giles, C.L.: Improving offline handwritten Chinese character recognition by iterative refinement. In: 2017 14th IAPR International Conference on Document Analysis and Recognition (ICDAR), vol. 1, pp. 5–10. IEEE (2017)
22. Zhang, J., Bi, H., Chen, Y., Wang, M., Han, L., Cai, L.: Smarthandwriting: handwritten Chinese character recognition with smartwatch. IEEE Internet Things J. **7**(2), 960–970 (2019)
23. Zhang, X., Zhang, L., Han, D., Bi, K.: Adaptive matching and retrieval for calligraphic character. J. Zhejiang Univ. (Eng. Sci.) **50**(4), 11 (2016)
24. Zhang, X.Y., Bengio, Y., Liu, C.L.: Online and offline handwritten Chinese character recognition: a comprehensive study and new benchmark. Pattern Recogn. **61**, 348–360 (2017)
25. Zhong, Z., Jin, L., Xie, Z.: High performance offline handwritten Chinese character recognition using googlenet and directional feature maps. In: 2015 13th International Conference on Document Analysis and Recognition (ICDAR), pp. 846–850. IEEE (2015)
26. Zhuang, Y.: TF-tree: an interactive and efficient retrieval of Chinese calligraphic manuscript images based on triple features. In: Proceedings of the ACM International Conference on Image and Video Retrieval, pp. 113–120 (2010)
27. Zhuang, Y., Zhuang, Y., Wu, F.: A hybrid-distance-tree-based index for large Chinese calligraphic characters database. J. Comput.-Aided Des. Comput. Graph. **19**(2), 7 (2007)

MHNA: Multi-Hop Neighbors Aware Index for Accelerating Subgraph Matching

Yuzhou Qin, Xin Wang$^{(\boxtimes)}$, and Wenqi Hao

College of Intelligence and Computing, Tianjin University, Tianjin, China
{yuzhou_qin,wangx,haowenqi}@tju.edu.cn

Abstract. With the proliferation of knowledge graphs in various domains, efficient processing of subgraph matching queries has become a crucial issue. However, the subgraph matching problem has been proven to be an NP-complete problem. While specific approaches aim to accelerate queries by leveraging favorable matching orders and pruning rules, they face limitations in handling large-scale graph data due to the exponential search space. Conversely, other methods employ graph indexes to enhance query efficiency, but these indexes provide limited acceleration capabilities or encounter challenges in widespread adoption due to their substantial size. In this paper, a novel index called MHNA (Multi-Hop Neighbors Aware) is proposed, which is devised to accelerate subgraph matching while minimizing the space overhead. Moreover, we introduce an efficient iterative index construction method that computes the MHNA index for each vertex by solely leveraging its neighboring information. Extensive experiments demonstrate that our methods reduce storage space and construction time by an order of magnitude compared to existing the state-of-the-art database indexing approaches. Regarding query processing time, MHNA achieves up to a 30-fold reduction in time overhead.

Keywords: Subgraph matching · Multi-hop neighbors · Bloom filter

1 Introduction

With the growing applications of knowledge graphs in diverse domains, numerous efforts have been made to efficiently analyze large graphs such as social networks and Resource Description Framework (RDF) [1]. One of the most famous problems for large graphs is *subgraph matching*. Given a data graph G and a query graph Q, the subgraph matching problem is to identify all matches of Q in G.

However, the problem of subgraph matching has been proven to be an NP-complete problem [2]. Extensive methods have been proposed to construct indexes that store the neighboring features of vertices, such as subgraphs or subtrees, to reduce the search space, thus accelerating the query process for subgraph matching [3–5]. However, the enumeration and storage of substructures incur exponential time and space complexity, which is infeasible for large-scale

X. Song et al. (Eds.): APWeb-WAIM 2023, LNCS 14331, pp. 283–298, 2024.
https://doi.org/10.1007/978-981-97-2303-4_19

graph data. At the same time, additional methods have been developed to mine frequent substructures from graphs to mitigate space overhead [6–9]. Nevertheless, extracting valuable frequent substructures from large-scale graph data has remained a challenging data mining task.

To address these issues, we propose a novel index called MHNA, which leverages the recorded multi-hop neighboring information to pre-filter a large number of intermediate results during the query process, thereby accelerating the query execution. To minimize the storage space cost, the *bloom filter* is employed to store the multi-hop neighboring information, substantially reducing storage overhead. Moreover, an index construction approach is devised to speed up the building process by effectively leveraging the information from neighboring vertexes. This approach involves γ rounds of iteration to construct γ-hop MHNA, where in each round, each vertex only needs to access its neighboring vertexes without further expansion, greatly enhancing the efficiency of index construction.

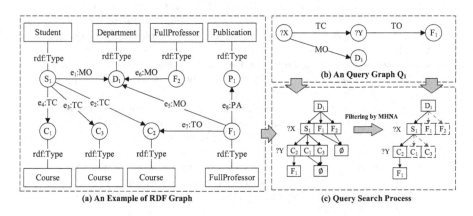

Fig. 1. An example of filtering intermediate results by MHNA. (The edge labels MO, TC, TO, and PA represent "MemberOf", "TakeCourse", "TeacherOf", and "publicatio-nAuthor", respectively.)

Example 1. For example, by executing Q_1 in Fig. 1(b) on the RDF graph shown in Fig. 1(a), the query search process illustrated in Fig. 1(c) can be obtained. The classical subgraph matching algorithm consists of two steps: (1) Determining the candidate sets for $?X$ and $?Y$, which are $\{S_1, F_1, F_2\}$ and $\{C_1, C_2, C_3\}$; (2) Based on these candidate sets, all possible results are enumerated and verified by employing a *backtracking* method to obtain the final results. However, by leveraging the information from multi-hop neighbors, it becomes possible to ensure in advance that F_1 and F_2 do not serve as 2-hop neighbors of F_1, while C_1 and C_3 are not 1-hop neighbors of F_1. After this filtering process, the candidate sets for $?X$ and $?Y$ are reduced to $\{S_1\}$ and $\{C_2\}$ correspondingly. This instance exemplifies the ability of MHNA in filtering out a significant number of intermediate results in advance to speed up the query process.

Our contributions in this paper can be summarized as follows:

(1) We propose a novel index method called MHNA, devised to improve the performance of subgraph matching by preserving multi-hop neighboring information while minimizing space overhead.
(2) An iterative index construction approach is proposed to construct MHNA effectively by leveraging the information from neighboring vertexes without further expansion.
(3) A subgraph matching algorithm is proposed that employs the MHNA within the backtracking based framework to improve the query efficiency.
(4) The extensive experiments have been conducted on both synthetic and real-world datasets to verify the efficiency of our methods. The experimental results show that MHNA outperforms the state-of-the-art methods by an order of magnitude.

The rest of this paper is organized as follows. Section 2 reviews related works. In Sect. 3, we introduce preliminary definitions. In Sect. 4, we describe the MNHA schema, its corresponding construction method, and the overall query processing in detail. Section 5 shows the experimental results, and we conclude in Sect. 6.

2   Related Work

In this section, the existing works on optimizing subgraph matching queries are concluded. Furthermore, we will review several specialized RDF systems.

2.1   Subgraph Matching

Building upon Ullman's backtracking based subgraph isomorphism algorithm [2] proposed in 1976, a multitude of algorithms such as Nauty [10], TurboISO [11], SQBC [12], and VF2++ [13] are devised to improve the efficiency of subgraph matching by selecting optimal matching orders and leveraging heuristic information to filter out intermediate results as early as possible. However, due to the limitation of the exponential search space, these algorithms are only applicable to small-scale graph data.

To accommodate large-scale graph data, extensive methods such as Closure-tree [4], CECI [3], and VEQS [5] have been employed to improve the performance of subgraph matching by constructing indexes in advance and utilizing the neighboring features (such as paths, subtrees, and subgraphs) stored in the indexes to accelerate queries. Nevertheless, as the size of the graph data increases, the scale of the index grows exponentially, making it unacceptable for large-scale graph data. To address this issue, several methods, such as gIndex [6], FG-index [7], Tree+Δ [8], and FPIRPQ [9], have been proposed to reduce the scale of the indexes by only indexing frequent substructures. However, mining valuable frequent substructures from large graph data has always been a challenging data mining problem. Therefore, MHNA is devised to preserve complete neighboring information. Meanwhile, at the expense of only sacrificing a small amount of accuracy, the index size of MHNA is reduced significantly.

2.2 Specialized RDF Systems

As RDF becomes increasingly popular, several specialized RDF systems have emerged to efficiently manage RDF graphs and answer complex queries. For example, Hexastore [14] and RDF-3X [15] employ a three-column relational table to store RDF triple data and accelerate queries by sextuple indexes. Virtuoso [16] is a high-performance multi-model database system in which graph data is organized in a single relational table, and query performance is ensured through indexes and compression techniques. In gStore [17], the neighbors of entity vertex are encoded as a binary string to construct a VS-Tree index, enabling efficient reduction of the search space for SPARQL queries. However, multi-hop neighboring information cannot be employed in these databases to speed up queries, which is one of our key contributions in MHNA.

This paper introduces MHNA, a novel index for large-scale graph data based on the bloom filter. The proposed index stores multi-hop neighboring information for each vertex and exploits it to filter intermediate results, thus improving the efficiency of the query.

3 Preliminaries

In this section, we present the concepts essential for understanding the content of this paper. Table 1 lists the notations frequently used in this paper.

Table 1. Frequently used notations

Notation	Description
G, Q	RDF graph and query graph
V, E	The set of vertices and edges
$\Delta, \mathcal{T}$	Properties of vertices and labels of vertices or edges
M	A subgraph match
$\mathbb{B}$	A bit vector, also referred to as bloom filter (BF)
H_k	Vector representation of k hash functions
$\mathbb{B}^{\leq\gamma}(v)$	BF stores edges of vertex v within γ hops
$\mathbb{B}^{\leq\gamma}(v, e)$	The BF stores edges of vertex v within γ hops, except those reachable from adjacent edge e of v
m, k	Size of bloom filters and hash function numbers
$[\![\mathbb{B}^{\leq\gamma}]\!]^{(m,k)}$	γ-hop MHNA

Definition 1 (RDF Graph). *Consider three disjoint infinite sets U, B, and L representing Uniform Resource Identifiers (URI), blank nodes, and literals, respectively. A triple $(s, p, o) \in (U \cup B) \times U \times (U \cup B \cup L)$ is called an RDF triple, in which s, p, o is the subject, predicate, and object, respectively.*

A finite set of RDF triples T is called an RDF dataset, and its corresponding RDF graph is $G = (V, E, \Delta, T)$, where (1) $V = \{s \mid (s, p, o) \in T\} \cup \{o \mid (s, p, o) \in T \wedge o \in (U \cup B)\}$ is a collection of vertices. (2) $E = \{(s, o) \mid (s, p, o) \in T \wedge s \in V \wedge o \in V\}$ is a collection of directed edges. (3) $\Delta : V \rightarrow U \times L$ is a mapping from vertices to their properties, for any $v \in V$, $\Delta : v \mapsto \{(p, o) \mid (s, p, o) \in T \wedge s = v \wedge o \in L\}$. (4) $T : (V \cup E) \rightarrow U$ is a mapping from vertices (edges) to their labels (URI), for any $v \in V$, $T : v \mapsto \{o \mid (s, p, o) \in T \wedge s = v \wedge p = \mathtt{rdf\!:\!type}\}$ for any $e \in E$, $T : e \mapsto \{p \mid (s, p, o) \in T \wedge (s, o) \in E\}$.

Definition 2 (BGP Query). *A BGP (Basic Graph Pattern) query is defined as $Q = (V_Q, E_Q, \Delta_Q, T_Q)$, where $V_Q \subseteq V \cup V_Q^{var}$ is a subset of vertices or variables, $E_Q \subseteq V_Q \times V_Q$ is edges in query Q, Δ_Q is a mapping from vertices V_Q to their properties, and T_Q is a mapping from vertices V_Q (edges E_Q) to their labels (URI). In particular, $e \in E_Q$ is a variable edge if $T_Q(e) = \emptyset$.*

Definition 3 (Subgraph Match). *For a connected query graph Q, a subgraph match $M : V_Q \rightarrow V_m$ is a function that maps V_m to V_Q ($|V_Q| \geq |V_m|$), where $V_m \subseteq V$. The subgraph match M can be regarded as a valid result of Q if and only if the following conditions are satisfied:*

1) *if $v_i \in V_Q^{var}$, $M(v_i) \in V_m$ must be satisfied; otherwise v_i must be as the same as $M(v_i)$. For any $v \in V_Q$, $\Delta_Q(v) \subseteq \Delta(M(v))$ and $T_Q(v) \subseteq T(M(v))$.*
2) *if there exists an edge $e_1 = (v_i, u_j) \in E_Q$ in query Q, there also exists an edge $e_2 = (M(v_i), M(v_j)) \in E$ in G and $T_Q(e_1) \subseteq T(e_2)$, then e_1 can match e_2. Note that e_1 can match any edge in G if $T_Q(e_1) = \emptyset$;*

Definition 4 (Bit-Vector). *A bit-vector $\mathbb{B}$ is an ordered sequence of bits, i.e., boolean values. The i-th bit of $\mathbb{B}$ is denoted as $\mathbb{B}[i]$, and the set of all bit-vectors with a size of m is represented as $\mathcal{B}_m$. For two bit-vectors $\mathbb{B}_1$ and $\mathbb{B}_2$ belonging to the set $\mathcal{B}_m$, the operations on $\mathbb{B}_1$ and $\mathbb{B}_2$ are defined as follows:*

1) *$\mathbb{B} = \mathtt{AND}(\mathbb{B}_1, \mathbb{B}_2)$, also denoted as $\mathbb{B} = \mathbb{B}_1 \wedge \mathbb{B}_2$, is defined as: $\forall 0 \leq i \leq m$, $\mathbb{B}[i] = 1$ holds iff $\mathbb{B}_1[i] = 1 \wedge \mathbb{B}_2[i] = 1$, otherwise $\mathbb{B}[i] = 0$.*
2) *$\mathbb{B} = \mathtt{OR}(\mathbb{B}_1, \mathbb{B}_2)$, also denoted as $\mathbb{B} = \mathbb{B}_1 \vee \mathbb{B}_2$, is defined as: $\forall 0 \leq i \leq m$, $\mathbb{B}[i] = 1$ holds iff $\mathbb{B}_1[i] = 1 \vee \mathbb{B}_2[i] = 1$, otherwise $\mathbb{B}[i] = 0$.*
3) *$\mathbb{B} = \mathtt{LR}(\mathbb{B}_1, k)$. Rotate the array to the left by k steps, where k is a non-negative integer. Formally, $\forall 0 \leq i \leq m, \mathbb{B}[i] = \mathbb{B}_1[(i - k + m) \mod m]$.*

Definition 5 (Bloom Filter). *BF (Bloom Filter) is an space-efficient probabilistic data structure. Mathematically, a bloom filter represents a set $S = \{y_1, y_2, ..., y_n\}$ of n elements by a bit-vector $\mathbb{B}$ of m bits, where each bit in $\mathbb{B}$ is initially set to zero. A BF employs a constant k of independent hash functions $h_1, h_2, ..., h_k$. Each hash function map elements to random numbers uniformly in the range $1, ..., m$ as an index in $\mathbb{B}$. For convenience, a mapping $H_k : S \rightarrow \mathcal{B}_m$ is employed to represent the k hash functions. For $\mathbb{B}' = H_k(y)$, where $y \in S$, the $\mathbb{B}'$ satisfies that $\forall 1 \leq i \leq k, \mathbb{B}'[h_i(y)] = 1$, and all other bits in $\mathbb{B}'$ are zero. For any element y, a BF supports the following operations:*

1) INS($\mathbb{B}, y$). *Insert a element y into BF $\mathbb{B}$, this operation set $\mathbb{B} = \mathbb{B} \vee H_k(y)$.*
2) Exist($\mathbb{B}, y$). *This operation returns true if y is possibly in $\mathbb{B}$, or false if y definitely not in $\mathbb{B}$. This operation returns true iff $H_k(y) = \mathbb{B} \wedge H_k(y)$. The possibility of false positive ε can be estimated by:*

$$\varepsilon = (1 - e^{-\frac{nk}{m}})^k$$

4   Multi-Hop Neighbors Aware Index

In this section, we first introduce the schema and construction method of MHNA. Subsequently, a query processing method based on MHNA is presented.

4.1   Index Schema of MHNA

Both enumerating and storing the neighboring information of each vertex in graph result in exponential complexity, making it unacceptable for large-scale graphs due to the excessive time and space overhead. To address this issue, the MHNA is introduced to reduce storage space cost significantly by incurring only a marginal accuracy trade-off (fewer than 5 bits per element are required for a 0.1 false positive rate). The novelty of MHNA lies in its ability to efficiently compute γ-hop MHNA indexes by leveraging the $\gamma - 1$ hop MHNA index information of neighboring vertices without any further expansion.

$\mathbb{B}^{\leq \gamma}(v)$, for $\gamma \geq 1$, is defined to be the BF that stores the set of all those neighboring edges of vertex v up to γ hops. The auxiliary data structure $\mathbb{B}^{\leq \gamma}(v, e)$ is used during the building process of MHNA, which represents the BF utilized to store all edges of vertex v up to γ hops, except those edges can be reached from adjacent edge e of v. The formal definition of MHNA is as follows.

Definition 6 (MHNA). *Let $G = (V, E, \Delta, \mathcal{T})$ be a graph, m, k, γ be a positive integer. $[\mathbb{B}^{\leq \gamma}]_G^{(m,k)} = \{\mathbb{B}^{\leq \gamma}(v) \mid v \in V\}$ is a γ-hop MHNA index on graph G, where m and k is the dimension of $\mathbb{B}^{\leq \gamma}(v)$ and the number of hash functions, respectively.*

For a vertex $v \in V$, E_v is the set of adjacent edges of v, where $E_v = \{(s, o) \mid (s, o) \in V \wedge (s = v \vee o = v)\}$. For any edge $e = (s, o)$, a mapping $\sigma : E \times V \to V$ is adopted to obtain the adjacent vertex of e with respect to v. $\sigma(e, v) = o$ holds iff $s = v$, otherwise, $\sigma(e, v) = o$. The $\mathbb{B}^{\leq \gamma}(v)$ can be defined recursively as follow:

1) *if $\gamma = 1$, then*
   a) *$\mathbb{B}^{\leq 1}(v) = \bigvee_{e_i \in E_v} H_k((e_i, \mathcal{T}(e_i)))$;*
   b) *for each $e \in E_v$, $\mathbb{B}^{\leq 1}(v, e) = \bigvee_{e_i \in E_v \wedge e_i \neq e} H_k((e_i, \mathcal{T}(e_i)))$;*
2) *if $\gamma > 1$, then*
   a) *$\mathbb{B}^{\leq \gamma}(v) = \mathbb{B}^{\leq 1}(v) \vee \text{LR}\left(\bigvee_{e_i \in E_v} \mathbb{B}^{\leq \gamma-1}(\sigma(e_i, v), e_i), 1\right)$;*
   b) *for each $e \in E_v$,*
   $$\mathbb{B}^{\leq \gamma}(v, e) = \text{LR}\left(\bigvee_{e_i \in E_v \wedge e_i \neq e} H_k((e_i, \mathcal{T}(e_i))) \vee \mathbb{B}^{\leq \gamma-1}(\sigma(e_i, v), e), 1\right);$$

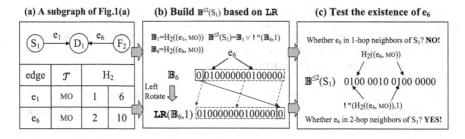

Fig. 2. An example of distinguishing the neighboring edges in different hops

To enhance efficiency, in Definition 6, the $\mathbb{B}^{\leq\gamma}(v)$ values are constructed based on the information of neighboring vertices, making it challenging to determine the distance of a multi-hop neighbor e from vertex v. To address this issue, the LR operation is introduced to preserve the depth information of multi-hop edges.

The working process of LR is depicted in Fig. 2 through an example. As shown in Fig. 2(b), for vertex S_1, its γ-hop MHNA index $\mathbb{B}^{\leq\gamma}(S_1)$ is equal to $\mathbb{B}_1 \vee \text{LR}(\mathbb{B}_6, 1)$. The LR operation performs a left rotation by 1 bit on $\mathbb{B}_6$, indicating that e_6 is a 2-hop neighbor of vertex S_1. As shown in Fig. 2(c), due to the bit position changes caused by the LR operation, when determining the presence of an edge e_6, the LR operation on the hash value of e_6 is required to match $\mathbb{B}^{\leq 2}(S_1)$.

Theorem 1. *Assuming E_v^i represents all i-hop neighboring edges for vertex $v \in V$, and $E_v^{\leq\gamma} = \{E_v^i \mid 1 \leq i \leq \gamma\}$. If there exists an edge e_j such that $\mathbb{B}_{e_j} = \mathbb{B}_{e_j} \wedge \mathbb{B}^{\leq\gamma}(v)$, where $\mathbb{B}_{e_j} = \text{LR}(H_2((e_j, \mathcal{T}(e_j)), i-1)$, it indicates that $e_j \in E_v^i$ satisfies with a probability of $1 - (1 - e^{-\frac{nk}{m}})^k$, where $n = |E_v^{\leq\gamma}|$, m is the bit size of $\mathbb{B}^{\leq\gamma}(v)$, and k is the hash function number. Otherwise, it can be inferred that $e_j \notin E_v^i$.*

Proof. (Sketch) Assuming there are k hash functions that uniformly select each position in $\mathbb{B}^{\leq\gamma}(v)$, the probability that the bit in $\mathbb{B}^{\leq\gamma}(v)$ is not selected when inserting e_i can be expressed as $(1 - \frac{1}{m})^k$. Based on the fact that $\lim_{m\to\infty}(1 - \frac{1}{m})^m = \frac{1}{e}$, we can approximate $(1 - \frac{1}{m})^k$ as $e^{-\frac{k}{m}}$ when m is large (error less than 10^{-2} when $m \geq 20$). After all edges in $E_v^{\leq\gamma}$ have been inserted into $\mathbb{B}^{\leq\gamma}(v)$, the probability that the bit in $\mathbb{B}^{\leq\gamma}(v)$ has been set to 1 is $1 - e^{-\frac{kn}{m}}$, where $n = |E_v^{\leq\gamma}|$. Thus, if there exists an edge e_j such that $\mathbb{B}_{e_j} = \mathbb{B}_{e_j} \wedge \mathbb{B}^{\leq\gamma}(v)$, the probability of $e_j \notin E_v^i$ is $(1 - e^{-\frac{nk}{m}})^k$. Hence, the probability of $e_j \in E_v^i$ is $1 - (1 - e^{-\frac{nk}{m}})^k$.

If e_j has already been inserted into $\mathbb{B}^{\leq\gamma}(v)$, it implies that the corresponding vector positions have been set to 1. However, in the case where $\mathbb{B}_{e_j} \neq \mathbb{B}_{e_j} \wedge \mathbb{B}^{\leq\gamma}(v)$, indicating that there exist vector positions corresponding to e_j that are still zero, we can infer that $e_j \notin E_v^i$. $\qquad\square$

Based on Definition 6, an efficient construction method for MHNA is presented in Algorithm 1. The optimal length of bit vectors and hash function

---

Algorithm 1: Constructing MHNA index

Input: The graph $G = (V, E, \Delta, \mathcal{T})$. The parameters ε, γ, and n which is the expected false positive rate, maximum hops, the estimated number of γ-hop neighbors, respectively.

Output: The MHNA index $[\![\mathbb{B}^{\leq \gamma}]\!]_G^{(m,k)}$

1   $m \leftarrow -\lfloor \frac{n \ln \varepsilon}{(\ln 2)^2} \rfloor$; $k \leftarrow \lfloor \frac{m}{n} * \ln 2 \rfloor$;   // count the optimal m and k

2   **foreach** $e = (v, u) \in E$ **do**   // initialize the auxiliary data structure

3     $\lfloor$ $\mathbb{B}^{\leq 0}(v, e) \leftarrow \{0\}^m$; $\mathbb{B}^{\leq 0}(u, e) \leftarrow \{0\}^m$;

4   **for** $i = 1, ..., \gamma$ **do** // build γ-hop MNHA index from bottom to top

5   **foreach** $v \in V$ **do**

6         $E_v \leftarrow \{(s, o) \mid (s, o) \in V \land (s = v \lor o = v)\}$;   // adjacent edges of v

7   **for** $j = 1, ..., |E_v|$ **do**

8           $\lfloor$ $\mathbb{B}_j \leftarrow H_k((e_j, \mathcal{T}(e))) \lor \mathrm{LR}\left(\mathbb{B}^{\leq i-1}(\sigma(e_j, v), e_j), 1\right)$;

9         $\mathbb{B}_{pre} \leftarrow \{0\}^m$; $\mathbb{B}_{suf} \leftarrow \{0\}^m$;   // the prefix and suffix result is kept to avoid redundant computation

10   **for** $j = 1, ..., |E_v|$ **do**   // traverse E_v, $\forall 1 \leq j \leq |E_v|$, $e_j \in E_v$

11           $l \leftarrow |E_v| - j$;

12           $\mathbb{B}^{\leq i}(v, e_j) \leftarrow \mathbb{B}^{\leq i}(v, e_j) \lor \mathbb{B}_{pre}$; $\mathbb{B}^{\leq i}(v, e_l) \leftarrow \mathbb{B}^{\leq i}(v, e_l) \lor \mathbb{B}_{suf}$;

13           $\mathbb{B}_{pre} \leftarrow \mathbb{B}_{pre} \lor \mathbb{B}_j$; $\mathbb{B}_{suf} \leftarrow \mathbb{B}_{suf} \lor \mathbb{B}_l$;

14         $\mathbb{B}^{\leq i}(v) \leftarrow \mathbb{B}_{pre}$;

15 **return** $\{\mathbb{B} \mid v \in V \land \mathbb{B} = \mathbb{B}^{\leq \gamma}(v)\}$;   // obtain γ-hop MHNA index on G

---

numbers, denoted by m and k, respectively, can be determined using the formulas presented in line 1 [18]. Algorithm 1 adopts a bottom-up index construction method. (1) First, the auxiliary data structures (lines 2–3) are initialized. (2) Next, for each vertex $v \in V$, a space-optimized dynamic programming approach is adopted to compute $\mathbb{B}^{\leq i}(v, e_j)$ and $\mathbb{B}^{\leq i}(v)$ efficiently (line 9–14). (3) Iteratively, the second step will be repeated until the γ-hop MHNA index is constructed. (line 4). The idea of this efficient construction algorithm is to fully exploit the information from neighboring vertices. In each iteration, only the directly adjacent vertices are required to be traversed without any further expansion, resulting in a substantial improvement in performance.

Complexity Analysis. The time complexity of MHNA index construction is $O(\gamma \lceil \frac{m}{W} \rceil |E|)$, where W denotes the number of bits operated together, typically 32 or 64 bits.

Proof. (Sketch) The time complexity of the MHNA index construction on graph G consists of three components: (1) Initialization of the auxiliary data structure, whose complexity of $O(|E|)$. (2) For each vertex $v \in V$, a dynamic programming approach are employed to compute the corresponding $\mathbb{B}^{\leq i}(v, e_j)$ and $\mathbb{B}^{\leq i}(v)$, in the time complexity of $(\lceil \frac{m}{W} \rceil |E_v|)$, where $\lceil \frac{m}{W} \rceil$ represents the time complexity of the AND operation on bit vector. Therefore, the total time complexity for this part is $O(\lceil \frac{m}{W} \rceil |E|)$. (3) The second step is repeated for $\gamma - 1$ times to construct the γ-hop MHNA index from bottom to top. Hence, the overall time complexity of the proposed algorithm is $O(\gamma \lceil \frac{m}{W} \rceil |E|)$.   □

Example 2. Figure 3(a) shows the auxiliary data structures corresponding to each vertex of the constructed γ-hop index, and Fig. 3(b) shows the hash values of each edge. In Fig. 1(a), the adjacent vertices of D_1 are $V_{D_1} = \{S_1, F_1, F_2\}$ and adjacent edges are $E_{D_1} = \{e_1, e_5, e_6\}$. The MHNA construction process is described here using D_1 as an example. (1) If $i = 1$, as shown in Fig. 3(c), $\mathbb{B}^{\leq 1}(D_1, e_1) = H((e_5, MO)) + H((e_6, MO))$. $\mathbb{B}^{\leq 1}(D_1, e_5)$, $\mathbb{B}^{\leq 1}(D_1, e_6)$, and $\mathbb{B}^{\leq 1}(D_1)$ are handled similarly. (2) When $i \geq 2$, as depicted in Fig. 3(d), B_1, B_2, and B_3 correspond to the auxiliary information of the three neighbors S_1, F_1, and F_2, respectively. Using prefix and suffix arrays to hold the intermediate results, all the auxiliary data structures and the i-hop MHNA corresponding to D_1 are computed in $O(|E_{D_1}|)$ time complexity. (3) By iterating the procedure in Fig. 3(d) for $\gamma - 1$ iterations, the γ-hop MHNA illustrated in Fig. 3(e) can be computed.

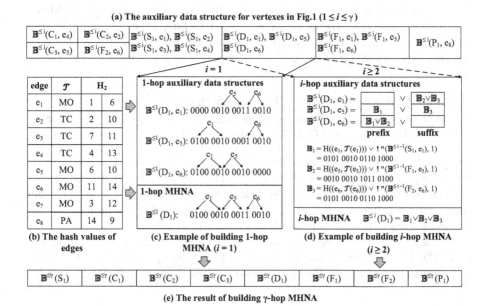

Fig. 3. Example of building γ-hop MHNA for the RDF graph in Fig. 1(a)

4.2 Query Processing

The query processing of subgraph matching is boosted by utilizing the γ-hop MHNA index, denoted as $[\![\mathbb{B}^{\leq\gamma}]\!]_G^{(m,k)}$, which acts as a filtering mechanism for intermediate results. The effective leverage of γ-hop reduces a large number of invalid intermediate results, thus speeding up the query.

In Algorithm 2, it is demonstrated that a backtracking based method that utilizes the MHNA index to speed up subgraph matching. As depicted in Algorithm 2, after all the vertices in query Q are successfully matched, M will be added to the query results $[\![M]\!]_G$ (line 2). Otherwise, an unmatched vertex v with minimal candidates is chosen from V_Q. Its candidate set is determined by its type and property information (line 4–6). For each valid candidate vertex $u \in C_M(v)$, (v, u) is added to the result M_{new} (line 9) and the `matching` function is invoked recursively (line 17).

However, our backtracking process differs from existing methods as follows. (1) On the one hand, for each matched vertex $v \in Q$ and its corresponding result u, a depth-first search algorithm is adopted to identify the neighboring edges within a γ-hop distance from v. The MHNA index $\mathbb{B}^{\leq\gamma}(u)$ and the corresponding distance from v are then assigned to these edges, thereby facilitating the subsequent filtering process (line 16, 18–24). (2) On the other hand, the information from neighbors up to γ-hop, which is stored in $[\![\mathbb{B}^{\leq\gamma}]\!]_G^{(m,k)}$, is leveraged to filter out invalid candidates at an early stage. All candidate edges associated with a candidate vertex u, corresponding to vertex v in query Q, are identified (line 6), and the MHNA index is utilized to filter them individually (line 10–14). Considering the false positive rate of MHNA, as defined in Theorem 1, the remaining candidates are verified to ensure the correctness of the results (line 15).

Complexity Analysis. Assume $D_Q = \frac{|E_Q|}{|V_Q|}$ is the average degree of query graph Q, the time complexity of filtering candidates by MHNA and setting filters to γ-hop neighboring edges is $O(k|E_Q|)$ and $O(\min\{(D_Q)^\gamma, |E_Q|\})$, respectively.

Proof. (Sketch) The time complexity of filtering candidates by MHNA consists of three components: (1) The number of adjacent edges for a vertex in Q is $O(D_Q)$; (2) The number of filters for each edge in Q is bound by $O(|V_Q|)$; (3) The time complexity of validating candidate edges is $O(k)$, where k is the number of hash functions. Hence, the overall complexity is $O(k|E_Q|)$;

For any vertex $v \in V_Q$, the number of γ-hop neighbors of v is bounded by $O((D_Q)^\gamma)$, and the maximum number of edges in Q is $O(|E_Q|)$. Therefore, the complexity of setting filters to γ-hop neighboring edges is $O(\min\{(D_Q)^\gamma, |E_Q|\})$. □

Algorithm 2: Computing subgraph matches

Input: Graph $G = (V, E, \Delta, \mathcal{T})$, query $Q = (V_Q, E_Q, \Delta_Q, \mathcal{T}_Q)$, subgraph match M, filters $F_Q : E_Q \to \mathbb{B}^{\leq \gamma} \times \mathbb{N}^+$, where $\mathbb{N}^+$ is the set of positive integers; γ-hop MNHA index $[\![\mathbb{B}^{\leq \gamma}]\!]_G^{(m,k)}$

Output: Query results $[\![M]\!]_G$

1 **Procedure** matching($G, Q, M, F_Q, [\![\mathbb{B}^{\leq \gamma}]\!]_G^{(m,k)}, [\![M]\!]_G$):
2   **if** $|M| = |V_Q|$ **then** $[\![M]\!]_G \leftarrow [\![M]\!]_G \cup \{M\}$;   // find a query result
3   **else**
4   **foreach** $v \in \{u \mid u \in V_Q \wedge u \notin M\}$ **do**
5        $\lfloor$ $C_M(v) \leftarrow \{u \mid u \in V \wedge \Delta_Q(v) \subseteq \Delta(u) \wedge \mathcal{T}_Q(v) \subseteq \mathcal{T}(u)\}$;
6   select next v from $\{u \mid u \in V_Q \wedge u \notin M\}$ with minimal $|C_M(v)|$;
7   **foreach** $u \in C_M(v)$ **do**
8          $E_v \in \{(s, o) \mid (s, o) \in E_Q \wedge (s = v \vee o = v)\}$;
9          $M_{new} \leftarrow M; M_{new}(v) \leftarrow u$;   // add (v, u) into subgraph match
10   **foreach** $e_Q = (s, o) \in E_v$ **do**
11            $e \leftarrow (M_{new}(s), M_{new}(o))$;   // e is the candidate edge
12   **foreach** $(\mathbb{B}, hops) \in F_Q(e_Q)$ **do**
13              $\lfloor$ $\mathbb{B}_e \leftarrow \mathrm{LR}(H_k((e, \mathcal{T}(e))), hops)$;
14   **if** $\mathbb{B}_e \neq \mathbb{B}_e \wedge \mathbb{B}$ **then continue**;   // filtered by MNHA
15   **if** $\mathcal{T}_Q(e_Q) \not\subseteq \mathcal{T}(e) \vee e \notin E$ **then continue**;
16          $F_Q^{new} \leftarrow F_Q$; setFilters($u, \mathbb{B}^{\leq \gamma}(u), F_Q^{new}, 1$);
17   matching($G, Q, M_{new}, F_Q^{new}, [\![\mathbb{B}^{\leq \gamma}]\!]_G^{(m,k)}, [\![M]\!]_G$) ;

18 **Function** setFilters($v, \mathbb{B}, F_Q, \gamma_v$):
   Input: vertex v, $\mathbb{B}$ is the MHNA index for v, current hops γ_v
   Output: filters F_Q
19    $visit(v) \leftarrow true$;   // mark the vertex v as visited
20   **if** $\gamma_v \geq \gamma$ **then return**;   // the exit of the recursion
21   **foreach** $e_i = (u, u_i) \in \{(s, o) \mid (s, o) \in E_Q \wedge s = u\}$ **do**
22   **if** $visit(u_i) = false$ **then**
23          $F_Q(e_i) \leftarrow F_Q(e_i) \cup \{(\mathbb{B}, \gamma_v)\}$;   // set filters for e_i
24   setFilters($u_i, \mathbb{B}, F_Q, \gamma_v + 1$);

---

5 Experiments

In this section, the efficiency and scalability of MHNA were evaluated by comparing MHNA with baseline methods. The experiment results show that MHNA can outperform the state-of-the-art methods by an order of magnitude.

5.1 Experimental Settings

The MHNA was implemented on the top of openGauss-graph[1]. The system was deployed on a server, which has a 64-core Intel Xeon Silver 4216@ 2.10 GHz CPU,

---

[1] https://gitee.com/opengauss/openGauss-graph.

with 512GB of RAM and 1920GB SSD, running a 64-bit CentOS 7.7 operating system.

Datasets and Queries. Our experiments were conducted on the real-world dataset DBpedia [19] and the synthetic datasets LUBM [20]. The statistics of the datasets are presented in Table 2. To evaluate the filtering effectiveness of MHNA, three different types of queries[2] were introduced, which included 2-hop queries (Q_1, Q_4), 3-hop queries (Q_2, Q_5), and 4-hop queries (Q_3, Q_6).

Table 2. Statistics of Datasets.

Dataset	#Triples	#Vertices	#Edges
LUBM10	1,316,700	207,429	630,757
LUBM20	2,782,126	437,558	1,332,030
LUBM30	4,109,002	645,957	1,967,309
LUBM40	5,495,742	864,225	2,630,657
LUBM50	6,890,640	1,082,821	3,298,814
DBpedia	23,445,441	2,257,499	6,876,041

Baselines. To verify the effectiveness and efficiency of MHNA, comparative experiments were conducted against three RDF database systems. Supporting SPARQL and Cypher query languages, openGauss-graph accommodates the data of RDF graph and property graph in a unified storage scheme. Virtuoso [16] is a high-performance RDBMS, which supports the storage of various data models including RDF graph. Five indexes including PSOG, POSG, SP, OP, and GS (S, P, O, and G stand for subject, property, object, and graph, respectively), were constructed in Virtuoso to accelerate query processing. As a specialized graph database system, gStore [17] improves the performance of SPARQL queries by employing VS-Tree index, which records the encoded neighboring features of each vertex to filter the invalid intermediate results.

5.2 Experimental Results

To evaluate the performance of MHNA, four different indexes including 2-hop MHNA, 3-hop MHNA, the VS-Tree in gStore, and the indexes in Virtuoso, are compared in terms of construction time and space overhead. Furthermore, the size of original dataset was introduced as a reference. The default value for the expected false positive rate is set to 0.1.

---

[2] https://github.com/rainboat2/MHNA.

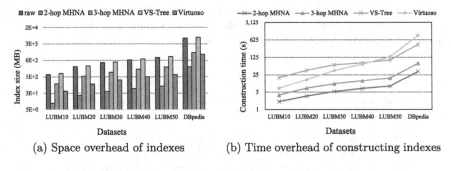

Fig. 4. Space and time overhead of constructing indexes

Exp 1. Index Size. As shown in Fig. 4(a), thanks to the space-efficient feature of the bloom filter, the storage overhead of 2-hop MHNA is an order of magnitude less than that of other methods. However, due to the inclusion of 3-hop neighboring information, 3-hop MHNA requires more space than 2-hop MHNA. While the bit vectors are also adopted to encode neighboring information for each vertex in VS-Tree, longer bit vectors are required to maintain the filtering effectiveness of non-leaf nodes. In the case of Virtuoso, although maintaining the five indexes, it employs row and page compression to reduce the size of indexes, which is the reason that the space overhead of Virtuoso is smaller than that of 3-hop MHNA.

Exp 2. Index Construction Time. As shown in Fig. 4(b), the index construction time of 2-hop MHNA is an order of magnitude on average less than that of VS-Tree and Virtuoso. The construction of the 3-hop MHNA takes twice the time required for the 2-hop MHNA. This efficiency can be attributed to the carefully designed iterative index construction method, which effectively leverages the information from neighboring vertices, thus minimizing redundant computations. In contrast, Virtuoso needs to build five different indexes and applies row and page compression on these indexes. Moreover, following the computation of the signature for each vertex, VS-Tree incurs a substantial time cost associated with constructing the tree structure and super edges [17].

Exp 3. Query Efficiency. As shown in Fig. 5 and Fig. 6, compared to Virtuoso, gStore, and openGauss-graph, 2-hop MHNA and 3-hop MHNA indexes can effectively improve query efficiency for most benchmark queries. For the Q_3 in Fig. 6, the 3-hop MHNA index can improve query efficiency up to 30 times. The reasons for the results are as follows. (1) The MHNA indexing approach effectively leverages the multi-hop, enabling the advanced filtering of a large number of intermediate results and accelerating queries. (2) Since more time-consuming JOIN operations of relation tables are required, openGauss-graph and Virtuoso take more time to handle complex queries. (3) In gStore, only 1-hop neighboring information is recorded, and it is impossible to filter intermediate results using multi-hop neighboring information. Nonetheless, for queries with few invalid intermediate results, e.g., Q_6 in Fig. 6, the filtering effectiveness of MHNA is limited and not as effective as the indexes of Virtuoso.

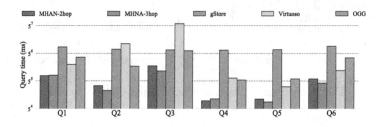

Fig. 5. The query processing time on DBpedia dataset

The 3-hop MHNA has an advantage in accelerating complex queries, while the 2-hop MHNA is more suitable for processing simple queries. As shown in Fig. 5 and Fig. 6, for simple queries like Q_1 and Q_4, the filtering effectiveness of 2-hop MHNA and 3-hop MHNA are similar. However, compared to the 2-hop MHNA, the 3-hop MHNA performs worse due to the additional computational cost. For the queries with higher complexity, the significant filtering effectiveness of 3-hop MHNA enables it to accelerate these queries more effectively.

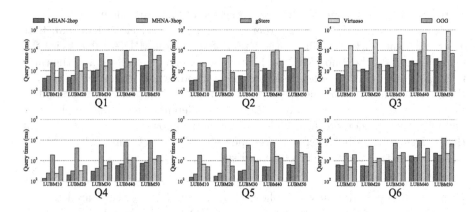

Fig. 6. The query processing time on different size of LUBM datasets

Exp 4. Impact of False Positive Rate. As shown in Fig. 7, the time and space overhead of constructing MHNA decreases significantly as the false positive rate increases, but the downward trend slows down and the query execution time increases gradually. It is worth noting that when the false positive rate is small ($\varepsilon = 0.01$), the query performance is poor. Since the lower false positive rate, the higher the space overhead of the MHNA, which increases the overall cost of utilizing the MHNA. Therefore, it is crucial to carefully select an appropriate false positive rate based on the specific circumstances.

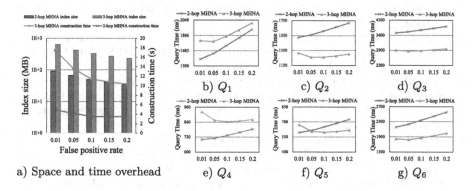

Fig. 7. Impact of different false positive rates (conducted on LUBM50)

6   Conclusion

In this paper, we introduce MHNA, a novel indexing approach devised to enhance the performance of subgraph matching queries. By leveraging the multi-hop neighboring information, MHNA effectively minimizes the number of intermediate results, reducing the computational overhead dramatically. The experimental results validate the effectiveness of MHNA, showing an order of magnitude reduction in both time and space overhead compared to the existing state-of-the-art database indexing approaches. Moreover, MHNA improves the query efficiency by up to 30 times compared to the state-of-the-art methods.

Acknowledgement. This work is supported by HUAWEI and the National Natural Science Foundation of China (61972275).

References

1. World Wide Web Consortium. RDF 1.1 concepts and abstract syntax (2014)
2. Ullmann, J.R.: An algorithm for subgraph isomorphism. J. ACM (JACM) **23**(1), 31–42 (1976)
3. Bhattarai, B., Liu, H., Huang, H.H.: CECI: compact embedding cluster index for scalable subgraph matching. In: Proceedings of the 2019 International Conference on Management of Data, pp. 1447–1462 (2019)

4. He, H., Singh, A.K.: Closure-tree: an index structure for graph queries. In: 22nd International Conference on Data Engineering (ICDE 2006), pp. 38–38. IEEE (2006)
5. Kim, H., Choi, Y., Park, K., Lin, X., Hong, S.-H., Han, W.-S.: Fast subgraph query processing and subgraph matching via static and dynamic equivalences. VLDB J. **32**(2), 343–368 (2023)
6. Yan, X., Yu, P.S., Han, J.: Graph indexing: a frequent structure-based approach. In: Proceedings of the 2004 ACM SIGMOD International Conference on Management of Data, pp. 335–346 (2004)
7. Cheng, J., Ke, Y., Ng, W., Lu, A.: FG-index: towards verification-free query processing on graph databases. In: Proceedings of the 2007 ACM SIGMOD International Conference on Management of Data, pp. 857–872 (2007)
8. Zhao, P., Yu, J.X., Yu, P.S.: Graph indexing: tree+ delta<= graph. In: Proceedings of the 33rd International Conference on Very Large Data Bases, pp. 938–949 (2007)
9. Wang, X., et al.: FPIRPQ: accelerating regular path queries on knowledge graphs. World Wide Web **26**(2), 661–681 (2023)
10. McKay, B.D., et al.: Practical graph isomorphism (1981)
11. Han, W.-S., Lee, J., Lee, J.-H.: Turboiso: towards ultrafast and robust subgraph isomorphism search in large graph databases. In: Proceedings of the 2013 ACM SIGMOD International Conference on Management of Data, pp. 337–348 (2013)
12. Zheng, W., Zou, L., Lian, X., Zhang, H., Wang, W., Zhao, D.: SQBC: an efficient subgraph matching method over large and dense graphs. Inf. Sci. **261**, 116–131 (2014)
13. Jüttner, A., Madarasi, P.: VF2++-an improved subgraph isomorphism algorithm. Discret. Appl. Math. **242**, 69–81 (2018)
14. Weiss, C., Karras, P., Bernstein, A.: Hexastore: sextuple indexing for semantic web data management. Proc. VLDB Endow. **1**(1), 1008–1019 (2008)
15. Neumann, T., Weikum, G.: The RDF-3X engine for scalable management of RDF data. VLDB J. **19**, 91–113 (2010)
16. Erling, O., Mikhailov, I.: RDF support in the virtuoso DBMS. In: Networked Knowledge-Networked Media: Integrating Knowledge Management, New Media Technologies and Semantic Systems, pp. 7–24 (2009)
17. Zeng, L., Zou, L.: Redesign of the gstore system. Front. Comput. Sci. **12**, 623–641 (2018)
18. Wikimedia Commons. Bloom filter (2023). https://en.wikipedia.org/wiki/Bloom_filter
19. Lehmann, J., et al.: DBpedia-a large-scale, multilingual knowledge base extracted from Wikipedia. Semant. Web **6**(2), 167–195 (2015)
20. Guo, Y., Pan, Z., Heflin, J.: LUBM: a benchmark for owl knowledge base systems. J. Web Semant. **3**(2–3), 158–182 (2005)

Keywords and Stops Aware Optimal Routes on Road Networks

JiaJia Li, Qiulin An$^{(\boxtimes)}$, Ying Zhao, Rui Zhu, and Na Guo

Shenyang Aerospace University, Shenyang, China
{lijiajia,zhurui}@sau.edu.cn, anqiulin0730@163.com

Abstract. Recently, the *keyword-aware routing* problem has been increasingly studied, which is to return the optimal route from the starting point s to the destination t, satisfying all the user-specified keyword requirements. Most existing solutions focus only on the travel cost (distance or time) of the route and do not take into account the number of stops required to meet the keyword service. However, parking more often can degrade the user experience, as finding a parking space in a city is difficult and time-consuming. In this paper, we study the *Keywords and Stops aware Optimal Routes (KSOR)* problem, which finds a set of non-dominated routes with minimum distance and minimum number of stops. It is non-trivial to search such routes because the newly added Point of Interest (POI) may reduce the stops when expanding the road network, which poses challenges for pruning the partial routes. To avoid this dilemma, we first utilize a spatial index to find the potential POIs around the shortest path from s to t. To efficiently combine these POIs for generating valid routes, we propose two route generation methods. The *light enumeration method* prune the range of enumerated route combinations by the calculated lower and upper bounds of the stops. And the *weighted calculation method* generates routes by calculating the optimal combination of POIs under different weights of distance and stops. Extensive experiments conducted on real road networks show that the proposed methods are more efficient.

Keywords: Keyword Aware Route · Skyline Route · Road Network

1 Introduction

With the development of global positioning technology and smart mobile devices, the routing nowadays on the road network can satisfy more user-specified needs. Among them, the *Optimal Routes with Collective Spatial Keywords(ORCSK)* [7,10,13] problem has been intensively studied. Given a road network G, where one Point of Interest (POI) may contain one or several keywords, a starting point s, a destination t, and a user-specified keywords list Q_{kw}, ORCSK returns the route with minimum distance from s to t covering all the keywords in Q_{kw}.

Such query has many applications in real-life, as shown in Fig. 1, a user travels from s to t and wants to go to a movie, buy some goods, and withdraw

© The Author(s), under exclusive license to Springer Nature Singapore Pte Ltd. 2024
X. Song et al. (Eds.): APWeb-WAIM 2023, LNCS 14331, pp. 299–314, 2024.
https://doi.org/10.1007/978-981-97-2303-4_20

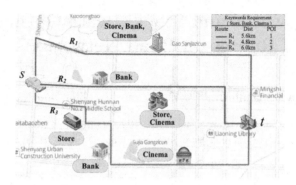

Fig. 1. Example of user route selection

money during the trip. The three routes R_1, R_2, R_3 are all valid routes since each of them can satisfy the *Store, Bank, Cinema* needs. The traditional *ORCSK* will return R_2 to the user because it has the shortest distance. However, R_1 is also competitive because it only requires one stop, which is important for users driving in the city. R_3, on the other hand, may not be preferred by most users, because it has no advantages in terms of distance and number of stops. Motivated by this, in this paper, we introduce a new route planning problem denoted as *Keywords and Stops aware Optimal Routes (KSOR)*, which returns routes that are either short distances or have few stops. It is worth noting that such a query is essentially a skyline query, so there may be more than one non-dominated route in the result.

As far as we know, there are two ways for solving *ORCSK*. One is expansion based [10,13], which traverses the network from s to the required POIs carefully until all the keywords are covered. It is non-trivial to solve *KSOR* by adopting these methods because the newly added Point of Interest (POI) may reduce the stops when expanding. As a result, plenty of partial routes cannot be pruned, thus reducing the query efficiency. Another way for solving *ORCSK* is based on the selection-and-generation mechanism [7], which searches for potential POIs around s and t first and then generates the final routes. Although we can obtain the POI candidate set easily by some spatial index, it is non-trivial to combine these candidate POIs to return the routes that *KSOR* is concerned about. This is because a larger number of enumerations are needed, and it is a challenge to balance the result quality and query efficiency.

In this paper, we adopt the selection-and-generation idea to solve *KOSR* but propose many strategies to reduce the enumerations. We first obtain the POI candidate set by using two spatial indexes (*IG-Tree* index and *Grid* index), then combine multiple routes based on the POIs in the candidate set, and finally filter the results by dominance relationships. To cover all the route combination cases as much as possible, we propose two route combination algorithms, one is a light enumeration method, which reduces the enumeration range by finding the upper and lower bounds of the number of stops. The other is based on a weighted calculation method. We define two costs for each POI, which represent its impact on route distance and number of stops respectively, and then perform a weighted calculation of the two costs. The POI with the smallest calculated value is selected for each query keyword category, and by varying the weights, multiple routes can be obtained. Moreover, our algorithm performs the shortest distance calculation several times in the process, and the traditional *Dijsktra* method is inefficient, so we combine the H2H index [14] to speed up the shortest distance calculation. The specific contributions of this paper are as follows:

- We propose a new keyword-aware routing problem *KSOR* and answer it using the spatial search approach and obtain the POI candidate set by means of *IG-Tree* and *Grid* index.
- To speed up the process of generating routes from POIs in the candidate set, we design bounds on the number of stops to limit the enumeration range and avoid enumeration of a large number of invalid paths, and propose a light enumeration method.
- We also propose another route enumeration method based on weighting calculation, which uses weighting calculation to evaluate the impact of each POI on the distance and number of stops of the route, and then select the more advantageous POI and cover all the stops of the route as much as possible by changing the weights.
- We conducted extensive experiments on the real large road network and verified the efficiency of the algorithm.

2 Related Work

In this section, we compare related work in three dimensions: number of keywords, metrics and network size, and summarise them in Table 1.

2.1 The Keyword-Aware Routing Problem Under Single Metric

Most of the existing work on keyword-aware routing problems considers route distance while some work focuses on travel times [5] and route recommendations [18]. Early research focused on single-keyword problems such as *Trip Planning Query(TPQ)* [8], *Optimal Sequenced Route(OSR)* [11,17] and *Keyword-aware Optimal Route(KOR)* [4], which plan an optimal result for the user. Later, to satisfy the travel preferences of different users, the *top-k OSR(KOSR)* [12] and *Rating Constrained Optimal Sequenced Route query(RCSOR)* [20] problems for finding top-k routes were proposed. However, none of these problems can be applied to solve *KSOR* because they generate routes without considering the problem of different combinations of POIs. They choose any POI that contains user-specified keyword categories, so they do not propose an efficient route combination algorithm to cover different POI combinations.

For route planning problems that contain multiple keywords per POI, we categorize them as *ORCSK* problems uniformly. The solutions to *ORCSK* can

Table 1. The keyword-aware routing problem

Problem	#kw in POI		Metrics		Network Size	
	1	≥1	Single	Mutiple	Small	Large
TPQ [9], OSR [11,17] KOSR [12], RCOSR [20]	✓		✓			✓
ORCSK [7,10]		✓	✓			✓
STMPC [3]	✓			✓		✓
KSR [15]		✓		✓	✓	
KSOR (This Paper)		✓		✓		✓

be categorized into two categories: based on route expansion and based on spatial search. The algorithms in [10,13] are both based on the route expansion methods. The main idea of the algorithm in [13] is greedy choice of POI, either the nearest POI or the POI that provides the most services. Since its goal is to plan an optimal route, and even if we compute a large number of sub-optimal routes by the same greedy way, we may not be able to cover all the cases of the number of stops and the computational cost will be very large. In [10], a best-first path expansion method *DA-CSK* based on deviation is proposed, which maintains a minimum priority queue that is ensured to generate top-k routes in increasing order of distance. However, *DA-CSK* exists the same problem, which may need to set a large k to cover all stops cases, and the computational cost will be very expensive as the number of query keywords increases. Therefore, based on route expansion way is not suitable to solve *KSOR*. Another popular solution to *ORCSK* is based on spatial search, which searches for POIs within a certain spatial range of the starting point and destination, and then combines the optimal routes based on these POIs. A new spatial index *IG-Tree* is proposed in [7] to solve *ORCSK*. *IG-Tree* combines $IR^2 - Tree$ [6] and *G-Tree* [19] to build a new index that contains both keywords and spatial road network information. Although the *IG-Tree* index is efficient in obtaining POI candidate sets, the query method proposed in [7] is only suitable for finding the optimal route. Because it does not consider the multiple route combination cases caused by the number of stops, so the route combination method suitable for *KSOR* is not proposed.

2.2   The Keyword-Aware Routing Problem Under Multiple Metrics

The *Skyline Trips of Multiple POIs Categories(STMPC)* [3] and *Keyword-aware Skyline Routes(KSR)* [15] both study the keyword-aware routing problem under two metrics, *STMPC* considers route distance and route cost, as well as different combinations of POIs not considered since a keyword per POI in its road network. Meanwhile the proposed *WPOIs* algorithm requires the calculation of global POIs, which is a particularly time-consuming process and some calculations that are far away from the origin and destination are meaningless. *KSR*

Table 2. Important notations

Notation	Definition
$K(\cdot)$	The union of keywords for a set of vertices or a route.
$D(R)$	The distance of route R.
$S(R)$	The number of stops of route R.
N	The Non-dominant valid route set.
P_{detour}	The detour distance of POI.
P_{cand}	A POI candidate set.
$sp(s,t)$	The shortest distance from s to t.
S^{upper}	The upper bound of stops.
S_{lower}	The lower bound of stops.

considers the route distance and the number of POIs visited, which is similar to *KSOR*. However, this problem is limited by the size of the map. It is only applicable to small indoor road networks. The *GMD* algorithm needs to traverse the global POI partition, and then combine these partitions. The calculation cost will be very large in the urban road network, so it is not applicable to *KSOR*.

3 Problem Definition

In this section, we will describe the detailed problem definition, as well as some related terminology, and we summarize some relevant notations in Table 2.

A road network is a graph $G(V, E)$, V is a set of vertices and E is a set of edges($E \subseteq V \times V$). For each vertex $v_i \in V$ contains zero or multiple keywords. We also call nodes with keywords as point-of-interest (POI), and let $K(\cdot)$ be the union of keywords for a set of vertices or a route, and $K(V)$ denote all the keyword categories of the graph G. For each edge $(u, v) \in E$ having a weight $w(u, v)$. A route R is a sequence of vertices $\langle v_0, \dots, v_n \rangle$ where $v_i \in V$. Note that some v_i are POI containing query keywords, and others are ordinary vertices on the shortest path between POI.

Definition 1 (Valid route). *Given a set S of keywords and a route R, R covers S if the collective keyword set of R contains S, i.e., $S \subseteq K(R)$. We define R as a valid path w.r.t. S. $D(R)$ denotes the total distance of route R.*

Definition 2 (Route stops). *Give a route $R\langle v_0, \dots, v_n \rangle$, the number of route stops is denoted as $S(R)$, which is the minimum number of POIs needed to cover all query keywords. These POIs come from the vertices in the route R.*

Definition 3 (Dominance relationship). *Given a road network, routes R_1 and R_1 are valid routes. If R_1 dominates R_2 (donated as $R_1 \succ R_2$), the following relationship needs to be satisfied: (i) $D(R_1) < D(R_2)$ and $S(R_1) \leq S(R_2)$ or (ii) $D(R_1) \leq D(R_2)$ and $S(R_1) < S(R_2)$.*

304   J. Li et al.

Definition 4 (Keywords and Stops aware Optimal Route (KSOR)). *G-iven a road network G(V, E), an KSOR query is a triple (s, t, Q_{kw}), where s is the query start point, t is the query end point, and Q_{kw} is the set containing the query keywords. KSOR is to return a set of non-dominated valid routes, denoted by N. The set N satisfies: (i)$\forall R' \notin N, \exists R \in N$ s.t.$R \succ R'$, and (ii) $\forall R \in N, \nexists R' \in N$ s.t. $R' \succ R$.*

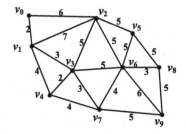

POI	Keywords
v_1	kw_1
v_2	kw_2,kw_3
v_4	kw_2
v_5	kw_1,kw_2,kw_3
v_6	kw_1,kw_2
v_7	kw_3

Fig. 2. Example road network with POIs and keywords

Figure 2 shows a road network, when given a $KSOR(v_0, v_9, \{kw_1, kw_2, kw_3\})$, the final result set $N = \{R_1, R_2, R_3\}$ is returned, where $R_1 = \langle v_0, \boldsymbol{v_1}, \boldsymbol{v_4}, \boldsymbol{v_7}, v_9 \rangle$ is the shortest route. $R_2 = \langle v_0, \boldsymbol{v_2}, \boldsymbol{v_6}, v_9 \rangle$ stops twice less than route R_1. $R_3 = \langle v_0, v_2, \boldsymbol{v_5}, v_8, v_9 \rangle$ need only stop once, but the distance is the longest.

4   Proposed Method for KSOR

In this section, we will introduce the solution method of *KSOR*, the framework of our algorithm is divided into two stages, i) **Obtain POI candidate set**: According to the query keywords specified by the user, the POI candidate set(denoted by P_{cand}) that contains these keywords is found by means of the spatial index *IG-Tree* and *Grid* index. ii) **Valid routes generation**: Multiple valid routes are combined by POIs in P_{cand}, and then the result set N is filtered from them by dominance relationship. To efficiently combine valid routes that cover all stops cases, we propose two route combination methods: the light enumeration method and the weighted calculation method. Details are shown in Sect. 4.2.

4.1   POI Candidate Set Generation

To solve *KSOR*, our first task is to obtain the POIs containing the query keywords. According to the most naive method, we directly search for all POIs containing the query keywords in graph. By using the keyword inverted index list established when reading the graph information, we can easily get all the POIs corresponding to each keyword. However, the method is not beneficial for subsequent route planning. Because combining POIs of the global graph will generate a huge number of routes, which is very expensive to compute. Moreover, this does not consider the spatial location of POIs, for some POIs that are particularly far from the origin and destination are not necessary to compute, because the routes planned by these POIs must be especially long in distance and will always be dominated by POIs that are closer. Therefore, to find the suitable POI candidate set, we use efficient spatial indexes to search for POIs within a certain spatial range from origin to destination. We used the *Grid* index and *IG-Tree* index respectively, the details will be shown below.

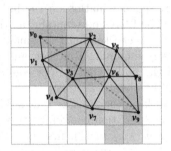

Fig. 3. Grid based index

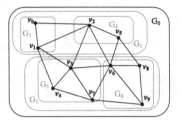

Fig. 4. IG-Tree based index

Grid Index: Build a grid index structure for a given graph $G(V, E)$. First, according to the latitude and longitude coordinates of the graph node, determine the minimum boundary rectangle (MBR) that can cover the map. The MBR is then divided into grid cells of uniform size and each grid cell is numbered. Meanwhile, record the POI number and corresponding keyword information contained in each grid. To promote the search of keyword information, we store binary code containing keywords in the grid.

After the grid index is established, we start to obtain the POI candidate set. We first connect the origin and destination into a straight line, and then record the grid cells that the line passes through. According to the binary code of the keywords in the grid, determine whether it has satisfied all the query keywords. If not satisfied, expand these grid areas outward to one layer and repeat the operation until it can satisfy all the query keyword. After determining the final grid cells, we record the POIs within these grid cells and delete the POIs that do not contain the query keywords to obtain the final POI candidate set.

As shown in Fig. 3, the road network uses grid index to obtain POI candidate set. First, v_0 and v_9 are connected into a straight line. The grey grid is the region through which the line passes. This region only contains v_3 nodes, and the v_3 does not contain the query keyword, so a layer of blue grid is continued to expand. The blue grid contains v_1 and v_6 nodes, but kw_3 is still not satisfied, so we continue to extend the green grid layer. At this time, the green grid contains nodes v_2, v_4, v_5 and v_7, all keywords are satisfied, POI candidate set $P_{cand} = \{v_1, v_2, v_4, v_5, v_6, v_7\}$.

IG-Tree Index: *IG-Tree* uses a multi-level partitioning algorithm [16] to divide the graph into subgraphs of equal size, forming multiple sub-graph partitions, and establishing a tree structure according to the relationship between partitions. Each leaf node corresponds to a map node at the bottom of the tree. From bottom to top, there are partition nodes composed of multiple leaf nodes. Each partition node records the keyword information of the node in the partition, which is also stored as binary code.

When using *IG-Tree* index to find a POI candidate set, it first finds the least common ancestor (LCA) node of the start and destination on the index structure, and checks whether the binary code of the LCA node contains all the query keywords. If not, the parent of the LCA node is checked until the smallest partition node that contains all the query keywords is found. After that, all leaf nodes under the partition node are stored, and the nodes that do not contain the query keyword are deleted to obtain the POI candidate set.

Figure 4 shows the partition of the road network by the *IG-Tree* index. The road network in Fig. 2 is evenly divided, with each partition containing two or three nodes. For a given query $KSOR(v_0, v_9, \{kw_1, kw_2, kw_3\})$, first find the minimum common ancestor of nodes v_0 and v_9, v_0 is in partition G_3, its parent node is G_1 node, v_9 is in partition G_6, its parent node is G_2, G_1 and $G_2's$ parent node are both G_0, so $LCA(v_0,v_9) = G_0$, and G_0 partition can satisfy all the query keywords. Therefore, all POIs containing query keywords in G_0 partition are POI candidate set, $P_{cand} = \{v_1, v_2, v_4, v_5, v_6, v_7\}$.

4.2 Valid Routes Generation

After obtaining the candidate set of POIs, we will use these POIs to generate valid routes. The most naive way to generate routes is to enumerate all POI combinations, but the computational cost of this method will increase exponentially as the number of query keywords increases, so we cannot enumerate all combinations of POIs directly. However, according to the properties of *KSOR*, we can find the boundary of the number of stops to limit the scope of enumeration and reduce a large number of unnecessary enumerations. According to this idea, we propose a light enumeration method. Moreover, we also propose another route generation method based on weighted calculation, which selects the optimal combination of POIs by weighting the distance and the number of POIs visited. In the following, we will introduce two methods in detail.

4.2.1 Enumeration Based Method

We can view the route generation process as a combination problem of POIs, where different routes are combined according to the POIs in the candidate set. To avoid enumerating all the combinations, we try to start with the property of *KSOR*. For the *KSOR*, N must be selected from the shortest routes with different the number of stops. Because of the routes with the same stops, it must be the shortest route that dominates the others. Meanwhile, the shortest route that satisfies all the query keywords must not be dominated by other routes, since it is better than any route in terms of distance, so the number of stops for this route can be considered as an upper bound. The other routes that can be the final result must have a number of stops that is less than the upper bound, because it is only then that they are not dominated by the shortest route.

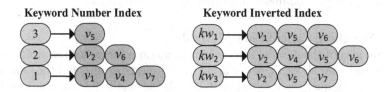

Fig. 5. Keyword number index and keyword inverted index

According to the property of the above two $KSOR$, we can enumerate the routes according to the number of stops. First, we find a shortest route in a greedy way. Find the nearest neighbor POI starting from source point s, then connect with s, regard the nearest neighbor POI as a new point s, iterate the operation until all query keywords are satisfied, and then connect the destination with t. In this way we already have a skyline result, which can be directly added to N, while obtaining the upper bound of the number of stops(S^{upper}). After that, we enumerate the routes whose the number of stops is under S^{upper}. In order to enumerate the routes with the different number of POI visited conveniently, after obtaining the POI candidate set, we build an index list according to the number of containing query keywords. According to the index list, we can similarly calculate the lower bound of the number of stops(S_{lower}) in a greedy way, each time the greedy selection contains the POI points that satisfy the query keyword at most. So our actual enumeration range is $[S_{lower}, S^{upper} - 1]$, which avoids enumerations outside the range. Note that we only enumerate a set of POIs, not a complete route, so we need to connect these POIs into a complete route in the same greedy way we used to find the shortest path. Finally, the result set N is obtained according to the dominance relationship between these valid routes.

According to the road network of Fig. 2, the keyword inverted index is established and shown in Fig. 5. For the query $KSOR(v_0, v_9, \{kw_1, kw_2, kw_3\})$, we obtain $P_{cand} = \{v_1, v_2, v_4, v_5, v_6, v_7\}$ in the stage of obtaining POI candidate set. The index of the number of keywords is established according to P_{cand}, and $S_{lower} = 1$ is calculated by the index. The shortest route $R_1 = \langle v_0, v_1, v_4, v_7, v_9 \rangle$ is obtained using the greedy algorithm. $D(R_1) = 15$, $S(R_1) = 3$, so $S^{upper} = 3$, and the enumeration range is $[1,2]$. First, enumerate the routes for $stop = 2$. The only POIs that can form a route with $stop = 2$ are those that contain at least two query keywords, so we enumerate from the nodes whose index key is

2 in the keyword count index. That is, v_2 and v_6 are used to combine, v_2 contains kw_2 and kw_3, so we combine v_2 with the node whose index key is kw_1 in the keywords inverted index, and combine v_6 with the node whose index key is kw_3. Finally, the set$\{\langle v_2, v_1 \rangle, \langle v_2, v_6 \rangle, \langle v_2, v_7 \rangle, \langle v_6, v_5 \rangle, \langle v_6, v_7 \rangle\}$ is obtained. After that, the POIs in the set are connected into routes in a greedy way, where the shortest route $R_2 = \langle v_0, v_2, v_6, v_9 \rangle$, $D(R_2) = 17$. Enumerating the routes with $stop = 1$ in the same way, we get $R_1 = \langle v_0, v_2, v_5, v_8, v_9 \rangle$ and $D(R_1) = 21$. R_1, R_2, and R_3 cannot dominate each other, so $N = \{R_1, R_2, R_3\}$.

4.2.2 Weighted Calculation Method

The main idea of the method is to weigh calculations for each POI, which calculates the cost that choosing it will bring to the route distance and the number of stops. For each query keyword, the POI with the smallest value obtained from the weighted calculation is selected, then an optimal set of POIs is obtained under the given weight. By varying the weights more optimal POI sets will be obtained, and the routes connected by each optimal POI set are the optimal routes under their corresponding weights. Finally, we filter the final result based on these routes.

Table 3. The cost of POI in Fig. 2

id	p_{detour}	p_{count}	$(p_{detour}\text{-}sp(s,t))/sp(s,t)$	$(Q_{count}\text{-}p_{count})/Q_{count}$
v_1	13	1	(13-13)/13 = 0	(3-1)/3 = 0.667
v_2	17	2	(17-13)/13 = 0.307	(3-2)/3 = 0.333
v_4	15	1	(15-13)/13 = 0.153	(3-1)/3 = 0.667
v_5	21	3	(21-13)/13 = 0.615	(3-3)/3 = 0
v_6	16	2	(16-13)/13 = 0.231	(3-2)/3 = 0.333
v_7	13	1	(13-13)/13 = 0	(3-1)/3 = 0.667

For each POI, we use the detour distance of POI (denoted as p_{detour}) to represent the impact on the route distance, and the number of keywords contained in POI (denoted as p_{count}) to represent the impact on the number of stops. p_{detour} is equal to the sum of the shortest distance from the source point s to the POI and the shortest distance from the POI to destination t. Considering that the number of visits of p_{detour} and p_{count} is not within a data range, it is necessary to calculate their normalization first. Considering that p_{detour} and p_{count} are not in a data range, they need to be normalized before weighting calculation. The final formula is as follows:

$$P_v = \frac{p_{detour} - sp(s,t)}{sp(s,t)} w + \frac{Q_{count} - p_{count}}{Q_{count}}(1 - w) \tag{1}$$

$sp(s,t)$ is the shortest distance from s to t, Q_{count} is the number of query keywords. Given a weight $w \in [0,1]$, each POI can compute a P_v value. For each query keyword, we select the POI that contains the keyword and has the small-est P_v value. Note that a POI may satisfy multiple query keywords at the same time. In order to avoid repeated calculation, each time we select a POI, we will judge whether it contains the remaining query keywords. If it contains, it does not need to calculate the keyword again. As we can see, a route is calculated for every value of w. To cover all possible routes, we can change the value of w and iterate the calculation process. For 1000 iterations, w is $(0, 0.001, 0.002, ..., 1)$. As in the light enumeration method, only the set of POIs is obtained after weighted calculation, so we connect the valid route in the same greedy way, the result set N is then filtered according to the dominant relationship (Table 3).

Table 2 shows the cost of each POI calculated according to Eq. 1. Taking $KSOR(v_0, v_9, \{kw_1, kw_2, kw_3\})$ as an example, when $w = 1$, for kw_1, contains kw_1 three POI for $\{v_1, v_5, v_6\} = \{0 * 1 + 0.667 * 0 = 0, 0.615 * 1 + 0 * 0 = 0.615, 0.231 * 1 + 0.333 * 0 = 0.231\}$. So select the v_1 node with the smallest P_v value, similarly for kw_2 and kw_3, v_4 and v_7 are selected, $R_1 = \{v_0, \boldsymbol{v_1}, \boldsymbol{v_4}, \boldsymbol{v_7}, v_9\}$. When $w = 0.5$, for kw_1, $\{v_1, v_5, v_6\} = \{0 * 0.5 + 0.667 * 0.5 = 0.3335, 0.615 * 0.5 + 0 * 0.5 = 0.3075, 0.231 * 0.5 + 0.333 * 0.5 = 0.282\}$, so choose the v_6, Since v_6 includes kw_2 at the same time, only the POI containing kw_3 needs to be calculated, and v_2 has the smallest P_v value, so $R_2 = \{v_0, \boldsymbol{v_2}, \boldsymbol{v_6}, v_9\}$. Similarly, when $w = 1$, $R_3 = \{v_0, v_2, \boldsymbol{v_5}, v_8, v_9\}$. Other $w \in [0, 1]$ will also calculate other possible routes, and eventually $S_R = \{R_1, R_2, R_3\}$ will also be obtained according to their dominance relationship.

Algorithm 1: $IG/Grid - enum(G, s, t, Q_{kw})$

Input: $G(V, E)$, $Q(s, t, Q_{kw})$
Output: The Non-dominant valid route N
1 $N \leftarrow \phi$;
2 $P_{cand} \leftarrow$ Obtained by $IG - Tree$ index / $Grid$ index ;
3 $L(k) \leftarrow EstablishList(P_{cand})$;
4 $S_{lower} \leftarrow ComputeLower(L(k))$;
5 $R_0 \leftarrow Shortpath(s, t, P_{cand})$;
6 $S^{upper} \leftarrow R_0.stops$;
7 $N \leftarrow R_0$;
8 **for** each v_i in $[S_{lower}, S^{upper} - 1]$ **do**
9     $C \leftarrow GetCombine(v_i)$;
10     $P_{set} \leftarrow Enumeration(C)$;
11     $R_{cand} \leftarrow ConnectRoute(P_{set})$;
12 $N \leftarrow Donminat(R_{cand})$;
13 **return** N;

4.3   Algorithm for KSOR

We now describe our algorithm based on the components introduced earlier, and since two strategies for each stage are described in Sects. 4.1 and 4.2

respectively, we can combine four algorithms, *IG-enum*, *IG-weight*, *Grid-enum*, and *Grid-weight*. we describe our algorithm according to the two categories of route combination methods, the details are provided in Algorithms 1 and 2.

$IG/Grid - enum$ is described in Algorithm 1, which first initialises the result set N,then obtains the POI candidate set P_{cand} according to the spatial index (Lines 1–2), and establishes the keyword count list $L(k)$ according to P_{cand}, calculates the lower bound on the number of stops by means of $L(k)$ (Lines 3–4). After that, the shortest route R_0 is calculated according to the greedy idea, and the number of stops of R_0 is the upper bound of the number of stops, and R_0 is added to N (Lines 5–7). Then all possible routes are enumerated according to the number of stops between the upper and lower bounds to obtain the route candidate set and R_{cand} (Lines 8–11), and finally the result set N is filtered according to the dominance relationship of routes in R_{cand} (Line 12).

Algorithm 2 shows the running process of *IG/Grid-weight*. The first steps are the same as Algorithm 1, initializing N and obtaining P_{cand} (Lines 1–2), then setting the value of w cyclically to iterative calculate the optimal route under 1000 different weights (Lines 3–5). For each weight, the P_v values of all POIs in P_{cand} are calculated (Lines 6–8), then the POIs with the smallest P_v values are selected according to each keyword and the set P_{set} is formed (Lines 9–11), the POIs in the set P_{set} are connected to generate valid routes and add them to R_{cand} (Lines 12–13), and the result set N is filtered from R_{cand} (Line 15).

Algorithm 2: $IG/Grid - weight(G, s, t, Q_{kw})$

Input: $G(V, E), Q(s, t, Q_{kw})$
Output: The Non-dominant valid route N
1 $N \leftarrow \phi$;
2 $P_{cand} \leftarrow$ Obtained by $IG - Tree$ index / $Grid$ index ;
3 $i \leftarrow 0$;
4 **while** $i \leq 1000$ **do**
5 $\quad$ $w \leftarrow i * 0.001$;
6 $\quad$ **for** *each p_i in P_{cand}* **do**
7 $\quad\quad$ $P_v(p_i) \leftarrow WeightedCalculate(p_i, w)$;
8 $\quad\quad$ $P.insert(P_v(p_i))$;
9 $\quad$ **for** *each kw in Q_{kw}* **do**
10 $\quad\quad$ $p_0 \leftarrow$ SelectMinimum(P,kw);
11 $\quad\quad$ $P_{set}.insert(p_0)$;
12 $\quad$ $R_0 \leftarrow ConnectRoute(P_{set})$;
13 $\quad$ $R_{cand}.insert(R_0)$;
14 $\quad$ $i \leftarrow i + 1$;
15 $N \leftarrow Donminat(R_{cand})$;
16 **return** N;

Table 4. Real world maps

Maps	$|V|$	$|E|$	#POI	$|K(V)|$
CAL	21K	43K	12K	63
NYC	264K	733K	20K	427
COL	435K	1057K	32K	548

Table 5. Parameters

Parameters	Values
Density	0.02,0.04,**0.08**,0.16,0.32
—K(V)—	**3%**,6%,9%,12%,15%
—Q_{kw}—	2,4,**6**,8,10

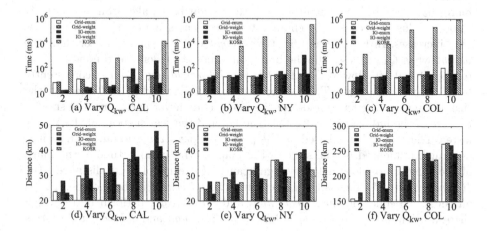

Fig. 6. Comparison of the number of various keywords

5 Experiment Evaluation

5.1 Experimental Setup

We compare the efficiency and accuracy of the four query processing algorithms from Sect. 4: *IG-enum*, *IG-weight*, *Grid-enum* and *Grid-weight*. Meanwhile, we modified the algorithm in *KOSR* to suit our problem and compared it with our algorithm. All algorithms are implemented in C++ and compiled with GNN GCC full optimization. We conducted the experiments on an Intel(R) Xeon(R) W-2245 CPU @3.2 GHz with 250 GB RAM under Linux(Ubuntu 18.04 LTS, 64bit).

Datasets: We use three real-world road networks [1] with the corresponding POI data [2] as shown in Table 4. To gain insight into the performance of the algorithm in different settings, we vary some map parameters, such as the type of keywords, POI density, etc. Meanwhile, we assign POIs to follow the zipfian distribution and then randomly fix keywords in the POIs, allowing each POI to contain a maximum number of keywords of 15. We choose NY as the default map to test the effect of the other parameters, which are listed in Table 5, with the default settings marked in bold.

5.2   Experimental Results

Effect of Q_{kw}: Figure 6 shows the running time of the algorithm and the average distance of the route while varying the number of query keywords. It is observed that our approach is consistently one to two orders of magnitude faster compared to *KOSR*. This is because the path expansion approach expands many invalid paths, which leads to inefficiency. The total running time of our algorithms increases with the increase of Q_{kw}, where the enum-based algorithm fluctuates more than one order of magnitude and the weight-based algorithm is relatively smooth. This is because the enumerated route cases increase dramatically as Q_{kw} increases. In Figs. 6b and 6c, *Grid-weight* will run slightly faster than *IG-weight* when Q_{kw} is less, but the both run almost the identical time when Q_{kw} reaches 8 and 10. However, in Fig. 6a, the *IG-weight* algorithm is consistently faster than *Grid-weight*. This is because the *IG-Tree* index is affected by the size of the map, the fewer nodes of the map divide fewer partitions, the smaller the size of the index tree finally constructed, and the search speed of the POI is faster. The grid index is based on the path to expand the grid outward, so it is not much influenced by the scale of the map, so the running time of the maps of the three maps does not vary significantly. In terms of route distances, the *IG-enum* route distance is the longest and the *IG-weight* is the shortest. Based on these, we conclude that *Grid-enum* and *Grid-weight* perform better when there are few query keywords, followed by *IG-enum*, and *IG-weight* performs best when there are more query keywords.

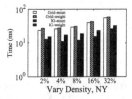

Fig. 7. POI densities

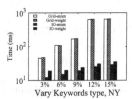

Fig. 8. Keyword types

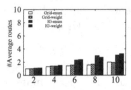

Fig. 9. Average number of routes

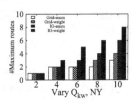

Fig. 10. Maximum number of routes

Effect of POI Density: As shown in Fig. 7, the running time of all algorithms increases with the increase of POI density. The increase of POI density means that there are more POIs, so there will be more combinations of route cases, and therefore more routes need to be calculated.

Effect of Keywords Type: In Fig. 8, the running time of all algorithms increases as the number of keyword types increases. Since the total number of POIs remains the same, the number of POIs containing each keyword type decreases. Therefore, a larger spatial search is needed to obtain the POIs, so the running time increases. In particular, grid index increase more time because the number of layers that need to be extended will be more.

Evaluation of the Number of the Routes: As shown in Fig. 9, among the average results of the 1000 sets of queries, *IG-enum* and *IG-weight* result sets have more routes and *Gird-enum* and *Grid-weight* have fewer. This indicates that in general *IG-Tree* index searches for a larger set of POI candidates, so a larger number of routes are generated and the final result is more likely to cover all the stops. However, this is also related to the distribution of the query keywords in the POIs, the more cases the POI contains the number of query keywords, the more cases the number of stops combined. As Fig. 10 illustrates, the maximum number of routes are all numerous.

6 Conclusion

In this paper, we propose a new keyword-aware routing problem *KSOR* and answer it efficiently using spatial search. Moreover, we propose two efficient route enumeration methods to generate routes that cover the overall stops case and reduce the generation of large scale useless routes. In future work, we plan to explore the possibility of addressing problems like *KSOR* through path expansion techniques. Subsequently, we will discuss the applicability of these two different methods in various scenarios.

Acknowledgements. The research work was partially supported by the Shenyang Young and Middle-aged Scientific and Technological Innovation Talent Support Plan under Grant No. RC220504; and the Natural Science Foundation of Liaoning Education Department under Grant No. LJKZ0205.

References

1. The 9th dimacs implementation challenge - shortest paths. http://users.diag.uniroma1.it/challenge9/download.shtml
2. Real datasets for spatial databases: Road networks and points of interest. http://www.cs.utah.edu/~lifeifei/SpatialDataset.htm
3. Aljubayrin, S., He, Z., Zhang, R.: Skyline trips of multiple POIs categories. In: Renz, M., Shahabi, C., Zhou, X., Cheema, M.A. (eds.) DASFAA 2015. LNCS, vol. 9050, pp. 189–206. Springer, Cham (2015). https://doi.org/10.1007/978-3-319-18123-3_12

4. Cao, X., Chen, L., Cong, G., Xiao, X.: Keyword-aware optimal route search. Proc. VLDB Endow. **5**(11), 1136–1147 (2012)
5. Fan, Y., Xu, J., Zhou, R., Li: Metaer-TTE: an adaptive meta-learning model for EN route travel time estimation. In: Proceedings of the Thirty-First International Joint Conference on Artificial Intelligence, vol. 23, pp. 2023–2029 (2022)
6. Felipe, I.D., Hristidis, V., Rishe, N.: Keyword search on spatial databases. In: IEEE International Conference on Data Engineering (2008)
7. Haryanto, A.A., Islam, M.S., Taniar, D., Cheema, M.A.: IG-Tree: an efficient spatial keyword index for planning best path queries on road networks. World Wide Web **22**, 1359–1399 (2019)
8. Kernighan, S.: An effective heuristic algorithm for the travelling-salesman problem. Oper. Res. **21**(2), 498–516 (1973)
9. Li, F., Cheng, D., Hadjieleftheriou, M., Kollios, G., Teng, S.-H.: On trip planning queries in spatial databases. In: Bauzer Medeiros, C., Egenhofer, M.J., Bertino, E. (eds.) SSTD 2005. LNCS, vol. 3633, pp. 273–290. Springer, Heidelberg (2005). https://doi.org/10.1007/11535331_16
10. Li, j., Xiong, x., Li, L., He, D., Zong, C., Xiaofang, Z.: Finding top-k optimal routes with collective spatial keywords on road networks. In: ICDE (2023)
11. Li, J., Yang, Y.D., Mamoulis, N.: Optimal route queries with arbitrary order constraints. TKDE **25**(5), 1097–1110 (2012)
12. Liu, H., Jin, C., Yang, B., Zhou, A.: Finding top-k optimal sequenced routes–full version. arXiv preprint arXiv:1802.08014 (2018)
13. Lu, H.C., Chen, H.S., Tseng, V.S.: An efficient framework for multirequest route planning in urban environments. TITS **PP**(4), 1–11 (2016)
14. Ouyang, D., Qin, L., Chang, L.: When hierarchy meets 2-hop-labeling: Efficient shortest distance queries on road networks. In: SIGMOD. pp. 709–724. ACM (2018)
15. Salgado, C.: Keyword-aware skyline routes search in indoor venues. In: the 9th ACM SIGSPATIAL International Workshop (2018)
16. Selvakkumaran, N., Karypis, G.: Analysis of multilevel graph partitioning. In: Supercomputing, IEEE/ACM Sc95 Conference (1995)
17. Sharifzadeh, M., Kolahdouzan, M., Shahabi, C.: The optimal sequenced route query. VLDB J. **17**, 765–787 (2008)
18. Xu, M., Xu, J., Zhou, R., Li, J., Zheng, K.: Empowering A* algorithm with neuralized variational heuristics for fastest route recommendation. TKDE **35**, 10011–10023 (2023)
19. Zhong, R., Li, G., Tan, K.L., Zhou, L., Gong, Z.: G-tree: an efficient and scalable index for spatial search on road networks. TKDE **27**(8), 2175–2189 (2015)
20. Zhu, H., Li, W., Liu, W., Yin, J., Xu, J.: Top k optimal sequenced route query with POI preferences. Data Sci. Eng. **7**(1), 3–15 (2022)

The Way to Success: A Multi-level Attentive Embedding Framework for Proposal Teamwork Analysis in Voting-Oriented System

Rui Zha[1], Ding Zhou[2(✉)], Le Zhang[3], and Tong Xu[1,2,3]

[1] University of Science and Technology of China, Hefei, China
zr990210@mail.ustc.edu.cn, tongxu@ustc.edu.cn
[2] Xiaohongshu Inc., Beijing, China
zhoudinglive@gmail.com
[3] Baidu Research, Baidu Inc., Beijing, China

Abstract. Effective teamwork is crucial for successful proposals, particularly in a competitive voting-oriented system. To increase the chances of success, teams should select collaborators with complementary expertise and broad influence to attract potential supporters. Traditional efforts in quantitative proposal analysis mainly focus on the voting decision process of individual voters from the perspective of social influence, while the teamwork effectiveness of proposals has been largely ignored. To bridge this gap, we propose a novel **M**ulti-level **A**ttentive **E**mbedding **F**ramework (**MAEF**) to reveal the secret behind a successful proposal from the perspective of teamwork, in which two levels of embedding have been learned to measure the teamwork effects. Specifically, at the individual level, expertise learning and a multi-channel Graph Attention Neural Network are developed for each collaborator to estimate the degree of expertise on different proposal topics and the attractiveness to supporters, respectively. At the team level, we aggregate the individual-level embeddings of collaborators to measure the overall teamwork effectiveness, taking into account the complementarity of expertise and closeness of collaboration. In this phase, the attention mechanism is introduced to distinguish individual contributions. Extensive experiments on a real-world legislative institution dataset clearly validate the effectiveness of our MAEF model compared with several state-of-the-art baseline methods, supporting the hypothesis that effective teamwork indeed improves the predictability of proposal success.

Keywords: Quantitative Proposal Analysis · Teamwork Effectiveness · Ideal Point Estimation · Graph Neural Network

1 Introduction

The trend towards collaborative work over solo efforts continues to rise, particularly in competitive, vote-driven environments such as legislative systems [6,15].

X. Song et al. (Eds.): APWeb-WAIM 2023, LNCS 14331, pp. 315–330, 2024.
https://doi.org/10.1007/978-981-97-2303-4_21

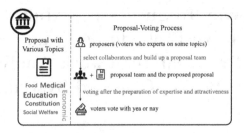

Fig. 1. Example of "Proposal-Voting" process.

Optimal teamwork integrates diverse expertise and skills, enhancing proposal quality and broadening appeal, thereby increasing success potential. Identifying key determinants of proposal success through measuring teamwork impact provides critical guidance for effective collaboration.

In voting systems, the "proposal-voting" process (Fig. 1) involves voters either affirm or deny proposals, with success contingent on the accumulation of affirmative votes. To optimize success, some voters transition into collaborator roles, forming an initial team based on their specific expertise and interests. They work collectively to ensure proposal success through comprehensive coverage of required expertise and by enticing potential supporters. Yet, forming a competent team is challenging, excessive collaborators can complicate and destabilize the collaborative process, elevating the failure risk. Therefore, maintaining balanced and effective teamwork necessitates a holistic solution.

Previous work in this area mainly relies on quantitative proposal analysis, employing preference-based [8] and influence-based [20,24] methods to estimate the matching degree of proposal topics. However, these techniques overlook the crucial role of teamwork, hindering the identification and selection of suitable collaborators. In contrast, some studies [11,13] have developed metrics to measure team expertise and collaboration strength, while others [5,25] explored team representation strategies for team-based analysis. Despite these efforts, existing methodologies frequently fail to model the complementary expertise of team members and their attractiveness to potential supporters. Therefore, rigorous analysis is required to discern the granular impact of teamwork on successful team formation.

Typically, selecting potential collaborators entails multi-level considerations. As individuals, potential collaborators are evaluated on their topical expertise and ability to garner support. As a team, collaborators must provide comprehensive coverage of necessary expertise and sustain robust collaboration for successful proposal implementation.

To this end, we propose **M**ulti-level **A**ttentive **E**mbedding **F**ramework (MAFE), a novel approach for proposal analysis and quantifying multi-level teamwork effects. This method predicts proposal success from a teamwork perspective. At the individual level, a multi-channel Graph Attention Neural Network is employed to estimate each collaborator's expertise degree in different proposal topics and their appeal to potential supporters. We then aggregate these individual-level embeddings to generate a final team-level embedding, considering complementary expertise and collaboration strength using an attention mechanism to distinguish individual contributions. Finally, a linear prediction head is applied to the team representation to assess proposal success. The main contributions of this work are summarized as follow:

- To the best of our knowledge, we are among the first ones to quantitatively analyze the teamwork effect of proposal success in a voting-oriented system.
- We propose a novel attentive embedding framework, MAEF, which jointly represents individual-level expertise and attractiveness, as well as team-level complementary expertise and collaboration effects.
- We conduct extensive experiments on the dataset obtained from a real-world legislative institution, which have clearly validated the effectiveness and interpretability of our proposed model.

2   Related Work

The related work of our task can be classified into three categories: proposal analysis, team representation, and graph neural networks.

Proposal Analysis. Early research in proposal analysis aimed to understand the individual voting behaviors. Specifically, preference-based methods attempt to estimate the ideal point of voters on different topics using topic models, such as labeled *Latent Dirichlet Allocation* (LDA) [8] and *Probabilistic Latent Semantic Analysis* (PLSA) [10]. Recently, based on the social influence theory which claims that decisions are usually mutually influenced [19], several influence-based methods have been proposed to incorporate social influence into individual decisions [20,24]. Different from these methods that mainly focus on the voting process of individual voters, we quantitatively analyze the proposal process from the perspective of teamwork, considering mutliple voter roles.

Team Representation. Previous efforts for team representation mainly focus on understanding the team effectiveness through manually defined metrics. For instance, *Lappas et al.* [16] designed a communication cost function to identify efficient teams, while *Anagnostopoulos et al.* [1] considered individual workloads, arguing more balanced teams are more effective. However, these artificial designs lack efficiency and generalizability. Recently, strategies considering group decision-making have been employed in group recommendation [2,5,12]. While these methods consider team-level representation, their optimization goal is team member satisfaction following specific strategies. Conversely, our teamwork task centers on voter satisfaction outside the team, evaluating complementary expertise and attractiveness to garner more potential supporters.

Graph Neural Networks. Graph Neural Networks (GNNs) are special types of neural networks capable of working with graph data structures, such as social networks and knowledge graphs [22]. In this direction, various propagation and

Table 1. Statistics of TLY Dataset.

Attribute	Number/Scope
# proposals ($|P|$)	7487
# voters ($|L|$)	245
# proposal topics ($|T|$)	251
# support channels ($|C|$)	3
# parties	8
# electoral districts	25
# departments in TLY	12
average size of a proposal team	12
average size of each session in TLY	123
proposal result	1 (Success), 0 (Failure)
# successful proposal	4115 (54%)
# failing proposal	3372 (46%)

aggregation functions have been designed for various downstream tasks, such as node representation and link prediction [23]. Early GNN variants, like Graph Convolutional Networks (GCNs) [14], use first-order neighbors to simplify the filter in CNNs. Subsequently, Graph Attention Networks (GATs) [21] apply the attention mechanism to aggregate the representations of neighbors and update node representations in parallel. Building on this, many works are emerging that employ GNNs with attention mechanisms other than the standard GATs to achieve better performance [3,4,7]. In this paper, inspired by the ability of GATs to extract parallel structural relationships in non-Euclidean domains, we also chose to modify GAT with a multi-channel approach to fit the support pattern in our investigated teamwork task.

3 Preliminaries and Problem Definition

In this section, we first introduce our real-world dataset with several intuitive pre-studies. On such basis, the preliminaries and the problem definition of our proposal analysis task will be formulated.

3.1 Dataset Description

To facilitate the quantitative study of our research task, we have collected our TLY dataset[1] from a real-world legislative institution. The dataset contains all proposal and voter records from 2009 to 2019, encompassing 7,487 proposals and 245 voters in total. The detailed statistics are presented in Table 1. The collected data consists of two parts:

---

[1] We will publish the dataset for reproducibility check after acceptance.

🪪 Proposal 2941	
💬 Proposal Topics	Typhoon Personnel Holiday Weather
🛕 Proposer	James Huang
👥 Collaborators	Xiaowei Qiu, David Gao, William Chen, Lisa Su ...
🏆 Voting Result	Success

Fig. 2. Example of proposal records.

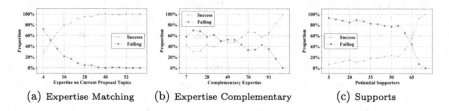

(a) Expertise Matching   (b) Expertise Complementary   (c) Supports

Fig. 3. Pre-study on TLY Dataset.

- **Proposal Records.** For each proposal, we collected its proposal team with all the *collaborators*. Also, the proposal content has been extracted as several standard *topics*. Additionally, we recorded the final result of each proposal, i.e., success or failure.
- **Voter Records.** For each voter, we collected her *historical proposal records*, where the individual acted as a collaborator. We also gathered data on the support channels used by each voter to attract potential supporters. In our study, we classified support channels into three types, including *party* (PC), *department* (DC), and *electoral district* (EDC), based on data characteristics.

Figure 2 depicts a proposal record, demonstrating the collaborative effort of a team of multiple collaborators, including the initial proposer, to achieve success. This motivates our research to investigate whether our basic assumption, that teamwork-related factors can affect proposal outcomes, is supported. We evaluated this hypothesis through a series of pre-studies.

3.2 Pre-study on Proposals with Teamwork

We conducted a pre-study to demonstrate the impact of teamwork on proposal success. Using the Pearson Correlation Coefficient (PCC), we measured the correlation between proposal outcomes and three teamwork-related factors, including expertise overlap, complementary expertise, and historical collaboration.

For expertise overlap, as shown in Fig. 3(a), a higher number of overlapping proposals among collaborators showed a positive correlation with success ($PCC = 0.4153, p < 0.001$). Complementary expertise, measured by the number of union topics divided by the number of collaborators, also positively correlated with success ($PCC = 0.3084, p < 0.001$), as depicted in Fig. 3(b). Furthermore,

Fig. 3(c) demonstrates that historical collaboration, representing potential support, strongly correlated with success ($PCC = 0.7201, p < 0.001$). The success rate significantly increased with more historical collaborators.

In the following sections, we will propose a comprehensive technical solution for analyzing proposals by integrating these factors.

3.3   Problem Formulation

Informed by preliminary findings, we mathematically formulate proposal analysis. We start from considering a set of historical proposals, P, and a set of voters, L. Each voter $l \in L$ votes potentially acts as a collaborator or supporter of a proposal team.

Along this line, to represent each proposal $p \in P$, we use a multi-hot vector $t_p \in \mathbb{R}^{|T|}$ to summarize the topics in the proposal content, where T denotes the topic set. A corresponding multi-hot vector $l_p \in \mathbb{R}^{|L|}$ records proposal p's collaborators. Besides, we assign a label $y_p \in \{1, 0\}$ to indicate the success or failure of the proposal p. Consequently, each historical proposal p can be represented as a triple $p = (l_p, t_p, y_p)$.

On such basis, to describe each voter $l \in L$, we use two attribute vectors to capture the expertise and support factors identified in our pre-study. The first vector $a_l \in \mathbb{R}^{|T|}$ measures the expertise of voter l in each topic t, with each element $a_{l,t} \in [0, 1]$ reflecting the level of expertise. The second vector $b_l \in \mathbb{R}^{|C|}$ captures the strategies used by voter l to attract supporters from different channels, where C denotes the set of support channels (e.g., PC, DC, EDC). Each element $b_{l,c} \in [0, 1]$ of b_l measures how much emphasis voter l places on supporters from channel $c \in C$. Notably, a_l and b_l are estimated from historical proposal records. Along this line, we formally define our proposal analysis task as follows:

Proposal Teamwork Analysis (PTA) Task. Given voters L and their proposal records P, each $p \in P$ corresponds to the triple $p = (l_p, tp, y_p)$. Our goal is to learn the attributes a_l and b_l of each voter l and predict the success potential of a new proposal $p^* = (lp^*, tp^*, yp^*)$.

4   Technical Solution for MAEF

In this section, we will introduce our technical solution to the PTA task. Generally, we will first give a brief overview of our MAEF model, and then explain the details of MAEF step-by-step.

4.1   An Overview of MAEF

As shown in Fig. 4, MAEF is composed of two layers of embedding: the *individual-level* and the *team-level* layers. In the individual-level embedding layer, we first estimate the expertise matching degree of each collaborator compared with the proposal topics. Then, a multi-channel Graph Attention Network

is utilized to learn the attractiveness of each collaborator to potential supporters based on historical collaboration adjacency graphs. Along this line, in the
team-level embedding layer, we use an attention mechanism to learn the complementary effects of expertise and support. We also include a constraint to ensure
that the team-level embedding satisfies the topic requirements of the proposal.
Finally, we use a two-hidden-layer MLP with softmax readout to make the final
prediction. For ease of reading, in this section, we will focus on a specific proposal
$p = (l_p, t_p, y_p)$, which can be easily generalized to other proposals.

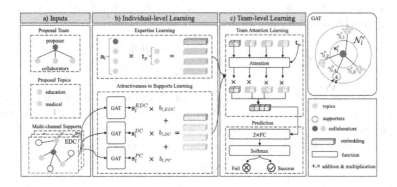

Fig. 4. Framework of Multi-level Attentive Embedding Framework (MAEF).

4.2 Individual-Level Embedding Layer

Expertise Learning. Since different voters may have different expertise, and
a team with more professional collaborators will succeed more easily, as proved
in Sect. 3.2, it will be beneficial to select the collaborators with better expertise
exactly on the proposal topics. To that end, as shown in Fig. 4b), for each collaborator $l \in l_p$, we estimate the expertise matching degree between l and the
proposal topics t_p as follows:

$$m_{l,p} = a_l \circ t_p, \tag{1}$$

where $\circ$ denotes the element-wise product, and $m_{l,p}$ is the expertise matching
degree between l and proposal p.

Attractiveness Learning. As stated in prior arts [8,10,20], voters could
be mutually influenced based on their similar interests/benefits. As proved in
Sect. 3.2 that potential support could be a crucial factor for the proposal's success, we designed a multi-channel GAT to quantitatively measure the attractiveness of each collaborator $l \in l_p$ to supports from different channels.

In detail, we first summarize the historical collaboration adjacency graph
$\mathcal{N}_l$ for each collaborator l, in which all nodes are the voters who have ever

collaborated with l in historical proposals. Then, for each channel c, we have the subgraph $\mathcal{N}_l^c$ to indicate the potential supporters from channel c. Afterwards, as shown in Fig. 4b), GAT is applied to each channel c to incorporate the mutuality between l and its potential supporters. Specifically, for collaborator l and its potential supporter $k \in \mathcal{N}_l^c$, the attractiveness from l to k in channel c can be formulated as:

$$s_{l,k}^c = \sigma(\boldsymbol{h}_1^c[\boldsymbol{a}_l \| \boldsymbol{a}_k]), \tag{2}$$

where $\sigma(\cdot)$ denotes the activation function, $\|$ denotes the vector concatenate operation, and $\boldsymbol{h}_1^c$ is the learnable weight matrix of channel c. Since the size of $\mathcal{N}_l^c$ for each collaborator l is different, we use the normalization operation to scale the attractiveness from l to k in channel c:

$$\gamma_{l,k}^c = \text{softmax}(s_{l,k}^c) = \frac{\exp(s_{l,k}^c)}{\sum_{j \in \mathcal{N}_l^c} \exp(s_{l,j}^c)}, \tag{3}$$

where $\exp(\cdot)$ denotes the exponential function. Finally, we obtain the normalized attractiveness coefficient $\gamma_{l,k}^c$ from l to its potential supporter k in channel c. Thus, the attractiveness of collaborator l via channel c can be aggregated by its supporters' expertise with a corresponding coefficient as follows:

$$\boldsymbol{s}_l^c = \sigma(\sum_{j \in \mathcal{N}_l^c} \gamma_{l,j}^c \times \boldsymbol{a}_j). \tag{4}$$

Besides, we extend attention utilized in $\boldsymbol{s}_l^c$ to multi-head attention so that the training process is more stable. Specifically, we repeat the calculation in Eq. (4) K times and summarize the learned embedding as the final embedding of potential support from channel c:

$$\boldsymbol{s}_l^c = \sum_{k=1}^{K}(\sigma(\sum_{j \in \mathcal{N}_l^c} \gamma_{l,j}^c \times \boldsymbol{a}_j)). \tag{5}$$

Since each channel has its own inherent characteristics, e.g., conservative or aggressive, it is necessary to incorporate these characteristics of each channel into a better representation of channel supports. Along this line, we design a character offset embedding ξ_c for each channel, and fuse them into the final embedding of potential support from channel c:

$$\tilde{\boldsymbol{s}}_l^c = \boldsymbol{s}_l^c + \xi_c. \tag{6}$$

Finally, we aggregate the embedding of potential support for collaborator l with the corresponding channel emphasis $\boldsymbol{b}_l$ described in Sect. 3.3, where different collaborators may emphasize different channels according to their own benefits. Specifically, the embedding of each channel is re-weighted via the personal preference of each collaborator as follows:

$$\boldsymbol{s}_{l,p} = \sum_{c \in C} b_{l,c} \times \tilde{\boldsymbol{s}}_l^c. \tag{7}$$

4.3 Team-Level Embedding Learning

Team Attention Learning. As analyzed in Sect. 3.2, we have realized that proposal success is related to not only individual-level expertise and attractiveness but also complementary effects and collaboration effectiveness. Thus, we utilize an attention mechanism to estimate how each collaborator contributes to team-level complementary expertise and collaboration. Specifically, to capture the effect of each collaborator on the whole proposal team, we first calculate the role coefficient of each collaborator $l \in l_p$ as follows:

$$r_{l,p} = \sigma(\boldsymbol{h}_2[\boldsymbol{t}_p\|\boldsymbol{m}_{l,p}\|\boldsymbol{s}_{l,p}]), \tag{8}$$

where $\boldsymbol{h}_2$ is a learnable weight matrix, and $r_{l,p}$ denotes how collaborator l contributes expertise and supports team l_p. As same as Eq. (3), we normalize $r_{l,p}$ with softmax function:

$$\beta_{l,p} = \text{softmax}(r_{l,p}) = \frac{\exp(r_{l,p})}{\sum_{j \in t_p} \exp(r_{j,p})}. \tag{9}$$

Thus, with the learned coefficient $\beta_{l,p}$, we aggregate the final team representation with both the complementary effect of expertise and the attractiveness of supports as follows:

$$\boldsymbol{g}_p = \sum_{l \in l_p} \beta_{l,p} \times [\boldsymbol{m}_{l,p}\|\boldsymbol{s}_{l,p}]. \tag{10}$$

Constraint for Complementary Effect. To further guarantee that the team representation can cover the topic requirements of the proposal as much as possible, we restrict to have a lower Euclidean Distance between our team representation $\boldsymbol{g}_p$ and topic requirements $\boldsymbol{t}_p$. Then, constraint $\mathcal{L}_1$ is designed as follows:

$$\mathcal{L}_1 = \frac{1}{|P|} \sum_{p \in P} \sqrt{(\boldsymbol{g}_p - \boldsymbol{t}_p)^2}. \tag{11}$$

4.4 Training Strategy

Finally, we will introduce the details of the training strategy. Since we formulate the PTA task as a binary classification problem, we use cross-entropy loss to measure the difference between the predicted value $\hat{y}_p$ and the ground truth y_p, defined as follows:

$$\mathcal{L}_2 = -\frac{1}{P} \sum_{p \in P} y_p \ln \hat{y}_p + (1 - y_p)\ln(1 - \hat{y}_p). \tag{12}$$

Then, the loss function of our task can be formulated as follows:

$$\mathcal{L} = \mathcal{L}_1 + \mathcal{L}_2 + \lambda\|\Theta\|_2^2, \tag{13}$$

where $\|\Theta\|_2^2$ is the L2-norm of Θ and λ is the regularization parameter.

4.5   Time Complexity and Convergence Analysis

The overall time complexity of our MAEF model is $O(|\Theta| \times |T|)$. The size of model parameters $|\Theta|$ is highly dependent on the number of topics $|T|$. Since the amount of topics $|T|$ can be pre-selected and controlled, the time complexity of our model is likely to be comparable or slightly higher than previous non-deep learning models, as observed in our dataset (Table 1). In addition, the loss function $\mathcal{L}$ comprises three parts: the cross-entropy constraint $\mathcal{L}_2$, L2-norm-like functions $\mathcal{L}_1$ and $\lambda\|\Theta\|_2^2$. Both the cross-entropy function and the L2-norm-like function satisfy the Lipschitz continuous condition, which guarantees that the gradient of $\mathcal{L}$ has an upper bound. Consequently, the SGD algorithm used for training will converge to the critical point, even for a non-convex and smooth loss function, as demonstrated in prior research [9]. In summary, the convergence of our MAEF model can be ensured.

5   Experiments

In this section, we present an evaluation of our proposed MAEF model using a real-world dataset. We will start by describing the experimental settings and then proceed to provide a comprehensive assessment of the overall performance of our model, including the results of the ablation studies. Moreover, we will analyze the learned embeddings from both individual- and team-level layers, which yielded several interesting insights.

5.1   Experimental Settings

A) Validation Task & Evaluation Metrics. The validations were conducted on the TLY data. To evaluate the effectiveness of MAEF, we evaluate the performance based on the prediction task of PTA, i.e., to predict whether one proposal will finally succeed or not. Accordingly, we adopt the metrics including Accuracy, Recall, and F1 to evaluate the performance.

B) Baselines. To evaluate the performance of MAEF on the proposal prediction task, we compare it with several baseline methods as follows:

- **Ruling Party Must Success (RPMS).** This heuristic method assumes that the majority party always wins the voting by attracting more than half of the votes, labeling their proposals as "success" and the rest as "failure."
- **Logistic Regression & Random Forests (LR & RF).** LR and RF are classic models for classification tasks, where we use the concatenation of the proposal team $\mathbf{l}_p$ and topics $\mathbf{t}_p$ in multi-hot form as inputs.
- **UltraGCN [17] & AGREE [5].** These two recommendation approaches treat each team as a virtual user and the topics combination as a virtual item. Totally, we have 5938 unique virtual users, 3101 virtual items, and 7487 interactions, resulting in a very sparse matrix for UltraGCN and AGREE.

– **Deepwalk** [18] & **GAT** [21] & **GATv2** [3]. We chose these three state-of-the-art network embedding methods to learn team member representations. As in MAFE, we used attention mechanism for team-level representation and MLP for final prediction. Note that we use DP-A to denote the Deepwalk combined with an attention mechanism.

C) **Implementation Details.** In MAEF implementation phase, we first initialize $\mathbf{a}_l, \mathbf{b}_l$ of each collaborator $l \in L$ through Gaussian distribution with mean equal to 0 and variance equal to 1. When learning attractiveness of each collaborator, a two-layer multi-channel GAT is designed, where each layer has $|T| = 251$ hidden neurons. Besides, $r_{l,p}$ in team attention learning is formulated as a two hidden layer MLP with 64 and 1 hidden neurons, respectively. Finally, $\hat{y}_p$ is a 2-dimensional vector predicted by MLP with $(2 \times |T|, 128, 32, 2)$ neurons from the input layer to the output layer. In the training phase, we randomly initialize parameters and optimize the model with Adam. Specifically, we set the mini-batch size to 16, the penalty parameter λ to $10e{-}6$ and the learning rate to $5e{-}5$ with half decay every 10 epochs. The source code of MAEF is publicly available at: https://anonymous.4open.science/r/MAEF-21CC.

Table 2. Overall performance on predicting the success of a proposal.

Methods	40% Training Data			60% Training Data			80% Training Data		
	Acc.	Recall	F1	Acc.	Recall	F1	Acc.	Recall	F1
RPMS	0.5314	0.5832	0.5480	0.5314	0.5832	0.5480	0.5314	0.5832	0.5480
LR	0.8245	0.8851	0.8317	0.8644	0.8830	0.8771	0.8711	0.8757	0.8846
RF	0.8718	0.8485	0.8928	0.8818	0.8687	0.8966	0.8918	0.8890	0.9038
UltraGCN	0.7891	0.8670	0.8384	0.8122	0.8800	0.8561	0.8350	0.9035	0.8705
AGREE	0.8083	0.8849	0.8925	0.8400	0.8914	0.9116	0.8511	0.8981	0.9192
DP-A	0.8517	0.8645	0.8669	0.8661	0.8544	0.8831	0.9178	0.9467	0.9239
GAT	0.8473	0.8834	0.8587	0.8738	0.8834	0.8966	0.8905	0.8738	0.9044
GATv2	0.8618	0.8800	0.8748	0.8895	0.8863	0.9018	0.8998	0.8824	0.9127
MAEF	**0.9225**	**0.9720**	**0.9272**	**0.9278**	**0.9681**	**0.9324**	**0.9452**	**0.9920**	**0.9483**

5.2 Quantitative Evaluations

Overall Performance. To validate the performance and robustness of the proposed MAEF model, we randomly selected 40%, 60%, and 80% of the TLY dataset as training samples. The overall performance of the model is summarized in Table 2. The results show that our MAEF consistently achieves the best performance, supporting the hypothesis that considering expertise and collaboration factors in teamwork effectiveness can enhance the predictability of proposal success.

Moreover, we have made some interesting observations in the experiments: 1) Firstly, it is worth noting that the ruling party may not always guarantee a victory in the voting process, as the RPMS achieved an F1-value of only 0.548. This highlights the fact that voters take various factors into consideration before making their decisions, and do not simply follow their party's orders. As a result, proposals from minor parties can also achieve success if the team demonstrates effective teamwork, such as complementary expertise and strong attractiveness. 2) Secondly, our proposed MAEF model takes into account comprehensive factors that are crucial for the success of a proposal. While general graph models, including DP-A, GAT, and GATv2, all incorporate the attention mechanism to measure social impact among collaborators, they do not consider factors such as expertise and attractiveness, resulting in poorer performance compared to our MAEF model. The MAEF model learns fine-grained representations by considering all factors, which undoubtedly benefits the overall performance. 3) Thirdly, we found that both the individual-level and team-level embedding layers contribute to the final performance. As shown in Table 2, the novel UltraGCN method has lower performance compared to traditional methods such as LR and RF. This may be due to UltraGCN's difficulty in handling both individual-level expertise and team-level effectiveness. Although AGREE can handle some of the limitations of UltraGCN, it is still challenging to deal with the highly sparse rate matrix without incorporating rich behavioral information for each collaborator. In contrast, LR and RF treat each team member and proposal topic as independent features and learn different weights for each feature to aggregate collaborators into the whole team-level representation. Even with this simple form, traditional methods can perform better than deeper models, indicating that a multi-layer embedding framework indeed benefits performance.

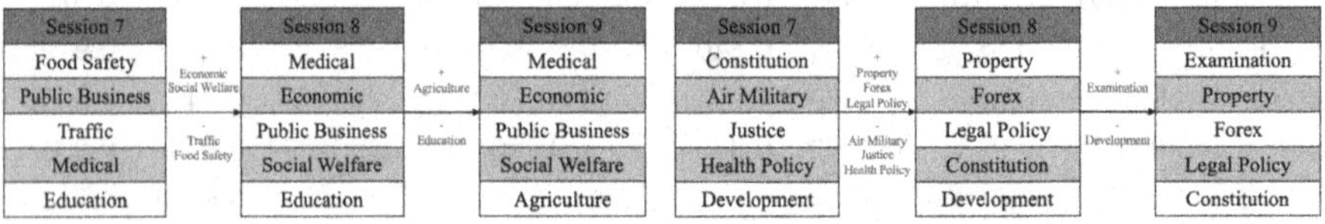

Session 7		Session 8		Session 9		Session 7		Session 8		Session 9
Food Safety	Economic Social Welfare	Medical	Agriculture	Medical		Constitution	Property Forex Legal Policy	Property	Examination	Examination
Public Business		Economic		Economic		Air Military		Forex		Property
Traffic	Traffic Food Safety	Public Business	Education	Public Business		Justice	Air Military Justice Health Policy	Legal Policy	Development	Forex
Medical		Social Welfare		Social Welfare		Health Policy		Constitution		Legal Policy
Education		Education		Agriculture		Development		Development		Constitution

(a) Evolution of top 5 topics.   (b) Evolution of last 5 topics.

Fig. 5. Trend evolution of topics from the 7th to 9th session in TLY.

Table 3. Ablation study on the success prediction task.

Methods	Metrics		
	Acc.	Recall	F1
MAEF-E	0.9239	0.9789	0.9377
MAEF-S	0.9339	0.9703	0.9382
MAEF-M	0.9379	0.9816	0.9415
MAEF	**0.9452**	**0.9920**	**0.9483**

Ablation Study. To further demonstrate and distinguish the effectiveness of each component in MAEF, three variants are proposed for the ablation study:

- **MAEF-E**, which masks the expertise learning component;
- **MAEF-S**, which masks the multi-channel GAT component;
- **MAEF-M**, which replaces the attention mechanism applied for team-level representation by average strategy, treating each collaborator equally;

Table 3 reveals that MAEF-E and MAEF-S exhibit comparable performance on most metrics, yet both fall short of MAEF. This demonstrates that both expertise matching learning and multi-channel GAT effectively model individual-level factors contributing to proposal success. Conversely, the inferior performance of MAEF-M confirms that collaborators contribute differently to the proposal team, underscoring the efficacy of our attention mechanism in team-level representation learning.

5.3 Qualitative Discussions

To further validate our basic assumption of teamwork's effect on proposal success and better understand the mechanism of proposal team formation strategy, we conducted several qualitative discussions on the attributes of voters. Along this line, four interesting rules have been revealed as follows:

1) *How to select suitable collaborators based on expertise, diversity, consistency?* First, we would like to summarize the collaborator selection strategy from the perspective of expertise. To do this, we measured the *cosine similarity* between each pair of collaborators and then calculated the average result as the *Consistency Index* (CI) for the team. Following this, we calculated the Pearson correlation coefficient (PCC) between the CI and the proposal result (1 or 0, using the same setting as the pre-study in Sect. 3.2). The resulting PCC score was 0.7908 with a p-value of less than 0.001, indicating a significant positive correlation. This finding suggests that while complementary expertise can be crucial for proposal success, it is still preferable for collaborators to be consistent in their expertise, which may facilitate collaboration.

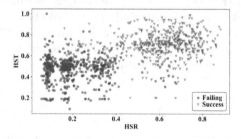

Fig. 6. A distribution of ruling parties' proposals with respect to HSR and HST.

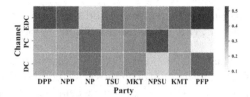

Fig. 7. A heatmap of how different parties emphasize on different channels, where darker colors mean more emphasis.

2) *Does collaboration strength influence proposal results?* Then, we would like to validate our basic assumption that the strength of collaboration may also affect the proposal results. Specifically, we normalized the attention coefficient between each pair of collaborators learned via Eq. (3). Then, the average coefficient has been compared with the proposal result to calculate the PCC score, which is estimated at 0.2766 with a p-value less than 0.0001. Moreover, we also summarized the average collaboration strength between each pair of collaborators, which has a PCC of 0.7790 and a p-value less than 0.0001 compared with the proposal result. According to these results, we concluded that more collaborators may not result in success as they may fail to unite and a united team with strong collaboration will definitely lead to better chance of success.

3) *Why ruling party still fail in the votings?* Despite the results presented in Sect. 5.2, voters consider factors beyond party order. To uncover why ruling party proposals may fail, we computed the Historical Success Rate (HSR) of the proposal team and the Historical Success Times (HST) of ruling party on current proposal topics (see Fig. 6). We can find that failures often corresponded to lower HSR and HST metrics, suggesting suboptimal topic proficiency or an unsuitable team assigned to the proposal.

4) *How channels and topics affect the proposal results?* Finally, we would like to discuss the effects of channels and topics. Specifically, we summarized the average attractiveness b_l of all voters from the same party, as shown in Fig. 7. This approach enabled us to identify different strategies employed by parties to attract support. For instance, major parties such as the KMT and the DPP typically focus on electoral district issues to secure re-election. Besides, we also analyzed the trend of proposal topics, as illustrated in Fig. 5. While the rankings changed across different sessions, topics related to people's livelihoods like agriculture and education, consistently remained the most popular, indicating that proposals on popular topics are more likely to succeed.

6   Conclusion

In this paper, we quantitatively analyze the success of proposals in a voting-oriented system from the perspective of the teamwork effect. Specifically, we propose a novel multi-level attentive embedding framework where individual-level and team-level layers of embedding are estimated and aggregated to comprehensively describe the proposal process with collaborator selection. Along this line, we measure and balance the complementary effect of expertise, its attractiveness to potential supporters, and the closeness of collaboration. Furthermore, we conducted extensive experiments on a dataset collected from a real-world legislative institution that clearly validate the effectiveness and interpretability of our MAEF model compared to several state-of-the-art baseline methods. These results support the hypothesis that teamwork effectiveness improves the predictability of proposal success.

Acknowledgements. This work was supported by the grants from National Natural Science Foundation of China (No. 62222213, 62072423), and the USTC Research Funds of the Double First-Class Initiative (No. YD2150002009).

References

1. Anagnostopoulos, A., Becchetti, L., Castillo, C., Gionis, A., Leonardi, S.: Power in unity: forming teams in large-scale community systems. In: Proceedings of the 19th ACM International Conference on Information and Knowledge Management, pp. 599–608. ACM (2010)
2. Boratto, L., Carta, S.: State-of-the-art in group recommendation and new approaches for automatic identification of groups. In: Soro, A., Vargiu, E., Armano, G., Paddeu, G. (eds.) Information Retrieval and Mining in Distributed Environments. Studies in Computational Intelligence, vol. 324, pp. 1–20. Springer, Heidelberg (2010). https://doi.org/10.1007/978-3-642-16089-9_1
3. Brody, S., Alon, U., Yahav, E.: How attentive are graph attention networks? In: International Conference on Learning Representations (2022)
4. Busbridge, D., Sherburn, D., Cavallo, P., Hammerla, N.Y.: Relational graph attention networks. arXiv preprint arXiv:1904.05811 (2019)
5. Cao, D., He, X., Miao, L., An, Y., Yang, C., Hong, R.: Attentive group recommendation. In: The 41st International ACM SIGIR Conference on Research & Development in Information Retrieval, pp. 645–654. ACM (2018)
6. Driskell, J.E., Salas, E., Driskell, T.: Foundations of teamwork and collaboration. Am. Psychol. **73**(4), 334 (2018)
7. Dwivedi, V.P., Bresson, X.: A generalization of transformer networks to graphs. arXiv preprint arXiv:2012.09699 (2020)
8. Gerrish, S., Blei, D.M.: How they vote: issue-adjusted models of legislative behavior. In: Advances in Neural Information Processing Systems (2012)
9. Ghadimi, S., Lan, G.: Stochastic first-and zeroth-order methods for nonconvex stochastic programming. SIAM J. Optim. **23**(4), 2341–2368 (2013)
10. Gu, Y., Sun, Y., Jiang, N., Wang, B., Chen, T.: Topic-factorized ideal point estimation model for legislative voting network. In: Proceedings of the 20th ACM SIGKDD International Conference on Knowledge Discovery and Data Mining, pp. 183–192. ACM (2014)
11. Guo, L., Yin, H., Wang, Q., Cui, B., Huang, Z., Cui, L.: Group recommendation with latent voting mechanism. In: 2020 IEEE 36th International Conference on Data Engineering (ICDE), pp. 121–132. IEEE (2020)

12. Huang, Z., Xu, X., Zhu, H., Zhou, M.: An efficient group recommendation model with multi-attention-based neural networks. IEEE Trans. Neural Netw. Learn. Syst. **31**(11), 4461–4474 (2020)
13. Kim, Y.J., Engel, D., Woolley, A.W., Lin, J.Y.T., McArthur, N., Malone, T.W.: What makes a strong team?: Using collective intelligence to predict team performance in league of legends. In: Proceedings of the 2017 ACM Conference on Computer Supported Cooperative Work and Social Computing (2017)
14. Kipf, T.N., Welling, M.: Semi-supervised classification with graph convolutional networks. arXiv preprint arXiv:1609.02907 (2016)
15. Kirkland, J.H., Gross, J.H.: Measurement and theory in legislative networks: the evolving topology of congressional collaboration. Soc. Netw. **36**, 97–109 (2014)
16. Lappas, T., Liu, K., Terzi, E.: Finding a team of experts in social networks. In: Proceedings of the 15th ACM SIGKDD International Conference on Knowledge Discovery and Data Mining, pp. 467–476. ACM (2009)
17. Mao, K., Zhu, J., Xiao, X., Lu, B., Wang, Z., He, X.: UltraGCN: ultra simplification of graph convolutional networks for recommendation. In: Proceedings of the 30th ACM International Conference on Information & Knowledge Management, pp. 1253–1262 (2021)
18. Perozzi, B., Al-Rfou, R., Skiena, S.: Deepwalk: Online learning of social representations. In: Proceedings of the 20th ACM SIGKDD International Conference on Knowledge Discovery and Data Mining, pp. 701–710. ACM (2014)
19. Qiu, J., Tang, J., Ma, H., Dong, Y., Wang, K., Tang, J.: Deepinf: social influence prediction with deep learning. In: Proceedings of the 24th ACM SIGKDD International Conference on Knowledge Discovery and Data Mining (2018)
20. Song, K., Lee, W., Moon, I.C.: Neural ideal point estimation network. In: Thirty-Second AAAI Conference on Artificial Intelligence (2018)
21. Veličković, P., Cucurull, G., Casanova, A., Romero, A., Lio, P., Bengio, Y.: Graph attention networks. arXiv preprint arXiv:1710.10903 (2017)
22. Wu, S., Sun, F., Zhang, W., Xie, X., Cui, B.: Graph neural networks in recommender systems: a survey. ACM Comput. Surv. **55**(5), 1–37 (2022)
23. Wu, Z., Pan, S., Chen, F., Long, G., Zhang, C., Philip, S.Y.: A comprehensive survey on graph neural networks. IEEE Trans. Neural Netw. Learn. Syst. **32**, 4–24 (2020)
24. Xu, T., Zhu, H., Zhong, H., Liu, G., Xiong, H., Chen, E.: Exploiting the dynamic mutual influence for predicting social event participation. IEEE Trans. Knowl. Data Eng. **31**(6), 1122–1135 (2018)
25. Yin, H., Wang, Q., Zheng, K., Li, Z., Yang, J., Zhou, X.: Social influence-based group representation learning for group recommendation. In: 2019 IEEE 35th International Conference on Data Engineering (ICDE), pp. 566–577. IEEE (2019)

SCS: A Structural Similarity Measure for Graph Clustering Based on Cycles and Paths

Jiayi Li, Lisong Wang, Zirui Zhang, and Xiaolin Qin[(✉)]

Nanjing University of Aeronautics and Astronautics, Nanjing, China
qinxcs@nuaa.edu.cn

Abstract. With the continuous development of business intelligence and scientific exploration, graphs have been extensively applied to various fields. Graph clustering has emerged as a crucial task for mining the structure and function of complex networks. However, existing clustering algorithms often overly emphasize the density and degree of vertices in the graph while neglecting the correlations and structural characteristics among vertices, resulting in poor performance when clustering graphs. In this paper, we propose a novel method called Structural and Cyclic Similarity (SCS) for structural graph clustering, aiming to improve the quality of clustering. Our method utilizes short-length cycles and paths, which are common graph motifs, to comprehensively capture the neighborhoods and graph motifs of connected vertices. This enables us to quantify the similarity between vertices effectively. The SCS is then applied to structural graph clustering algorithms, thereby improving the clustering quality. To efficiently compute the SCS, we give an algorithm of subgraph counting, which rapidly counts all short-length cycles in the graph. Experimental results conducted on six real-world datasets demonstrate that the clustering algorithm based on SCS outperforms other similarity measures in terms of clustering quality and can improve the effectiveness of graph clustering.

Keywords: Structural Graph Clustering · Structure Similarity · Graph Motif

1 Introduction

With the rapid development of society, much of the data in today's world can be modeled as graphs. A graph is composed of a series of vertices connected by edges, representing relationships between objects. For example, in a social network, vertices represent individuals, and edges represent friendship between individuals [30]. The surge in graph applications has shifted the major direction of graph research towards managing and analyzing graph data, leading to graph clustering being recognized as a fundamental problem and receiving significant attention and research.

X. Song et al. (Eds.): APWeb-WAIM 2023, LNCS 14331, pp. 331–345, 2024.
https://doi.org/10.1007/978-981-97-2303-4_22

Graph clustering is the process of grouping vertices in a graph such that the edges between vertices within the same cluster are dense, while the edges between vertices belonging to different clusters are sparse. Graph clustering has extensive applications in various fields. For example, it can be used to detect hidden structures in a graph [31]. In social networks(e.g., Facebook), clusters in the graph can represent communities [9]. In collaboration networks (e.g., DBLP), clusters may represent groups of researchers with similar research interests.

However, there is still much room for research on graph clustering. In graphs, different graph motifs play different roles in representing the local clustering structure of the graph. Cycles, as one of the most basic graph motifs, are the simplest position-insensitive graph motifs and occur in much higher numbers in real-world graphs than in random graphs [20]. Edges within cycles are more likely to be intra-cluster edges. As basic structures in graphs, cycles possess important functionalities and significance. Therefore, cycles can serve as important features for assessing the significance of vertex or edge clustering. Existing literature has proposed various graph clustering methods, including graph partitioning, density-based methods, and so on. However, they have certain limitations: they often overly emphasize the density and degree of vertices in the graph while neglecting the correlations and structural characteristics among vertices, leading to poor performance for graph clustering. The application of graph motifs in graph clustering methods offers the following advantages: (1) it can meet the diverse clustering demands for graphs, (2) it can improve the efficiency and effectiveness of graphs clustering, and (3) it can identify noise in complex graphs to avoid impacting the clustering results.

To overcome the limitations mentioned above and further improve the quality of graph clustering, we propose a new similarity measure named Structural and Cyclic Similarity (SCS) based on existing structural graph clustering algorithms. SCS maps the number of cycles and paths traversed by each edge $e = (u, v)$ to the degree of association similarity between u and v. It can be embedded as a similarity measure in various types of structural graph clustering algorithms to improve the quality of clustering for graphs, such as social networks, collaborative networks, and more.

For example, Fig. 1 shows a club network with 28 vertices and 43 edges. If we apply the pSCAN [4] with Cosine similarity to cluster this network, the modularity score [11] is 0.4 in the best case. There are 10 vertices that are not assigned to any cluster, including 5, 19, and 20, which are hubs and outliers. Additionally, 7 vertices are not assigned to the desired clusters. However, if we apply the pSCAN with SCS for clustering, the modularity score is 0.58. Only 3 vertices, including 5, 19, and 20, which are hubs and outliers, are not assigned to any cluster. This demonstrates that considering only the degree of vertices or the number of common neighborhoods is insufficient for achieving satisfactory clustering results in certain graphs. By incorporating frequently occurring subgraph structures like cycles into the consideration, further improvements in clustering results can be achieved.

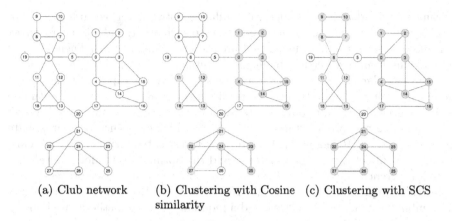

(a) Club network   (b) Clustering with Cosine   (c) Clustering with SCS
                        similarity

Fig. 1. Clustering with different similarity measures

Therefore, the significance of this paper lies in combining cycles and paths as a new clustering similarity measure named Structural and Cyclic Similarity on top of existing similarity measures. We give an algorithm of subgraph counting to efficiently obtain cycles from graphs for computing SCS. The objective is to improve the quality of clustering results and explore potential applications. The key contributions of our work can be summarized as follows:

1. We propose a similarity measure based on graph motifs (cycles and paths) to improve the clustering quality of structural graph clustering algorithms in real-world graphs, and discuss the effectiveness of cycle in capturing local cluster information in graphs;
2. We give an algorithm of subgraph counting, which not only computes the number of cycles, but also obtains edges of them for computing SCS quickly;
3. We embed SCS into two classical structural graph clustering algorithms to cluster real-world graphs. Experimental results show that the SCS combining neighborhoods and subgraph structures can take advantage of clustering on graphs and get better clustering results.

2 Related Work

Graph Clustering, also known as community detection in the social network, refers to the division of a graph into a set of subgraphs, called clusters. Given a graph $G = (V, E)$, where V is the set of vertices and E is the set of edges between vertices, the goal of graph clustering is to partition G into k disjoint subgraphs, where the value of k can be predetermined or undetermined. Scholars have been studying the clustering of networks for decades, especially in the fields of computer science and physics. Here we review some of the more common methods.

Structural Graph Clustering. Reference [31] proposed the original algorithm SCAN, which defines a structural similarity between adjacent vertices. Given parameters ε and μ, if the structural similarity between vertices u and v is not less than ε, two vertices are structure-similar to each other, and u is a core vertex if it has at least μ neighborhoods that are structure-similar to it. Then, clusters are growing from core vertices by including into each cluster all vertices that are structure similar to core vertices in the cluster. SCAN is capable of identifying hubs and outliers. In addition to the precise structural graph clustering studied in References [25,31], an approximate method that improves efficiency through edge sampling was proposed in Reference [17], which can generate approximate clustering results. References [14,28] further investigated structural graph clustering without setting the parameter ε, and the optimal results for ε were computed. Reference [4] proposed the algorithm pSCAN, which prior processes vertex with large probabilities as cores. SHRINK achieves the best clustering results by combining structural similarity and a modularity-based method based on the principle of greed [27], and no parameter input was required.

Other Graph Clustering. In addition, there are clustering methods based on various principles, including spectral clustering [18], graph partitioning [6,19,23], modularity-based methods [21,24], and density-based method [26]. Recent research has also focused on effective cohesive subgraph detection. Most existing graph clustering solutions aim to optimize intra-cluster density or inter-cluster sparsity based on various definitions of "density" and "sparsity" of graph clusters, or a combination of both. Few approaches focus on the original structure of the graph, let alone apply it to graph clustering.

Graph motifs, also known as graphlets, are small subgraphs that occur more frequently than expected in random graphs [20]. As elementary structures in graphs, graph motifs have important implications and represent a broad range of natural phenomena. Triangles [3] and related clustering coefficients [15] have been used for spam detection. Graph motifs have also been employed in analyzing protein-protein interaction networks [5]. However, there has been no research on utilizing graph motifs to improve the effectiveness of graph clustering. Enumerating graph motifs in large-scale graphs is challenging [10]. Specifically, triangle enumeration has been extensively studied, including memory-resident algorithms [16] with optimal complexity of $O(|E|^{\frac{3}{2}})$ in the worst case. Counting short cycles can be accomplished in $O(|V|^\omega)$, where $\omega < 2.376$ is the exponent of matrix multiplication [1]. Reference [22] developed ESCAPE, the first software package aimed at exact counting of all five-vertex subgraphs on moderate-sized graphs. Applying graph motifs to graph clustering is a novel and promising research direction.

In summary, graph clustering methods offer high flexibility and scalability, and selecting the appropriate clustering algorithm is crucial for different types of graph. Among various structural graph clustering algorithms, the similarity measure used to evaluate the similarity between vertices plays a key role in determining the quality of clustering results. Currently, most algorithms employ Cosine similarity as the measure, which is mainly based on the common neighborhoods between two vertices. When two vertices are connected, the common neighborhoods are the number of triangles that both vertices are involved in.

However, SCS relies not only on adjacency but also combines adjacency with the multiple motifs, mapping it into the similarity between vertices. Using SCS can further leverage the correlation and structural features between vertices, enhancing the effectiveness of structural graph clustering algorithms and improving the quality of clustering results.

3 Structural and Cyclic Similarity

Most existing structural graph clustering algorithms aim to find the optimal clustering of a network based on the neighborhoods of two connected vertices. And the mapping of the neighborhoods of two connected vertices into a numerical value represents their similarity. Generally, the more common neighborhoods two vertices have, the more similar they are, and the higher the probability that they belong to the same cluster.

Definition 1. *The structural neighborhood of a vertex u, denoted by $N[u]$, is defined as the closed neighborhood of u, that is $N[u] = \{v \in V \mid (u,v) \in E\} \cup \{u\}$.*

Note that the open neighborhood of u is $N(u) = N[u] \setminus \{u\}$. The closed degree of the vertex u is $d[u] = |N[u]|$.

Definition 2. *The structural similarity between u and v is denoted by $\sigma(u,v)$, is defined as the number of common structural neighborhoods between u and v, normalized by the geometric mean of their cardinalities of the structural neighborhood [31].*

$$\sigma(u,v) = \frac{|N[u] \cap N[v]|}{\sqrt{d[u] \cdot d[v]}} \tag{1}$$

Intuitively, for two vertices, the more common vertices in their structural neighborhoods, the larger the structural similarity value.

In comparison to global features of graphs, such as degree distributions, diameters, and community structures [13], graph motifs are small, connected graphs captured in the local vicinity of vertices or edges, and are primarily used as elementary features representing key functionalities of graphs [32]. Graph clusters are relatively dense subgraphs, and therefore, the occurrence of small-sized graph motifs within them is highly probable. The edge e is likely to be frequently involved in different motifs within the cluster. Therefore, we evaluate the similarity between two vertices connected by edge e by examining the distribution of graph motifs that include e as a constituent edge. The higher the similarity, the closer the relationship between the two connected vertices.

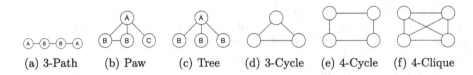

| (a) 3-Path | (b) Paw | (c) Tree | (d) 3-Cycle | (e) 4-Cycle | (f) 4-Clique |

Fig. 2. Graph motifs with 3 and 4 vertices

Figure 2 illustrates graph motifs with 3 and 4 vertices, which can be broadly classified into two categories. The position-sensitive graph motifs ((a), (b) and (c)) contain vertices in different equivalence classes w.r.t. graph isomorphism. In contrast, in position-insensitive graph motifs ((d), (e) and (f)), all vertices are isomorphic. However, not all these graph motifs play a crucial role in clustering. Path motifs (a) and tree motifs (c) are often used to identify cuts spanning across different clusters [29]. Therefore, we consider short-length cycle motifs as the prime features for similarity assessment due to the following reasons:

1. A high density of edges in the cluster will form a large number of short-length cycles within;
2. Cycles are the simplest position-insensitive graph motifs, so edges within a cycle can be treated uniformly w.r.t. graph isomorphism;
3. Cycles can form other complex position-insensitive graph motifs, such as cliques.

In this paper, we only consider simple, undirected and unweighted graphs, but based on the principle of measure, it can be simply extended to directed graphs or other types. For given graph $G = (V, E)$, where V is the set of vertices and E is the set of edges, we use n to denote the number of vertices $|V|$ and m to denote the number of edges $|E|$. $(u, v) \in E$ represents an edge between vertex u and v.

Definition 3. *Given a path* $p = (v_1, \cdots, v_{l+1})$, *where* $(v_i, v_{i+1}) \in E$, $1 \leq i \leq l$. *A path containing* $l + 1$ *vertices, or* l *edges, is length* l, *denoted as an* l-*path for brevity. If* $v_1 = v_{l+1}$, *the path p is closed and it turns out to be a cycle, denoted as an* l-*cycle.*

An l-cycle is a cycle that consists of l vertices (or edges). The shortest cycles are 3-cycles, which is commonly known as triangles. Similarly, 4-cycles are rectangles, and so on.

Definition 4. *Given an edge* $e = (u, v) \in E$ *of the graph* $G(V, E)$, $c^l(e)$ *denotes the number of* l-*cycles* $(l \geq 3)$, *each of which includes e as a constituent edge, indicating the number of* l-*cycles that both u and v pass through.* $p^l(e)$ *denotes the number of* l-*paths containing e as the central edge, allowing for closed paths where vertices in the path can be repeated.*

Definition 5. *Given an edge* $e = (u, v) \in E$, *a cycle length L and an aggregation function F, the Cyclic Similarity between the connected vertices u, v is defined as follows.*

$$CS(u,v) = F\left(\frac{c^l(e)}{p^l(e)}\right)(l = 3, \cdots, L) \tag{2}$$

$F(\cdot)$ is an aggregation function such as SUM or AVG. When F is SUM, $CS(u,v) = \sum_{l=3}^{L}\frac{c^l(e)}{p^l(e)}$. When F is AVG, $CS(u,v) = \frac{\sum_{l=3}^{L}\frac{c^l(e)}{p^l(e)}}{L-2}$. In addition, the cycle length L is typically not set too large. Setting a large L would result in higher computational costs and may have minimal positive effects on clustering. In this paper, we only consider cases where $L \le 5$. Intuitively, for two vertices, the more short-length cycles they have in common, the lager their cycle similarity will be.

Definition 6. *Given an edge $e = (u,v) \in E$, a cycle length L and an aggregation function F, the Structural and Cyclic Similarity (SCS) between the connected vertices u, v is defined as follows.*

$$CSC(u,v) = \frac{|N[u] \cap N[v]|}{\sqrt{d[u] \cdot d[v]}} + F\left(\frac{c^l(e)}{p^l(e)}\right)(l = 3, \cdots, L) \tag{3}$$

By quantifying the similarity, it enables the measurement of vertex neighborhoods while capturing and utilizing the intrinsic structural properties of the graph, leading to improved clustering results.

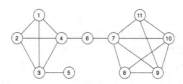

Fig. 3. Example graph

For example, in Fig. 3, we consider a simple graph with $L = 5$ and the aggregation function F as SUM. For the edge $(1,2)$, its $c^3(1,2)$, $c^4(1,2)$, and $c^5(1,2)$ are 2, 2, and 0 respectively, indicating that it passes through two 3-cycles and two 4-cycles. Its $p^3(1,2)$, $p^4(1,2)$, and $p^5(1,2)$ are 4, 12, and 36 respectively, indicating that it serves as the central segment for four 3-paths, twelve 4-paths, and thirty-six 5-paths. Therefore, $CS(1,2) = 0.625$, and $SCS(1,2) = 1.625$. If we consider the edges $(3,5)$ or $(4,6)$, then $CS(3,5) = 0$, $CS(4,6) = 0$, $SCS(3,5) = \frac{2}{\sqrt{10}}$, and $SCS(4,6) = \frac{2}{\sqrt{15}}$, indicating that these two edges are not included in any short-length cycles, suggesting they may be inter-cluster edges, and their endpoints likely belong to different clusters. Thus, combining graph motifs with graph clustering shows feasibility, as graph motifs help reveal the relationships within a graph.

4   Counting Short-Length Cycles

Counting the frequency of small subgraphs occurrence in large graphs is a classical research problem. ESCAPE [22] is an algorithmic framework that can accurately compute the counts of 4-vertex and 5-vertex subgraphs in a graph. It leverages the direction of graph degrees to eliminate duplicate counting, employs cutting and combining for precise counting. However, it only outputs the counts of specified subgraphs in the graph and does not provide vertices or edges of them. We use wedge sets to obtain the edges where all 4-cycles and 5-cycles occur in the graph.

Input an undirected graph G, the degree ordering of G by $\prec$. For vertices u and v, we say $u \prec v$, if either $d(u) < d(v)$ or $d(u) = d(v)$ and $u < v$ (comparing vertex id). ESCAPE constructs the degree ordered DAG $G^{\rightarrow}$ by orienting all edges in G according to $\prec$.

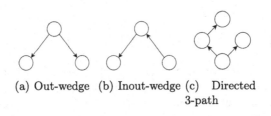

(a) Out-wedge   (b) Inout-wedge   (c)   Directed
3-path

Fig. 4. Fundamental patterns for 4-cycle and 5-cycle counting

Wedge is the most basic pattern in subgraph counting. When a wedge has a direction, it can be categorized into two types: out-wedge and inout-wedge (as shown in Fig. 4 (a) and (b)). For counting 4-cycles, ESCAPE performs cutting along any set of opposite vertices. Considering the acyclic orientations of the 4-cycle as shown in Fig. 5 (a), (b) and (c). The key observation is that wedges between i and j are either 2 out-wedges, 2 inout-wedges, or one of each. Therefore, the number of 4-cycles, $C_4(G)$, is as follows.

$$C_4(G) = \sum_{j \prec i} \binom{W_{++}(i,j) + W_{+-}(i,j)}{2} \tag{4}$$

where $W_{++}(i,j)$ denotes the number of out-wedges between i and j, and $W_{+-}(i,j)$ denotes the number of inout-wedges between i and j.

For counting 5-cycle, ESCAPE utilizes the directed 3-paths in Fig. 4 (c). The 5-cycles can be formed by combining a directed 3-path and either an out-wedge or an inout-wedge (Fig. 5 (d) and (e)), with i and j as the cut vertices. When there exists a directed tailed-triangle (Fig. 5 (f)), it also satisfies the combination rule but does not form a 5-cycle. Therefore, let $P(i,j)$ be the number of directed 3-paths between vertex i and j, and Z be the number of directed tailed-triangles, the number of 5-cycles, $C_5(G)$, is as follows.

$$C_5(G) = \sum_{j \prec i} P(i,j) \cdot (W_{++}(i,j) + W_{+-}(i,j)) - Z \tag{5}$$

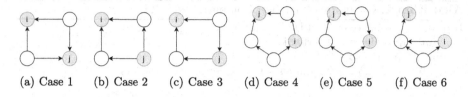

(a) Case 1   (b) Case 2   (c) Case 3   (d) Case 4   (e) Case 5   (f) Case 6

Fig. 5. All acyclic orientations of 4-cycle and 5-cycle

To obtain the edges of 4-cycle and 5-cycle, we use wedge sets U_i to store all the intermediate vertices of out-wedges and inout-wedges between i and j. For directed graph $G^{\rightarrow}$, given a vertex v, $N^{\leftarrow}(v)$ represents its in-neighborhoods and $N^{\rightarrow}(v)$ represents its out-neighborhoods.

The pseudo-code is given in Algorithm 1. First, we initialize the counters and orient the graph G (line 1–2). Then, we compute the wedge set and directed 3-path set of vertex i by Algorithm 2 and Algorithm 3, respectively (line 4–5). After that, we start counting 5-cycles (line 6–12) and 4-cycles (line 13–16) as described above.

Algorithm 1. Short-length Cycle Counting: CountingSC(G)

Input: G: an undirected graph
Output: C_4: 4-cycle number of edges, C_5: 5-cycle number of edges
1: Initialize short-length cycle number C_4, C_5 of edges
2: Orient G using degree orientation to produce $G^{\rightarrow}$
3: **for** $i \leftarrow 0$ to $n - 1$ **do**
4:     $U_i \leftarrow$ ComputeWS$(G^{\rightarrow}, i)$; //Compute wedge set by Algorithm2
5:     $D_i \leftarrow$ ComputeD3S$(G^{\rightarrow}, i)$; //Compute directed 3-path set by Algorithm3
6:     **for each** path $p = (i, k, l, j) \in D_i$ **do** //Start counting 5-cycles
7:         $C_5[(i, k)], C_5[(k, l)]$ and $C_5[(l, j)]$ plus $|U_i[j]|$ respectively;
8:         **for each** $x \in U_i[j]$ **do**
9:             **if** $x = k$ or $x = l$ **then** //Form a directed tailed-triangle
10:                 $C_5[(i, k)], C_5[(k, l)]$ and $C_5[(l, j)]$ minus 1 respectively;
11:             **else**
12:                 $C_5[(i, x)]$ and $C_5[(j, x)]$ plus 1 respectively;
13:     **for** $j \leftarrow i + 1$ to $n - 1$ **do** //Start counting 4-cycles
14:         **if** $|U_i[j]| \geq 2$ **then**
15:             **for each** $x, y \in U_i[j]$ and $x \neq y$ **do**
16:                 $C_4[(i, x)], C_4[(x, j)], C_4[(i, y)]$ and $C_4[(y, j)]$ plus 1 respectively;
17: **return** C_4, C_5

Algorithm 2. Compute Wedge Set: ComputeWS($G^{\rightarrow}, i$)

Input: $G^{\rightarrow}$: a degree ordered DAG, i: a vertex
Output: U_i: wedge set of vertex i
1: **for** $x \in N^{\leftarrow}(i)$ **do**
2:     **for** $j \in N(x)$ **do** //Form an out-wedge or an inout-wedge
3:         $U_i[j] \leftarrow U_i[j] \cup x$;
4: **return** U_i

Algorithm 3. Compute Directed 3-path Set: ComputeD3S($G^{\rightarrow}, i$)

Input: $G^{\rightarrow}$: a degree ordered DAG, i: a vertex
Output: D_i: directed 3-path set of vertex i
1: **for** $k \in N^{\leftarrow}(i)$ **do**
2:     **for** $l \in N^{\leftarrow}(k)$ **do**
3:         **for** $j \in N^{\rightarrow}(l)$ **do**
4:             **if** $j \neq i$ and $j \neq k$ **then** //Form a directed 3-path start from vertex i
5:                Add the path $p = (i, k, l, j)$ to D_i;
6: **return** D_i

5   Experiments

In this section, we evaluate the SCS using six real-world datasets. We embed it into two representative structural graph clustering algorithms: pSCAN and Den-Shrink [7], and compare it with three typical similarity measures: Cosine similarity, Jaccard similarity [2] and Dice similarity [2]. For pSCAN, two parameters, ε and μ, need to be set manually. To ensure fair comparison, we choose the values of ε and μ that yield the best clustering results. All the algorithms are implemented in C++ and compiled using Clion. The experiments were conducted on a macOS machine with a frequency of 3.2 GHz, 8 cores, and 16 GB of memory.

5.1   Datasets

We consider six real-world datasets, including social networks, co-authorship networks, and shopping affiliation networks, among others. These datasets are obtained from the Stanford Network Analysis Platform and are preprocessed before conducting the clustering experiments. The information of datasets is shown in Table 1, with the last two columns indicating the average degree and average clustering coefficient. The datasets are sorted in ascending order based on the number of vertices.

5.2   Evaluation Quality

Clustering validity checking is an important issue in cluster analysis [12]. Here, we evaluate our clustering results using the modularity [8,11,21] and adjusted Rand index (ARI) [13] measures. These quality measures have been widely used in practice due to their effectiveness and computational efficiency, and they are consistent with existing literature on graph clustering.

Table 1. Datasets ($\bar{d}$: Average degree, c: Average clustering coefficient)

Graph	Vertex	Edge	$\bar{d}$	c
facebook-tvshow	3892	17239	8.85	0.412
ca-GrQc	4158	13422	6.46	0.557
musae_facebook	22470	170823	15.20	0.232
Deezer_HR	54573	498202	18.25	0.215
DBLP	317080	1049886	6.62	0.632
amazon2017	334863	925872	5.52	0.397

The modularity is an internal evaluation metric that assesses the compactness of clusters without requiring real-world community data. It can be applied to evaluate clustering on all datasets, while the ARI is an external evaluation metric that can only be applied to datasets with ground-truth clusters, such as DBLP and amazon2017.

Higher modularity scores suggest better clusterings, and higher ARI scores suggest a better match with the ground-truth clustering. Neither the modularity nor the ARI can exceed 1, and the ARI may be negative if the clustering is "worse than random."

5.3 Experimental Results

Short-Length Cycles Counting. Table 2 shows the time cost of counting short cycles in different datasets. We consider different values of the cycle length L ($3 \leq L \leq 5$). When $L = 3$, we only count the number of 3-cycles each edge belongs to. When $L = 4$, we count the number of 3-cycles and 4-cycles each edge belongs to, and so on. The time cost of cycle counting is not significantly influenced by the size of the graph but rather by its density. When the average degree of the graph is high, the counting time is longer. However, this counting process only needs to be performed offline once, and the tiny time cost is affordable compared to the improving of clustering quality.

Table 2. The number of short-length cycles and counting time

Graph	$|c^3|$	Time (L = 3)	$|c^4|$	Time (L = 4)	$|c^5|$	Time (L = 5)
facebook-tvshow	87090	0.001	2889901	0.313	120296318	22.917
ca-GrQc	48260	0.001	1054030	0.054	29812384	3.271
musae_facebook	794953	0.009	33942241	4.392	1859381211	532.452
Deezer_HR	664332	0.012	11208502	4.665	248371391	94.267
DBLP	2224385	0.071	210902219	21.998	19676700606	4245.32
amazon2017	667129	0.022	18228789	8.765	348371391	201.312

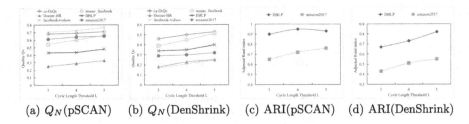

(a) Q_N(pSCAN)   (b) Q_N(DenShrink)   (c) ARI(pSCAN)   (d) ARI(DenShrink)

Fig. 6. Clustering quality under different cycle length L

Cycle Length L. Figure 6 shows that when more different short-length cycles are considered, it is more likely to identify the intra-cluster vertices. However, when $L = 5$, more time is required for counting the 5-cycles. Considering that the short-length cycles counting can be performed offline, it is possible to allocate more time for graph clustering in order to obtain better clustering quality.

Aggregation Function F. The choice of aggregation function can result in different measurement values, which in turn may affect the final clustering results. In this experiment, we selected two aggregation functions, SUM and AVG, and examined their impact on the clustering quality of the graph. Figure 7 shows that in most cases, AVG provides better clustering results compared to the SUM. This is because the SUM combines the occurrences of vertices through different cycle lengths, emphasizing the connections between similar vertices. On the other hand, AVG averages the occurrences of vertices through different cycle lengths, resulting in an average ratio of occurrences. Therefore AVG may be more suitable for other graph operations, such as sparse graphs.

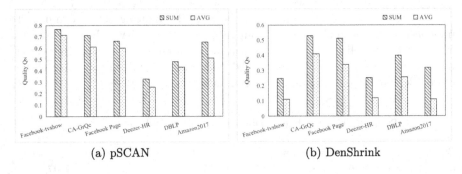

(a) pSCAN   (b) DenShrink

Fig. 7. Q_N under different aggregation function F

Evaluation Under Different Clustering Algorithms. This section shows the clustering results obtained using Cosine similarity, Jaccard similarity, Dice similarity and SCS under two different structural graph clustering algorithms, pSCAN and DenShrink, respectively. In which, the cycle length L is 5, and the aggregation function F is SUM. Figure 8 (a) and (c) show that when using pSCAN, the modularity Q_N obtained using the SCS is significantly better than that obtained using the Jaccard similarity and Dice similarity for all the datasets. The improvement ranges from 9% to 25%. In some datasets, the SCS performs slightly better than the Cosine similarity. For example, in the Facebook-tvshow, the clustering quality is improved by 6%; in the ca-GrQc and Deezer-HR, it is improved by 3% and 4% respectively. In the DBLP and amazon2017 with ground-truth community, the SCS also shows improvements ranging from 6% to 30% in most cases. This is because the SCS takes into account the adjacency between vertices and further considers the connectivity between vertices in subgraphs, leading to better clustering results.

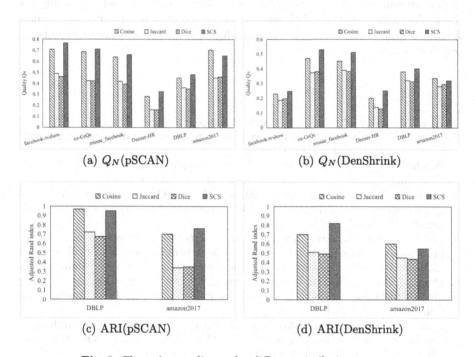

(a) Q_N(pSCAN)   (b) Q_N(DenShrink)

(c) ARI(pSCAN)   (d) ARI(DenShrink)

Fig. 8. Clustering quality under different similarity measures

In the case of the DenShrink, the SCS improves the modularity Q_N by 8% to 29% compared to the Jaccard similarity and Dice similarity. For some datasets, it improves by 3% to 9% compared to the Cosine similarity. For the ARI, the SCS improves by 10% to 33% compared to the Jaccard similarity and Dice similarity. In the DBLP, it improves by 12% compared to the Cosine similarity. This is because, compared to pSCAN, the DenShrink does not require the setting of two important parameters and can adapt to different similarity measures while selecting the best clustering quality based on a greedy criterion. From Table 1, it can be seen that the SCS performs best on datasets with high density, high average degree, and large number of cycles, maximizing the advantages of clustering.

6   Conclusion

In this paper, we propose a novel similarity measure called Structural Cycle Similarity that utilizes cycles and paths to measure the similarity between vertices, which is then applied to the structural graph clustering. To efficiently compute the SCS, we give an algorithm of subgraph counting, which rapidly counts all short-length cycles in the graph. We embed the SCS into two structural graph clustering algorithms, pSCAN and DenShrink, and demonstrate its advantages in real-world graph clustering through extensive experiments.

In future research, we plan to explore the application of the SCS in overlapping community detection or graph sparsification. By further expanding the application domains of the SCS, we can gain deeper insights into its effectiveness in various graph analysis tasks and provide more choices and improvements for practical applications.

References

1. Alon, N., Yuster, R., Zwick, U.: Finding and counting given length cycles (1998)
2. Augsten, N., BHlen, M.H.: Similarity joins in relational database systems. Synth. Lect. Data Manage. **5**(5), 1–124 (2013)
3. Becchetti, L., Boldi, P., Castillo, C., Gionis, A.: Efficient algorithms for large-scale local triangle counting. ACM Trans. Knowl. Disc. Data. **4**, 1–8 (2010)
4. Chang, L., Wei, L., Lin, X., Lu, Q., Zhang, W.: pSCAN: fast and exact structural graph clustering. In: 2016 IEEE 32nd International Conference on Data Engineering (ICDE) (2016)
5. Chen, J., Hsu, W., Lee, M.L., Ng, S.K.: Nemofinder: dissecting genome-wide protein-protein interactions with Meso-scale network motifs. In: Twelfth ACM SIGKDD International Conference on Knowledge Discovery & Data Mining (2006)
6. Ding, C., He, X., Zha, H., Ming, G., Simon, H.D.: A min-max cut algorithm for graph partitioning and data clustering. In: Proceedings 2001 IEEE International Conference on Data Mining (2001)
7. Feng, H.B.: Density-based shrinkage for revealing hierarchical and overlapping community structure in networks. Statist. Mech. App. Phys. A. **390**, 2160–2171 (2011)
8. Feng, Z., Xu, X., Yuruk, N., Schweiger, T.A.J.: A novel similarity-based modularity function for graph partitioning (2007)
9. Fortunato, S.: Community detection in graphs. Phys. Rep. **486**(3–5), 75–174 (2010)
10. Grochow, J.A., Kellis, M.: Network motif discovery using subgraph enumeration and symmetry-breaking. In: Speed, T., Huang, H. (eds.) RECOMB 2007. LNCS, vol. 4453, pp. 92–106. Springer, Heidelberg (2007). https://doi.org/10.1007/978-3-540-71681-5_7
11. Guimerà, R., Nunes Amaral, L.A.: Functional cartography of complex metabolic networks. Nature **433**, 895–900 (2005)
12. Halkidi, M., Batistakis, Y., Vazirgiannis, M.: Cluster validity methods: part i. ACM SIGMOD Record **31**(2), 40–45 (2002)
13. Hubert, L., Arabie, P.: Comparing partitions. J. Classif. **2**(1), 193–218 (1985)

14. Kim, M.S., Han, J.: A particle-and-density based evolutionary clustering method for dynamic networks. Proc. VLDB Endow. **2**(1), 622–633 (2009)
15. Kutzkov, K., Pagh, R.: On the streaming complexity of computing local clustering coefficients. In: Proceedings of the Sixth ACM International Conference on Web Search and Data Mining (2013)
16. Latapy, M.: Main-memory triangle computations for very large (sparse (power-law)) graphs. Theoret. Comput. Sci. **407**, 458–473 (2008)
17. Lim, S., Ryu, S., Kwon, S., Jung, K., Lee, J.G.: LinkSCAN*: overlapping community detection using the link-space transformation. In: IEEE International Conference on Data Engineering (2014)
18. Liu, J., Chi, W., Danilevsky, M., Han, J.: Large-scale spectral clustering on graphs. In: Proceedings of the Twenty-Third International Joint Conference on Artificial Intelligence (2013)
19. Lu, W., Xiao, Y., Shao, B., Wang, H.: How to partition a billion-node graph. In: IEEE International Conference on Data Engineering (2014)
20. Milo, R., Shen-Orr, S., Ltzkovitz, S., Kashtan, N., Alan, U.: Network motifs: Simple building blocks of complex networks (2011)
21. Newman, M., Girvan, M.: Finding and evaluating community structure in networks. Phys. Rev. E **69**(2 Pt 2), 026113 (2004)
22. Pinar, A., Seshadhri, C., Vishal, V.: Escape: efficiently counting all 5-vertex subgraphs. In: The Web Conference (2017)
23. Shi, J., Malik, J.M.: Normalized cuts and image segmentation. IEEE Trans. Pattern Anal. Mach. Intell. **22**, 888–905 (2000)
24. Shiokawa, H., Fujiwara, Y., Onizuka, M.: Fast algorithm for modularity-based graph clustering. In: National Conference on Artificial Intelligence (2013)
25. Shiokawa, H., Fujiwara, Y., Onizuka, M.: Scan++: efficient algorithm for finding clusters, hubs and outliers on large-scale graphs. VLDB Endow. **8**, 1178–1189 (2015)
26. Singh, M.: SPICi: a fast clustering algorithm for large biological networks. Bioinformatics **26**(8), 1105–11 (2010)
27. Sun, H., Huang, J., Han, J., Deng, H., Sun, Y.: Shrink: a structural clustering algorithm for detecting hierarchical communities in networks. In: ACM International Conference on Information & Knowledge Management (2010)
28. Sun, H., Huang, J., Han, J., Deng, H., Zhao, P., Feng, B.: gSkeletonClu: density-based network clustering via structure-connected tree division or agglomeration (2010)
29. Tsourakakis, C., Bonchi, F., Gionis, A., Gullo, F., Tsiarli, M.: Denser than the densest subgraph: extracting optimal quasi-cliques with quality guarantees. In: ACM SIGKDD International Conference on Knowledge Discovery & Data Mining (2013)
30. Wasserman, S., Faust, K.: Social Network Analysis: Methods and Applications (1994)
31. Xu, X., Yuruk, N., Feng, Z., Schweiger, T.A.: Scan: a structural clustering algorithm for networks. In: Proceedings of the 13th ACM SIGKDD International Conference on Knowledge Discovery and Data Mining, pp. 824–833 (2007)
32. Zhao, P.: gSparsify: Graph motif based sparsification for graph clustering. ACM (2015)

Time Series Model Interpretation via Temporal Feature Sampling

Zhaoyang Liu[1] , Xiaodong Li[1(✉)] , and Yanping Cui[2]

[1] College of Computer and Information, Hohai University, Nanjing 210098, China
{zyliu,xiaodong.li}@hhu.edu.cn
[2] Jiangsu Province Hydrology and Water Resources Investigation Bureau,
Shanghai Road, Nanjing, China

Abstract. Model interpretation methods play a critical role in enhancing the applicability of time series neural networks in high-risk domains. However, existing model interpretation methods, primarily designed for static data like images, do not yield satisfactory results while dealing with time series data. Although some studies have explored the time dimension and evaluated feature importance at each time point through feature removal, they neglect the potential correlations among multiple features that impact the model's predictive outcomes. To address this limitation, we introduce the concept of Shapley value into the process of time series model interpretation and propose the TFS (Temporal Feature Sampling) algorithm. This algorithm calculates the importance scores of feature subsets, which include the removed features, during the model interpretation process. Additionally, it models the distribution of features by sampling within the time series data. We conducted comparative experiments between TFS and several baseline methods on two synthetic datasets and one real-world dataset, and the experimental results confirmed the efficiency and performance of our algorithm.

Keywords: Time series model · Explainable intelligence · Shapley value

1 Introduction

Due to the rapid development of neural network models, deep learning has achieved significant breakthroughs in various fields [1–3]. The effectiveness of various neural networks mainly relies on complex network structures, models as well as parameter settings. However, these factors also contribute to the low transparency of neural networks, making it difficult for people to assess the reliability of their decisions. The "black-box" properties of deep neural networks severely hinder their widespread application in domains with high security requirements, such as medical diagnosis [4], finance [5], and autonomous driving [6]. Therefore, comprehensible deep learning models are extremely important, which has led to

X. Song et al. (Eds.): APWeb-WAIM 2023, LNCS 14331, pp. 346–360, 2024.
https://doi.org/10.1007/978-981-97-2303-4_23

the emergence and development of many model interpretability methods based on feature importance assessment [7].

Essentially, in the field of artificial intelligence, eXplainable Artificial Intelligence (XAI) refers to the presentation of model decision outcomes to humans in an understandable way. Existing explanation methods for neural networks often focus on extracting the importance or contribution of input features to the model prediction [7]. However, most model interpretation methods were originally designed for static data, such as image data, and are not directly applicable to time series datasets and models. This is because, on one hand, the dynamic nature of temporal data introduces time dependencies in feature importance [8]. On the other hand, recurrent neural network (RNN) models, which are commonly used for time series modeling, have more complex structures. Existing methods for interpreting time series neural network models can be mainly divided into two categories. In addition to the aforementioned static model interpretation methods, another category is attention-based interpretation methods. Many studies have suggested that attention scores can be interpreted as importance scores that change over time, but the usability and accuracy of such interpretation method are still debatable [9]. To address the limitations of the above two types of methods, existing methods for achieving dynamic interpretation in time series neural networks primarily focus on removing certain features in different ways to measure the changes in model predictions and provide explanations. For example, Zeiler et al. [10] removes individual features from the feature vector at each time step and measures the change in model predictions before and after the feature modification as an explanation. Tonekaboni et al. [11] propose the FIT (Feature in Time) algorithm. This algorithm utilizes a generator to produce the remaining features outside a specified subset in the feature vector. It measures the impact of including or excluding this feature subset on the model predictions using Kullback-Leibler Divergence (KL divergence) at the same time step. We can observe that, first, these explanation methods, which compute local feature importance by removing specified features or their complements from the feature vector, do not take into account the correlations between these features and the remaining ones. Second, the approach of using a generator to generate data, as in [11], requires a high demand for the data feature distribution, and if the data distribution is complex, it may not achieve optimal results.

Regarding the above two issues, we propose the TFS (Time Feature Sampling) method, which incorporates the idea of Shapley values [12] when removing features in time series models. The TFS method calculates the importance of all feature subsets containing the feature being removed by sampling. This approach has the advantage of considering the correlations between the specified feature and other features. Additionally, it is suitable for datasets with complex data distributions. Firstly, the objective of this method is to compute the Shapley values of feature subsets as importance explanations. For a specified feature subset, its Shapley value represents its contribution to the prediction and can be obtained by weighted summation over all possible combinations of features. Secondly, to mitigate the computational complexity arising from the number of exponential-level feature combinations, we introduce a sampling algorithm to approximate the final results. Lastly, considering the dynamic nature of time series data, we calculate such importance scores at each time step and aggregate the final results. The TFS method is model-agnostic, allowing its application to any black-box model in time series analysis. In our experiments, we validated our ideas and proposed methods on synthetic datasets with varying degrees of complexity and the real medical dataset MIMIC III. The experimental results demonstrate that the TFS algorithm outperforms existing baseline methods to a certain extent in terms of performance and stability. Our contributions are as follows:

- We propose applying the method of sampling Shapley values to feature removal in model interpretation to evaluate the impact of feature subsets on model predictions.
- We conduct experiments on several datasets, and the results demonstrate that taking into account the correlations between features enhances the effectiveness of explanations for time series models and data.
- Through experimental validation, we find that using a generator to generate the removed features doesn't yield satisfactory results for datasets with complex data distributions. However, sampling features from the original dataset is not affected by this issue.

The remainder of this paper is organized as follows. In Sect. 2, we introduce the development of traditional static model interpretation methods and existing dynamic interpretation methods for time series models. In Sect. 3, we present the concept of model interpretation based on Shapley values and TFS algorithm we proposed. In Sect. 4, we discuss the experimental settings and present the experimental results. Finally, Sect. 5 provides a summary of the entire paper.

2   Related Work

In this section, we first categorize commonly used static model interpretability methods, followed by a summary of existing interpretability methods in the field of time series. Finally, we briefly describe the shortcoming that the current interpretability methods for time series data do not take feature correlations into account.

2.1   Traditional Model Interpretation Methods

Existing model interpretability methods based on importance attribution can be mainly categorized into three types [7]: gradient-based methods [9,13–17], perturbation-based methods [10,18], and propagation-based methods [19–21]. Gradient-based methods calculate the sensitivity of the target variable to input features through backpropagation to assess the importance of input features. Perturbation-based methods determine the most influential features on the results by perturbing feature permutations or removing features. Propagation-based methods involve propagating and decomposing model prediction scores through multi-layer networks to locate features with high contributions. However, these methods were originally designed for processing static data, such as image data, and have limited applicability to recurrent neural network models which are commonly used for time series data. On one hand, this is due to the vanishing gradient in recurrent structures, which affects the quality of feature importance allocation [22]. Bento et al. [23] proposed the TimeSHAP method, which perturbs sequence to explain the decision process of recurrent neural networks, enhancing the applicability of static interpretation methods to recurrent models. However, this method is more suitable for neural network models with specific structures. On the other hand, as highlighted by Ismail et al. [8], these methods fundamentally overlook the temporal property of time series data and do not explicitly model thier time dependency, leading to poor performance in such applications.

In addition to the aforementioned attribution-based static methods, attention mechanisms originally designed for machine translation tasks [24] are also commonly used for interpreting time series models [25,26]. These methods enhance model performance by learning the long-term dependencies in time series and interpret the model through its parameters. Guo et al. [27] pointed out that recurrent models have complex mappings with their latent space, which makes it difficult to directly attribute attention weights to individual predictions in time series. Therefore, Choi et al. [25] proposed the use of separate parameter sets to enhance the attention mechanism and obtain importance scores that vary with time and features.

2.2   Model Interpretation Methods for Time-Series Data

It is observed that there are limitations when applying static model interpretation methods to time series data. Due to the dynamic nature of time series data compared to image data, only a few model interpretation methods are currently applicable to this type of data. Ismail et al. [8] proposed a two-stage temporal saliency rescaling (TSR) method to address the issue of static attribution methods not being able to identify important features synchronously on the temporal axis. Suresh et al. [28] applied the feature occlusion (FO) method [14] to time series data. Instead of setting the specified feature to zero as in the original method, they replaced the target feature with a baseline value and calculated the importance of the replaced feature based on the difference in model outputs. Tonekaboni et al. [11] introduced the FIT (Feature Importance in Time) method, which generates new features using a generator and replaces the complement of the specified feature. The importance of each feature in predicting changes at the same time step is measured using scores based on KL divergence. Crabbe et al. [29] applied masking techniques to the interpretation of time series data. They used a dynamic mask module to perturb the values of different features by learning the input feature matrix mask, thereby highlighting important features.

In fact, the aforementioned model interpretability methods for time series data can all be viewed as feature removal related methods. Covert et al. [30] suggested that existing model interpretability methods can be described within a unified framework based on feature removal. Feature removal based methods can be categorized from three dimensions: feature removal approach, model interpretability behavior, and feature summarization approach. Regarding the feature removal approach, the aforementioned methods [11,28,29] respectively employ methods such as replacing with baseline values, generating data using a generator, and dynamic masking to substitute the removed features in the data. In terms of model interpretability behavior, these methods analyze the changes in model predictions after feature removal. As for the feature summarization approach, FO [10] considers using the L1 norm to measure the similarity or dissimilarity of model predictions, while FIT [11] uses KL divergence to describe the difference in model outputs when a specific feature is present or absent at the same time step. We believe that neither of the two categories of algorithms mentioned above, from the perspective of feature removal, takes into account the impact of correlations between different features on model predictions during model interpretation. Moreover, the method employed by FIT, which generates features using a generator, may not be suitable for datasets with complex distributions. Therefore, leveraging the idea of Shapley values [31], we propose the TFS algorithm for interpreting time series models. By sampling feature distributions in the data, our approach considers the correlations between multiple features and calculates the Shapley values of different features as feature importance measures. This can compensate the deficiencies in the above studies to some extent.

3   Model Interpretation Methods

The method proposed in this paper is primarily based on Shapley values and interpretation methods for feature removal in the context of time series data. In this section, we will provide a detailed explanation of this paper's research approach and introduce the TFS algorithm that we have developed.

3.1   Preliminary Notation

We frame the problem of model interpretation within the context of time series classification. Assume that we are given a trained black-box model f_θ, it takes a multivariate time series $X \in R^{D \times T}$ as input, as well as $Y \in y$ as output, where T is the number of time steps and N is the number of features. We use uppercase symbols (e.g., X) to denote random variables and lowercase ones (e.g., x) to denote their values. Further, $X_t \in R^D$ is the set of all features at time $t \in 1, ..., T$, denoted by the vector $[x_{1,t}, x_{2,t}, ..., x_{D,t}]$ and $X_{0,t} \in R^{D \times t}$ is the matrix $[\mathbf{X}_0; \mathbf{X}_1; \cdots ; \mathbf{X}_t]$. Finally, let $X_{S,t}$ denotes the subset of all features at time t, where $S \in \{1, 2, ..., D\}$, and $X_{S^c,t}$ represents the feature complement corresponding to the subset S, where $S^c = \{1, 2, ..., D\} \backslash S$.

3.2 Introduction to Shapley Values

The key component of the general approach to model interpretability is to calculate the contribution of individual input features to the predictions of the model. For instance, consider the following linear model:

$$f_\theta(x) = f_\theta(x_1, \ldots, x_n) \approx y = \beta_0 + \beta_1 x_1 + \ldots + \beta_n x_n, \qquad (1)$$

The contribution of the i-th feature to the model $f_\theta(x)$ can be defined in the following way:

$$\varphi_i(x) = f_\theta(x_1, \ldots, x_n) - E\left[f(x_1, \ldots, X_i, \ldots, x_n)\right], \qquad (2)$$

where $E(\beta_i)$ denotes the average impact estimate of feature i. Then equation (2) represents the difference in predictions for the same instance when feature i is unknown. In contrast, Strumbelj et al. [32] pointed out that perturbing one feature at a time would result in a situation where the feature contributions are all zero after perturbation for non-additive models. For similar scenarios involving correlations among sample feature subsets during model interpretation, the concept of Shapley values can be applied. This involves calculating the individual contributions of each feature subset to the output and then aggregating them. Expressed in terms of Shapley's formula as:

$$\varphi_i(v) = \sum_{S \subseteq N \setminus \{i\}} \frac{|S|! \, (|N| - |S| - 1)!}{|N|!} \left(v\left(S \cup \{i\}\right) - v\left(S\right)\right), \qquad (3)$$

where N is the total number of features, S is a subset of N, and v denotes the black-box function to be explained, then the calculation result of the above formula $\varphi_i(v)$ denotes the contribution of feature i to this model considering all feature subsets.

However, as described in Eq (3), the number of feature subsets is usually exponential, which leads to a high computational complexity when the algorithm is applied in practice. For this reason, there are related studies that discuss methods to reduce the computational complexity [18,32]. In the study of Strumbelj et al. [32], a Monte Carlo sampling method is used to form the distribution of feature i to be interpreted in the sample, which in turn avoids the computation of all possible sets of features with and without the use of feature i. This approximation reduces the computational complexity to a large extent. The final Shapley value sampling formula proposed by Strumbelj et al. [32] is expressed as:

$$\hat{\varphi}_i = \frac{1}{M} \sum_{m=1}^{M} \left(\hat{f}_\theta\left(x_{+i}^m\right) - \hat{f}_\theta\left(x_{-i}^m\right)\right), \qquad (4)$$

where M denotes the number of samples, and $+i$ and $-i$ denote whether the current feature instance contains the specified feature i, respectively.

3.3   TFS: Temporal Feature Sampling

In the traditional image domain, the interpretation of a feature can usually be achieved by simply removing the feature for perturbation [10,12] and observing the change in model predictions before and after the perturbation to measure the feature importance. However, in the field of time series, this feature removal approach is usually not applicable. This is because, first, most models require complete set of features for training, and second, for structured data such as time-series data, feature 0 often carries specific meanings [18]. Using the interpretation method of setting features to 0 can be confusing for the results. Therefore, when dealing with the intricate scenario of feature removal in time-series models, it becomes necessary to differentiate existing model interpretation methods within the time-series domain based on the various approaches used for feature removal. For example, Suresh et al. [28] replace the features to be interpreted with pre-defined baseline values to achieve the effect of feature removal so as not to the interference of feature 0 on the prediction results; Tonekaboni et al. [11] introduced the FIT algorithm, which deviates from the notion of removing the features to be interpreted and instead removes its complement at a specific moment. This ensures that the model prediction value at that moment is solely influenced by the features being explained.

Further, we explore the internal mechanism of the FIT, which is based on the idea of comparing the impact of the model on its predictives by a given subset of features, S, at moment t. When a new observation $\mathbf{x}_{\mathbf{S},t}$ is obtained at time point t, its significance can be assessed based on whether it adds additional information that is typically used to explain the change in the predicted distribution compared to the distribution prior to t. To ensure that features at moment t are only affected by a subset S, the distribution $p\left(\mathbf{x}_t \mid X_{0:t-1}\right)$ based on past-conditional measurements is estimated using the recursive generator G to approximate the counterfactual distribution $p\left(\mathbf{x}_{S^c,t} \mid X_{0:t-1}, \mathbf{x}_{S,t}\right)$ and model the non-smoothness in the time series, and finally use the cross-entropy function to calculate the importance score I of the feature subset S at moment t. The formula is as follows:

$$I\left(\mathbf{x}_{\mathbf{S}}, t\right) = KL(p\left(y \mid X_{0:t}\right) \| p\left(y \mid X_{0:t-1}\right)) - \\ KL(p\left(y \mid X_{0:t}\right) \| p\left(y \mid X_{0:t-1}, \mathbf{x}_{\mathbf{S},t}\right)), \tag{5}$$

the overall score I measures how important observing $\mathbf{x}_{\mathbf{S},t}$ would be for estimating the output distribution at time t. To ensure that the denominator at moment t in the second fractional equation is related only to the feature subset S, the authors use generators and sampling methods to generate the complement of the feature set, denoted as S^c.

However, it is worth noting that, similar to FO, this approach still only considers the contribution of the features to be interpreted to the prediction itself, without taking into account the correlation of multiple features to the prediction results (Fig. 1). In addition, the role of the generator in FIT is to generate feature complements to model non-stationarity in time series. We believe that this approach may not be applicable for handling data with more complex distribution.

Therefore, we propose TFS, a model interpretation method for time-series data, which incorporates the idea of Shapley value sampling when removing features and samples data directly from the original dataset as a subset of features.

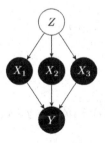

Fig. 1. Multi-feature prediction structure, where the observations of the conditional distribution $p(y|x_1)$ cannot be used as the contribution of X_1 to Y because of the (potential) interference of the common cause Z among multiple features.

Considering the correlation of multiple features on the influence of predicted values when removing features, we modeled feature correlation using Shapley value sampling instead of individually removing or retaining features to be interpreted in the original feature vector. First, a feature instance z : $(z_{(1)}, \ldots, z_{(S)}, \ldots, z_{(D)})$ is randomly sampled in the original data sequence, and for the interested feature subset S, at time t, we construct two instances with different feature sequences, one including the feature subset S and the other excluding S:

$$x_{+s} = [x_1, \ldots, x_S, z_{S+1}, \ldots, z_D], \tag{6}$$

$$x_{-s} = [x_1, \ldots, z_S, z_{S+1}, \ldots, z_D]. \tag{7}$$

where the first instance x_{+S} is an instance including the feature subset S, where all values after S (excluding the values of the feature subset S) are replaced by the feature values in the sample z. The second instance x_{-S} is similar to the former, where all values are in the previous order, but all values after the instance x including the feature subset S are replaced by the values corresponding to the feature subset S in the sample z. Then, the average of difference values of the two feature sequences at moment t after multiple sampling is calculated by Eq. 4.

In time-series data, similar to FIT, the importance of the features at each moment needs to be considered and therefore a similar Monte Carlo sampling calculation needs to be performed at each moment in all time period T. For a detailed description of this method see Algorithm 1 (where j denotes the feature to be interpreted):

Since TFS also needs to calculate the feature importance on each moment, both FIT and TFS have the same time complexity of $O(n^2)$. However, since the FIT algorithm requires additional training of the feature generator before the

Algorithm 1. TFS

Input: f_θ: Trained Black-box model, time series $\boldsymbol{X}_{0:T}$, where T is the max time, L: Number of Monte Carlo samples, j: feature of interest.
Output: Importance score matrix I

1: **for** each $t \in [T]$ **do**
2:   **for** each $l \in [L]$ **do**
3:     Draw random instance $z \in X_{0:T}$
4:     Set instance $z : z = [z_1, \dots, z_D]$
5:     With feature f_j :
6:     $x_{+j} = [x_1, \dots, x_j, z_{j+1}, \dots, z_D]$
7:     Without feature f_j :
8:     $x_{-j} = [x_1, \dots, z_j, z_{j+1}, \dots, z_D]$
9:     $p\left(\hat{y}_{+j}^l\right) = f_\theta\left(X_{0:t-1}, X_{+j,t}\right), p\left(\hat{y}_{-j}^l\right) = f_\theta(X_{0:t-1}, X_{-j,t})$
10:    **end for**
11:   $I(X_j, t) = \frac{1}{L}\sum_{l=1}^{L}\left(p\left(\hat{y}_{+j}^l\right) - p\left(\hat{y}_{-j}^l\right)\right)$
12: **end for**
13: **return** $I(X_j, t)$

features can be interpreted, this will consume a significant amount of time. In addition, TFS samples data distribution directly from the data sequence, which can serve as a reliable approximation of the distribution of the original features when the data lacks clear distribution patterns or is challenging to be fitted by the generator.

4   Experimental Results and Analysis

We evaluated the proposed model interpretation method on several simulated and real datasets while comparing it with several baseline methods including FIT.

4.1   Data Set and Metrics

Different from image data, evaluating interpretation methods on time series datasets is often challenging due to the absence of labels. Although we can perform different forms of saliency analysis (e.g., heat maps) on time series with traditional XAI methods, such analysis is often incomprehensible due to the fact that time-series data usually lack the inherent semantics present in image data (e.g., explaining the classification of a rooster based on the rooster comb is easily) [33]. To address this problem, two common treatments are currently used. The first approach involves artificially synthesizing time series data with semantic information (i.e., labels) to facilitate interpretation. Secondly, the evaluation of interpretation methods on real datasets involves analyzing the fidelity properties of these methods. This is achieved by perturbing samples based on the interpretation results obtained from the models and observing the deviation in predictions made by the black-box model.

Based on the variations in evaluation approaches aforementioned, our experiments primarily utilize two types of datasets: synthetic simulated datasets and real datasets. The following sections outline the characteristics of each dataset type and the corresponding evaluation methods employed.

Simulated Data

- **Spike Data:** As in Tonekaboni et al. [11], we used a multivariate data consisting of 3 random NARMA time series where the outcome (label) changes to 1 as soon as a spike is observed in the relevant feature (and is 0 otherwise). We employed the generator to generate a dataset comprising 10,000 such instances, with 8,000 instances allocated for training and 2,000 instances for testing purposes. The task at hand is relatively straightforward, aiming to ensure that the black-box classifier can effectively learn the accurate relationship between the crucial features and the corresponding outcomes.
- **State Data:** To illustrate the limitations of FIT's approach in generating interpretable features for datasets with more intricate state distributions, we utilized a dataset with a more intricate state distribution [29] for validation. Specifically, we use state data consisting of multivariate time series signals with 3 features to model the intricate dynamics in real-time data. A non-stationary Hidden Markov Model (HMM) with 2 hidden states and a linear transfer matrix is used to generate time-varying observations. Moreover, at each time step t, the input feature vectors are sampled from a multivariate normal distribution determined by the hidden states of the HMM. According to this approach, the state transfer probabilities are modeled as a function of time to model non-stationarity. We generated 1000 such data sequences of time length 200, of which 800 were used for training and the remaining 200 for testing.

Clinical Data. MIMIC-III is a comprehensive multivariate clinical time series dataset comprising electronic medical records of around 40,000 ICU patients from Beth Israel Deaconess Medical Center in Boston, Massachusetts [34]. This dataset is widely used in healthcare and medical artificial intelligence related research and contains several time series related tasks, such as mortality and length-of-stay prediction. We used the preprocessing step described by Tonekaboni et al. [11] to predict patient mortality using 8 vital signs and 20 laboratory group measurements per hour over a 48-hour period. Finally, 229,888 eligible admissions data were extracted and we divided them into 65%, 15%, and 20% training, validation, and test sets.

Evaluation Metrics. For simulated data, we can assess the performance of saliency methods by leveraging the metrics already existed for classification problems, since the simulated data explicitly defines the features that have a significant impact on the interpretation results. Therefore, we use AUROC (Area Under the Receiver Operating Characteristic curve) and AUPRC (Area Under the Precision-Recall Curve) to evaluate the quality of feature importance assignment across baselines compared to ground-truth.

For real data, we primarily focus on evaluating the fidelity properties of the model interpretation method. The underlying concept is that removing a feature from the data should adversely affect the overall prediction performance of the black-box model, if such feature is important for the outcome prediction. Therefore, a greater performance degradation indicates a more accurate importance attribution. Similar to Tonekaboni et al. [11], the performance of the method is measured from two perspectives using global and local evaluation methods. The global approach corresponds to measuring the impact of removing the observations with the highest importance scores (top 5%) in the entire test set. The local approach corresponds to removing the observations with the highest importance scores (top $K = 50$) within each instance. For both experiments, we present the decrease in the metric AUC (Area Under the Curve) and the average predicted change when important features are masked.

4.2 Model Setup and Baseline Methods

For all experiments, we use a 1-layer GRU model with 200 hidden units as the black-box model f_θ to be interpreted. For a subset of features S, which can be expressed as an individual feature or a collection of features chose beforehand. For simplicity, we restrict the set of features to one separate feature, i.e., $S = i$. To ensure the stability of the results, we perform a 5-fold cross-validation on the training set and calculate the mean and standard deviation of the results. All experiments were reconducted on a device with a 32-core AMD EPYC 7601 CPU and an Nvidia GeForce RTX 3070.

We utilize several widely used static model interpretation methods and dynamic methods designed for time-series models (e.g., FO, FIT) as baseline methods, combined with TFS to conduct experiments on the aforementioned datasets, respectively. According to the previous section, the baseline methods used are classified into the following categories according to their categories:

1. Gradient-based methods: in this category, we used Deep-LIFT [16], Grad-SHAP [18]. For the consistency of the algorithm implementation, we used the implementation provided in the Captum library [35].
2. Perturbation-based approach: Feature Occlusion (FO) [28] - the feature importance is determined by calculating the differences in predictions when each feature i is substituted with a random sample from a uniform distribution. The augmented feature occlusion (AFO) method introduced by Tonekaboni et al. [11] is also used. Unlike FO, AFO replaces features from a uniform distribution, this method selects features directly from the original data for replacement to avoid generating out-of-distribution noise samples.
3. Attention-based approach (RETAIN) [25]: this is an attention-based model that provides feature importance over time by learning dual attention scores (over time and features).

4. Feature importance in time (FIT) [11]: assigns importance to observations based on the KL divergence between the predicted distribution and counterfactuals (the remaining features were not observed at the last time step).

4.3 Results

We restricted the model evaluation to a subset of features consisting of a single feature and compared it to the baseline approach described above. The experimental results for the synthetic data are shown in Table 1 and results for real data are shown in Table 2. Between different Tables, the optimal results produced by the algorithm are highlighted in bold and the suboptimal results are underlined.

We can find that for the simpler synthetic data of Spike data, almost all methods are able to correctly identify the features most relevant to the prediction results. Compared to static XAI methods, dynamic model interpretation methods (TFS, FO, AFO, FIT) perform more consistently in both types of evaluation metrics. TFS which incorporates the concept of Shapley value into perturbation-based methods (FO, AFO) and takes into account the impact of feature sets on prediction results, resulting in improved performance compared to previous methods. In comparison to the FIT algorithm, TFS demonstrates superior performance in the AUPRC metric, while achieving comparable results in other evaluation metrics.

Table 1. Experimental results on synthetic data.

Datasets	SPIKE		STATE	
Method	AUROC	AUPRC	AUROC	AUPRC
TFS	0.948 ± 0.001	$\mathbf{0.946 \pm 0.002}$	$\mathbf{0.87 \pm 0.02}$	$\mathbf{0.68 \pm 0.01}$
FIT	$\mathbf{0.968 \pm 0.020}$	0.866 ± 0.022	0.73 ± 0.02	0.48 ± 0.01
AFO	0.942 ± 0.002	$\underline{0.932 \pm 0.006}$	0.84 ± 0.01	0.63 ± 0.01
FO	0.943 ± 0.001	0.894 ± 0.026	$\underline{0.85 \pm 0.01}$	0.64 ± 0.01
Deep-Lift	0.941 ± 0.002	0.520 ± 0.098	0.78 ± 0.00	0.61 ± 0.01
RETAIN	0.249 ± 0.168	0.001 ± 0.000	0.80 ± 0.02	$\underline{0.65 \pm 0.03}$
GradSHAP	0.933 ± 0.004	0.516 ± 0.039	0.79 ± 0.00	0.55 ± 0.00

For state data, which is designed more complex, our algorithm achieves the best results among all the interpretation methods. Additionally, the experimental results validate our hypothesis that the FIT algorithm may not be suitable for complex data distributions. In contrast, TFS, which leverages sampling from the original data, exhibits superior ability to capture the underlying data distribution, leading to accurate interpretation results.

For the real data, i.e. MIMIC III data, we show the results in Table 2. It is evident that TFS achieves either the best or second-best performance across all evaluation metrics. Compared with FIT, TFS outperforms FIT in terms of the average predicted change metric in the global evaluation method. This difference may arise from the MIMIC III dataset's distinct data distribution, where the FIT algorithm effectively captures feature changes and simulates the removal of features at fixed moments. However, when considering the actual algorithm runtime, FIT takes significantly longer time due to the large number of features in this dataset. It requires approximately 15 h to generate the missing features using the generator, whereas the TFS algorithm runs faster (approximately 2.5 h) without this step. Consequently, our algorithm achieves comparable experimental results to FIT while running more efficiently.

Table 2. Experimental results on MIMIC III data.

Criteria	PERFORMANCE (TOP 5%)		PERFORMANCE (K = 50)	
Method	AUC DROP	PRED.CHANGE	AUC DROP	PRED.CHANGE
TFS	0.054 ± 0.002	$\mathbf{0.06 \pm 0.002}$	0.068 ± 0.009	0.064 ± 0.002
FIT	$\mathbf{0.066 \pm 0.007}$	0.056 ± 0.003	$\mathbf{0.073 \pm 0.007}$	$\mathbf{0.067 \pm 0.003}$
AFO	0.049 ± 0.009	0.05 ± 0.002	0.066 ± 0.007	0.06 ± 0.003
FO	0.047 ± 0.004	0.045 ± 0.003	0.067 ± 0.008	0.057 ± 0.004
Deep-Lift	0.025 ± 0.002	0.031 ± 0.002	0.021 ± 0.003	0.029 ± 0.002
RETAIN	0.020 ± 0.014	0.022 ± 0.0019	0.032 ± 0.019	0.036 ± 0.002
GradSHAP	0.022 ± 0.004	0.03 ± 0.002	0.021 ± 0.004	0.027 ± 0.002

5   Conclusion

For time series data, existing model interpretation methods designed for image data exhibit poor performance due to the absence of dynamics. Similarly, interpretation methods designed for time series data overlook the correlation between the features of interest and the remaining features. To address these limitations, this paper proposes a novel model interpretation method called TFS, specifically designed for time-series neural networks. TFS incorporates the concept of Shapley values and considers the collective influence of different features on predictions at each moment. To evaluate its effectiveness, we conducted experiments on synthetic and real datasets, comparing TFS with several existing methods. The experimental results show that considering feature correlations when removing features can improve the accuracy of the interpretation results to some extent, while also making them less susceptible to data quality. Furthermore, when compared to FIT, TFS achieves comparable performance on real data while demonstrating superior computational efficiency. Future research work can consider not only the correlation between features when interpreting, but also the problem of feature continuity in time series into the XAI algorithm.

References

1. Yuan, Y., Zhou, X., Pan, S., Zhu, Q., Song, Z., Guo, L.: A relation-specific attention network for joint entity and relation extraction. In: IJCAI. vol. 2020, pp. 4054–4060 (2020)
2. Ho, N.H., Yang, H.J., Kim, S.H., Lee, G.: Multimodal approach of speech emotion recognition using multi-level multi-head fusion attention-based recurrent neural network. IEEE Access **8**, 61672–61686 (2020)
3. Caruana, R., Lou, Y., Gehrke, J., Koch, P., Sturm, M., Elhadad, N.: Intelligible models for healthcare: Predicting pneumonia risk and hospital 30-day readmission. In: Proceedings of the 21th ACM SIGKDD International Conference on Knowledge Discovery and Data Mining, pp. 1721–1730 (2015)
4. Zhang, Z., Xie, Y., Xing, F., McGough, M., Yang, L.: MDNet: a semantically and visually interpretable medical image diagnosis network. In: Proceedings of the IEEE Conference on Computer Vision and Pattern Recognition, pp. 6428–6436 (2017)
5. Wang, D., Quek, C., Ng, G.S.: Bank failure prediction using an accurate and interpretable neural fuzzy inference system. AI Commun. **29**(4), 477–495 (2016)
6. Kim, J., Rohrbach, A., Darrell, T., Canny, J., Akata, Z.: Textual explanations for self-driving vehicles. In: Ferrari, V., Hebert, M., Sminchisescu, C., Weiss, Y. (eds.) ECCV 2018. LNCS, vol. 11206, pp. 577–593. Springer, Cham (2018). https://doi.org/10.1007/978-3-030-01216-8_35
7. Linardatos, P., Papastefanopoulos, V., Kotsiantis, S.: Explainable AI: a review of machine learning interpretability methods. Entropy **23**(1), 18 (2020)
8. Ismail, A.A., Gunady, M., Corrada Bravo, H., Feizi, S.: Benchmarking deep learning interpretability in time series predictions. Adv. Neural. Inf. Process. Syst. **33**, 6441–6452 (2020)
9. Serrano, S., Smith, N.A.: Is attention interpretable? arXiv preprint arXiv:1906.03731 (2019)
10. Zeiler, M.D., Fergus, R.: Visualizing and understanding convolutional networks. In: Fleet, D., Pajdla, T., Schiele, B., Tuytelaars, T. (eds.) ECCV 2014. LNCS, vol. 8689, pp. 818–833. Springer, Cham (2014). https://doi.org/10.1007/978-3-319-10590-1_53
11. Tonekaboni, S., Joshi, S., Campbell, K., Duvenaud, D.K., Goldenberg, A.: What went wrong and when? Instance-wise feature importance for time-series black-box models. Adv. Neural. Inf. Process. Syst. **33**, 799–809 (2020)
12. Petsiuk, V., Das, A., Saenko, K.: Rise: randomized input sampling for explanation of black-box models. arXiv preprint arXiv:1806.07421 (2018)
13. Sundararajan, M., Taly, A., Yan, Q.: Axiomatic attribution for deep networks. In: International Conference on Machine Learning, pp. 3319–3328. PMLR (2017)
14. Selvaraju, R.R., Cogswell, M., Das, A., Vedantam, R., Parikh, D., Batra, D.: Grad-cam: visual explanations from deep networks via gradient-based localization. In: Proceedings of the IEEE International Conference on Computer Vision, pp. 618–626 (2017)
15. Montavon, G., Lapuschkin, S., Binder, A., Samek, W., Müller, K.R.: Explaining nonlinear classification decisions with deep Taylor decomposition. Pattern Recogn. **65**, 211–222 (2017)
16. Shrikumar, A., Greenside, P., Kundaje, A.: Learning important features through propagating activation differences. In: International Conference on Machine Learning, pp. 3145–3153. PMLR (2017)

17. Erion, G., Janizek, J.D., Sturmfels, P., Lundberg, S.M., Lee, S.I.: Learning explainable models using attribution priors (2019)
18. Lundberg, S.M., Lee, S.I.: A unified approach to interpreting model predictions. In: Advances In Neural Information Processing Systems, vol. 30 (2017)
19. Bach, S., Binder, A., Montavon, G., Klauschen, F., Müller, K.R., Samek, W.: On pixel-wise explanations for non-linear classifier decisions by layer-wise relevance propagation. PLoS ONE **10**(7), e0130140 (2015)
20. Kindermans, P.J., et al.: Learning how to explain neural networks: Patternnet and pattern attribution. arXiv preprint arXiv:1705.05598 (2017)
21. Zhang, J., Bargal, S.A., Lin, Z., Brandt, J., Shen, X., Sclaroff, S.: Top-down neural attention by excitation backprop. Int. J. Comput. Vision **126**(10), 1084–1102 (2018)
22. Ismail, A.A., Gunady, M., Pessoa, L., Corrada Bravo, H., Feizi, S.: Input-cell attention reduces vanishing saliency of recurrent neural networks. In: Advances in Neural Information Processing Systems, vol. 32 (2019)
23. Bento, J., Saleiro, P., Cruz, A.F., Figueiredo, M.A., Bizarro, P.: Timeshap: explaining recurrent models through sequence perturbations. In: Proceedings of the 27th ACM SIGKDD Conference on Knowledge Discovery & Data Mining, pp. 2565–2573 (2021)
24. Vaswani, A., et al.: Attention is all you need. In: Advances in Neural Information Processing Systems, vol. 30 (2017)
25. Choi, E., Bahadori, M.T., Sun, J., Kulas, J., Schuetz, A., Stewart, W.: Retain: an interpretable predictive model for healthcare using reverse time attention mechanism. In: Advances In Neural Information Processing Systems, vol. 29 (2016)
26. Song, H., Rajan, D., Thiagarajan, J., Spanias, A.: Attend and diagnose: clinical time series analysis using attention models. In: Proceedings of the AAAI Conference on Artificial Intelligence, vol. 32 (2018)
27. Guo, T., Lin, T., Lu, Y.: An interpretable LSTM neural network for autoregressive exogenous model. arXiv preprint arXiv:1804.05251 (2018)
28. Suresh, H., Hunt, N., Johnson, A., Celi, L.A., Szolovits, P., Ghassemi, M.: Clinical intervention prediction and understanding using deep networks. arXiv preprint arXiv:1705.08498 (2017)
29. Crabbé, J., Van Der Schaar, M.: Explaining time series predictions with dynamic masks. In: International Conference on Machine Learning, pp. 2166–2177. PMLR (2021)
30. Covert, I.C., Lundberg, S., Lee, S.I.: Explaining by removing: a unified framework for model explanation. J. Mach. Learn. Res. **22**(1), 9477–9566 (2021)
31. Shapley, L.S., et al.: A value for n-person games (1953)
32. Štrumbelj, E., Kononenko, I.: Explaining prediction models and individual predictions with feature contributions. Knowl. Inf. Syst. **41**, 647–665 (2014)
33. Schlegel, U., Arnout, H., El-Assady, M., Oelke, D., Keim, D.A.: Towards a rigorous evaluation of XAI methods on time series. In: 2019 IEEE/CVF International Conference on Computer Vision Workshop (ICCVW), pp. 4197–4201. IEEE (2019)
34. Johnson, A.E., et al.: MIMIC-III, a freely accessible critical care database. Sci. Data **3**(1), 1–9 (2016)
35. Kokhlikyan, N., et al.: Captum: a unified and generic model interpretability library for PyTorch. arXiv preprint arXiv:2009.07896 (2020)

K-PropNet: Knowledge-Enhanced Hybrid Heterogeneous Homogeneous Propagation Network for Recommender System

Fenghang Li⬡, Chunyang Ye(✉)⬡, Keqi Li⬡, Yongyue Yang⬡, and Hui Zhou⬡

Hainan University, Haikou, HN 570203, China
{fhli,cyye,kqli,yyy,zhouhui}@hainanu.edu.cn

Abstract. In order to address the cold start problem and enhance model interpretability, recommender systems commonly incorporate knowledge graphs as supplementary information. However, existing knowledge graph-based recommendation methods primarily focus on exploring users' deeper-level preferences or establishing item associations based on implicit knowledge graph information. Unfortunately, these approaches tend to overlook the crucial interactions between users and items, resulting in the underutilization of the rich information available from both parties. To overcome these challenges, this paper introduces a recommender model called the K-PropNet: Knowledge-Enhanced Hybrid Heterogeneous Homogeneous Propagation Network for Recommender System, which combines heterogeneous and homogeneous propagation techniques. Heterogeneous propagation leverages explicit user-item interaction information to broaden user interests, while homogeneous propagation employs convolutional neural networks to incorporate the structural and semantic information inherent to items themselves. Furthermore, the model utilizes two neural networks with attention-like mechanisms to discern the respective contributions of various neighbors in the two propagation paths within the knowledge graph. Ultimately, the information from both heterogeneous and homogeneous propagation is integrated to make the final prediction. The effectiveness of the proposed K-PropNet model is evaluated on four public datasets, demonstrating a significant performance improvement over state-of-the-art baseline methods.

Keywords: Recommender Systems · Heterogeneous Propagation · Homogeneous Propagation · Knowledge Graph

1 Introduction

Recommender systems have become a solution to tackle information overload, with Collaborative Filtering (CF) [8] being a popular approach. However, CF-based methods encounter challenges like the cold start problem and sparse user-item interactions [23], which can result in reduced recommendation quality. To

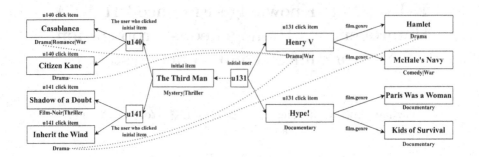

Fig. 1. An example of KG spreading in a movie scene, labels like "u131" representing user ID in KG.

address these challenges, knowledge graphs (KGs) have been integrated into recommender systems [5], enhancing their ability to understand users' interests and behaviors and leading to more accurate recommendations.

In recent years, numerous approaches have emerged that aim to enhance user preferences by leveraging the structural relationships between head and tail entities in knowledge graphs [11,15]. These methods iteratively propagate along the links in the knowledge graph, with the goal of unveiling users' latent interests in potential items. By incorporating these latent interests as auxiliary information for recommendations, the objective is to enhance recommendation accuracy. However, these approaches often overlook the valuable insights hidden within user-item interactions, leading to a limited exploration of the interaction information between users and items.

Let's illustrate this issue with an example in the movie domain, as shown in Fig. 1. Suppose User u131 has previously clicked on the movies "Henry V" and "Hype!". By propagating the "film.genre" information from the knowledge graph triplets, we obtain four movies (Hamlet, McHale's Navy, Paris Was a Woman, Kids of Survival) that share the genres "Drama," "War,", "Documentary" and "Comedy". However, we cannot confidently determine u131's preference for movies in the Mystery or Thriller category, and we fail to predict u131's next click on the movie "The Third Man" in the Mystery and Thriller genre.

To address the aforementioned challenges, we propose K-PropNet (Hybrid Heterogeneous Homogeneous Propagation Network for Recommender System), which combines heterogeneous and homogeneous propagation techniques to effectively explore explicit interaction information between users and items. In the heterogeneous propagation phase, we leverage explicit interaction signals and integrate them with the knowledge graph to expand users' interests within the graph. To capture diverse contributions from different neighbors, we employ a neural network with knowledge attention mechanisms to learn weighted contributions in a biased manner. For instance, in Fig. 1, the movie "The Third Man" was clicked by users u140 and u141, who have shown interest in movies with similar genres to the four movies related to user u131. By exploring such user-item interactions, we enhance the accuracy of predicting user u131's likelihood of clicking on "The Third Man".

However, we acknowledge that relying solely on interaction signals may overlook certain semantic and structural characteristics of the initial items. Therefore, in the homogeneous propagation phase, we directly aggregate the initial items with the user interests obtained after expanding them within the knowledge graph. By selectively aggregating the neighborhood information of the initial items, we aim to learn their semantic and structural information in a biased manner. Finally, we aggregate the embeddings obtained from both heterogeneous and homogeneous propagation to derive the final user embeddings and item embeddings, which are utilized for the ultimate prediction task.

We conducted comprehensive experiments to evaluate the effectiveness of K-PropNet in four real-world recommendation scenarios: music, book, movie, and restaurant recommendations. The experimental results revealed significant improvements in both the AUC (Area Under the Curve) [21] and ACC (Accuracy) [18] metrics compared to the latest baseline methods for the music, book, and movie datasets. Specifically, our model achieved an average AUC improvement of 1.07%, 2.13%, and 0.54%, as well as an ACC improvement of 1.99%, 2.41%, and 0.94% on the music, books, and MovieLens-1M datasets, respectively.

To summarize, this paper has made the following key contributions: First, we propose K-PropNet, a novel recommendation approach that combines the strengths of heterogeneous and homogeneous propagation techniques. By effectively leveraging interaction signals and integrating it with the knowledge graph, K-PropNet captures explicit interaction signals between users and items. Additionally, it incorporates the semantic and structural information of initial items through convolutional neural networks in the knowledge graph. The aggregation of embeddings from these two pathways enhances the accuracy of the final predictions. Second, we introduce two neural networks inspired by attention mechanisms in the propagation processes. These networks effectively handle the contributions of different neighbors during knowledge graph propagation, ensuring that both heterogeneous and homogeneous paths are adequately considered. By taking into account the respective impacts of these paths, K-PropNet significantly improves the overall recommendation performance.

The rest of this paper is organized as follows: Section 2 reviews literature on knowledge graph for recommender systems. Section 3 presents our K-PropNet model. Section 4 evaluates our proposal using extensive experiments. Section 5 concludes the work and highlights some future research directions.

2   Related Work

To mitigate the issues of data sparsity and the cold start problem in collaborative filtering (CF), the integration of knowledge graphs as auxiliary information in recommendation systems has gained significant importance. Knowledge graph-based recommendation methods can be broadly categorized into embedding-based methods, path-based methods, and propagation-based methods.

Embedding-based approaches (KGE) [6] encode knowledge graph information into low-dimensional vector embeddings to improve recommender models. Traditional KGE methods, including TransE [1], TransH [16], TransR [7], and TransD [4], DistMult [20] and RESCAL [10], emphasize semantic matching or translation distance. However, these methods often overlook the interconnectedness of information in the knowledge graph, making them more suitable for tasks like link prediction rather than recommender systems.

Path-based methods, on the other hand, focus on the multiple connections between entities in the knowledge graph. These methods require algorithms to select the most relevant paths or define meta-paths according to the specific recommendation context, enabling propagation along these paths. However, this approach is limited by domain knowledge, as a deep understanding of the recommendation domain is needed to design appropriate meta-paths. Moreover, manual path design can be time-consuming and complex, particularly in complex recommendation scenarios involving intricate knowledge graphs, resulting in significant time and human resource costs.

To overcome the limitations of embedding-based and path-based methods, Wang et al. introduced RippleNet [11] as a propagation-based recommendation model specifically designed for click-through rate prediction. RippleNet expands users' latent interests iteratively within the knowledge graph, enabling preference distributions over candidate items. However, RippleNet lacks consideration for the varying contributions of different neighbors during propagation and exhibits increased computational complexity in large-scale knowledge graphs, leading to potential memory overhead and computational costs.

To address these challenges, Wang et al. proposed KGCN [14] and KGNN-LS [12], which utilize convolutional neural networks to capture item correlations and integrate neighbor information for recommendation systems. However, these models overlook the explicit interactions between users and items within the entire system and fail to fully exploit the item information. To address this limitation, CKAN [15] presents an approach that leverages collaborative information to fully harness the explicit interaction information between users and items. However, by replacing the initial item with the collaborative information set, the model loses the semantic and structural information associated with the initial item during the subsequent propagation process, thereby impacting the accuracy of the recommendation results.

3   K-PropNet

3.1   Overview

In this section, we present a detailed description of the framework of the proposed K-PropNet model, as depicted in Fig. 2. The K-PropNet model comprises two primary components: heterogeneous propagation and homogeneous propagation. In the heterogeneous propagation component, we leverage the interaction information between users and items and utilize the knowledge graph to propagate user interests. On the other hand, in the homogeneous propagation

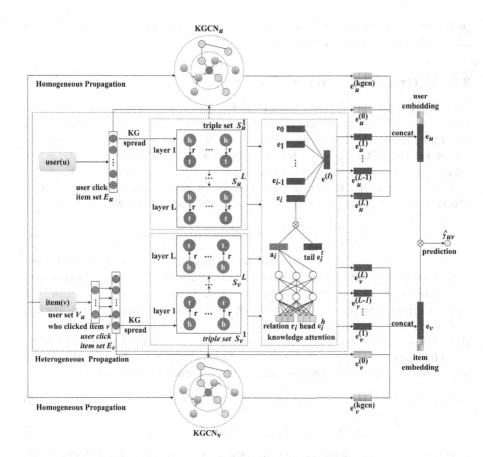

Fig. 2. The overall framework of the K-PropNet.

component, our focus is on learning the semantic and structural information of the initial item's neighborhood within the knowledge graph. These two propagation mechanisms are subsequently aggregated using separate neural networks that incorporate attention-like mechanisms. This aggregation process generates the final user embedding and item embedding, which are utilized for prediction tasks aimed at determining the probability of users expressing interest in items.

In the following sections, we will provide a detailed description of the underlying principles of these two components. To ease the presentation, we define a group of m users as $U = \{u_1, u_2, \ldots, u_m\}$, and a set of n items as $V = \{v_1, v_2, \ldots, v_n\}$. We define an interaction matrix, $Y \in R^{m \times n}$, based on the implicit feedback between users and items, where $y_{uv} = 1$ indicates the presence of implicit interaction between user u and item v. The knowledge graph is defined as G, which comprises triplets (h, r, t) representing the relationships among head entities, relations, and tail entities, $\{(h, r, t) \mid h, t \in E, r \in R\}$. We use $A = \{(v, e) \mid v \in V, e \in E\}$ to establish the correspondence between entities e in the knowledge graph and items v. Our mainly goal is to learn the prediction function $\hat{y}_{uv} = F(u, v \mid \theta, Y, G)$, which calculates the probability of user u showing interest in item v.

3.2 Heterogeneous Propagation

Users' preferences are inclined towards items with which they have had previous interactions. Hence, we utilize the historical explicit interactions between users and items to obtain the corresponding set of items, which serves as a representation for the initial user. In this way, we effectively leverage the interaction information between users and items. Specifically, as illustrated in Eq. 1, we transform user u into the initial item set E_u for heterogeneous propagation by utilizing the user's historical interaction set A within the knowledge graph.

$$E_u = \{e \mid (v, e) \in A \text{ and } v \in \{v \mid y_{uv} = 1\}\} \tag{1}$$

Similarly, for item v, users who have previously interacted with it in the historical interaction set A demonstrate similar interaction behaviors towards that item. Consequently, we can represent the characteristics of item v by using the corresponding user set V_u. Subsequently, employing the same transformation, each user in the V_u set is converted into the corresponding item set based on their historical interaction set A. These sets are then merged to obtain the E_v set. This approach allows us to utilize the E_v set as a representation of the initial item v for heterogeneous propagation within the knowledge graph.

$$V_u = \{v_u \mid u \in \{u \mid y_{uv} = 1\} \text{ and } y_{uv_u} = 1\} \tag{2}$$

$$E_v = \{e \mid (v_u, e) \in A \text{ and } v_u \in V_u\} \tag{3}$$

In heterogeneous propagation, we utilize these initial entity sets, which effectively capture the interactions between users and items, to represent the initial user u and item v. This approach enables us to enhance the information of users and items, thereby improving the performance of the recommendation system.

KG Spread. During KG propagation, the strong correlations between neighboring entities in the KG are leveraged. The items within the E_u and E_v sets are propagated along KG pathways individually to generate expanded entity sets and their corresponding triple sets. The contributions of these sets are influenced by the propagation distance, enabling an effective expansion of the embedding representations for users and items. The propagation process is outlined below:

$$E_o^l = \{t \mid (h, r, t) \in G \text{ and } h \in E_o^{l-1}\}, l = 1, 2, \ldots, L \tag{4}$$

$$S_o^l = \{(h, r, t) \mid (h, r, t) \in G \text{ and } h \in E_o^{l-1}\}, l = 1, 2, \ldots, L \tag{5}$$

Here, the subscript "o" acts as a placeholder for the symbols "u" or "v", and the subsequent subscripts "o" in the article have the same significance. By incorporating KG propagation as auxiliary information in the model, we can effectively utilize the KG to expand the latent preferences of users and the characteristics of items, as illustrated in Fig. 2. The entity sets obtained through heterogeneous propagation resemble ripples on the water's surface, propagating layer by layer along the pathways in the KG, from close proximity to further distances. This ripple-like propagation allows us to deeply extend user preferences within the KG, capturing high-order interaction information between users and items based on the KG. As a result, users and items are more accurately represented as embedding vectors in the model.

Knowledge Attention. During KG propagation, when different head entities and relations lead to the same tail entity, their meanings and embedding representations can vary. For example, films like "Forrest Gump" and "Cast Away" share many commonalities when viewed from the perspective of directors, actors, or countries, resulting in a high degree of similarity between them. However, when considering movie genres and ratings, they may not exhibit the same level of similarity. Therefore, it is not sufficient to simply aggregate the obtained triplets from KG propagation, as they carry diverse weights. To address this issue, we introduce knowledge attention to assign distinct attention weights to tail entities, enabling the differentiation of their connections with various head entities and relations, and attributing them with distinct meanings.

Let (h, r, t) be the set of triplets for the i-th triplet in the l-th layer, we utilize the following approach to generate the attention weight a_i for the tail entity.

$$a_i = \pi(e_i^h, r_i) \tag{6}$$

where e_i^h and r_i represent the embedding representations of the head entity and the relation in the triplet. We employ the function $\pi(e_i^h, r_i)$ to regulate the association between the head entity and the tail entity, thereby influencing the computed attention weight a_i by the head entity. The function $\pi()$ is defined using a neural network architecture similar to an attention mechanism, and its specific formulation is as follows.

$$\pi(e_i^h, r_i) = \sigma(w_2 ReLU(w_1 z_0 + b_1) + b_2) \tag{7}$$

$$z_0 = ReLU(w_0 [e_i^h, r_i]_{cat} + b_0) \tag{8}$$

We employ the ReLU layer as the non-linear activation function and utilize the Sigmoid function as the activation function in the neural network's final layer. The concatenation of e_i^h and r_i is denoted as $[e_i^h, r_i]_{cat}$. The trainable weights and biases are represented by w and b, respectively, with their subscripts indicating the corresponding layers in the neural network. Ultimately, we normalize the attention weight coefficients of the triplets using the softmax function.

$$\widetilde{\pi}\left(e_i^h, r_i\right) = \frac{\exp\left(\pi\left(e_i^h, r_i\right)\right)}{\sum_{(h',r',t')\in\mathcal{S}_o^l} \exp\left(\pi\left(e_i^{h'}, r_i'\right)\right)} \tag{9}$$

Based on the attention weights, we can identify the adjacent tail entities that require more attention, thereby effectively capturing the interrelationships within the KG. Through a dot product operation between the attention weights and the corresponding tail entities, we obtain the embedding representation e_i for the current triplet. Based on the embedding of e_i, we derive the embedding representation $e_u^{(l)}$ or $e_v^{(l)}$ for the triplets in the l-th layer of user u or item v.

$$e_i^{(o)} = a_i e_i^t \qquad (10)$$

$$e_o^{(l)} = \sum_{i=1}^{|\mathcal{S}_o^l|} e_i^{(o)}, l = 1, 2, \ldots, L \qquad (11)$$

Considering that the entities in the initial entity set bear the closest resemblance to the original representations of users and items, and thus possess a strong correlation with the initial user and item, we augment the embeddings of the initial user and item by including the embedding representations of users and items from the initial entity set. This augmentation effectively amplifies the impact of the initial information throughout the entire recommendation process.

$$e_o^{(0)} = \frac{\sum_{e \in E_o^0} e}{|E_o^0|} \qquad (12)$$

3.3 Homogeneous Propagation

During heterogeneous propagation, we replace the propagation of initial items and users within the KG with collaborative information sets derived from the interactions between users and items. However, in doing so, we unintentionally overlook the semantic information embedded in the initial items themselves, as well as their structural characteristics within the KG triplets. Figure 3 illustrates an example in the movie domain where the initial item "Die Haed 2" is initially categorized as Action and Thriller, undergoes heterogeneous propagation and is likely to be misidentified as Comedy and Children's genres. This deviation from the users' genuine expectations ultimately leads to suboptimal recommendations.

To amplify the influence of the initial items on the entire model, we perform convolutions on the initial items using the S_u^l and S_v^l sets derived from KG propagation in homogeneous propagation, where $l = 1, 2, \ldots, L$. By conducting separate convolutions on each layer of user and item sets, we effectively capture the semantic and structural information of the initial items and transform it into embedding representations for model predictions. This compensates for the information loss encountered in heterogeneous propagation. Consequently, the impact of the initial items is better incorporated into the model and leveraged during the prediction phase of the recommendation process.

To capture the higher-order semantic and structural information of the initial items within the KG, we start by retrieving the neighborhood information of the initial items. This includes entity information e_v and relation information r_v. Since there is a correspondence between entities and items, we perform convolution operations on the entity nodes ho within the S_u^l and S_u^l sets and the neighborhood information of the items.

$$I_{r_v}^{h_o} = g(h_o, r_v) \qquad (13)$$

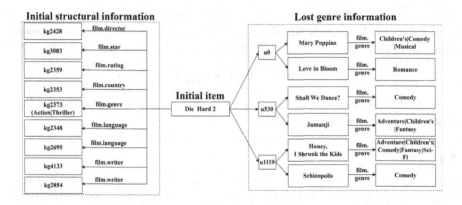

Fig. 3. Loss of semantic structural information in heterogeneous propagation. Using movie genres as an example, the label "kg2373" on the left side represents the genre encoding of "Die Hard 2" within the knowledge graph.

The function $g(a, b)$ represents the dot product between a and b. In this case, h_o acts as a unified representation of the entity nodes h_u and h_v within the S_u^l and S_v^l sets. We employ the function g to calculate the scores between entity nodes and neighborhood relation information, thereby determining the significance of the relation r_v to the entity node h_o. Through these convolution operations, we can leverage the neighborhood information of the initial items to explore their semantic and structural characteristics within the KG. Consequently, the influence of the initial items can be better incorporated into the model and utilized during the prediction phase of the recommendation process.

$$\widetilde{I}_{r_v}^{h_o} = \frac{\exp\left(g(h_o, r_v)\right)}{\sum_{(h_o' \in S_o^l, r_v' \in S_v^{orgin})} \exp\left(g(h_o', r_v')\right)} \tag{14}$$

$$V_{N(v)}^{h_o} = \sum_{e_v \in N(v)} g(\widetilde{I}_{r_v}^{h_o}, e_v) \tag{15}$$

We normalize the scores $I_{r_v}^{h_o}$ between the entity h_o and relation r_v by applying the softmax function. Then, we integrate the normalized scores with the entity's neighborhood information e_v using the function g to obtain the set of topological neighborhood nodes $V_{N(v)}^{h_o}$. Here, S_v^{orgin} represents the collection of KG neighborhood triplets for the initial item, and $N(v)$ represents the set of KG neighborhood entity information for the initial item. Since the size of $N(v)$ can vary significantly in different practical scenarios, we ensure computational consistency by uniformly sampling the entity's neighborhood. Specifically, we denote the neighborhood of entity e as $v_{s(v)}^{h_o}$, where $S(v)$ is defined as follows:

$$S(v) = \{e \mid e \in N(v), \mid S(v) \mid = 2\} \tag{16}$$

The condition $\mid S(v) \mid = 2$ indicates that we choose the 2-th layer neighborhood of the entity as the aggregated neighborhood. Subsequently, we aggregate the entity representation v with its corresponding neighborhood representation $V_{S(v)}^{h_o}$. This aggregation involves vector addition and a subsequent non-linear transformation, resulting in the final embedding $e_o^{(kgcn)}$. The process is represented by Eq. 17:

$$e_o^{(kgcn)} = \sigma(W^{(kgcn)}(v + v_{N(v)}^{h_o}) + b^{(kgcn)})$$ (17)

where $W^{(kgcn)}$ and $b^{(kgcn)}$ represent the weight and bias parameters of the fully connected layer, respectively.

3.4   Forecast

Based on the preceding calculations, we aggregate the obtained embeddings, which include $e_o^{(kgcn)}$, $e_o^{(0)}$, and $e_o^{(l)}$, $(l = 1, 2, \ldots, L)$, using Eq. 18 to derive the ultimate user embedding (e_u) and item embedding (e_v). Subsequently, we predict the likelihood of a user's potential interest in an item by computing the inner product between the user embedding (e_u) and the item embedding (e_v).

$$e_o = \sigma(W_o(e_o^{kgcn} \mid\mid e_o^{(0)} \mid\mid e_o^{(1)} \mid\mid e_o^{(2)} \mid\mid \ldots \mid\mid e_o^{(L)}) + b_o)$$ (18)

$$\hat{y}_{uv} = e_u^\top e_v$$ (19)

3.5   Loss Function

The K-PropNet model's loss includes RS module loss, Het module loss, and Hom module loss, as defined below:

$$Loss = L_{RS} + L_{Het} + L_{Hom}$$ (20)

$$L_{RS} = \sum_{u \in U, v \in V} \mathcal{F}(\hat{y}_{u'v'}, y_{uv})$$ (21)

$$L_{Het} = \frac{\lambda_{l2}}{2}\left(\sum_{(h,r,t) \in S_o^l} (\|\,h\,\|_2^2 + \|\,r\,\|_2^2 + \|\,t\,\|_2^2)\right)$$ (22)

$$L_{Hom} = \frac{\lambda_{l2}}{4}\left(\sum_{h_o \in S_o^l} (\|\,h_o\,\|_2^2 + \|\,V_{N(v)}^{h_o}\,\|_2^2)\right)$$ (23)

Here, $\mathcal{F}$ denotes binary cross-entropy for recommendation, the Het module loss involves λ_{l2}-controlled l_2 regularization for heterogeneous propagation, and the Hom module loss integrates regularization from homogeneous propagation.

4   Evaluation

In this section, we conducted experiments using four real-world recommendation scenarios to answer the following research questions (RQ):

RQ 1: How does the performance of the K-PropNet model compare to state-of-the-art knowledge graph-based recommendation methods?

RQ 2: What is the effectiveness of the homogeneous propagation and heterogeneous propagation modules within the overall model?

4.1 Experimental Setup

Datasets. To assess the effectiveness of K-PropNet, we chose four public datasets that exhibit varying degrees of sparsity across the domains of music, books, movies, and restaurants. The statistics of these datasets are shown in Table 1. To construct the knowledge graph (KG) for the MovieLens-1M, Last.FM, and Book-Crossing datasets, we followed the approaches employed in related works such as [12–14] and utilized Microsoft's Satori to create the subgraph. The knowledge graph construction for the Dianping-Food dataset was performed using an internal toolkit provided by Meituan-Dianping Group. It is important to note that due to the high sparsity of the Last.FM and Book-Crossing data, we did not set any threshold for data filtering.

Table 1. Basic statistics settings for the four datasets.

Dataset	users	items	interactions	relations	KG triples
Last.FM	1872	3846	42346	60	15518
Book-Crossing	17860	14967	139746	25	151500
Movie-1M	6036	2347	753772	7	20195
Dianping-Food	2298698	1362	23416418	7	160519

Baseline. We conducted a performance comparison of our K-PropNet model with several state-of-the-art methods, including RippleNet [11], MKR [13], KGCN [14], CKAN [15], KGARA [24], HKC [19], and EMKR [2]. The hyperparameter settings for these methods were aligned with the configurations reported in their respective original papers or official source code.

Parameters. The datasets from the four scenarios were randomly split into training, validation, and testing sets in a ratio of 6:2:2. In the context of CTR prediction [17,22], our models were trained on the training set and evaluated on the testing set. AUC [21] and ACC [18] were employed as performance metrics to assess the quality of the CTR prediction models. Hyperparameters were optimized based on the AUC and ACC results obtained from the testing set. During model training, the learning rate was set to 2×10^{-3}, and the $l2$ regularization coefficient was set to 10^{-5}. The attention network in the homogeneous propagation comprised two layers, with the hidden layer dimension set to the embedding size. In the heterogeneous propagation, the attention network consisted of one layer. The Adam optimizer [3] was utilized for all models. Each experiment was repeated five times, and the best result was selected as the final outcome. If the error range exceeded 5×10^{-3} among the five experiments, an additional five experiments were conducted, and the results of ten experiments were averaged. The implementation of K-PropNet was based on Python 3.6, utilizing the PyTorch 1.3.1 and NumPy 1.19.2 environments.

4.2   Results

Performance Comparison (RQ1). As shown in Table 2, we have gathered and analyzed the results of CTR prediction across four datasets.

Table 2. The results of AUC and ACC in CTR prediction.

Model	Last.FM		Book-Crossing		MovieLens-1M		Dianping-Food	
	AUC	ACC	AUC	ACC	AUC	ACC	AUC	ACC
MKR	0.796	0.755	0.732	0.702	0.916	0.842	0.867	0.790
RippleNet	0.780	0.698	0.701	0.640	0.919	0.845	0.855	0.756
KGCN	0.799	0.726	0.692	0.632	0.911	0.838	0.847	0.764
CKAN	0.845	0.742	0.752	0.663	0.923	0.849	0.877	0.800
KGARA	0.715	0.648	0.678	0.621	0.901	0.829	0.837	0.759
HKC	0.805	0.748	0.741	0.697	0.922	0.848	/	/
EMKR	0.808	0.753	0.746	0.705	0.923	0.848	**0.890**	**0.814**
K-PropNet	**0.854**	**0.768**	**0.766**	**0.722**	**0.928**	**0.854**	**0.889**	0.812

Table 2 reveals notable improvements of models such as KGCN, HKC, EMKR, and CKAN compared to RippleNet, particularly on sparse datasets like music and book. This indicates the effectiveness of considering the contributions of different item neighbors in knowledge graph propagation for sparse datasets. Additionally, CKAN demonstrates notable improvement over KGCN, indicating that leveraging collaborative signals from both users and items to assist knowledge graph-based recommendations enhances the recommendation outcomes on sparse datasets. The HKC and EMKR models, which combine MKR and KGCN, also demonstrate noticeable performance improvements compared to using MKR or KGCN alone. This suggests that alternate learning with item neighborhood information surpasses using knowledge graph embeddings alone on sparse datasets, implying that aggregating item neighborhood information captures more valuable insights for recommendations. For datasets with richer user interactions like MovieLens-1M and Dianping-Food, CKAN, HKC, and EMKR models perform better due to the utilization of additional information, underscoring the importance of aggregating neighborhood information from the knowledge graph in recommendation models.

Overall, our K-PropNet model achieves the best performance on the music, book, and MovieLens-1M datasets. On the Dianping-Food dataset, our approach is comparable to the state-of-the-art EMKR method and outperforms other baseline models significantly, demonstrating the effectiveness of our approach on this dataset. Compared to the best baselines, K-PropNet yields an average AUC improvement of 1.07%, 2.13%, and 0.54%, as well as an ACC improvement of 1.99%, 2.41%, and 0.94% on the music, book, and MovieLens-1M datasets, respectively. On the Dianping-Food dataset, the differences are 0.1% and 0.2%

compared to the best baseline EMKR, which are statistically insignificant. This indicates that our model achieves a similar performance level to the best baseline on the Dianping-Food dataset. Compared to the overall performance of EMKR, our model significantly outperforms it. Thus, it can be concluded that our K-PropNet model demonstrates significant improvements over the state-of-the-art baselines across these four datasets.

Ablation Experiment (RQ2). To assess the impact of homogeneous and heterogeneous propagation modules, we conducted ablation experiments, with results presented in the following tables.

Table 3. Validation of the Effectiveness of Heterogeneous Propagation and Homogeneous Propagation through AUC.

Category	$K - \text{PropNet}_{w/oHet}$	$K - \text{PropNet}_{w/oHom}$	K-PropNet
music	0.845	0.828	**0.854**
book	0.750	0.743	**0.768**
movie	0.912	0.920	**0.928**
restaurant	0.878	0.879	**0.889**

Table 4. Validation of the Effectiveness of Heterogeneous Propagation and Homogeneous Propagation through ACC.

Category	$K - \text{PropNet}_{w/oHet}$	$K - \text{PropNet}_{w/oHom}$	K-PropNet
music	0.742	0.741	**0.768**
book	0.657	0.660	**0.722**
movie	0.843	0.845	**0.854**
restaurant	0.801	0.796	**0.812**

We compare the performance of the complete model (K-PropNet) with the models where either the heterogeneous propagation module is removed ($K - \text{PropNet}_{w/oHet}$) or the homogeneous propagation module is removed ($K - \text{PropNet}_{w/oHom}$) across four datasets in terms of AUC, ACC, and F1 [9]. The results in Tables 3, 4, and 5 clearly indicate that the complete model consistently outperforms the models without the heterogeneous propagation module and the models without the homogeneous propagation module on all datasets. This compelling evidence underscores the pivotal role of heterogeneous propagation, effectively harnessing collaborative information via explicit interaction signals between users and items, and seamlessly integrating them into the knowledge graph. Additionally, the incorporation of convolutional neural networks in homogeneous propagation modules to exploit semantic and structural information from initial items within the knowledge graph significantly contributes to elevating the holistic model performance. This accumulation of substantial evidence serves to firmly affirm the effectiveness of both modules.

Table 5. Validation of the Effectiveness of Heterogeneous Propagation and Homogeneous Propagation through F1.

Category	$K - PropNet_{w/oHet}$	$K - PropNet_{w/oHom}$	K-PropNet
music	0.771	0.760	**0.791**
book	0.670	0.661	**0.683**
movie	0.845	0.848	**0.857**
restaurant	0.800	0.784	**0.818**

5 Conclusions

This paper presents the K-PropNet model, which integrates the features of both heterogeneous propagation and homogeneous propagation. K-PropNet utilizes explicit interaction information between users and items to facilitate heterogeneous propagation, expanding their interests within the knowledge graph. It employs a neural network with an attention-like mechanism to discern the contributions of different neighbors to users and items. Additionally, in the homogeneous propagation phase, K-PropNet leverages another neural network with a similar attention mechanism to selectively aggregate the extended interests and neighborhood information of initial items within the knowledge graph, thus enhancing the impact of initial items in recommendations. Extensive experiments demonstrate that K-PropNet outperforms the best baseline models significantly.

In the future, we plan to further enhance our approach by refining the handling of initial user and item information, as well as optimizing information propagation within the knowledge graph.

Acknowledgments. This work was supported in part by the National Key Research and Development Program of China (No. 2018YFB2100805), National Natural Science Foundation of China (No. 61962017), and the Key Research and Development Program of Hainan Province (No. ZDYF2020008 and No. ZDYF2022GXIS230).

References

1. Bordes, A., Usunier, N., Garcia-Duran, A., Weston, J., Yakhnenko, O.: Translating embeddings for modeling multi-relational data. In: Advances in Neural Information Processing Systems, vol. 26 (2013)
2. Gao, M., Li, J.Y., Chen, C.H., Li, Y., Zhang, J., Zhan, Z.H.: Enhanced multi-task learning and knowledge graph-based recommender system. IEEE Trans. Knowl. Data Eng. (2023)

3. Huang, F., Wu, X., Hu, Z.: AdaGDA: faster adaptive gradient descent ascent methods for minimax optimization. In: International Conference on Artificial Intelligence and Statistics, pp. 2365–2389. PMLR (2023)
4. Ji, G., He, S., Xu, L., Liu, K., Zhao, J.: Knowledge graph embedding via dynamic mapping matrix. In: Proceedings of the 53rd Annual Meeting of the Association for Computational Linguistics and the 7th International Joint Conference on Natural Language Processing (Volume 1: Long papers), pp. 687–696 (2015)
5. Khan, N., Ma, Z., Ullah, A., Polat, K.: Categorization of knowledge graph based recommendation methods and benchmark datasets from the perspectives of application scenarios: a comprehensive survey. Expert Syst. Appl., 117737 (2022)
6. Li, Z., Liu, H., Zhang, Z., Liu, T., Xiong, N.N.: Learning knowledge graph embedding with heterogeneous relation attention networks. IEEE Trans. Neural Netw. Learn. Syst. **33**(8), 3961–3973 (2021)
7. Lin, Y., Liu, Z., Sun, M., Liu, Y., Zhu, X.: Learning entity and relation embeddings for knowledge graph completion. In: Proceedings of the AAAI Conference on Artificial Intelligence, vol. 29 (2015)
8. Linden, G., Smith, B., York, J.: Amazon.com recommendations: item-to-item collaborative filtering. IEEE Internet Comput. **7**(1), 76–80 (2003)
9. Lipton, Z.C., Elkan, C., Narayanaswamy, B.: Thresholding classifiers to maximize F1 score. arXiv preprint arXiv:1402.1892 (2014)
10. Nickel, M., Tresp, V., Kriegel, H.P., et al.: A three-way model for collective learning on multi-relational data. In: ICML, vol. 11, pp. 3104482–3104584 (2011)
11. Wang, H., et al.: RippleNet: propagating user preferences on the knowledge graph for recommender systems. In: Proceedings of the 27th ACM International Conference on Information and Knowledge Management, pp. 417–426 (2018)
12. Wang, H., et al.: Knowledge-aware graph neural networks with label smoothness regularization for recommender systems. In: Proceedings of the 25th ACM SIGKDD International Conference on Knowledge Discovery & Data Mining, pp. 968–977 (2019)
13. Wang, H., Zhang, F., Zhao, M., Li, W., Xie, X., Guo, M.: Multi-task feature learning for knowledge graph enhanced recommendation. In: The World Wide Web Conference, pp. 2000–2010 (2019)
14. Wang, H., Zhao, M., Xie, X., Li, W., Guo, M.: Knowledge graph convolutional networks for recommender systems. In: The World Wide Web Conference, pp. 3307–3313 (2019)
15. Wang, Z., Lin, G., Tan, H., Chen, Q., Liu, X.: CKAN: collaborative knowledge-aware attentive network for recommender systems. In: Proceedings of the 43rd International ACM SIGIR Conference on Research and Development in Information Retrieval, pp. 219–228 (2020)
16. Wang, Z., Zhang, J., Feng, J., Chen, Z.: Knowledge graph embedding by translating on hyperplanes. In: Proceedings of the AAAI Conference on Artificial Intelligence, vol. 28 (2014)
17. Wu, C., Wu, F., Lyu, L., Huang, Y., Xie, X.: FedCTR: federated native ad CTR prediction with cross-platform user behavior data. ACM Trans. Intell. Syst. Technol. (TIST) **13**(4), 1–19 (2022)
18. Wu, L., He, X., Wang, X., Zhang, K., Wang, M.: A survey on accuracy-oriented neural recommendation: from collaborative filtering to information-rich recommendation. IEEE Trans. Knowl. Data Eng. (2022)
19. Xiaowang, G., Hongbin, X., Yuan, L.: Hybrid recommendation model of knowledge graph and graph convolutional network. J. Front. Comput. Sci. Technol. **16**(6), 1343 (2022)

20. Yang, B., Yih, W.T., He, X., Gao, J., Deng, L.: Embedding entities and relations for learning and inference in knowledge bases. arXiv preprint arXiv:1412.6575 (2014)
21. Yang, T., Ying, Y.: AUC maximization in the era of big data and AI: a survey. ACM Comput. Surv. **55**(8), 1–37 (2022)
22. Yang, Y., Zhai, P.: Click-through rate prediction in online advertising: a literature review. Inf. Process. Manag. **59**(2), 102853 (2022)
23. Yang, Y., Huang, C., Xia, L., Li, C.: Knowledge graph contrastive learning for recommendation. In: Proceedings of the 45th International ACM SIGIR Conference on Research and Development in Information Retrieval, pp. 1434–1443 (2022)
24. Zhang, Y., Yuan, M., Zhao, C., Chen, M., Liu, X.: Aggregating knowledge-aware graph neural network and adaptive relational attention for recommendation. Appl. Intell., 1–13 (2022)

Multi-patch Adversarial Attack for Remote Sensing Image Classification

Ziyue Wang, Jun-Jie Huang, Tianrui Liu, Zihan Chen, Wentao Zhao[✉],
Xiao Liu, Yi Pan, and Lin Liu

College of Computer Science, National University of Defense Technology, Changsha,
Hunan, China
{wangzy13,jjhuang,trliu,chenzihan21,wtzhao,liuxiao13a,panyi_jsjy,
liulin16}@nudt.edu.cn

Abstract. Deep Neural Networks (DNNs) have shown excellent image classification performance both in accuracy and efficiency. Therefore, it is of great value to deploy adversarial patch to protect critical facilities from DNNs-based scene classification in Remote Sensing Image (RSI). However, adversarial patch attack for RSI scene classification has not been investigated. The existing adversarial patch attack methods are designed for natural images and need to generate a single large adversarial patch which is of too large size to be physically feasible for RSI applications. In this paper, we propose a Multi-Patch Adversarial Attack (MPAA) method for RSI scene classification task. We propose to deploy multiple small adversarial patches on key locations and formulate the problem as a constrained optimization problem which jointly optimize patch locations and adversarial patches. The proposed MPAA takes a searching and optimization strategy to tackle it and consists of an Effective Location Selection module and a Patch Optimization module. Extensive experimental results on Aerial Image Dataset show that the proposed MPAA achieves 96.98% attacking success rate by using 16 small patches where each patch only occupies 0.0625% of the image size, which significantly outperforms other adversarial patch methods.

Keywords: Remote sensing image · Scene classification · Adversarial patch attack

1 Introduction

Deep Neural Networks (DNNs) have achieved superior performance on Remote Sensing Image (RSI) classification [6,19,25,36]. However, DNNs are vulnerable to adversarial attacks. The adversarial attack methods [7,22,24,30,31] generate adversarial examples by adding minor and dense adversarial noise over the benign image. Therefore, adversarial attack methods can be of great value in protecting critical facilities from DNNs-based RSI scene classification.

X. Song et al. (Eds.): APWeb-WAIM 2023, LNCS 14331, pp. 377–391, 2024.
https://doi.org/10.1007/978-981-97-2303-4_25

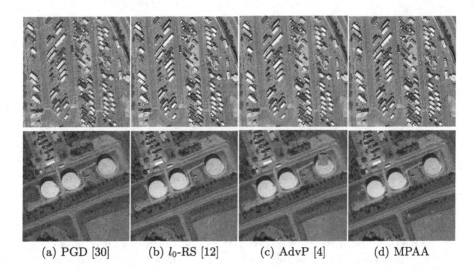

(a) PGD [30]   (b) l_0-RS [12]   (c) AdvP [4]   (d) MPAA

Fig. 1. Visualization of the adversarial examples generated by different methods. (a) PGD attack with default setting, (b) l_0 − RS changes 1000 pixels (0.27%), (c) AdvP uses 1% of image size to deploy a single patch, and (d) the proposed MPAA utilizes 16 small patches.

The existing RSI adversarial attack methods [1,5,33,34] mainly investigated effective approaches to deceive the DNNs-based RSI scene classification methods. Xu *et al.* [33] study the threat of adversarial examples on RSI scene classification and notice that the existing adversarial attack methods are also applicable on RSI scene classification. Though crafting RSI adversarial examples by adding adversarial noise is effective, it could be impractical to deploy the dense and additive adversarial noise in physical world.

In order to avoid the undeployable dense and additive adversarial noise, sparse adversarial attack methods [8,12,23] have been proposed. The l_0-norm is used to restrict the perturbation on as few pixels as possible. Specifically, Modas *et al.* [23] propose a geometry inspired sparse attack method and approximate the decision boundary as an affine hyperplane to compute the sparse perturbations. Around 2.7% pixels are modified to mislead the classifiers. Croce *et al.* [8] make the sparse perturbation more imperceptible by considering the color and edge changes. Hein *et al.* [12] propose to use Random Search (RS) strategy to optimize the sparse perturbation, termed as l_0 − RS. By designing specific sampling distributions, l_0 − RS method only need to change 0.1% to 0.3% pixels to fool the classifiers. But modifying 0.1% pixels also equals to hundreds of points changing in RSI, so the sparse attack approaches are also difficult to be physically implemented for RSI applications.

Brown *et al.* [4] propose adversarial patch attack (AdvP) which is a seminal physical world adversarial attack method by pasting an adversarial patch on the image to fool the DNNs-based image classifier. They do not limit the perturbation with l_2 or l_∞-norm, but restrict the perturbation to small area by using a binary mask matrix, so the attack can be deployed in physical world. The adversarial patch approaches have then been applied in different applications, including autonomous driving [28,29], pedestrian detection [14,15,18], RSI object detection [21,37], etc. There are advanced works [28,29] that utilize the adversarial patch to attack the autonomous driving system by pasting special stickers on traffic signs. The classifier can be successfully mislead by searching the suitable position to set stickers which may lead to severe safety accidents. Adversarial patch attack also has been applied on pedestrian detection [14,15,18] by placing the learned adversarial patch on clothes or hat to mislead the pedestrian detectors. In addition to improve the effectiveness of attack, Bao *et al.* [2] and Hu *et al.* [14] pay more attention on generating more natural looking adversarial patch by using Generative Adversarial Networks [3]. Adversarial patch has also been applied on RSI object detection [21,37]. To the best of our knowledge, there is no prior work that investigated physically realizable adversarial patches for RSI scene classification.

There are two challenges that hinder the direct application of adversarial patch methods for natural images on RSI scene classification: (i) It is more difficult to attack a RSI scene classifier since RSI typically contains a smaller number of scene categories than natural images; (ii) The adversarial patch for RSI should be as small as possible in order to have a small physical size when deploying on real scene. However, the existing adversarial patch methods are mainly designed for natural images, and need to use a relative large adversarial patch in order to ensure high attacking success rate. Therefore, for RSI applications, there is a dilemma that a small adversarial patch cannot ensure high attacking success rate, while a large adversarial patch can be difficult to deploy on real scene.

Figure 1 shows visual comparison results of different adversarial attack methods on two exemplar remote sensing images. Projected Gradient Descent (PGD) method, which is a traditional adversarial attack method, [30] generates adversarial examples with dense noise. The $l_0 - RS$ method [12] generates the most imperceptible examples, but it needs to change hundreds of single pixels, which still can not be deployed in real sense. AdvP method [4] uses a patch of 1% image size which has huge physical size (over 30×30 m^2) compared to the building. Therefore, the existing adversarial attack methods face difficult when deploying on remote sensing applications.

In this paper, we propose a novel Multi-Patch Adversarial Attack (MPAA) method for RSI scene classification to tackle the aforementioned problems. In order to achieve practically deployable adversarial attack, we propose to select n key locations on the remote sensing image and optimize n small adversarial patches to achieve robust adversarial attack, at the same time, keep the total patch area small.

The contribution of this work is mainly three-fold:

- We propose a novel Multi-Patch Adversarial Attack (MPAA) method for practically deployable RSI scene classification by optimizing multiple small patches rather than a single large one.

- We use a searching and optimizing strategy to generate the adversarial patches: an Effective Location Selection (ELS) module is first used to determine the patch locations according to the effectiveness map, a Patch Optimization (PO) module is then used to optimize the adversarial patches given the selected locations.
- From extensive experimental results, the proposed MPAA method achieves significantly higher attacking success rate when compared to other methods, and as a consequence of deploying multiple small adversarial patches, MPAA achieves less visually perceptible attack and is more practically feasible for RSI scene.

The rest of the paper is organized as follows: Sect. 2 briefly reviews the related work. Section 3 introduces our proposed Multi-Patch Adversarial Attack method. Section 4 presents experimental results, finally, Sect. 5 draws conclusions.

2   Related Work

2.1   Adversarial Attack on RSI

The existing adversarial attack methods for nature images can be transferred to the RSI applications. Burnel et al. [5] generate untargeted natural adversarial examples for remote sensing images based on Wasserstein Generative Adversarial Networks [1] and achieve high transferability over different DNN classification models. Xu et al. [34] use a surrogate model to extract the shallow feature of clean images and mix-up images, and generate universal adversarial examples in RSI dataset by adding perturbations to the benign images. These works can only apply in digital world and are impracticable in physical world.

More recent works propose to improve the practicality of attacks on RSI with object detection. Zhang et al. [37] find that the size of objects in RSI usually vary a great deal, so they propose to generate an universal adversarial patch that can adapt to multi-scale objects. They formulate a joint optimization problem to attack as many objects as possible and use a scale factor to adapt to objects with various size. Lian et al. [21] propose Adaptive-Patch-based Physical Attack (AP-PA) method for aerial detection to place the adversarial patch both on the object and outside the object. And there are still few works studying applying adversarial patch on RSI scene classification.

2.2   Adversarial Patch Attack

Brown et al. [4] propose the Adversarial Patch (AdvP) method to fool the classifier. The proposed AdvP generates an universal patch which can be pasted on almost all images by iterating in a continuous loop over different input benign images. But AdvP needs to utilize a patch of 5% image size which is too large for attacking RSI in reality. Fu et al. [10] prove the effectiveness of adversarial patch attack on vision transformers [9] and select the location of patch with guidance of corresponding saliency map termed as Patch Fool (PFool). Li et al. [20] propose

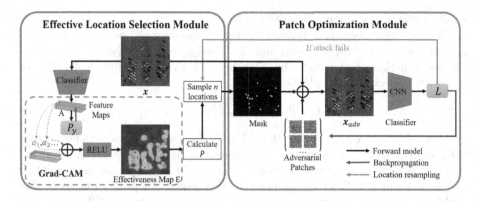

Fig. 2. Overview of the proposed Multi-Patch Adversarial Attack method for remote sensing images. In Effective Location Selection module, benign image x is firstly fed into the target classifier to calculate the effectiveness map $\mathcal{E}$ using Gradient-weighted Class Activation Map (Grad-CAM) method. Then we calculate the selecting probability P with respect to $\mathcal{E}$ and sample n locations to settle the mask. In Patch Optimization module, a gradient descent with backpropagation is set to update the adversarial patches with the given mask. And if attack fails, a new set of patch locations will be resampled with respect to $\mathcal{E}$.

an end-to-end differentiable adversarial patch attack method termed as Generative Dynamic Patch Attack (GDPA). They employ a generator to generate the patch pattern and decide the location instead of using an optimizer, therefore reduce the inference time. As a price, GDPA can not achieve a high attacking success rate with a small patch size.

3   Proposed Method

3.1   Problem Formulation

We aim to achieve feasible and robust adversarial patch attack for RSI scene classification and we propose to generate an adversarial remote sensing image $x_{adv}(m, p)$ by optimizing n small adversarial patches on n feasible locations in the benign remote sensing image x.

In this paper, we mainly focus on untargeted attack, the optimization objective for MPAA can be expressed as follows:

$$\arg\min_{m,p} L\left(f\left(x_{adv}(m, p), y\right)\right), \text{ s.t. } S_p/S_x \leq \varepsilon, \tag{1}$$

with

$$x_{adv}(m, p) = (1 - m) \odot x + m \odot p, \tag{2}$$

where $\odot$ denotes the Hadamard product, p denotes the adversarial patches and m is the same size of x denoting the corresponding binary mask matrix to constrain the shape and position of the adversarial patches, y denotes the ground-truth class. Moreover, $f(\cdot)$ denotes the target classifier, S_p and S_x denote the area of a single adversarial patch and clean image, respectively. And L denotes the loss function of $f(\cdot)$, following [38], here we use the output probability of $f(\cdot)$ with respect to y to calculate L:

$$L = \Pr\left(y \mid f\left(x_{adv}(m, p)\right)\right). \tag{3}$$

From Eq. (1), we can see that the objective function involves both optimizing the patch locations m as well as the adversarial patches p, and the adversarial patches depend on patch locations. This is a bi-level optimization problem and is in general difficult to optimize. In this paper, we propose to solve this problem with a search and optimization strategy, *i.e.*, using an Effective Location Selection (ELS) Module to determine m and using a Patch Optimization (PO) Module to optimize p. Figure 2 shows the overview of the proposed MPAA method.

3.2   Effective Location Selection Module

Different from the single patch attack methods, we propose to deploy n small patches on benign image x. It is a combinatorial optimization problem and difficult to solve. We propose a Effective Location Selection (ELS) module which is used to guide the optimization of the mask m by considering an attack effectiveness map $\mathcal{E}$. Gradient-weighted Class Activation Map (Grad-CAM) [27] method is widely used to generate a heatmap of class activation attention which measures the contribution of different parts to the classified class label. We leverage Grad-CAM method to generate an effectiveness map $\mathcal{E}$ to guide the selection of the key locations to deploy the adversarial patches. The effectiveness map $\mathcal{E}$ can be expressed as:

$$\mathcal{E} = \text{ReLU}\left(\sum_k \alpha_k A^k\right), \tag{4}$$

with

$$\alpha_k = \frac{1}{Z}\sum_i\sum_j \frac{\partial P_y}{\partial A_{ij}^k}, \tag{5}$$

where A denotes a feature layer, follow the general setting, we use the last convolutional layer to calculate $\mathcal{E}$. k denotes the $k-th$ channel in layer A, α denotes the neuron importance weights, Z denotes the length and width, P_y denotes the output score of ground-truth label y.

Directly selecting the top-n values of the effectiveness map $\mathcal{E}$ cannot ensure the optimum solution, since all the n adversarial patches should cooperate rather than working independently. Here, we propose a class activation map guided random sampling based patch location selection (CGRSamp) method. Let us denote $\mathcal{V} = \mathbf{1}_{M \times M} \otimes \mathcal{E}$ where $\mathbf{1}_{M \times M}$ represents an all ones convolution kernel of size $M \times M$ and $\otimes$ represents the convolutional operator. We hereafter treat the values of $\mathcal{V}$ as a probabilistic guidance for selecting patch locations. That is, the

---

Algorithm 1: MPAA

Input : benign image x, classifier $f(\cdot)$, ground-truth label y, confidence
  threshold T, number of resampling attempts K, number of patches n,
  size of a single patch $M \times M$, maximum iterations N;

Output: Adversarial image x_{adv};

1 Generate Effectiveness Map $\mathcal{E} \leftarrow \text{ReLU}\left(\sum_k \alpha_k A^k\right)$;

2 Calculate $\mathcal{V} \leftarrow \mathbf{1}_{M \times M} \otimes \mathcal{E}$;

3 Calculate probability $P_{i,j} \leftarrow \dfrac{\exp(\mathcal{V}_{i,j}/t)}{\sum_{u,v} \exp(\mathcal{V}_{u,v}/t)}$;

4 **for** $j = 0 \rightarrow K - 1$ **do**

5 $\quad$ Randomly initialize n patches p;

6 $\quad$ Sample mask m *w.r.t.* probability P;

7 $\quad$ **for** $l = 0 \rightarrow N - 1$ **do**

8 $\quad\quad$ Update adversarial image $x_{adv}(m, p) \leftarrow (1 - m) \odot x + m \odot p$;

9 $\quad\quad$ Update the confidence of y: $P_y \leftarrow \Pr\left(y \mid f\left(x_{adv}(m, p)\right)\right)$;

10 $\quad\quad$ **if** $P_y > T$ **then**

11 $\quad\quad\quad$ | Update p by gradient descent with backpropagation;

12 $\quad\quad$ **else**

13 $\quad\quad\quad$ | Break;

14 $\quad\quad$ **end**

15 $\quad$ **end**

16 $\quad$ **if** x_{adv} *is adversarial* **then**

17 $\quad\quad$ | Break;

18 $\quad$ **end**

19 **end**

20 **return:** x_{adv}.

---

location with larger value has a higher probability to be selected. The proposed CGRSamp method uses softmax with temperature [13] to soft the contribution and calculate the probability as follows:

$$P_{i,j} = \frac{\exp\left(\mathcal{V}_{i,j}/t\right)}{\sum_{u,v} \exp\left(\mathcal{V}_{u,v}/t\right)}, \tag{6}$$

where (i, j) denotes the top left location of patch, $\mathcal{V}_{i,j}$ denotes the sum of value within the patch at (i, j), $P_{i,j}$ denotes the probability of selecting (i, j), and t represents the temperature hyper-parameter. We will discuss the value of t in detail in Sect. 4.4.

Since a single random sampling attempt cannot ensure successful attack, we propose a patch location resampling strategy. The patch locations will be resampled *w.r.t.* Eq. (6) if the previously sampled locations did not lead to a successful attack. With this patch location resampling strategy, the proposed MPAA can achieve improved attacking success rate with better patch locations.

Fig. 3. Visualizations of the generated adversarial examples by MPAA w/ TV loss (row 1) and w/o TV loss (row 2).

3.3   Patch Optimization Module

Given the mask m generated by Effective Location Selection module, Patch Optimization (PO) module is to optimize the adversarial patches p to fool the classifiers.

The adversarial patches p can be optimized $w.r.t.$ Eq. (1) with m being fixed. The values of patch p is randomly initialized within the range of $[0, 1]$. Then gradient descent with backpropagation is used to update p. Specifically, the optimizer is Adam [17] with learning rate 2/255. We set the maximum number of iterations to $N = 2000$. The adversarial attack will be considered as a successful attack and terminated if the output confidence of y is lower than a threshold $T = 10\%$, otherwise a new set of patch locations will be resampled with respect to the effectiveness map $\mathcal{E}$.

The pseudo-code of the proposed MPAA is shown in Algorithm 1. Lines 1–3 correspond to the Effective Location Selection (ELS) module, and lines 4–19 correspond to the Patch Optimization (PO) module.

3.4   Imperceptible MPAA

The visual imperceptibility of the adversarial patches is also essential. This section further shows the scalability of MPAA to be imperceptible. We add a Total Variation (TV) [26] loss to reduce the visibility of the adversarial patches, the TV loss function can be expressed as:

$$L_{TV} = \sum_{i,j} \sqrt{(x_{i,j-1} - x_{i,j})^2 + (x_{i+1,j} - x_{i,j})^2}. \tag{7}$$

So the total loss function can be expressed as:

$$L_{total} = L\left(f\left(x_{adv}\right), y\right) + \lambda L_{TV}, \tag{8}$$

where λ is a hyper-parameter to balance the imperceptibility and attacking success rate (ASR), and we set $\lambda = 0.01$ by default. We will disscuss the impact of λ in Sect. 4.4.

The TV loss can help generate adversarial examples with a smoother patch pattern. The visualizations of examples with and without TV loss are shown in Fig. 3. It is clear to see that with TV loss constraint, the adversarial patches will be less perceptible and more closer to the background. In the following of the paper, let us denote the proposed MPAA method learned with the TV loss as MPAA-IP.

4   Experiments

4.1   Experimental Setup

Dataset and Classification Model: Aerial Image Dataset (AID) [32] is used to evaluate the proposed method. AID has 10,000 images within 30 classes, all images are with the same resolution of 600×600. The dataset has been randomly splitted into a training set and a testing set with a ratio of 7:3. The pre-trained ResNet50[1] [11] is used as the target classifier, which achieves 3.83% top-1 error by fine-tuning. We have also fine-tuned ResNet101 [11] and DenseNet121 [16] with 4.43% and 3.80% top-1 error, respectively, to evaluate the proposed MPAA method. All experiments are performed on a computer with a NVIDIA RTX 3090 GPU with 24 GB memory.

Evaluation Metrics: We utilize the Attacking Success Rate (ASR) and processing time to evaluate the effectiveness and efficiency of MPAA. During testing, the images which cannot be correctly classified are discarded.

Baselines: To validate the effectiveness of the proposed MPAA method, three comparison patch attack methods have been included: AdvP [4], GDPA [20] and PFool [10]. For fair comparison, all methods are evaluated in AID and untargeted setting, the total patch size is set to 1% of image size. For MPAA, we set patch number $n = 16$, the maximum number of resampling attempts $K = 3$ by default.

4.2   Comparisons

To evaluate the effectiveness of the proposed method, we compare our MPAA method with baseline methods on three different classifiers. Table 1 shows the adversarial attack performance of different methods. We can see that, proposed

---

[1] https://download.pytorch.org/models.

Table 1. The attack performance of different adversarial patch attack methods against the commonly used classification models including ResNet50, ResNet101 and DenseNet121.

Methods	Metrics	Classifiers		
		ResNet50	ResNet101	DenseNet121
AdvP [4]	ASR (%)	71.48	66.32	61.15
	Time (s)	12.1	12.7	13.2
GDPA [20]	ASR (%)	63.10	48.80	40.61
	Time (s)	9.6	**10.1**	**10.3**
PFool [10]	ASR (%)	90.20	66.47	70.79
	Time (s)	**6.3**	10.8	11.8
MPAA-IP	ASR (%)	88.18	88.50	78.86
	Time (s)	8.4	11.4	12.9
MPAA	ASR (%)	**96.98**	**94.42**	**88.74**
	Time (s)	6.4	10.7	12.3

Table 2. The attacking performance of MPAA without different components. W/o CGRSamp means randomly select the location of patches without any guidance.

Settings		Metrics	Patch Number				
CGRSamp	Resampling		1	2	4	8	16
√	×	ASR(%)	73.80	77.65	82.66	86.36	92.48
		Time(s)	7.4	7.2	6.8	5.7	4.0
×	√	ASR(%)	75.39	81.92	87.89	90.12	92.84
		Time(s)	13.5	12.7	10.2	8.5	6.6
√	√	ASR(%)	79.42	84.19	90.05	94.18	96.98
		Time(s)	13.1	12.4	10.2	8.0	6.4

MPAA method achieves the highest ASR as well as an acceptable speed against all classifiers. The MPAA-IP with TV loss also has a satisfactory attack performance and generates more imperceptible adversarial examples. And the three comparison methods get a lower ASR especially when attacking classifiers with deeper layers. AdvP [4] method uses one patch and selects the location randomly which leads to a huge gap (over 25% ASR on all three classifiers) with MPAA using multiple patches, although the total size of patches are the same. GDPA [20] uses a generator to generate the adversarial patch instead of gradient iteration approach, so achieves a lower time complexity against the more complex classifiers. However, GDPA achieves the lowest ASR among all methods. PFool [10] method divides the image into a fixed number of blocks by the size of patches, then uses a saliency map to select the blocks and place the patches which restricts the flexibility of patch locations. And the results also indicate that PFool achieves a lower ASR than our location selection method using effectiveness map information as a probabilistic guidance.

Table 3. The ASR and LPIPS of different λ for MPAA-IP.

λ	0.1	0.05	0.01	0.005	0
ASR (%)	79.61	82.16	88.18	90.19	96.98
LPIPS$\downarrow$	0.0425	0.0436	0.0459	0.0501	0.0560

4.3 Ablation Study

To investigate the properties of MPAA method, we conduct ablation study to investigate the effect of different components to the overall performance. As it is shown in Table 2, the proposed location selection method CGRSamp indeed gets a better ASR than randomly selecting the location of patches without any guidance. And the resampling strategy leads to an improved ASR which also results in higher processing time cost. It is also interesting to note that, by increasing the number of patches, MPAA not only achieves a higher ASR, but also generates adversarial examples with a reduced processing time. This validates that the proposed *Multi-Patch Adversarial Attack* leads to both improved effectiveness and efficiency. If we keep increasing the number of patches, MPAA will be closer to a sparse attack, with a higher ASR, but harder to be deployed for the excessive patch number. Considering practical deployment difficulties and attacking success rates, we set patch number $n = 16$ by default.

4.4 Analysis

In the imperceptible version MPAA-IP, the hyper-parameter λ is of great importance to balance the imperceptibility and attacking effect. Table 3 shows the different performance of MPAA-IP with different value of λ. The Learned Perceptual Image Patch Similarity (LPIPS) [35] is widely used to valuate the perceptual quality of the generated adversarial image, the lower value means that the adversarial examples are more similar to the benign images. It is obvious that with λ increases, the ASR get lower but the adversarial examples become more imperceptible. To balance the ASR and imperceptibility, we use $\lambda = 0.01$ by default.

In CGRSamp method, temperature t is an essential variable to soft the distribution. Table 4 shows the results of CGRSamp method with different t. When t equals to 1, it becomes a normal softmax function which leads to extreme probability of the maximum value of effectiveness map. This makes the selection lose randomness, and the patches may overlap together which leads to a lower ASR. When t is larger than 40, the ASR begins to decline. Furthermore, when the softmax function degenerates to a weighted mean algorithm with the values as weighting, the ASR keep declining to 93.29%. Therefore, we choose $t = 40$ as our default setting.

Table 4. The ASR of different temperatures t in Eq. (6) to calculate the probability in CGRSamp method.

Temperature	1	10	30	40	50	Weighted mean
ASR (%)	66.91	74.63	96.57	**96.98**	96.95	93.29
Time (s)	13.9	10.4	7.6	**6.4**	6.5	7.9

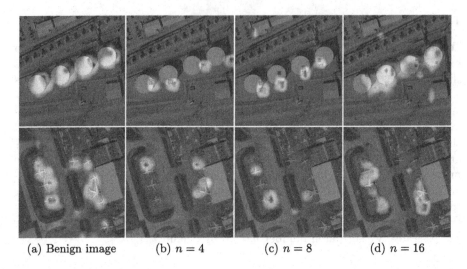

(a) Benign image   (b) $n = 4$ (c) $n = 8$ (d) $n = 16$

Fig. 4. The heatmaps of the generated adversarial examples by MPAA with different number of adversarial patches n.

We follow Grad-CAM [27] method to visualize the attention transfer of the classifier between the benign images and adversarial images. The visualization results are shown in Fig. 4. In the first column, we show the heatmap of two benign images of the ground-truth label storage tanks and the airport and we can see that the classifier mainly focuses on the tanks and the aircrafts. But in the adversarial images generated by MPAA, most attention has been attracted to the deployed patches. As the number of patches increases, the attention becomes more scattered. We conjecture that it is the reason why the proposed MPAA becomes easier to success with an increasing number of adversarial patches.

In the effective map, we find although most patches attract attention from the classifier, there are still certain adversarial patches attract less attention. Therefore, we have further investigated whether these patches are necessary or not. We try to remove these least effective adversarial patches (Marked in red box in (b)) as shown in Fig. 5. The result shows that after removing the least effective adversarial patch, the classification result changes from *Resort* with 75% confidence to *Airport* with 77% confidence and the heatmap shows that the attention transfers back to the aircrafts. This indicates that different adversarial patches are collaborated to form an adversarial example and all contribute to the success of adversarial attack.

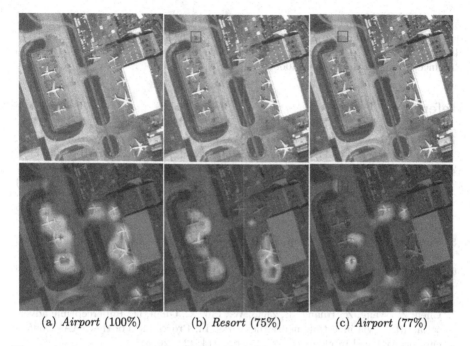

(a) *Airport* (100%)   (b) *Resort* (75%)   (c) *Airport* (77%)

Fig. 5. Analysis on the effect of the adversarial patch with the least attention. The first row shows the benign image, the adversarial image and the patch removed adversarial image. The second row shows the corresponding heatmap, and the numbers in brackets represent the top-1 classification confidence.

5   Conclusions

In this paper, we propose a novel adversarial attack method for remote sensing images scene classification termed as Multi-patch Adversarial Attack (MPAA) which utilizes multiple small and less perceptible adversarial patches instead of a single one to perform physically feasible adversarial attack. Within the proposed MPAA, an Effective Location Selection (ELS) module is used for selecting the optimum location set and a Patch Optimization (PO) module is used for optimizing the adversarial patches. In ELS module, we propose a class activation map guided random sampling based patch location selection (CGRSamp) method to select locations for pasting the adversarial patches according to Grad-CAM [27]. In PO module, we use a gradient descent to update the adversarial patches and a resampling strategy to improve the attacking success rate. From the experimental results, using multiple adversarial patches can not only lead to a higher attacking success rate with smaller patch area, but also can accelerate the attacking process. From extensive experiments on Aerial Image Dataset (AID) [32], the proposed MPAA with CGRSamp method has been proved to be both efficient and effective.

Acknowledgments. This work is supported by the National Natural Science Foundation of China under Project 62201600, 62102430, 62201604 and U1811462, NUDT Research Project ZK22-56 and ZK22-50, Natural Science Foundation of Hunan Province, China Project 2021JJ40688.

References

1. Arjovsky, M., Chintala, S., Bottou, L.: Wasserstein generative adversarial networks. In: International Conference on Machine Learning, pp. 214–223. PMLR (2017)
2. Bao, G.D., Xue, M., Ma, S., Abbasnejad, E., Ranasinghe, D.C.: Tnt attacks! universal naturalistic adversarial patches against deep neural network systems (2021)
3. Bengio, Y., et al.: Generative adversarial networks. Statistics (2014)
4. Brown, T.B., Mané, D., Roy, A., Abadi, M., Gilmer, J.: Adversarial patch. In: NIP (2017)
5. Burnel, J.C., Fatras, K., Flamary, R., Nicolas, C.: Generating natural adversarial remote sensing images. IEEE Trans. Geosci. Remote Sens. **60**, 1–14 (2021)
6. Chen, Y., Lin, Z., Zhao, X., Wang, G., Gu, Y.: Deep learning-based classification of hyperspectral data. IEEE J. Sel. Topics Appl. Earth Obs. Remote Sens. **7**(6), 2094–2107 (2014)
7. Chen, Z., Wang, Z., Huang, J., Zhao, W., Liu, X., Guan, D.: Imperceptible adversarial attack via invertible neural networks. In: Proceedings of the AAAI Conference on Artificial Intelligence, vol. 37, pp. 414–424 (2023)
8. Croce, F., Hein, M.: Sparse and imperceivable adversarial attacks. In: International Conference on Computer Vision (2019)
9. Dosovitskiy, A., et al.: An image is worth 16×16 words: transformers for image recognition at scale. arXiv preprint arXiv:2010.11929 (2020)
10. Fu, Y.F., Wu, S., Lin, Y., et al.: Patch-fool: are vision transformers always robust against adversarial perturbations? In: ICLR 2022 (2021)
11. He, K., Zhang, X., Ren, S., Sun, J.: Deep residual learning for image recognition. In: Proceedings of the IEEE Conference on Computer Vision and Pattern Recognition, pp. 770–778 (2016)
12. Hein, F.C.A.D.S.F.: Sparse-rs: a versatile framework for query-efficient sparse black-box adversarial attacks. In: AAAI-22 Technical Tracks 6 (2022)
13. Hinton, G., Vinyals, O., Dean, J.: Distilling the knowledge in a neural network. Comput. Sci. **14**(7), 38–39 (2015)
14. Hu, Y.C.T., Kung, B.H., Tan, D.S., Chen, J.C., Hua, K.L., Cheng, W.H.: Naturalistic physical adversarial patch for object detectors. In: Proceedings of the IEEE/CVF International Conference on Computer Vision, pp. 7848–7857 (2021)
15. Hu, Z., Huang, S., Zhu, X., Sun, F., Zhang, B., Hu, X.: Adversarial texture for fooling person detectors in the physical world. In: Proceedings of the IEEE/CVF Conference on Computer Vision and Pattern Recognition, pp. 13307–13316 (2022)
16. Huang, G., Liu, Z., Van Der Maaten, L., Weinberger, K.Q.: Densely connected convolutional networks. In: Proceedings of the IEEE Conference on Computer Vision and Pattern Recognition, pp. 4700–4708 (2017)
17. Kingma, D., Ba, J.: Adam: a method for stochastic optimization. Comput. Sci. (2014)
18. Komkov, S., Petiushko, A.: Advhat: real-world adversarial attack on arcface face id system. In: 2020 25th International Conference on Pattern Recognition (ICPR), pp. 819–826. IEEE (2021)

19. Kussul, N., Lavreniuk, M., Skakun, S., Shelestov, A.: Deep learning classification of land cover and crop types using remote sensing data. IEEE Geosci. Remote Sens. Lett. **14**, 778–782 (2017)

20. Li, X., Ji, S.: Generative dynamic patch attack. In: British Machine Vision Conference (BMVC) (2021)

21. Lian, J., Mei, S., Zhang, S., Ma, M.: Benchmarking adversarial patch against aerial detection. IEEE Trans. Geosci. Remote Sens. **60**, 1–16 (2022)

22. Luo, C., Lin, Q., Xie, W., Wu, B., Xie, J., Shen, L.: Frequency-driven imperceptible adversarial attack on semantic similarity. In: Proceedings of the IEEE/CVF Conference on Computer Vision and Pattern Recognition, pp. 15315–15324 (2022)

23. Modas, A., Moosavi-Dezfooli, S.M., Frossard, P.: Sparsefool: a few pixels make a big difference. In: 2019 IEEE/CVF Conference on Computer Vision and Pattern Recognition (CVPR) (2019)

24. Moosavi-Dezfooli, S.M., Fawzi, A., Frossard, P.: Deepfool: a simple and accurate method to fool deep neural networks. In: Proceedings of the IEEE Conference on Computer Vision and Pattern Recognition, pp. 2574–2582 (2016)

25. Rezaee, M., Mahdianpari, M., Zhang, Y., Salehi, B.: Deep convolutional neural network for complex wetland classification using optical remote sensing imagery. IEEE J. Sel. Topics Appl. Earth Obs. Remote Sens. **11**(9), 3030–3039 (2018)

26. Rudin, L.I., Osher, S., Fatemi, E.: Nonlinear total variation based noise removal algorithms. Physica D **60**(1–4), 259–268 (1992)

27. Selvaraju, R.R., Cogswell, M., Das, A., Vedantam, R., Parikh, D., Batra, D.: Gradcam: visual explanations from deep networks via gradient-based localization. In: Proceedings of the IEEE International Conference on Computer Vision (2017)

28. Shabtai, A.Z.K.E.: The translucent patch: a physical and universal attack on object detectors. In: 2021 IEEE/CVF Conference on Computer Vision and Pattern Recognition (CVPR) (2021)

29. Song, D., et al.: Robust physical-world attacks on machine learning models. Learning (2017)

30. Vladu, A., et al.: Towards deep learning models resistant to adversarial attacks. Statistics (2017)

31. Wu, J., et al.: Substitute meta-learning for black-box adversarial attack. IEEE Signal Process. Lett. **29**, 2472–2476 (2022)

32. Xia, G.S., et al.: Aid: a benchmark data set for performance evaluation of aerial scene classification. IEEE Trans. Geosci. Remote Sens. **55**(7), 3965–3981 (2017)

33. Xu, Y., Du, B., Zhang, L.: Assessing the threat of adversarial examples on deep neural networks for remote sensing scene classification: attacks and defenses. IEEE Trans. Geosci. Remote Sens. **59**(2), 1604–1617 (2020)

34. Xu, Y., Ghamisi, P.: Universal adversarial examples in remote sensing: methodology and benchmark. IEEE Trans. Geosci. Remote Sens. **60**, 1–15 (2022)

35. Zhang, R., Isola, P., Efros, A.A., Shechtman, E., Wang, O.: The unreasonable effectiveness of deep features as a perceptual metric. In: Proceedings of the IEEE Conference on Computer Vision and Pattern Recognition, pp. 586–595 (2018)

36. Zhang, W., Tang, P., Zhao, L.: Remote sensing image scene classification using CNN-capsnet. Remote Sens. **11**(5), 494 (2019)

37. Zhang, Y., et al.: Adversarial patch attack on multi-scale object detection for UAV remote sensing images. Remote Sens. **14**(21), 5298 (2022)

38. Zhao, Z., Liu, Z., Larson, M.A.: On success and simplicity: a second look at transferable targeted attacks. In: Neural Information Processing Systems (2021)

Multi-branch Residual Fusion Network for Imbalanced Visual Regression

Zhirong Huang[1,2], Shichao Zhang[1,2(✉)], Debo Cheng[1,2,3], Rongjiao Liang[1,2], and Mengqi Jiang[1,2]

[1] Key Lab of Education Blockchain and Intelligent Technology, Ministry of Education, Guangxi Normal University, Guilin 541004, China
zhangsc@mailbox.gxnu.edu.cn
[2] Guangxi Key Lab of Multi-Source Information Mining and Security, Guangxi Normal University, Guilin 541004, China
[3] UniSA STEM, University of South Australia, Mawson Lakes, Adelaide, Australia

Abstract. Imbalance visual regression is an important and challenging task, and research on it is still in its early stages. Among the existing solutions, either additional calibration layers need to be added to calibrate the label distribution or feature distribution of the imbalanced data, or a multi-stage training method with a loss function is used to mitigate the imbalance of the data, all of which cause significant time loss. In this paper, we propose an end-to-end multi-branch dynamic residual fusion network (MBDRFN) for imbalance visual regression. MBDRFN consists of a multi-branch residual fusion module (MBRF) and a dynamic balance factor (DBF) that works in concert to overcome the challenge of imbalanced visual regression. The MBRF module has three specialized branches corresponding to three magnitudes of sample labels, and can optimize each branch in a targeted manner to ensure overall fitting performance. The DBF builds on the MBRF and guides the model to gradually focus on learning and optimizing rare sample labels during the training process. Then, we use residual fusion to aggregate the loss functions of dedicated and master branches. The proposed MBDRFN method not only ensures that the model performs well on frequent and regular sample labels, but also improves its performance on rare sample labels with no significant time loss. Finally, we conduct extensive experiments on two real-world datasets to demonstrate the effectiveness and superiority of the proposed MBDRFN method.

Keywords: Imbalanced data · Imbalanced visual regression · Multi-branch network · Dynamic balance factor · End-to-end

1 Introduction

In modern deep learning [1,29], training data with balanced categories can enhance a model's generalization and fairness. However, imbalanced data is

---

Z. Huang and D. Cheng—Equal contribution.

X. Song et al. (Eds.): APWeb-WAIM 2023, LNCS 14331, pp. 392–406, 2024.
https://doi.org/10.1007/978-981-97-2303-4_26

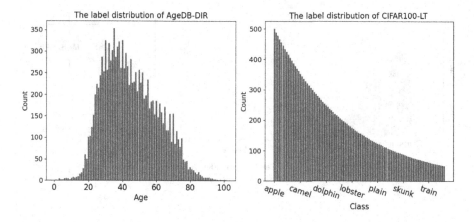

Fig. 1. The comparison of label distributions for imbalanced visual regression (left) and classification (right). On the left is the label distribution of AgeDB-DIR [27], which is a common dataset for imbalanced regression, and on the right is the label distribution of CIFAR100-LT [4] (with an imbalance ratio of 0.1), which is a common dataset for imbalanced classification. It is evident that the label distribution of imbalanced visual regression is more complex and diverse than imbalanced image classification.

quite common in many scenarios [8,13,14,25,32]. Existing research on imbalanced data has primarily focused on classification tasks with discrete labels, whereas many real-world applications involve regression tasks with continuous labels [7,10,27,28,30,31]. For example, age estimation [3,16] requires the prediction of a person's age based on their facial image, where age is a continuous value. However, most of the training images depict middle-aged and young individuals, resulting in an overall imbalanced distribution of the training data.

Currently, imbalanced image classification has received extensive and in-depth research [4,5,12,26], while the study on imbalanced visual regression is still in its early stages. In imbalanced image classification, the labels of the training data are usually discrete, and there are often clear boundaries between them. For example, there are distinct appearance differences between dogs and birds, which make it easy for models to learn and differentiate between them. Unlike classification, imbalanced visual regression has to deal with consecutive labels that have different degrees of similarity and there is no clear dividing line between adjacent labels. For instance, the appearance of a young person between the ages of 23 and 24 is visually similar and lacks a clear distinction. As shown in Fig. 1, due to the continuous nature of imbalanced visual regression labels, its label distribution cannot present a long-tail shape by adjusting the arrangement of labels as in classification, and the label distribution is more complex and diverse. Therefore, it is challenging to deal with imbalanced visual regression.

The research on imbalanced regression typically involves two levels: the data level and the model level. At the data level, resampling techniques [2,23] have been used to generate new samples for rare labels, thus changing the label

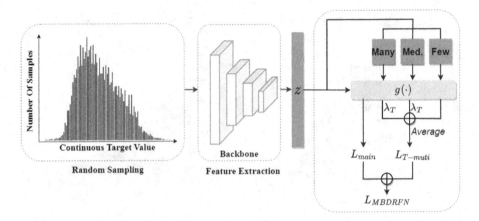

Fig. 2. The overall of the proposed MBDRFN method, which consists of a multi-branch residual fusion module (referred to as MBRF) and a dynamic balancing factor (referred to as DBF). The MBRF comprises three dedicated branches: "Many", "Med", and "Few", optimized for fitting the performance of their respective samples. The DBF module represented as (λ_T), works together with MBRF to learn the features of frequent labels while gradually focusing on the features of rare labels. Thus, DBF enables MBDRFN to learn from all samples effectively and improves its overall performance.

distribution of the dataset to make the number of each label more balanced. At the model level, Ren et al. [17] reexamined the widely used loss function of mean squared error (MSE) from a statistical perspective in the context of regression tasks. Yang et al. [27] found that rare sample features inherit the prior of frequent samples, leading to unreasonable similarities between sample features. They proposed the feature distribution smoothing (FDS) method to estimate and correct the potential bias in feature distribution.

Traditional models typically achieve better fitting performance for frequent labels but have poorer performance for rare labels. This is because traditional models fail to effectively capture the features of rare samples. To address this issue, we propose the MBDRFN method, which includes a dynamic balancing factor (DBF) and a multi-branch residual fusion module (MBRF) that can learn from all samples effectively and improve overall performance. Figure 2 illustrates the overall architecture of the proposed MBDRFN method.

As shown in Fig. 2, the MBRF comprises three branches: "Many", "Med", and "Few", each of which is optimized for the fitting performance of the respective samples. This allows the model to effectively learn the features of rare labels. However, it is also important for the model to learn the features of frequent labels in the early stages. Therefore, the output results of each branch and the main branch are aggregated using a residual fusion mechanism with a DBF to obtain the final results of the MBDRFN method. To the best of our knowledge, our proposed MBDRFN method is the first to handle imbalanced visual regression tasks using MBRF and a DBF. Our main contributions are summarized as follows:

- We propose a novel end-to-end multi-branch dynamic residual fusion network for handling imbalanced visual regression tasks.
- The proposed MBDRFN method employs a dynamic balancing factor to learn from all samples effectively and improves its overall performance.
- Extensive experiments on the two real-world datasets demonstrate the effectiveness and excellence of the proposed MBDRFN method.

2   Related Works

In recent years, the task of imbalanced visual regression has received more and more attention. In contrast to imbalanced image classification, imbalanced visual regression has not been fully explored, and some researchers have adapted methods for solving imbalanced image classification tasks (such as Focal Loss [11], Two-stage training [9], etc.) into regression scenarios. We summarize the existing studies on imbalanced visual regression as follows:

2.1   Data Level Solutions

The data level solution refers to pre-processing training data that is to balance the distribution before model training. Typically, there are two strategies: oversampling the minority class or undersampling the majority class. Branco et al. [2] proposed the SMOGN algorithm, which creates synthetic samples for predefined minority classes by augmenting them with Gaussian noise. However, the SMOGN algorithm uses an empirical label distribution to divide minority and majority classes and does not consider the dependency between data samples with continuous neighboring labels. Yang et al. [27] argue that empirical label distribution cannot accurately reflect the true label density distribution and propose label distribution smoothing, which uses kernel density estimation to extract effective label distributions. Stocksieker et al. [20] introduce a data augmentation algorithm that combines the procedures of weighted resampling (WR) and data augmentation (DA) procedures. This algorithm aims to address the issue of imbalanced continuous target covariate distribution by adjusting the sample distribution. On the other hand, Tian et al. [22] propose the deep imbalanced regression variational autoencoder (DIRVAE) algorithm to mitigate the disparity in the number of regression samples across different intervals. The proposed DIRVAE algorithm achieves this by generating missing samples and minority samples.

2.2   Model Level Solutions

The model level methods are primarily used to address the issue of imbalance in the feature space or loss of space. Yang et al. [27] applied classic reweighting algorithms, such as inverse frequency weighting (INV) and its square-root weighting variant (SQINV), to imbalanced regression tasks and propose a method that features distribution smoothing (FDS) to calibrate the potential deviation estimate of the feature distribution in the feature space. Ren et al. [17] revisited the widely used mean error (MSE) from a statistical perspective and found that MSE brings label imbalance into the prediction phase, resulting in poor prediction performance for rare labels. Therefore, they proposed a new loss function, called balanced MSE (BMSE), to accommodate training samples with imbalanced label distribution. Steininger et al. [19] propose a sample weighting method, named DenseWeight, to tackle the problem of unbalanced regression. They have also introduced a cost-sensitive learning method called DenseLoss, which is based on the DenseWeight weighting method. DenseWeight utilizes Kernel Density Estimation (KDE) to assign weights to data points based on the rarity of their corresponding target values. This weighting allows rare data points to have a higher influence on model training compared to ordinary data points. DenseLoss adjusts the impact of each data point on the loss function by considering the DenseWeight values. As a result, the training process gives more emphasis to rare data points, enabling the model to better handle the challenges posed by imbalanced regression datasets.

3   The Proposed MBDRFN Method

We present an overview of our proposed MBDRFN method in Fig. 2. The MBDRFN method comprises one main branch and three dedicated branches, namely "Many", "Median", and "Few". The main branch is responsible for learning common features for all samples, while the "Many" branch handles samples with frequent labels, the "Median" branch handles samples with regular labels, and the "Few" branch handles samples with rare labels. It is important to note that the dynamic balancing factor is designed for the three dedicated branches. In the next two subsections, we will introduce the details of the multi-branch residual fusion module and the dynamic balancing factor in our MBDRFN method.

3.1   Multi-branch Residual Fusion Module

To efficiently aggregate the branches, we have adopted the idea of residual learning and proposed the multi-branch residual fusion module (MBRF), as shown in Fig. 3. We have divided the visual features extracted from the backbone network into three groups, namely "Many", "Median", and "Few", based on the number of labeling samples in the training set. We then allow the dedicated branches to independently compute their losses on each group of features. Finally, we perform residual fusion on the losses of both the main branch and the dedicated branches.

The MBRF can optimize each branch in a targeted manner to ensure overall fitting performance while improving the performance of fitting rare labeling samples. By dividing the visual features into groups based on the number of labeling

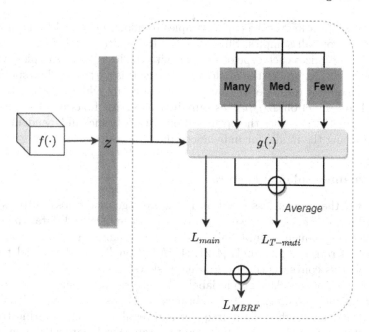

Fig. 3. The overall of the multi-branch residual fusion module, it has not yet added the dynamic balance factor(λ_T). z dentoes the $f(\cdot)$ the obtained features. The "Many", "Median" and "Few" are responsible for processing labels with more than 100 training samples, between 100 and 20, and less than 20, respectively. The output of each branch is fed into a regressor, denoted by $g(\cdot)$.

samples, the MBRF provides a more fine-grained way of handling imbalanced data, which has shown promising results in various tasks.

In the multi-branch residual fusion module, the loss function is designed during training as follows:

$$L_{main} = L\left(N_{h+m+t}(X), Y\right), \tag{1}$$

$$L_{muti} = \frac{\sum_{i \in \{h,m,t\}} L(N_i(Q_i), P_i)}{\sum_{i \in \{h,m,t\}} num(P_i)}, \tag{2}$$

$$L_{MBRF} = L_{main} + L_{muti}, \tag{3}$$

where X and Y represent the images and corresponding labels in each batch of training data, respectively. Q_h and P_h are subsets of X and Y, respectively, which contain the images and corresponding labels of frequent samples. Similarly, (Q_m, P_m) and (Q_t, P_t) represent subsets containing only regular and rare samples, respectively. N_h, N_m, and N_t represent the "Many", "Median", and "Few" branches, respectively, with N_{h+m+t} being equivalent to the main branch. $num(P_i)$ represents the number of labels for the corresponding sample. The loss function L_{main} optimizes the entire samples to allow the model to learn universal features from all samples. Subsequently, the multi-branch fusion loss function L_{multi} optimizes each type of sample, with the loss of rare samples typically being larger. L_{multi} guides the model to optimize this part of the loss specifically. Finally, the output results of the two loss functions are added together to obtain the total loss function L_{MBRF} as our final objective function. Our proposed MBDRFN method enables the optimization of each branch independently, while also optimizing the model for imbalanced data.

3.2   Dynamic Balance Factors

In Sect. 3.1, the branch loss function is the average of the loss outputs of the three dedicated branches. However, in the case of imbalanced data, the convergence effect of frequent sample labels and regular sample labels is often better than that of rare sample labels [4, 21, 24]. Additionally, as the model training, the overall loss convergence rate gradually slows down, especially for frequent sample labels and regular sample labels. To address this issue, we introduce a dynamic balancing factor, λ_T, to the branch of the loss function, which dynamically adjusts the weights of the frequent and regular branches during training. Specifically, we use a decreasing function to obtain this dynamic balancing factor at epoch λ:

$$\lambda_T = \min\left\{\frac{\lambda}{t}, 1\right\}, \tag{4}$$

$$L_{T-muti} = \frac{\sum_{i\in\{h,m\}} \lambda_T L\left(N_i\left(Q_i\right), P_i\right) + L\left(N_t\left(Q_t\right), P_t\right)}{\sum_{i\in\{h,m,t\}} num\left(P_i\right)}, \tag{5}$$

$$L_{MBDRFN} = L_{main} + L_{T-muti}, \tag{6}$$

We substitute the dynamic balancing factor of Eq. (4) into the multi-branch fusion loss function L_{muti} of Eq. (2) to obtain a new branch loss function L_{T-muti}. Therefore, we have the final loss function of the proposed MBDRFN method as presented in Eq. (6). At the beginning of training, the new branch loss function trains all branches with equal weights since learning from the head and middle data is equally important in the early stages of training. Then, at epoch λ-th, the balance factor is introduced, and the weight of the losses in the "Many" and "Median" branches dynamically decays as the training period t progresses until the end of training. The dynamic balancing factor enables the model to learn the features of frequent and regular labels fully while gradually emphasizing the learning of features with rare labels, resulting in a more balanced performance.

4   Experiments

4.1   The Details of Two Real-World Datasets

We demonstrate the effectiveness and performance of the proposed MBDRFN method using two real-world datasets, namely AgeDB-DIR and IMDB-WIKI-DIR. Figure 4 shows the label density distribution of both datasets, providing insight into their respective levels of imbalance.

IMDB-WIKI-DIR is an age estimation dataset derived from the IMDB-WIKI dataset [18] using specific rules by Yang et al. [27]. The dataset consists of 19.15K training images with images labeled with ages ranging from 0 to 186 in increments of 1. The number of images per age ranges from 1 to 7,149, exhibiting a significant imbalance in the label distribution. In addition, the dataset also includes 11.0K images with balanced label distribution for validation and testing.

AgeDB-DIR is a dataset derived from AgeDB [15] using a similar approach to IMDB-WIKI-DIR. It consists of 12.2K training set images labeled with ages from 0 to 101 in increments of 1 and the number of images per age from 1 to 353, showing a lighter label distribution imbalance compared to IMDB-WIKI-DIR. In addition, 2.1K label distribution balance images were used as validation and test sets.

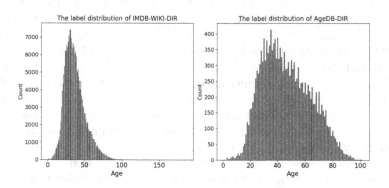

Fig. 4. The training set label distribution for two real-world datasets, IMDB-WIKI-DIR (left) and AgeDB-DIR (right), is presented in an overview. It is evident that both datasets demonstrate a noticeable imbalance in their label distributions.

4.2   BaseLines

Below, we provide a description of the baseline algorithms used for the comparison in the experiments:

- Vanilla: It is a model that does not incorporate any specific techniques for handling the issue of class imbalance. The backbone network used in Vanilla is ResNet-50 [6].

- Focal-R: It is a regression version of Focal loss that has been adapted to handle regression tasks. It replaces the scale factor of the original Focal loss with a continuous function that maps the absolute error to the range $[0, 1]$. The Focal-R loss function, specifically based on the L_1 distance, is defined as follows: $\frac{1}{n} \sum_{i=1}^{n} \sigma \left(|\beta e_i|\right)^{\gamma} e_i$, where e_i represents the L_1 error of the i-th sample in the regression task. The function $\sigma(\cdot)$ refers to the Sigmoid activation function. The hyperparameters β and γ are used to control the behavior of the loss function. In the experiments conducted by the authors, β and γ were set to 0.2 and 1, respectively.
- Square-root inverse-frequency weighting (SQINV): It is a traditional re-weighting method based on the label distribution and is referred to as a square-root weighting variant of inverse-frequency weighting (INV).
- Feature distribution smoothing(FDS): It is designed to address imbalanced regression problems by smoothing the feature distribution and calibrating potential bias estimates in the feature space. It aims to alleviate the impact of imbalanced distributions on regression tasks.

4.3 Experimental Setting

We used Resnet-50 as the backbone network to randomly extract feature representations of the training sets from IMDB-WIKI-DIR and AgeDB-DIR. Adam was used as the optimizer to update the weights of the backbone network, with an initial learning rate of 0.001. The model was trained for 90 epochs on an NVIDIA A100 GPU, with the learning rate being decayed by a factor of 0.1 at the 60-th and 80-th epochs, and a batch size of 256. In addition, we set the hyperparameter of the dynamic balance factor to 30, which means that the weights of the "Many" and "Median" branches will start to decay smoothly after 30 epochs. To demonstrate the generality and superiority of our method, we refer to the method of Yang et al. [27] and combine Focal-R and SQInv with a multi-branch residual fusion module and a dynamic balance factor for experiments.

We use Mean Absolute Error (MAE) and Mean Squared Error (MSE) as the evaluation indicators, which are commonly used in regression tasks for evaluating the performance of the developed novel model. The MAE is defined as $\frac{1}{N} \sum_{i=1}^{N} |y_i - \hat{y}_i|$, which calculates the average absolute difference between the true and predicted values for all samples, providing a measure of the model's average error. MSE is defined as $\frac{1}{N} \sum_{i=1}^{N} (y_i - \hat{y}_i)^2$, which calculates the average squared difference between the true and predicted values for all samples, providing a measure of the model's average squared error. Following the approach of [12], we divide the validation and test sets into three distinct subsets based on the number of training samples: the "Many" sample labels (labels with more than 100 training samples), the "Median" sample labels (labels with 20 to 100 training samples), and the "Few" sample labels (labels with fewer than 20 training samples). To provide a more comprehensive evaluation of the effectiveness of our method, we not only report the overall performance of the model in fitting the entire dataset, but also the performance of the model in fitting these subsets.

4.4   Experimental Results

In this section, the experimental results on the AgeDB-DIR and IMDB-WIKI-DIR datasets are reported in Tables 1, 2, 3 and 4, respectively.

Table 1. The experimental results of MAE on AgeDB-DIR dataset.

Method	MAE			
	All	Many	Med.	Few
Vanilla	8.026	7.101	9.009	14.039
Vanilla+FDS	7.814	6.899	8.935	13.357
Vanilla+MBRF	7.806	6.963	**8.758**	13.136
Vanilla+MBRF+DBF	**7.679**	**6.685**	9.116	**13.093**
Focal-R	7.74	6.784	9.032	13.194
Focal-R+FDS	7.609	6.848	8.533	12.241
Focal-R+MBRF	7.563	6.787	**8.478**	12.358
Focal-R+MBRF+DBF	**7.508**	**6.675**	8.706	**12.057**
SQINV	7.710	7.110	**12.478**	23.723
SQINV+FDS	7.906	7.297	12.872	23.113
SQINV+MBRF	7.858	7.273	12.621	22.495
SQINV+MBRF+DBF	**7.688**	**7.084**	12.616	**22.777**
Ours(BEST)VS.VANILLA	**+0.518**	**+0.426**	**+0.877**	**+3.797**

Table 2. The experimental results of MSE on AgeDB-DIR dataset.

Method	MSE			
	All	Many	Med.	Few
Vanilla	108.905	84.510	128.559	284.955
Vanilla+FDS	101.706	79.917	121.518	252.707
Vanilla+MBRF	104.283	83.437	**121.068**	255.596
Vanilla+MBRF+DBF	**98.913**	**75.018**	128.945	**241.536**
Focal-R	101.807	78.493	129.341	245.845
Focal-R+FDS	99.506	**75.362**	133.272	234.156
Focal-R+MBRF	99.392	80.160	**116.895**	232.627
Focal-R+MBRF+DBF	**96.876**	76.604	121.635	**219.855**
SQINV	104.169	85.996	122.579	224.900
SQINV+FDS	**100.889**	**85.595**	122.491	**185.596**
SQINV+MBRF	105.254	89.739	125.081	196.952
SQINV+MBRF+DBF	102.079	89.151	**112.878**	194.311
Ours(BEST)VS.VANILLA	**+12.029**	**+9.492**	**+12.681**	**+90.644**

First of all, we explain the meaning of the terms and method abbreviations in Tables 1, 2, 3 and 4. "Vanilla" represents the original model without using any techniques for handling imbalanced data. Focal-R refers to the model

402   Z. Huang et al.

that uses the Focal-R method (the regression version of Focal Loss), while SQINV refers to the model that uses the square root inverse frequency weighting method. FDS, MBRF, and DBF represent feature smooth distribution, multi-branch residual fusion module, and dynamic balance factor, respectively, where MBRF and DBF are the methods proposed in this paper. "Vanilla+FDS" represents the use of feature smooth distribution on the Vanilla model, and similarly, "Vanilla+MBRF+DBF" represents the simultaneous use of multi-branch residual fusion mechanism and dynamic balance factor on the Vanilla model, and so on. "Focal-R+" and "SQINV+" represent the use of certain methods on these models, respectively.

Table 3. The experimental results of MAE on IMDB-WIKI-DIR dataset.

Method	MAE			
	All	Many	Med.	Few
Vanilla	8.018	7.152	15.372	27.369
Vanilla+FDS	7.929	7.137	**14.689**	**25.401**
Vanilla+MBRF	8.069	7.199	15.641	26.054
Vanilla+MBRF+DBF	**7.900**	**7.088**	14.849	25.644
Focal-R	8.041	7.226	15.049	25.594
Focal-R+FDS	7.963	7.281	**13.546**	24.859
Focal-R+MBRF	8.006	7.231	14.669	24.727
Focal-R+MBRF+DBF	**7.827**	**7.129**	13.662	**24.185**
SQINV	7.710	7.110	**12.478**	23.723
SQINV+FDS	7.906	7.297	12.872	23.113
SQINV+MBRF	7.858	7.273	12.621	22.495
SQINV+MBRF+DBF	**7.688**	**7.084**	12.616	**22.777**
Ours(BEST)VS.VANILLA	+0.330	+0.068	+2.756	+4.592

Table 4. The experimental results of MSE on IMDB-WIKI-DIR dataset.

Method	MSE			
	All	Many	Med.	Few
Vanilla	136.392	105.836	368.435	1039.164
Vanilla+FDS	**134.254**	**104.633**	365.150	961.803
Vanilla+MBRF	136.978	106.570	378.509	950.512
Vanilla+MBRF+DBF	135.157	106.529	**358.579**	**932.853**
Focal-R	137.640	108.855	363.584	929.341
Focal-R+FDS	134.235	107.740	**331.343**	949.774
Focal-R+MBRF	138.212	110.153	359.406	902.353
Focal-R+MBRF+DBF	**131.831**	**105.482**	333.096	**901.006**
SQINV	129.839	106.072	**301.655**	901.369
SQINV+FDS	135.505	110.712	323.373	871.277
SQINV+MBRF	132.480	109.549	306.596	**810.098**
SQINV+MBRF+DBF	**128.627**	**105.019**	307.861	826.506
Ours(BEST)VS.VANILLA	+7.765	+0.817	+61.839	+229.066

From Tables 1, 2, 3 and 4, we have the following observations: (1) the proposed MBRF and DBF methods have demonstrated enhanced performance on the underlying models for all three different strategies. Notably, the "Few" strategy exhibits the most significant improvement, surpassing the state-of-the-art (SOTA) methods by a substantial margin. (2) the improvement effect of the proposed methods on the IMDB-WIKI-DIR dataset is superior to that on the AgeDB-DIR dataset. This difference in performance could be attributed to the availability of a larger amount of training data in the IMDB-WIKI-DIR dataset. (3) our method exhibits good generality and can be used in conjunction with other techniques for handling imbalanced data. It is worth noting that we excluded label distribution smoothing (LDS) from the comparison since it addresses imbalanced data at the data level, whereas our method operates at the model level.

After analyzing Tables 1, 2, 3 and 4, we concluded that combining MBRF with the base model significantly improved various performance metrics of the base model, especially for "Few". MBRF provides dedicated branches for different magnitudes of sample labels, allowing targeted optimization of the poorly performing branches during training. When DBF and MBRF are used together with the base model, the model's performance improvement is even more significant. DBF gradually guides the model to focus on dedicated branches with poor performance based on MBRF and better optimize these branches during training, thereby enhancing the model's performance.

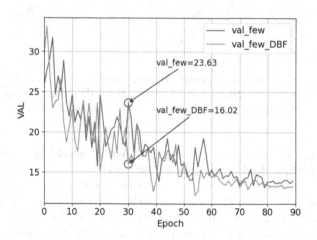

Fig. 5. The MAE change curve of the "Few" sample labels in the AgeDB-DIR dataset under different λ conditions is shown. The blue curve represents the MAE change curve without introducing the DBF, while the green curve represents the MAE change curve with the introduction of DBF (with λ set to 30). (Color figure online)

Figure 5 shows the MAE change curve of the Few labels under different λ conditions. The blue curve represents the MAE change curve without introducing DBF, while the green curve represents the MAE change curve with DBF introduced (with λ set to 30). By observing Fig. 4, it can be seen that when introducing the dynamic balance factor (DBF), the loss convergence of rare labels is significantly better than when DBF is not introduced after the 30th epoch. This indicates that DBF can guide the model to gradually focus on the optimization of rare labels during the training process.

This phenomenon may be attributed to the fact that in the early stages of training, the model learns features from frequent labels but fails to effectively learn features from rare labels. In the later stages of training, these features learned from frequent and regular labels have a significant impact on the model's ability to fit rare labels, leading to poor convergence of the loss function for rare labels. However, it is also crucial for the model to learn features from frequent and regular labels in the early stages of training. Therefore, it shows that our proposed dynamic balancing factor smoothly increases the model's focus on learning features from rare labels during training.

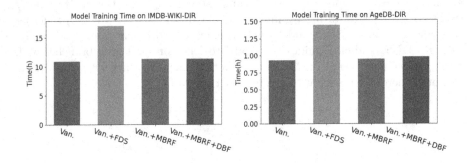

Fig. 6. The training times of the Vanilla model, combined with the FDS, MBRF, and DBF methods on the AgeDB-DIR and IMDB-WIKI-DIR datasets, respectively, were analyzed. The "Van." in the figure represents the Vanilla model.

Figure 6 displays the training time of the Vanilla model in conjunction with the FDS, MBRF, and DBF methods on the AgeDB-DIR and IMDB-WIKI-DIR datasets. The outcomes indicate that the "Van.+FDS" model takes significantly longer training time than the other methods. The reason for this is that FDS needs an additional feature calibration layer to calibrate the feature distribution of the training samples for each epoch, which substantially increases the model's training time. In contrast, the "Van.+MBRF" and "Van.+MBRF+DBF" models do not require additional training time and achieve excellent fitting performance. This is because MBRF and DBF operate directly on the features and loss functions during the training process, eliminating the need for extra time on re-balancing operations.

5 Conclusions

In this work, we propose a novel MBDRFN method with the multi-branch residual fusion module and a dynamic balancing factor to deal with imbalanced visual regression tasks. MBRF classifies the sample labels into frequent, regular, and rare categories based on the number of training samples and optimizes the model's fitting performance using dedicated branches. DBF, which is based on MBRF, guides the model to focus on optimizing rare labels gradually during the training process. These methods are general and can be combined with other techniques for handling imbalanced data. We conducted experiments on AgeDB-DIR and IMBD-WIKI-DIR datasets to demonstrate the effectiveness and superiority of our proposed MBDRFN method.

Acknowledgments. This research was supported in part by the Project of Guangxi Science and Technology (GuiKeAB23026040).

References

1. Bengio, Y., Lecun, Y., Hinton, G.: Deep learning for AI. Commun. ACM **64**(7), 58–65 (2021)
2. Branco, P., Torgo, L., Ribeiro, R.P.: SMOGN: a pre-processing approach for imbalanced regression. In: First International Workshop on Learning with Imbalanced Domains: Theory and Applications, pp. 36–50. PMLR (2017)
3. Cao, D., Zhu, X., et al.: Domain balancing: face recognition on long-tailed domains. In: Proceedings of the IEEE/CVF Conference on Computer Vision and Pattern Recognition, pp. 5671–5679 (2020)
4. Cao, K., Wei, C., et al.: Learning imbalanced datasets with label-distribution-aware margin loss. In: Advances in Neural Information Processing Systems, vol. 32 (2019)
5. Cui, J., Liu, S., et al.: Reslt: residual learning for long-tailed recognition. IEEE Trans. Pattern Anal. Mach. Intell. (2022)
6. He, K., Zhang, X., et al.: Deep residual learning for image recognition. In: Proceedings of the IEEE Conference on Computer Vision and Pattern Recognition, pp. 770–778 (2016)
7. Hu, R., Cheng, D., et al.: Low-rank feature selection for multi-view regression. Multimedia Tools Appl. **76**, 17479–17495 (2017)
8. Kang, B., Li, Y., et al.: Exploring balanced feature spaces for representation learning. In: International Conference on Learning Representations (2021)
9. Kang, B., Xie, S., et al.: Decoupling representation and classifier for long-tailed recognition. arXiv preprint arXiv:1910.09217 (2019)
10. Liang, R., Zhang, G., et al.: Bilateral-branch network for imbalanced visual regression. In: 2022 IEEE 34th International Conference on Tools with Artificial Intelligence (ICTAI), pp. 140–147. IEEE (2022)
11. Lin, T.Y., Goyal, P., et al.: Focal loss for dense object detection. In: Proceedings of the IEEE International Conference on Computer Vision, pp. 2980–2988 (2017)
12. Liu, Z., Miao, Z., et al.: Large-scale long-tailed recognition in an open world. In: Proceedings of the IEEE/CVF Conference on Computer Vision and Pattern Recognition, pp. 2537–2546 (2019)

13. Ma, Y., Tian, Y., et al.: Class-imbalanced learning on graphs: a survey. arXiv preprint arXiv:2304.04300 (2023)
14. Menon, A.K., Jayasumana, S., et al.: Long-tail learning via logit adjustment. arXiv preprint arXiv:2007.07314 (2020)
15. Moschoglou, S., Papaioannou, A., et al.: AgeDB: the first manually collected, in-the-wild age database. In: proceedings of the IEEE Conference on Computer Vision and Pattern Recognition Workshops. pp. 51–59 (2017)
16. Punyani, P., Gupta, R., Kumar, A.: Neural networks for facial age estimation: a survey on recent advances. Artif. Intell. Rev. **53**, 3299–3347 (2020)
17. Ren, J., Zhang, M., et al.: Balanced MSE for imbalanced visual regression. In: Proceedings of the IEEE/CVF Conference on Computer Vision and Pattern Recognition, pp. 7926–7935 (2022)
18. Rothe, R., Timofte, R., Van Gool, L.: Deep expectation of real and apparent age from a single image without facial landmarks. Int. J. Comput. Vision **126**(2–4), 144–157 (2018)
19. Steininger, M., Kobs, K., et al.: Density-based weighting for imbalanced regression. Mach. Learn. **110**, 2187–2211 (2021)
20. Stocksieker, S., Pommeret, D., Charpentier, A.: Data augmentation for imbalanced regression. arXiv preprint arXiv:2302.09288 (2023)
21. Tan, J., Wang, C., et al.: Equalization loss for long-tailed object recognition. In: Proceedings of the IEEE/CVF Conference on Computer Vision and Pattern Recognition, pp. 11662–11671 (2020)
22. Tian, H., Tian, C., et al.: Unbalanced regression sample generation algorithm based on confrontation. Inf. Sci. 119157 (2023)
23. Torgo, L., Ribeiro, R.P., Pfahringer, B., Branco, P.: SMOTE for regression. In: Correia, L., Reis, L.P., Cascalho, J. (eds.) EPIA 2013. LNCS (LNAI), vol. 8154, pp. 378–389. Springer, Heidelberg (2013). https://doi.org/10.1007/978-3-642-40669-0_33
24. Wang, X., Lian, L., et al.: Long-tailed recognition by routing diverse distribution-aware experts. arXiv preprint arXiv:2010.01809 (2020)
25. Wu, O.: Rethinking class imbalance in machine learning. arXiv preprint arXiv:2305.03900 (2023)
26. Yang, Y., Xu, Z.: Rethinking the value of labels for improving class-imbalanced learning. In: Advances in Neural Information Processing Systems, vol. 33, pp. 19290–19301 (2020)
27. Yang, Y., Zha, K., et al.: Delving into deep imbalanced regression. In: International Conference on Machine Learning, pp. 11842–11851. PMLR (2021)
28. Zha, K.: Deep Imbalanced Regression: Challenges, Methods, and Applications. Ph.D. thesis, Massachusetts Institute of Technology (2022)
29. Zhang, K., Lu, G., et al.: Personalized headline generation with enhanced user interest perception. In: Pimenidis, E., Angelov, P., Jayne, C., Papaleonidas, A., Aydin, M. (eds.) ICANN 2022. LNCS, vol. 13530, pp. 797–809. Springer, Cham (2022). https://doi.org/10.1007/978-3-031-15931-2_65
30. Zhang, S., Cheng, D., et al.: Supervised feature selection algorithm via discriminative ridge regression. World Wide Web **21**, 1545–1562 (2018)
31. Zhang, S., Yang, L., et al.: Leverage triple relational structures via low-rank feature reduction for multi-output regression. Multimedia Tools Appl. **76**, 17461–17477 (2017)
32. Zhang, Y., Kang, B., et al.: Deep long-tailed learning: a survey. IEEE Trans. Pattern Anal. Mach. Intell. (2023)

Learning Temporal Graph Representation via Memory-Aware Autoencoder

Jingyu Chen, Chengxin He$^{(\boxtimes)}$, Yuening Qu, Zhenyang Yu, Yuanhao Zhang, and Lei Duan$^{(\boxtimes)}$

School of Computer Science, Sichuan University, Chengdu, China
{chenjingyu1025,hechengxin,quyuening,yuzhenyang,
zhangyuanhao}@stu.scu.edu.cn, leiduan@scu.edu.cn

Abstract. Representation learning over temporal graphs has garnered significant attention from researchers due to the prevalence of real-world graphs that naturally evolve over time. However, some existing methods that discretize temporal graphs into snapshots often result in the loss of temporal information. On the other hand, some continuous-time methods ensure temporal continuity but neglect the intrinsic correlations among events and the influence of node features on their behavior. Therefore, we propose a novel temporal graph representation learning model with memory-aware autoencoder, called MATE, which consists of a node-wise memory unit and a temporal representation learning module. The temporal representation learning module is on the basis of autoencoder. Specifically, a temporal GNN-based encoder is constructed to aggregate temporal topological neighborhood features, while a dynamic event decoder is employed to learn the dynamic evolution of the temporal graph. Additionally, a static feature decoder is included to guide the encoder to maximize the retention of node features in its temporal representation. As for the node-wise memory unit, MATE utilizes it to store the historical information of nodes. Through the interactive iteration of the node-wise memory unit and temporal representation learning module, MATE is able to capture the intrinsic correlations among events occurring in a sequence. Extensive experiments conducted on various real-world datasets demonstrate the superiority of MATE.

Keywords: Temporal graph · Graph neural network · Graph representation learning

1 Introduction

Graph-structured data are ubiquitous in real-world scenarios. Correspondingly, numerous graph representation learning methods have been proposed to extract

This work was supported in part by the National Natural Science Foundation of China (61972268), and the Joint Innovation Foundation of Sichuan University and Nuclear Power Institute of China.

Fig. 1. An example of an academic collaboration network. Researchers P_1 and P_2 are both supervised by the same mentor. Researchers P_1 and P_3 collaborate on a large interdisciplinary project at time t_2, which arouses P_1's interest in the research field of P_3. As a result, P_1 seeks out more related topics, which may have led to the collaboration between P_1 and P_4 (whose research direction is the same as P_3) at time t_4.

valuable insights from such data. The majority of these methods focus on static graphs, i.e., the graph structures are assumed to remain fixed over time. However, graphs in the real world often exhibit complex dynamics [2,12,20]. For example, in social networks, new connections between users are continuously established over time, while in co-author networks, researchers collaborate with different peers as time progresses. Therefore, temporal graph representation learning has gradually become a research hotspot.

Temporal graph representation learning aims to model the evolution process of temporal graphs. By decomposing a temporal graph into a sequence of snapshots, static graph representation learning methods can be extended to temporal graphs [11,16,19,26,27], resulting in what is known as discrete-time methods. However, discrete-time methods have a coarse modeling granularity for temporal graph evolution because they ignore temporal information within the same snapshot. Hence, some researchers have proposed continuous-time methods [15,18,25], which consider the establishment of temporal edges as interaction events between nodes and view each event as an individual training instance. These methods can preserve the original time information of each temporal edge, and model the evolution process of temporal graphs at the finest granularity.

Most of the current temporal graph representation learning methods require retraining the model to obtain representations for new nodes that are not seen in the training phase, making it difficult to directly apply these methods to real-world scenarios [13,28]. Moreover, most existing methods either fail to preserve the original static features of nodes while capturing the dynamic nature of temporal graphs, or ignore the sequence of events in temporal graphs [25]. Solving representation learning problems on temporal graphs is highly challenging. We summarize the challenges into the following three points:

- *(Dynamic Evolution)* How to capture the dynamic changes in the topological structure of temporal graphs? As new nodes are continually added and new links are established, the topological structure of a temporal graph is no longer static. As shown in Fig. 1, from time t_1 to t_2, researcher P_3 gradually joins the graph. At time t_3, existing researchers P_1 and P_2 also establish new connections.

- (**Static Correlation**) *How to preserve the original static features of each node in the encoded temporal representations?* Node features imply patterns of temporal graph evolution, and some features that certain nodes possess may lead them to interact frequently with other nodes. For example, in Fig. 1, researchers P_1 and P_2 are more likely to collaborate frequently because they have the same advisor.
- (**Event Sequence**) *How to fully utilize the sequence of events?* The historical events in which a node participates can influence its future behavior, thereby affecting the direction of temporal graph evolution. Therefore, capturing the intrinsic connections between events is helpful for modeling the evolution process of temporal graphs. For example, in Fig. 1, researcher P_1's research interests become more diverse at time t_2 due to his interaction with P_3, which leads to his interaction with researcher P_4 at time t_4.

To address these challenges, we propose a novel model named MATE (short for learning temporal graph representation via <u>m</u>emory-<u>a</u>ware auto<u>e</u>ncoder). Specifically, for **Challenge 1**, a temporal GNN-based encoder with a functional time encoding is used to encode the temporal topological structure of the graph. For **Challenge 2**, a dynamic event decoder and a static feature decoder are used in combination with a reconstruction loss to guide the encoder to learn the dynamic nature of the graph while preserving the original static features of nodes. For **Challenge 3**, a node-wise memory unit is used to store the historical information of nodes in a compressed form. And the iterative interaction between the node-wise memory unit and the temporal representation learning module enables MATE to learn the potential connections between sequentially occurring temporal events.

The contributions of this paper are summarized as follows:

- We propose a novel temporal graph representation learning model, named MATE. It consists of a node-wise memory unit and a temporal representation learning module.
- MATE can preserve the original static features of nodes while learning the dynamic nature of the graph. Moreover, MATE is able to make full use of the sequential information of events.
- We conduct comprehensive experiments on public real-world datasets. The results demonstrate the effectiveness of MATE in solving the problem of temporal graph representation learning.

2   Related Work

Our work is related to static graph representation learning and temporal graph representation learning. We briefly introduce the related work as follows.

2.1   Static Graph Representation Learning

Graph representation learning aims to learn a mapping function which can map nodes in the graph to a low-dimensional space. DeepWalk [17] and Node2vec [3] perform random walks on the graph and use skip-gram [14] to train the model. These random walk-based methods directly learn embeddings of nodes, while GNN-based methods learn an aggregation function which can help produce representations of nodes. GCN [8] first generalizes convolution operation to graph-structured data. GraphSAGE [4] generates representations of nodes by sampling and aggregating features from their local neighborhood. GAT [22] leverages masked self-attention mechanism to specify different weights to different nodes when aggregating the features of neighboring nodes. Simply applying static graph representation learning approaches to temporal graphs can lead to severe loss of temporal information.

2.2   Temporal Graph Representation Learning

Representation learning methods on temporal graphs can be categorized into discrete-time methods and continuous-time methods.

Discrete-Time Methods. These methods are extended from static graph representation learning methods, which divide temporal graphs into multiple snapshots. DySAT [19] uses structural self-attention mechanism to compute representations of each node in each snapshot. Then it uses temporal self-attention mechanism to update representations of nodes. EvolveGCN [16] does not depend on the historical representations of nodes. It uses GCN to process each snapshot, then adopts RNN to evolve the parameters of GCN. These methods model temporal graphs at a coarse granularity. It should be noticed that the time interval between snapshots is user-specified. Thus an inappropriate selection of time interval may lead to sub-optimal performance [15].

Continuous-Time Methods. Recent studies have shown the superiority of continuous-time methods which aim at modeling the continuous evolution process of temporal graphs. CTDNE [15] performs random walk with temporal constraints. CAW-N [23] is another random walk-based approach with a novel anonymization strategy which is capable of encoding temporal network motifs. HTNE [28] integrates Hawkes process [6] and attention mechanism to capture the influence of historical events on the occurrence of current events. MMDNE [13] and TREND [24] also employ Hawkes process, and both of which consider characteristics of events at an individual and collective scale. DyRep [21] proposes a two-time scale deep temporal point process model to capture the dynamics of graph. JODIE [9] utilizes two recurrent neural networks to update representations of nodes each time they are involved in a new event. TGAT [25] is based on GNN and it proposes a functional time encoding to replace positional encodings in self-attention mechanism. TGN [18] is another GNN-based framework which is computationally efficient.

3   Preliminaries

Definition 1 (Temporal Graph). *A temporal graph is denoted as* $\mathcal{G} = (\mathcal{V}, \mathcal{E}, \mathcal{T})$, *where* $\mathcal{V}$ *indicates a set of nodes and* $\mathcal{T}$ *is a set of timestamps. Each*

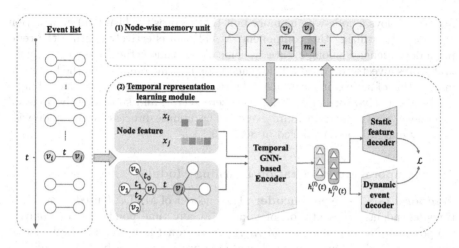

Fig. 2. Illustration of the proposed model MATE.

temporal edge $e = (v_i, v_j, t) \in \mathcal{E}$ refers to a connection between node v_i and node v_j at timestamp t. Note that there can be multiple edges between two nodes at different timestamps.

For instance, the academic collaboration network shown in Fig. 1 can be abstracted as a temporal graph, where the node set is $\{v_1, v_2, v_3, v_4, v_5\}$ and the edge set is $\{(v_1, v_2, t_1), (v_1, v_3, t_2), (v_1, v_2, t_3), (v_1, v_4, t_4), (v_2, v_5, t_5)\}$. There exist two edges (v_1, v_2, t_1) and (v_1, v_2, t_3), which means nodes v_1 and v_2 forming two different links at two different time t_1 and t_3.

Definition 2 (Temporal Graph Representation Learning). *Given a temporal graph $\mathcal{G} = (\mathcal{V}, \mathcal{E}, \mathcal{T})$, a node feature matrix $X \in \mathbb{R}^{|\mathcal{V}| \times d_0}$ (d_0 represents the feature dimension), and a temporal edge feature matrix $E \in \mathbb{R}^{|\mathcal{E}| \times d_0}$, the goal of temporal graph representation learning is to learn a parameterized model $\mathcal{F}(X, E; \theta)$ containing multiple variables θ, such that any node in the feature space X can be mapped to a latent feature space at any time. In this space, the representation vector of each node contains rich temporal information.*

4 The Design of MATE

The proposed model MATE consists of two parts: (1) a node-wise memory unit that stores the historical information of nodes, and (2) a temporal representation learning module that includes a temporal GNN-based encoder, a dynamic event decoder, and a static feature decoder. The two modules interact with each other iteratively. Figure 2 illustrates the overall architecture of MATE.

4.1 Node-Wise Memory Unit

The characteristics of a node can be altered by the historical events it participates in, consequently affecting its future behavior. To capture this information,

MATE employs a node-wise memory unit that records the historical information of nodes. By utilizing this component, MATE can effectively capture the temporal dependencies among events. Specifically, at time t, the node-wise memory unit stores a vector $\mathbf{m}_i(t) \in \mathbb{R}^{d_0}$ for each node v_i, which represents the historical information of node v_i in a compressed format. Upon the first appearance of node v_i, its memory vector is initialized as a zero vector, and subsequently updated as node v_i participates in more events. The update process of the node-wise memory unit will be introduced in Sect. 4.3.

4.2 Temporal Representation Learning Module

Temporal GNN-based Encoder. The encoder of MATE is used to generate time-dependent representation of node v_i at any time point t. The proposed temporal GNN-based encoder in MATE relies on the temporal graph neural network layer. The temporal graph neural network layer can be considered as a local aggregation operator that efficiently integrates node features, edge features, time information, and the temporal topological structure of graph.

We use $\mathbf{h}_i^{(l)}(t) \in \mathbb{R}^{d_l}$ to denote the output representation of node v_i at time t in the l-th layer, which can be derived as,

$$\mathbf{h}_i^{(l)}(t) = \mathbf{W}_{self}^{(l)} \left(\mathbf{h}_i^{(l-1)}(t) \, \| \, \Phi(0) \| \, \tilde{\mathbf{h}}_i^{(l)}(t) \right) \tag{1}$$

$$\tilde{\mathbf{h}}_i^{(l)}(t) = \sigma \left(\sum_{(v_i, v_{j'}, t') \in \mathcal{H}_i(t)} \mathbf{W}_{hist}^{(l)} \left(\mathbf{h}_{j'}^{(l-1)}(t') \, \| \, \mathbf{e}_{i,j'}(t') \| \, \Phi(t-t') \right) \right) \tag{2}$$

where $\mathcal{H}_i(t) = \{(v_i, v_{j'}, t') \in \mathcal{E} : t' < t\}$ is the historical events set of node v_i at time t. We call $v_{j'}$ a temporal neighbor of node v_i at time t. The temporal neighbors can be sampled uniformly or just reserve the most recent neighbors. $\mathbf{e}_{i,j'}(t')$ indicates the edge feature of event $(v_i, v_{j'}, t')$. $\|$ means the concatenation operation. And σ is an activation function. $\left\{ \mathbf{W}_{self}^{(l)}, \mathbf{W}_{hist}^{(l)} : l = 1, 2, ..., L \right\}$ are weight matrices need to be learned. $\Phi : T \to \mathbb{R}^{d_T}$ stands for a continuous functional mapping from time domain to the d_T-dimensional vector space based on Bochner's theorem [25]. More formally,

$$t \mapsto \Phi(t) = \sqrt{\frac{1}{d_T}} \left[\cos(\omega_1 t), \sin(\omega_1 t), \dots, \cos(\omega_{d_T} t), \sin(\omega_{d_T} t) \right] \tag{3}$$

where $\{\omega_i, 1 \le i \le d_T\}$ are learnable parameters.

By stacking L layers of temporal graph neural network, nodes can receive and aggregate information from their L-hop temporal neighbors. In the first layer, we not only initialize node information using the node's original static features, but also consider the node's historical information. We use $\mathbf{x}_i \in \mathbb{R}^{d_0}$ to represent the original features of node v_i, and $\mathbf{m}_i(t) \in \mathbb{R}^{d_0}$ to represent the memory vector of node v_i that can be accessed in the memory unit at time t. Taking into account the distinct impacts of node features and memory vectors, the first-layer input of node v_i is calculated as follows,

$$\mathbf{h}_i^{(0)}(t) = \mathbf{x}_i + \text{MLP}(\mathbf{m}_i(t)) \tag{4}$$

Dynamic Event Decoder. The dynamics of events are reflected by the establishment of temporal edges. The dynamic event decoder aims to reconstruct temporal edges based on the node representations output by the encoder. Considering that different nodes have different importance in the occurrence of events, MATE adopts an adaptive method to calculate the influence of nodes on edge establishment. Specifically, MATE designs a function to calculate the probability of each temporal edge (v_i, v_j, t) existing. Given the node representations of node v_i and v_j obtained by l layers of temporal graph neural networks at time t, represented by $\mathbf{h}_i^{(l)}(t)$ and $\mathbf{h}_j^{(l)}(t)$ respectively, the probability of the temporal edge (v_i, v_j, t) existing is calculated as follows,

$$\lambda_{i,j}(t) = \text{FFN}\left(\mathbf{h}_i^{(l)}(t) \| \mathbf{h}_j^{(l)}(t)\right) \tag{5}$$

$$= \text{ReLU}\left(\left[\mathbf{h}_i^{(l)}(t) \| \mathbf{h}_j^{(l)}(t)\right] \mathbf{W}_0 + \mathbf{b}_0\right) \mathbf{W}_1 + \mathbf{b}_1 \tag{6}$$

where $\mathbf{W}_0$ and $\mathbf{W}_1$ represent the weight matrices of two fully connected neural networks. $\mathbf{b}_0$ and $\mathbf{b}_1$ represent the bias vectors respectively. Given a temporal edge (v_i, v_j, t), we minimize the following reconstruction loss:

$$\mathcal{L}_e(v_i, v_j, t) = -\log\left(\sigma\left(\lambda_{i,j}(t)\right)\right) - Q \cdot \mathbb{E}_{v_k \sim P_n} \log\left(\sigma\left(1 - \lambda_{i,k}(t)\right)\right) \tag{7}$$

where σ is the Sigmoid function. A negative sample (v_i, v_k, t) is generated by randomly corrupting the target node of edge (v_i, v_j, t). P_n is the negative sampling distribution over the node space, and Q is the negative sample size.

Static Feature Decoder. The node's original static features are closely related to its behavior, and therefore should be preserved in its temporal representation. MATE utilizes a MLP to reconstruct a node's original static features, and uses a reconstruction loss to guide the encoder to maximize the preservation of the node's original static features. Taking node v_i as an example, its reconstructed feature is denoted as $\tilde{\mathbf{x}}_i(t)$, and the process can be represented as,

$$\tilde{\mathbf{x}}_i(t) = \text{MLP}\left(\mathbf{h}_i^{(l)}(t)\right) = \mathbf{h}_i^{(l)}(t) \mathbf{W}_2 + \mathbf{b}_2 \tag{8}$$

where $\mathbf{W}_2$ and $\mathbf{b}_2$ stand for weight matrix and bias vector respectively. For a temporal edge (v_i, v_j, t), the feature reconstruction loss is defined as:

$$\mathcal{L}_a(v_i, v_j, t) = \|\tilde{\mathbf{x}}_i(t) - \mathbf{x}_i\|_2^2 + \|\tilde{\mathbf{x}}_j(t) - \mathbf{x}_j\|_2^2 \tag{9}$$

Overall Loss. Based on the aforementioned process, the overall loss function of MATE can be obtained as:

$$\mathcal{L} = \sum_{(v_i, v_j, t) \in \mathcal{E}} \mathcal{L}_e(v_i, v_j, t) + \eta \mathcal{L}_a(v_i, v_j, t) \tag{10}$$

where $\eta > 0$ is a hyperparameter, controlling the weight of static feature reconstruction loss.

4.3   Iterative Interactive Process

The node-wise memory unit and temporal representation learning module of MATE interact with each other in an iterative manner. When a node participates in a new event, i.e., establishes new links, information related to the new event will be used to update its memory vector. Specifically, for a node v_i participating in an event (v_i, v_j, t) and encoded by the temporal GNN-based encoder, a message can be computed to update node v_i's memory vector. The message is computed as follows,

$$\text{msg}_i(t) = \mathbf{h}_i^{(l)}(t)||\mathbf{h}_j^{(l)}(t)||\Phi(\Delta t)||\mathbf{e}_{i,j}(t) \tag{11}$$

where $\mathbf{h}_i^{(l)}(t)$ and $\mathbf{h}_j^{(l)}(t)$ respectively represent the representations of node v_i and v_j after being encoded by the temporal GNN-based encoder at time t. Δt denotes the difference between the current time t and the last update time of node v_i's memory vector. $\mathbf{e}_{i,j}(t)$ is the feature of event (v_i, v_j, t), and $||$ denotes the concatenation operation.

In practice, MATE does not adopt the strategy of processing events between nodes one by one in chronological order and updating memory vectors of nodes immediately when nodes are involved in new events. For the sake of efficiency, MATE uses batch training. Intuitively, in a training batch, there may be multiple events related to the same node v_i, and each event can generate a message. These messages are represented as $\text{msg}_i(t_1), \text{msg}_i(t_2), \cdots, \text{msg}_i(t_n), \text{msg}_i(t)$, where $t_1, \cdots, \leq t_n \leq t$. MATE uses a message aggregator to aggregate these messages and update the node's memory vector using the aggregated message. The message aggregation process can be described as follows:

$$\overline{\text{msg}}_i(t) = \text{agg}(\text{msg}_i(t_1), \text{msg}_i(t_2), \cdots, \text{msg}_i(t_n), \text{msg}_i(t)) \tag{12}$$

There are multiple choices for the implementation of message aggregator. For instance, *mean* (keep the average of all messages for a node) and *most recent* (only keep the most recent message for a node). Considering the time complexity, MATE adopts the strategy of most recent aggregation.

After computing the messages, MATE updates the node's memory vector using GRU [1]. The update process is as follows,

$$\mathbf{m}_i(t^+) = \text{GRU}(\overline{\text{msg}}_i(t), \mathbf{m}_i(t)) \tag{13}$$

where $\mathbf{m}_i(t)$ represents the memory vector of node v_i obtained from the memory unit at the current time t. $\mathbf{m}_i(t^+)$ represents the memory vector after being updated at time t. Since the memory vector of node v_i is updated using the message at current time t, the update time of the memory vector of node v_i is also updated to t.

Algorithm 1. Training Procedure of MATE

Input: $\mathcal{G} = (\mathcal{V}, \mathcal{E}, \mathcal{T})$: training graph; X: node feature matrix; E: temporal edge feature matrix; K: the total number of training epochs; $\mathcal{B}$: batch size; α: learning rate; η: the coefficient of feature reconstruction loss

Output: θ: learning parameters of MATE

1: initialize θ;
2: arrange temporal edges in $\mathcal{E}$ in chronological order;
3: **for** $k = 1 \rightarrow K$ **do**
4:     initialize memory vector of each node with a zero vector;
5:     $msg \leftarrow \{\}$;
6:     **while** not converged **do**
7:         $\overline{msg} \leftarrow$ aggregator (msg)
8:         update node-wise memory unit using aggregated message $\overline{msg}$;
9:         sample a batch of temporal events with batch size $\mathcal{B}$ from $\mathcal{E}$;
10:        **for** each interaction (v_i, v_j, t) in the batch **do**
11:            get memory vector $\mathbf{m}_i(t), \mathbf{m}_j(t)$ of node v_i and v_j;
12:            calculate temporal representations $\mathbf{h}_i^{(l)}(t), \mathbf{h}_j^{(l)}(t)$ for v_i, v_j;
13:            decode latent representations and calculate overall loss;
14:            $msg \leftarrow$ calculate message of nodes v_i, v_j;
15:        **end for**
16:        $\theta \leftarrow$ backpropagation of overall loss with learning rate α;
17:    **end while**
18: **end for**
19: **return** θ.

We present the pseudo-code for the training procedure of MATE in Algorithm 1. From the training process, it can be seen that MATE does not update the memory vector of a node when processing the current batch, i.e., the memory vector of the same node accessed at different times in the same batch is the same. The batch size affects the granularity of model updates, and an appropriate batch size can reduce computation time without affecting model performance.

5 Experiments

5.1 Experimental Setup

Datasets. We adopt three public datasets, namely cit-HepTh, Movielens and Wikipedia. The statistics of these datasets are summarized in Table 1.

- **cit-HepTh** [10]: a citation graph on the theory of high energy physics. If a paper v_i cites paper v_j at time t, there exists a temporal edge (v_i, v_j, t). The abstracts of each paper are trained by word2vec [14] and the node features are initialized as the output vectors.
- **Movielens** [5]: a dataset of users rating for movies. A temporal edge (v_i, v_j, t) indicates a user v_i rates movie v_j at time t. The textual features of users and movies are preprocessed to be used as node features.

Table 1. Statistics of datasets.

Dataset	# Nodes	# Temporal edges	# Node features	Timespan
cit-HepTh	7,577	51,315	128	124 months
Movielens	2,625	100,000	24	7 months
Wikipedia	8,227	157,474	172	30 days

- **Wikipedia** [9]: a temporal graph where the temporal edges represent interactions of users editing on the Wikipedia pages in one month. The textual features of user edits are extracted and transformed into 172-dimensional LIWC feature vectors. Edit vectors for each user are added and normalized as node features.

Main Task. We use *temporal link prediction* as our main task. We chronologically split the dataset into training set, validation set and test set by 70%–15%–15% according to timestamps of temporal edges. In testing, we treat edges in test set as positive ones, and randomly sample the same number of negative node pairs. For a candidate triple (v_i, v_j, t) in testing time, the goal is to predict whether nodes v_i and v_j will form a link at time t. To be more specific, we first generate representations of nodes v_i and v_j at time t based on the trained encoder, then calculate a score of (v_i, v_j, t) by using the dynamic event decoder. We adopt *accuracy* (ACC) and *average precision* (AP) as evaluation metrics. We run our model for ten times and report the mean and standard deviation of these metrics.

Baselines. In our experiments, several state-of-the-art graph representation learning methods are selected as baselines, which can be categorized into two groups. (1)*Static methods*: VGAE [7], GAT [22] and GraphSAGE [4]. They are trained on the static graph formed from the training edges, without considering any time information. (2)*Temporal methods*: DySAT [19], we train this discrete-time model by splitting all data into 10 static snapshots. JODIE [9], DyRep [21], TGAT [25], TGN [18] and TREND [24]. They are trained by using the same data partitioning as MATE.

Implementation Details. The parameters of MATE are set as follows: for cit-HepTh, we set the number of layers of the temporal GNN-based encoder to 1 and the number of sampled temporal neighbors to 15. For Movielens, we set the number of layers of the temporal GNN-based encoder to 2 and the number of sampled temporal neighbors to 5. For Wikipedia, we set the number of layers of the temporal GNN-based encoder to 1 and the number of sampled temporal neighbors to 10. For all datasets, we fix the output dimension of each layer of the temporal graph neural network and the dimension of the time encoding to be the same as that of the node features. We use most recent sampling instead

Table 2. Overall performance on temporal link prediction. All results are converted to percentage by multiplying by 100. The best results are **bold**.

Model	cit-HepTh		Movielens		Wikipedia	
	ACC	AP	ACC	AP	ACC	AP
VGAE [7]	52.00 (0.66)	56.12 (0.65)	52.79 (1.00)	64.41 (2.01)	51.04 (0.59)	52.69 (0.43)
GAT [22]	66.39 (3.55)	78.94 (2.24)	63.76 (3.79)	66.62 (3.27)	72.71 (2.00)	70.45 (2.35)
GraphSAGE [4]	65.25 (2.05)	82.59 (2.70)	58.84 (2.64)	68.97 (2.44)	76.57 (0.35)	75.39 (0.52)
DySAT [19]	77.03 (1.50)	79.31 (2.26)	50.86 (0.88)	61.68 (1.73)	61.23 (5.87)	84.87 (1.24)
JODIE [9]	63.20 (0.90)	66.94 (1.96)	73.26 (0.24)	76.63 (0.53)	81.65 (1.21)	89.75 (2.04)
DyRep [21]	65.25 (1.13)	68.85 (1.36)	73.23 (0.22)	76.43 (0.40)	83.40 (1.13)	93.13 (0.72)
TGAT [25]	80.04 (1.81)	86.11 (2.22)	71.62 (0.68)	77.02 (0.68)	80.77 (0.98)	90.98 (0.82)
TGN [18]	82.46 (0.93)	89.00 (0.87)	75.87 (1.56)	81.01 (2.08)	88.69 (0.27)	96.21 (0.14)
TREND [24]	75.68 (1.27)	80.88 (1.55)	65.62 (2.69)	63.72 (2.12)	77.56 (0.80)	87.77 (0.62)
MATE	**86.54 (0.41)**	**92.93 (0.28)**	**80.24 (0.36)**	**86.10 (0.63)**	**89.56 (0.69)**	**96.75 (0.32)**

of random uniform sampling for temporal neighbor sampling. For all datasets, we set the batch size to 200. The number of negative samples for all datasets is set to 1. We employ Adam optimizer for training with the learning rate 0.0002 for cit-HepTh and Wikipedia, 0.0001 for Movielens. To avoid overfitting, we set the maximum number of epoch to 150 and use early stopping strategy with patience set to 10. We implement MATE by Python 3.8.13 with PyTorch 1.7.1. The source code is publicly available on Github[1].

5.2 Overall Performance Comparisons

We present the evaluation results in Table 2. The experimental results show that MATE achieves the best performance on all metrics for the three datasets.

In general, temporal graph representation learning methods outperform static methods, which validates the importance of time information. Applying static methods to temporal graphs ignores temporal constraints, which may result in the model erroneously utilizing future information for training on past interaction events, consequently leading to inaccurate inferences during the testing phase. For cit-HepTh, most of the continuous-time methods are superior to discrete-time methods. For Movielens and Wikipedia, all continuous-time methods outperform discrete-time methods. This demonstrates the superiority of continuous-time methods in modeling temporal graph evolution. Among continuous-time methods, most GNN-based methods (TGAT, TGN, and MATE) achieve good performance, indicating the strong ability of GNN in capturing graph structure evolution. However, TGAT and TGN fail to fully capture the driving factors of temporal graph evolution as they do not consider the influence of the static features of nodes on their behavior. Among GNN-based approaches, MATE performs best, which verifies the significance of node-wise memory unit and static feature decoder.

---

[1] https://github.com/scu-kdde/HGA-MATE-2023.

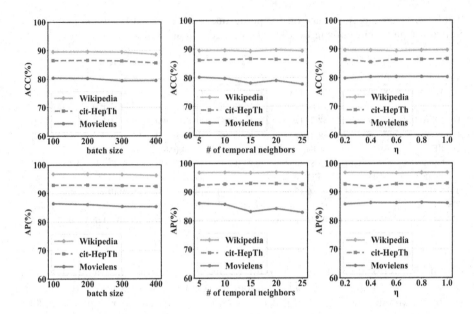

Fig. 3. Results of hyperparameter study.

5.3 Hyperparameter Study

To investigate the robustness of MATE, we analyze the impact of some important hyperparameters on model performance. As shown in Fig. 3, for cit-HepTh, the model performs consistently well when the batch size is 100 and 200. As the batch size increases beyond 200, the model performance gradually decreases. As mentioned earlier, MATE does not update the memory vectors of related nodes immediately after each event occurs, but updates their memory vectors after processing each batch. For the first event in the current batch, the information in the node-wise memory unit is the latest because it involves all the information about previous events. For the last event in the current batch, the information in the node-wise memory unit is outdated because it does not contain information about previous events in the current batch. Therefore, selecting an appropriate batch size requires a trade-off between model performance and speed. For the number of sampled temporal neighbors, it can be seen from the results that a small number of temporal neighbors cannot fully express the temporal topological structure of the graph, while sampling too many temporal neighbors can introduce redundant information. As for η, in general, when η is set to 1, the model performs the best. The static feature decoder is used to guide the encoder to retain the static features of the nodes when learning their temporal representations. The results show that the original static features of nodes contain rich information in all datasets, indicating a strong correlation between the behavior of the nodes and their intrinsic features.

Table 3. Results of ablation study.

Model	cit-HepTh		Movielens		Wikipedia	
	ACC	AP	ACC	AP	ACC	AP
MATE_d	86.23 (0.60)	92.59 (0.46)	80.10 (0.49)	85.94 (0.71)	88.68 (0.97)	96.27 (0.48)
MATE_m	85.72 (0.41)	92.17 (0.48)	65.21 (1.78)	69.09 (2.08)	81.56 (0.51)	90.98 (0.60)
MATE_uni	86.23 (0.49)	92.69 (0.45)	77.00 (0.63)	82.39 (0.81)	89.41 (0.33)	96.62 (0.20)
MATE	**86.54 (0.41)**	**92.93 (0.28)**	**80.24 (0.36)**	**86.10 (0.63)**	**89.56 (0.69)**	**96.75 (0.32)**

5.4 Ablation Study

To investigate the impact of each component of MATE on its performance, we design three variants based on MATE by removing/replacing one component at a time: (1) MATE_d: removing the static feature decoder, in which case only the temporal edge reconstruction loss is retained in the overall loss; (2) MATE_m: removing the node-wise memory unit; (3) MATE_uni: replacing the most recent sampling by uniform sampling. The results are listed in Table 3.

Clearly, MATE outperforms its variants on all metrics in all datasets. Among them, MATE_m performs the worst, indicating the importance of learning from the chronological order of events. Secondly, the experimental results of MATE_d show that the original static features of the nodes have a significant impact on their behavior. Finally, the experimental results of MATE_uni indicate that the information contained in temporal neighbors closer to the current moment is more important. In conclusion, each component of MATE is indispensable for modeling temporal graphs.

5.5 Inductive Capability Analysis

The model's inductive ability refers to its ability to process nodes not seen in the training set. This ability is particularly critical for temporal graph representation models, as new nodes will continue to be added over time and existing nodes will interact with new ones.

We randomly select 10% of nodes from dataset and remove the edges related to these nodes from the training set. We also remove the edges that do not contain these nodes from the validation and test sets. We select TGAT and TGN, which are both inductive temporal graph representation learning models, as comparative models.

As shown in Table 4, MATE outperforms the comparative models in all metrics. The experimental results confirm that MATE has strong inductive ability.

Table 4. Results of inductive learning.

Model	cit-HepTh		Movielens		Wikipedia	
	ACC	AP	ACC	AP	ACC	AP
TGAT [25]	75.39 (1.25)	80.05 (1.69)	68.60 (0.59)	74.50 (0.80)	71.97 (1.54)	81.94 (1.27)
TGN [18]	77.38 (3.39)	82.68 (4.05)	71.73 (2.61)	76.61 (3.88)	79.40 (1.08)	90.89 (0.63)
MATE	**83.87 (0.54)**	**89.62(0.51)**	**77.39 (0.61)**	**83.58 (0.79)**	**80.33 (0.72)**	**91.29 (0.46)**

6 Conclusion

In this paper, we propose a novel method for temporal graph representation learning, called MATE. By introducing a node-wise memory unit to store the historical information of nodes, MATE can learn from the sequentiality of events. Thanks to the temporal GNN-based encoder, dynamic event decoder as well as static feature decoder, MATE can capture the temporal dynamics of the graph while preserving the original static features of nodes. Comparative experimental results on real-world datasets have verified the effectiveness of MATE.

References

1. Cho, K., et al.: Learning phrase representations using RNN encoder-decoder for statistical machine translation. In: EMNLP, pp. 1724–1734 (2014)
2. Dai, H., Wang, Y., Trivedi, R., Song, L.: Deep coevolutionary network: embedding user and item features for recommendation. arXiv preprint arXiv:1609.03675 (2016)
3. Grover, A., Leskovec, J.: node2vec: Scalable feature learning for networks. In: SIGKDD, pp. 855–864 (2016)
4. Hamilton, W.L., Ying, Z., Leskovec, J.: Inductive representation learning on large graphs. In: NeurIPS, pp. 1024–1034 (2017)
5. Harper, F.M., Konstan, J.A.: The movielens datasets: history and context. ACM Trans. Interact. Intell. Syst. **5** (2016)
6. Hawkes, A.G.: Spectra of some self-exciting and mutually exciting point processes. Biometrika **58**, 83–90 (1971)
7. Kipf, T.N., Welling, M.: Variational graph auto-encoders. CoRR abs/1611.07308 (2016)
8. Kipf, T.N., Welling, M.: Semi-Supervised classification with graph convolutional networks. In: ICLR (2017)
9. Kumar, S., Zhang, X., Leskovec, J.: Predicting dynamic embedding trajectory in temporal interaction networks. In: SIGKDD, pp. 1269–1278 (2019)
10. Leskovec, J., Kleinberg, J.M., Faloutsos, C.: Graphs over time: densification laws, shrinking diameters and possible explanations. In: SIGKDD, pp. 177–187 (2005)
11. Li, J., Dani, H., Hu, X., Tang, J., Chang, Y., Liu, H.: Attributed network embedding for learning in a dynamic environment. In: CIKM, pp. 387–396 (2017)
12. Li, Z., et al.: Temporal knowledge graph reasoning based on evolutional representation learning. In: SIGIR, pp. 408–417 (2021)
13. Lu, Y., Wang, X., Shi, C., Yu, P.S., Ye, Y.: Temporal network embedding with micro- and macro-dynamics. In: CIKM, pp. 469–478 (2019)
14. Mikolov, T., Sutskever, I., Chen, K., Corrado, G.S., Dean, J.: Distributed representations of words and phrases and their compositionality. In: NeurIPS, pp. 3111–3119 (2013)
15. Nguyen, G.H., Lee, J.B., Rossi, R.A., Ahmed, N.K., Koh, E., Kim, S.: Continuous-time dynamic network embeddings. In: WWW, pp. 969–976 (2018)
16. Pareja, A., et al.: EvolveGCN: evolving graph convolutional networks for dynamic graphs. In: AAAI, pp. 5363–5370 (2020)
17. Perozzi, B., Al-Rfou, R., Skiena, S.: DeepWalk: online learning of social representations. In: SIGKDD, pp. 701–710 (2014)

18. Rossi, E., Chamberlain, B., Frasca, F., Eynard, D., Monti, F., Bronstein, M.M.: Temporal graph networks for deep learning on dynamic graphs. CoRR abs/2006.10637 (2020)
19. Sankar, A., Wu, Y., Gou, L., Zhang, W., Yang, H.: DySAT: deep neural representation learning on dynamic graphs via self-attention networks. In: WSDM, pp. 519–527 (2020)
20. Trivedi, R., Dai, H., Wang, Y., Song, L.: Know-evolve: deep temporal reasoning for dynamic knowledge graphs. In: ICML, vol. 70, pp. 3462–3471 (2017)
21. Trivedi, R., Farajtabar, M., Biswal, P., Zha, H.: DyRep: learning representations over dynamic graphs. In: ICLR (2019)
22. Velickovic, P., Cucurull, G., Casanova, A., Romero, A., Liò, P., Bengio, Y.: Graph attention networks. In: ICLR (2018)
23. Wang, Y., Chang, Y., Liu, Y., Leskovec, J., Li, P.: Inductive representation learning in temporal networks via causal anonymous walks. In: ICLR (2021)
24. Wen, Z., Fang, Y.: TREND: temporal event and node dynamics for graph representation learning. In: WWW, pp. 1159–1169 (2022)
25. Xu, D., Ruan, C., Körpeoglu, E., Kumar, S., Achan, K.: Inductive representation learning on temporal graphs. In: ICLR (2020)
26. Zhou, L., Yang, Y., Ren, X., Wu, F., Zhuang, Y.: Dynamic network embedding by modeling triadic closure process. In: AAAI, pp. 571–578 (2018)
27. Zhu, D., Cui, P., Zhang, Z., Pei, J., Zhu, W.: High-order proximity preserved embedding for dynamic networks. IEEE Trans. Knowl. Data Eng. 30, 2134–2144 (2018)
28. Zuo, Y., Liu, G., Lin, H., Guo, J., Hu, X., Wu, J.: Embedding temporal network via neighborhood formation. In: SIGKDD, pp. 2857–2866 (2018)

HoME: Homogeneity-Mining-Based Embedding Towards Detecting Illicit Transactions on Bitcoin

Zitian Chen[✉], Guang Li, Danyang Xiao, Weigang Wu, and Jieying Zhou

School of Computer Science and Engineering, Sun Yat-sen University,
Guangzhou, China
{chenzd8,liguang7,xiaody}@mail2.sysu.edu.cn,
{wuweig,isszjy}@mail.sysu.edu.cn

Abstract. Cryptocurrencies have brought booming economic innovations in recent years, but they have also given rise to many intractable problems related to financial crimes, urgently requiring for some effective countermeasures. Transaction network analysis based on machine learning methods has been widely used to detect illicit transactions. In this paper, we aim at mining the homogeneity of transaction patterns, with the inspiration that homogeneous transactions share similar dominant features and such information can be used to enhance illicit detection. Accordingly, we propose the **H**omogeneity-Mining-based **E**mbedding (HoME) framework. The main idea of HoME is to convert homogeneity among transactions into topological connectivity for representation learning. First, we design the Pure Cluster Search algorithm based on agglomerative hierarchical clustering to find out groups of homogeneous transactions. Then, we represent homogeneity using virtual vertices and add them into the transaction network. Further, we design a Two-Phase Walk method to capture neighborhood similarity in the new transaction network. As shown in evaluations on the well-known Elliptic dataset, HoME can achieve good effectiveness and robustness, outperforming popular baselines.

Keywords: Blockchain · Bitcoin · Illicit Transaction Detection · Data Mining · Node Embedding

1 Introduction

In recent years, the public has observed a remarkable surge in the popularity of cryptocurrencies. Despite booming innovations, the prosperity of cryptocurrencies has also introduced new challenges in terms of illicit activities within the global financial ecosystem. The inherent anonymity of cryptocurrencies offers convenience for conducting illicit activities, including but not limited to phishing [3] and Ponzi scams [5], dark web trading [9] and money laundering [20]. It is reported that illicit transaction volume hit an all-time high of $20.6 billion in

X. Song et al. (Eds.): APWeb-WAIM 2023, LNCS 14331, pp. 422–436, 2024.
https://doi.org/10.1007/978-981-97-2303-4_28

2022[1]. Nowadays, regulators around the world face a significant dilemma: How to harness the benefits of advanced decentralized technology to promote efficiency in trading and investment while ensuring financial security? Developing practical and effective techniques for detecting illicit activity is essential to steer the situation towards a positive direction.

Bitcoin [14] is the pioneer of cryptocurrencies as well as the hotbed of illicit activities since its birth. For instance, a notorious darknet marketplace SilkRoad used it for trading a large number of illicit substances, amounting to around 10 million Bitcoins between 2011 and 2013[2]. Some of existing works on detecting illicit activities focus on specific illicit types, such as dark web [9], ransomware [1,6] and Ponzi scams [5], while other works do not distinguish specific types. These works on detecting illicit activities of generic type are commonly referred to as Anti-Money Laundering, since different types of illicit activities are often interconnected through money laundering services to evade detection and tracking.

Taking into account the association among transactions, Bitcoin data can be effectively represented by a transaction network, wherein the nodes represent transactions and the edges denote the flow of payments. Given the format of network, it is practical to introduce graph learning that mines topological information to facilitate illicit activity detection. In existing works on detecting illicit activities of general type [2,18,20], however, machine learning techniques, particularly ensemble learning methods, have demonstrated satisfactory performance when leveraging statistical features, while graph learning methods which aim to aggregate neighbor information counter-intuitively have either yielded inferior performance or only shown marginal improvement.

In order to adequately exploit topological information of the transaction network to improve detection, our work focuses on the homogeneity of transaction patterns. We define a group of transactions as homogeneous if they share similar dominant features and possess the same label. Then, **homogeneous transactions** are regarded as having the **same pattern**. Within a transaction network, a connection between two transactions indicates a dependency between their respective patterns. In other words, if transaction t_1 and t_2 are homogeneous, then t_1's neighbor is expected to exhibit a similar pattern to t_2's neighbor.

Our work aims to address the following question: **How can we capture the similarity among the neighbors of homogeneous transactions and utilize this information to enhance illicit transaction detection?**

Accordingly, we propose the Homogeneity-Mining-based Embedding (HoME) framework to convert homogeneity among transactions into topological connectivity in the Bitcoin transaction network for representation learning. First, we design a Pure Cluster Search algorithm based on agglomerative hierarchical clustering. It adopts Depth-First-Search to extract groups of homogeneous transactions that share the same pattern on an agglomerative hierarchy. Then, we represent homogeneity using virtual vertices in the Bitcoin transaction network. Each virtual vertex corresponds to a group of homogeneous transactions.

---

[1] https://go.chainalysis.com/2023-Crypto-Crime-Report.html.
[2] https://archive.org/details/UlbrichtCriminalComplaint_201310.

Moreover, we add new edges to connect homogeneous transactions to their respective virtual vertices. This arrangement places a virtual vertex as a shared neighbor for a group of homogeneous transactions. Subsequently, we employ a Two-Phase Walk method to capture neighborhood similarity within the modified transaction network. The Two-Phase Walk consists of two steps: uniform walking and biased walking.

The union of vertex series derived from the Two-Phase Walk is ultimately used as input for a SkipGram [13] model to learn embeddings for Bitcoin transactions. The performance of HoME is evaluated with different base classifiers (Random Forest, Gradient Boosting, AdaBoost) on the Elliptic [20] data set, the most popular data set for detecting illicit activity on Bitcoin. We also provide sensitive analysis on key parameters of HoME.

The main contribution of this paper can be summarized as follows:

– We propose a HoME framework that exploits homogeneity of transaction patterns for representation learning towards illicit transaction detection on Bitcoin.
– We design a novel Pure Cluster Search algorithm to find out homogeneous transactions and a Two-Phase Walk method to obtain transaction embeddings.
– We evaluate the effectiveness and robustness of HoME using the Elliptic benchmark dataset. The experimental results indicate that HoME is superior to nine popular baselines.

The remainder of this paper is arranged as follows. Section 2 summarizes prior researches related to our work; consequently Sect. 3.2 illustrates the details of our proposed HoME framework. Section 4 presents experimental settings and results with corresponding analysis. Finally, Sect. 5 concludes our work.

2   Related Work

Existing works on detection of illicit activities in cryptocurrencies can be divided into two main categories. (1) Works focusing on detection performance, which aim at improving the performance of detection by extracting relevant features or developing effective classification models. (2) Works focusing on utilization of a limited amount of labeled data, designing heuristic expansion strategy or improving training efficiency.

The former can be further divided into two sub-categories: statistical feature based and graph topology based. Statistical features can be effectively leveraged by traditional machine learning methods. In [20], the authors provided the Elliptic dataset and compare effectiveness of various methods, presenting the superiority of Random Forest. In [9], a voting-based classification method is employed to identify Bitcoin addresses involved in darknet markets, with conclusion that transactions related to darknet tend to provide higher fee than others. In [18], the authors proposed a detection framework based on Adaptive Stacked eXtreme Gradient Boosting (ASXGB) to handle evolving data streams, and adopted a data-sampling approach NCL-SMOTE to achieve additional improvement, using the Elliptic dataset as well. In [4], account features and contract opcode features are extracted and fed into an XGBoost classifier to detect Ponzi scheme of smart contracts in Ethereum. In [17], the authors studied 2-motifs in Bitcoin transaction network and proposed a pattern of short thick band (STB) as a potential laundering pattern. Furthermore, [8] considered more generally N-motifs and found that 3-hop graph neighborhood features dominate. In [3], a graph based hierarchical feature extraction method is proposed to assist detection of phishing scams in Ethereum. As for network embedding, [21] designed an embedding method that samples the Ethereum network based on transaction amount and timestamp, and used one-class SVM to identify phishing activities. In [15], the authors extracted features based on the distance of one node to a known illicit target. Graph learning method such as Graph Convolution Network (GCN) is applied as well. In [20], the authors also adopted EvolveGCN to capture the dynamics of transaction networks. In [2], a modified GCN with random walk normalisation is used to improve performance of detection.

The latter category deals with the challenge of working with a limited amount of labeled data in the detection task. In [22], the authors uncovered two types of Bitcoin mixing mechanisms, namely exchange and confusion. They also develop a heuristic algorithm to extend the addresses of mixing services and trace the flow of funds in a real attack scenario. In [11], the authors analyzed illicit activities on Ethereum and proposed an improved Smart Local Moving algorithm to expand an initial illicit address set. To meet the challenge of label scarcity, [12] studied the application of active learning on the illicit transaction detecting task, where only 10% of tags are utilized to achieve performance close to the best supervised baseline.

In summary, existing works on illicit activity detection encompass traditional machine learning, network embedding and heuristic methods. Similarly to [15, 21], our work designs a specialized embedding method to enhance detection. The key distinction lies in our method's modification of transaction network through the mining of homogeneous patterns for a more comprehensive integration of statistical features and topological information.

3   Methodology

3.1   Problem Definition

The transaction detection task is regarded as a graph node binary classification problem. Given a transaction network $G = (V, E, X, L)$, $V = \{v_1, \cdots, v_n\}$ is a set of transactions, $E \subseteq V \times V$ is a set of edges and $\epsilon = (v_i, v_j) \in E$ indicates that an input of v_j points to an output of v_i. $X \in \mathbb{R}^{|V| \times S}$ represents transaction features, where S is the size of the feature space. $L \in \mathbb{R}^{M \times 2}$ is a set of labels. For simplicity, G is treated as indirected. Our framework aims to learn representation $\Theta \in \mathbb{R}^{|V| \times \gamma}$ with embedding size γ for transactions in V.

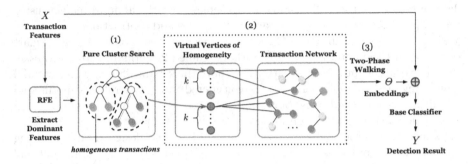

Fig. 1. Overall design of HoME. We use red and blue to denote the classes of illicit and licit, respectively. Within the tree-view of the Pure Cluster Search algorithm, leaf nodes enclosed by a dashed line indicate a group of homogeneous transactions, which corresponds to a virtual vertex. The virtual vertices are connected to their respective groups of transactions. The labels of gray nodes in the transaction network are unknown. (Color figure online)

3.2 Framework Design

Figure 1 demonstrates the overall design of our Homogeneity-Mining-based Embedding (HoME) framework. Initially, we utilize RFE (recursive feature elimination) to extract a set of dominant features. During each round of RFE, the candidate features are ranked based on their importance scores, and subsequently the least important feature(s) are eliminated. The recursive procedure terminates when the number of candidate features reaches a predefined threshold λ ($\ll S$). The dominant feature space is denoted as X_d. Following the main process of HoME, the embedding vectors Θ are concatenated with the statistical features X and fed into a base classifier, such as Random Forest, to generate the classification result Y. Algorithm 1 demonstrates the main process of HoME, emphasizing three essential steps as follows.

(1) Obtain homogeneous transactions in the form of *pure clusters*. For conciseness, we use the term *pure* as an adjective to indicate that samples within a cluster share the same label. Additionally, it is worth noting that samples grouped together within a cluster also possess similar features. As a result, a pure cluster represents a group of homogeneous transactions that exhibit the same pattern. We propose the Pure Cluster Search (see Sect. 3.3) algorithm which offers a certain number ($2 \times k$) of pure clusters.

(2) Add virtual vertices into the transaction network. The essence of this step is a conversion from the abstract homogeneity relationship to explicit topological connectivity among transaction vertices. Each pure cluster corresponds to a virtual vertex. Additionally, to address the issue of the class-imbalance in the data, a balancing strategy is applied (see Sect. 3.4). As a result, we obtain a new transaction network denoted as G'.

Algorithm 1: HoME

Input: transaction graph $G(X, V, E, L)$
   dominant features X_d, pure clusters per class k
   uniform walks per vertex r_u, biased walks per virtual vertex r_b
   walk length l, window size w, embedding size γ
Output: matrix of transaction embedding $\Theta \in \mathbb{R}^{|V| \times \gamma}$
   preliminary identification result L'

1 /* (1) Obtain homogeneous transactions */
2 $C \Leftarrow \text{PureClusterSearch}(X_d, L, k)$
3 /* Identify partial unseen samples preliminarily */
4 Initialize L' as an empty set.
5 **for** $x_i \in X_{unseen}$ **do**
6   Initialize A as an empty list.
7   **for** $c_j \in C$ **do**
8   **if** $\|\mu_j - x_i\|_2 \leq \delta_j$ **then**
9   Append j to A.
10   **if** $|A| > 0$ **then**
11   Sort A in ascending order of $\|\mu_j - x_i\|_2$.
12   Put i into $c_{A[0]}$ and $x_i \rightarrow c_{A[0]}$'s label into L'.
13 /* (2) Add virtual vertices into the transaction network */
14 Initialize V', E' as empty sets.
15 **for** $c_j \in C$ **do**
16   /* Create a new virtual vertex v_j corresponding to c_j */
17     $V' \Leftarrow V' \cup \{v_j\}$
18   **for** $i \in c_j$ **do**
19         $E' \Leftarrow E' \cup \{(v_i, v_j)\}$
20 Balance E' to obtain E''.
21 $G' \Leftarrow (X, V \cup V', E \cup E'', L)$
22 /* (3) Execute Two-Phase Walk for representation learning */
23 $W \Leftarrow \text{TwoPhaseWalk}(G', r_u, r_b, l)$
24 $\Theta \Leftarrow \text{SkipGram}(W, w, \gamma)$

---

(3) Execute Two-Phase Walk on G' for representation learning (see Sect. 3.5).
It is a combination of uniform walk and biased walk within a variable ratio.

3.3 Pure Cluster Search

Algorithm 2 demonstrates the detail of Pure Cluster Search. It performs agglomerative hierarchical clustering based on the dominant feature space X_d and traverses the resulting hierarchy in a Depth-First-Search manner to extract pure
clusters. As shown in Fig. 1, a group of homogeneous transactions forms a sub-
tree within the hierarchy. Each pure cluster is classified as either licit or illicit
based on the labels of its constituent transactions.

Algorithm 2: PureClusterSearch

Input: X_p, L, k
Output: C /* a set of pure clusters */
1 Initialize C as an empty set.
2 **Function** *DFS(x)* :
3   **if** *x is leaf* **then**
4         $c \Leftarrow \{x\}$
5   /* retrieve x's label in terms of L */
6   **Return** (c, y)
7      $c_L, y_L \Leftarrow DFS(x.left)$
8      $c_R, y_R \Leftarrow DFS(x.right)$
9   **if** $y_L \neq NaN$ **and** $y_L = y_R$ **then**
10   /* merge c_L and c_R */
11        $C \Leftarrow C \setminus \{c_L, c_R\}$
12        $C \Leftarrow C \cup \{c_L \cup c_R\}$
13   **Return** $(c_L \cup c_R, y_L)$
14   **Return** $(\emptyset, NaN)$
15 **End Function**
16 $H \Leftarrow$ AgglomerativeHierarchicalClustering(X_p)
17 $DFS(H.root)$
18 /* 0 as licit and 1 as illicit */
19 Divide C into C_0, C_1 and remain top-k respectively in terms of cluster size.
20 $C \Leftarrow C_0^k \cup C_1^k$

Based on the pure cluster set, HoME uses Euclidean distance to measure affinity between pure clusters and unseen samples, and further performs preliminary identification to assist the next step of constructing virtual vertices.

Let μ_j be the centroid of pure cluster c_j. Let δ_j be the median of $\|\mu_j - x_i\|_2$, where $x_i \in c_j$. As shown in Algorithm 1, if an unseen transaction sample x_i is closer to the centroid of c_j than a half of samples in c_j ($\|\mu_j - x_i\|_2 \leq \delta_j$), then c_j is regarded as a candidate cluster for x_i. Let c' be the most closest cluster for x_i among its candidate targets. Finally, x_i is assigned to c' and labeled correspondingly. For consistency, HoME uses UPGMC (Unweighted Pair Group Method using Centroids) to measure distance during hierarchical clustering:

$$D_{UPGMC}(c_i, c_j) = \|\frac{1}{|c_i|} \sum_{x_p \in c_i} x_p - \frac{1}{|c_j|} \sum_{x_q \in c_j} x_q\|_2 = \|\mu_i - \mu_j\|_2 \quad (1)$$

3.4 Constructing Virtual Vertices

Given $C = \{c_{01}, ..., c_{0k}, c_{11}, ..., c_{1k}\}$, where $c_{(0,\cdot)}$ represents a licit cluster and $c_{(1,\cdot)}$ represents an illicit cluster, HoME generates additional vertices:

$$V' = \{v_{ij} \mid 0 \leq i \leq 1, 1 \leq j \leq k\} \quad (2)$$

where v_{ij} corresponds to c_{ij}, with additional edges:

$$E' = \bigcup_{\substack{0 \leq i \leq 1 \\ 1 \leq j \leq k}} E'_{ij} \tag{3}$$

where $\epsilon = (v_p, v_{ij}) \in E'_{ij}$ indicates that transaction v_p is clustered in c_{ij}. The order of subscripts is adjusted to meet the following condition:

$$|c_{ip}| \geq |c_{iq}|, \ 0 \leq i \leq 1, \ \forall \, p < q \tag{4}$$

Let s_j be the minimum of $\{|c_{0j}|, |c_{1j}|\}$. Function $\phi(A, s)$ randomly samples from A to obtain a new set of size s. HoME further adjusts the additional edges:

$$E''_{ij} = \begin{cases} E'_{ij}, & |E'_{ij}| = s_j \\ \phi(E'_{ij}, s_j), & |E'_{ij}| > s_j \end{cases} \tag{5}$$

So far, we achieve a balance of new edges for licit versus illicit transactions:

$$|E''_{0j}| = |E''_{1j}|, \ \forall \, 1 \leq j \leq k \tag{6}$$

3.5 Two-Phase Walk

The first phase of Two-Phase Walk is uniform walking that performs random walks for each vertex in $V \cup V'$, leading to a vertex series $(v_i^1, v_i^2, ..., v_i^t)$, where v_i^1 is the starting vertex and v_i^{j+1} is a neighbor of vertex v_i^j chosen randomly. For robustness, we perform r_u independent walks for each vertex. The second phase is biased walking that performs r_b walks only for virtual vertices in V'. Let $d(\cdot)$ denote the shortest path distance between a vertex and the starting vertex. In biased walking, the transition probability of visiting a neighbor n from v is:

$$\pi_{vn} = \frac{exp(-d(n))}{\sum_{n_i \in neighbor(v)} exp(-d(n_i))} \tag{7}$$

Accordingly, a neighbor of the current vertex that is closer to the starting virtual vertex is more likely to be chosen as the next vertex in a walk, so as to enhance capture of homogeneity.

3.6 Illicit Transaction Detection

Given vertex series from the union of uniform walking and biased walking, HoME obtains embedding vectors via SkipGram [13], which can be described as an optimization problem:

$$\min_{\Theta} \ - logPr(\{v_{i-w}, ..., v_{i-1}, v_{i+1}, ..., v_{i+w}\} \mid \Theta(v_i)) \tag{8}$$

Finally, the transaction features and embeddings are concatenated and fed into a base classifier for detection. Moreover, the preliminary identification result L' is used to amend the detection.

$$Y(x) = \begin{cases} L'(x), & if \ x \ in \ L' \\ Classifier([x, \theta]), & else \end{cases} \tag{9}$$

4   Experiments

4.1   Experimental Setup

The proposed HoME is compared with the following popular baselines, including traditional machine learning and graph-base learning methods.

- Random Forest constructs a multitude of decision trees.
- Gradient Boosting iteratively combines weak prediction models into a single strong model based on boosting optimization method.
- AdaBoost tweaks weak prediction models towards good adaptiveness.
- MLP composes of multiple fully-connected layers of perceptrons.
- GCN [10] passes feature information based on Laplacian matrix.
- Skip-GCN inserts skip connection(s) to enhance GCN.
- GAT [19] applies attention mechanism to aggregate neighbor information.
- GraphSAGE [7] exploits neighborhood sampling for inductive learning.
- DeepWalk [16] obtains embedding based on random walk and SkipGram.

HoME is respectively combined with three different base classifiers (Random Forest, Gradient Boosting, AdaBoost) for robustness evaluation. The same combinations are applied on DeepWalk to provide comprehensive comparisons.

Implementations. The traditional machine learning methods are implemented by the scikit-learn Python package. For assemble learning methods, we applies Random Forest with 50 estimators and 50 max features, Gradient Boosting with binary cross-entropy loss function, and AdaBoost with 50 estimators and a learning rate of 0.1. The remaining parameters of the above methods are officially default. MLP is set with one hidden layer of 100 neurons and trained for 500 epochs using the Adam optimizer with a learning rate of 0.001. The graph-base learning methods excluding DeepWalk are implemented by PyTorch package. The numbers of message passing layers in GCN, Skip-GCN, GAT and GraphSAGE are set to 2, 2, 3, 3 respectively. These graph-base learning models are all set with 64 hidden channels and a dropout rate of 0.5 and trained for 1000 epochs using the same setting of Adam optimizer with MLP. For both HoME and DeepWalk, the walk length, window size and embedding size are set to 10, 5 and 128 respectively. The threshold of RFE is set to 10.

Dataset. We use the well-known Elliptic dataset [20], which is the largest publicly available data set for detecting illicit activities in Bitcoin. In this dataset, 203,769 transaction nodes are divided into 49 independent sub-graphs, of which 4,545 (2%) are labelled as illicit and 42,019 (21%) are labelled as licit. Each transaction has 166-dimensional custom designed features, of which the first 94 represent local information and the last 72 represent aggregated information. All features are standardized with zero mean and unit variance. We use the first 30 sub-graphs with 26,905 transactions for training and the remaining 19 sub-graphs with 19,659 transactions for testing (of a 6:4 ratio).

Table 1. Preliminary identification results in homogeneity mining. Given $k = 5, 10, 15,$ 20, the number of identified transactions increases distinctly and the accuracy metrics remains relatively stable.

k	5		10		15		20	
class	licit	illicit	licit	illicit	licit	illicit	licit	illicit
#transactions	902 (5.0%)	425 (26.7%)	1207 (6.7%)	585 (36.7%)	1542 (8.5%)	598 (37.6%)	2047 (11.3%)	604 (38.0%)
(preliminary) accuracy	0.999	1.000	0.999	1.000	0.999	0.995	1.000	0.985

4.2 Experimental Results

Preliminary Identification. Table 1 shows preliminary identification results after Pure Cluster Search (see Sect. 3.3). The number of licit/illicit clusters k is set to 5, 10, 15, 20 respectively. The third row (#transactions) shows different numbers of identified transactions for licit/illicit class when varying k. Numbers in brackets are percentages with respect to unseen samples. We illustrate two main observations: (1) is for correctness and (2) is for robustness with different k.

(1) The accuracy metrics are close to 1 for both classes in all cases, showing notable reliability of our homogeneity mining method. Note that the numbers of identified transactions for licit class are larger than the numbers for illicit class in all cases, due to the imbalance of the data. From the perspective of percentage, 26.7%–38.0% illicit transactions are identified but only 5.0%–11.3% for licit transactions. This discrepancy proves the effectiveness of the balance mechanism in homogeneity mining.

(2) With increase of k, the number of identified transactions increases distinctly. Although a higher $k(= 20)$ identifies more transactions, it meanwhile slightly decreases accuracy, especially for illicit class (1.000 to 0.985). Note that the increase for illicit transactions is not significant (36.7% to 38.0%) given k from 10 to 20, compared with increasing k from 5 to 10. In consideration of the relatively stable accuracy given a proper interval of k and the nonlinear contribution of its increase, we conclude that k is an important but not determining factor in our homogeneity mining method.

Comparative Results. Table 2 shows the effectiveness results of detecting illicit transactions using different methods. The table is divided into three regions. The methods in the first region (from Random Forest to GraphSAGE) only consider statistical features. The second region and the third region provide results of DeepWalk and the proposed HoME combined with three base classifiers (Random Forest, Gradient Boosting, AdaBoost) respectively. Since the goal of our work is to detect illicit transactions, we display three widely-used evaluation metrics: Precision, Recall and F1-score $= (2 \times P \times R)/(P + R)$ only for

Table 2. Effectiveness Results of Detecting Illicit Transactions. Random Forest (RF), Gradient Boosting (GB) and AdaBoost (AB) are employed as base classifiers for both DeepWalk and HoME respectively.

Method	Illicit			WeightedAVG
	Precision	Recall	F1	F1
Random Forest	0.9021	0.7938	0.8445	0.9757
Gradient Boosting	0.8985	0.7957	0.8440	0.9755
AdaBoost	0.9398	0.5984	0.7312	0.9607
MLP	0.6315	0.5902	0.6101	0.9380
GCN	0.6603	0.4451	0.5317	0.9443
Skip-GCN	0.6959	0.4986	0.5810	0.9496
GAT	0.6232	0.7788	0.6924	0.9468
GraphSAGE	0.9473	0.5143	0.6667	0.9619
DeepWalk w/ RF	0.9412	0.7882	0.8579	0.9780
DeepWalk w/ GB	0.8930	0.7910	0.8389	0.9747
DeepWalk w/ AB	0.9357	0.6223	0.7475	0.9628
HoME w/ RF	**0.9637**	**0.8014**	**0.8751**	**0.9807**
HoME w/ GB	0.9147	0.7951	0.8507	0.9767
HoME w/ AB	0.9574	0.6788	0.7944	0.9693

illicit class. For completeness, we also display WeightedAVG F1-score which provides a comprehensive evaluation for both licit and illicit class. The comparative results demonstrate that HoME has superior effectiveness for detecting illicit transactions. Details are concluded as follows:

(1) Ensemble learning methods demonstrate strong performance in this task. Random Forest has the best comprehensive performance (F1-score = 84.5%) for illicit class among the 8 methods with only original features. Despite similar mechanism, AdaBoost achieves higher precision score (94.0%) but has a much lower recall rate (59.8%), after trying various learning rates to achieve a relatively high F1-score (73.1%). This observation suggests that the adaptive strategy employed by AdaBoost may not be well-suited for the data distribution in this particular task.

(2) Graph-base learning methods (GCN, Skip-GCN, GAT and GraphSAGE) are surprisingly inapposite for this task as all have F1-scores below 70%. Note that a degree of aggregated information of transaction nodes has been gathered by the last 72 original features in the data set. Thus, although these methods make good effort to exploit topology information and aggregate neighbor features, the redundancy of data may impede they to capture dominant features, reducing the effect of classification. GAT achieves better comprehensive performance (F1-score = 69.2%) than other 3 methods, reflecting a relative advantage of attention mechanism.

(3) Node embedding via DeepWalk improves F1-score (+1.3%) for Random Forest with an evident increase of precision (+3.9%) and a slight decrease of recall (−0.5%). It works negatively for Gradient Boosting with an decreased F1-score (−0.5%). For AdaBoost, it slightly decreases precision (−0.4%). It indicates that contribution of node embedding via direct DeepWalk is quite limited for detecting illicit transactions.

(4) Our proposed HoME framework exhibits substantial and robust improvement across different classifiers. It achieves a significant increase in F1-score (+3.1%) for Random Forest, accompanied by improvements in precision precision (+6.1%) and recall (+0.8%), surpassing all other methods. For Gradient Boosting, HoME shows a positive impact with an increased F1-score (+0.7%), which distinguishes from DeepWalk that lacks homogeneity information. Particularly noteworthy is the notable enhancement in recall rate (+8.0%) and F1-score (+6.3%) for AdaBoost. Additionally, HoME demonstrates adequate ability to identify licit transactions, as evident from the modest improvements in WeightedAVG F1-score (+0.5%, +0.1%, +0.9%). From a comprehensive perspective, these results highlight two key facts: (i) Homogeneity information significantly contributes to the task of illicit transaction detection, (ii) our mining mechanism effectively exploits valuable homogeneity information, in the presence of imbalanced data.

Parameter Sensitivity. To analyze the parameter sensitivity of HoME, we conducted experiments by varying the number of licit/illicit pure clusters (k), embedding size and walking ratio. The results are presented in Figs. 2, 3 and 4, respectively.

- Figure 2 illustrates that combining the results of HoME with three different base classifiers, a moderate value of k (=10) achieves the best overall performance. This observation aligns with the preliminary identification results, as indicated in Table 1, wherein a higher value of k (=15) leads to a slight decrease in accuracy.
- Figure 3 reveals that there is a marginal diminishing effect when increasing the embedding size. Considering a comprehensive perspective, an embedding size that is closer to the size of the original statistical features in the data (166) tends to yield better results. This can be attributed to the equilibrium of information achieved by having a similar dimensionality between the embeddings and the original features.
- Figure 4 shows the effect of biased walking in HoME. A walking ratio around one thousand provides good performance. It is reasonable since the number of transaction vertices (200K) in the data is around 10,000 times than the number of virtual vertices added by HoME. A walking ratio of zero leads to poor performance, indicating the superior of biased walking.

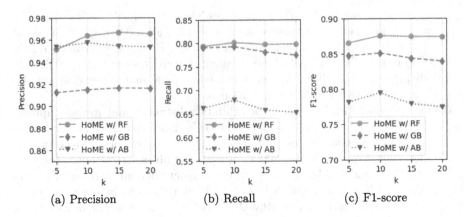

Fig. 2. Performance of HoME with different k (number of licit/illicit clusters)

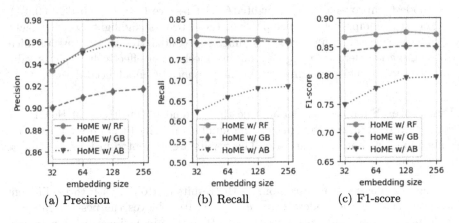

Fig. 3. Performance of HoME with different embedding sizes

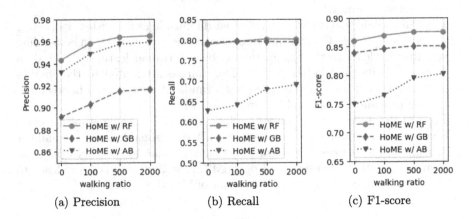

Fig. 4. Performance of HoME with different walking ratios

5   Conclusion and Future Work

In this paper, we design a novel framework HoME for detecting illicit transactions on Bitcoin. HoME consists of three main steps. Firstly, we introduce the Pure Cluster Search algorithm, which utilizes agglomerative hierarchical clustering to extract homogeneous transactions. Secondly, we represent homogeneity as virtual vertices and additional edges within the transaction network. Lastly, we employ a Two-Phase Walk method, combining uniform walking and biased walking, on the modified transaction network to generate embedding vectors. HoME captures similarity among the neighbors of homogeneous transactions, facilitating a more comprehensive integration of statistical features and topological information to enhance illicit transaction detection. Experiments demonstrate that HoME outperforms popular baseline methods and exhibits good robustness.

We hope that our work offers a fresh perspective on cryptocurrency transaction analysis. Moving forward, we will aim at extending the ground-truth data and enhancing the efficiency of representation learning. Meanwhile, we urge the implementation of more effective countermeasures to foster a healthier blockchain ecosystem.

Acknowledgement. This work is supported by The Key-Area Research and Development Program of Guangdong Province (2020B0101090005).

References

1. Akcora, C.G., Li, Y., Gel, Y.R., Kantarcioglu, M.: BitcoinHeist: topological data analysis for ransomware prediction on the bitcoin blockchain. In: Proceedings of the Twenty-Ninth International Joint Conference on Artificial Intelligence (2021)
2. Alarab, I., Prakoonwit, S., Nacer, M.I.: Competence of graph convolutional networks for anti-money laundering in bitcoin blockchain. In: ICMLT '20, pp. 23–27. Association for Computing Machinery, New York, NY, USA (2020)
3. Chen, W., Guo, X., Chen, Z., Zheng, Z., Lu, Y.: Phishing scam detection on ethereum: towards financial security for blockchain ecosystem. In: Proceedings of the Twenty-Ninth International Joint Conference on Artificial Intelligence (2021)
4. Chen, W., Wu, J., Zheng, Z., Chen, C., Zhou, Y.: Market manipulation of bitcoin: evidence from mining the mt. gox transaction network. In: IEEE INFOCOM 2019 - IEEE Conference on Computer Communications, pp. 964–972 (2019)
5. Chen, W., Zheng, Z., Cui, J., Ngai, E., Zheng, P., Zhou, Y.: Detecting Ponzi schemes on ethereum: towards healthier blockchain technology. In: Proceedings of the 2018 World Wide Web Conference, pp. 1409–1418 (2018)
6. Conti, M., Gangwal, A., Ruj, S.: On the economic significance of ransomware campaigns: a bitcoin transactions perspective. Comput. Secur. **79**(C), 162–189 (2018)
7. Hamilton, W.L., Ying, R., Leskovec, J.: Inductive representation learning on large graphs. In: Proceedings of the 31st International Conference on Neural Information Processing Systems. NIPS'17, pp. 1025–1035. Curran Associates Inc., Red Hook, NY, USA (2017)

8. Jourdan, M., Blandin, S., Wynter, L., Deshpande, P.: Characterizing entities in the bitcoin blockchain. In: 2018 IEEE International Conference on Data Mining Workshops (ICDMW), pp. 55–62 (2018)
9. Kanemura, K., Toyoda, K., Ohtsuki, T.: Identification of darknet markets' bitcoin addresses by voting per-address classification results. In: 2019 IEEE International Conference on Blockchain and Cryptocurrency (ICBC), pp. 154–158 (2019)
10. Kipf, T.N., Welling, M.: Semi-supervised classification with graph convolutional networks. In: International Conference on Learning Representations (2017)
11. Li, J., et al.: Measuring illicit activity in DeFi: the case of ethereum. In: Bernhard, M., et al. (eds.) FC 2021. LNCS, vol. 12676, pp. 197–203. Springer, Heidelberg (2021). https://doi.org/10.1007/978-3-662-63958-0_18
12. Lorenz, J., Silva, M.I., Aparício, D., Ascensão, J.a.T., Bizarro, P.: Machine learning methods to detect money laundering in the bitcoin blockchain in the presence of label scarcity. In: Proceedings of the First ACM International Conference on AI in Finance. ICAIF '20. Association for Computing Machinery, New York, NY, USA (2021)
13. Mikolov, T., Chen, K., Corrado, G., Dean, J.: Efficient estimation of word representations in vector space. In: Bengio, Y., LeCun, Y. (eds.) 1st International Conference on Learning Representations, ICLR 2013, Scottsdale, Arizona, USA, 2–4 May 2013, Workshop Track Proceedings (2013)
14. Nakamoto, S.: Bitcoin: a peer-to-peer electronic cash system (2008)
15. Oliveira, C., Torres, J., Silva, M.I., Aparício, D., Ascensão, J.T., Bizarro, P.: GuiltyWalker: distance to illicit nodes in the bitcoin network. arXiv preprint arXiv:2102.05373 (2021)
16. Perozzi, B., Al-Rfou, R., Skiena, S.: DeepWalk: online learning of social representations. In: Proceedings of the 20th ACM SIGKDD International Conference on Knowledge Discovery and Data Mining. KDD '14, pp. 701–710. Association for Computing Machinery, New York, NY, USA
17. Ranshous, S., et al.: Exchange pattern mining in the bitcoin transaction directed hypergraph. In: Financial Cryptography Workshops (2017)
18. Vassallo, D., Vella, V., Ellul, J.: Application of gradient boosting algorithms for anti-money laundering in cryptocurrencies. SN Comput. Sci. **2**(3) (2021)
19. Veličković, P., Cucurull, G., Casanova, A., Romero, A., Liò, P., Bengio, Y.: Graph attention networks. arXiv preprint arXiv:1710.10903 (2018)
20. Weber, M., et al.: Anti-money laundering in bitcoin: experimenting with graph convolutional networks for financial forensics. arXiv preprint arXiv:1908.02591 (2019)
21. Wu, J., et al.: Who are the phishers? phishing scam detection on ethereum via network embedding. IEEE Trans. Syst. Man Cybern. Syst. **52**(2), 1156–1166 (2022)
22. Wu, L., et al.: Towards understanding and demystifying bitcoin mixing services. In: Proceedings of the Web Conference 2021. WWW '21, pp. 33–44. Association for Computing Machinery, New York, NY, USA (2021)

BoundEst: Estimating Join Cardinalities with Tight Upper Bounds

Jia Yang[1], Yujie Zhang[1], Bin Wang[1,2,3], and Xiaochun Yang[1(✉)]

[1] School of Computer Science and Engineering, Northeastern University, Shenyang, China
`yangxc@mail.neu.edu.cn`
[2] National Frontiers Science Center for Industrial Intelligence and Systems Optimization, Shenyang, China
[3] Key Laboratory of Data Analytics and Optimization for Smart Industry (Northeastern University), Ministry of Education, Shenyang, China

Abstract. Cardinality estimation is a critical component of query optimization. Despite extensive research, achieving efficient and accurate estimation for join queries remains challenging. Estimating tight upper bounds for join cardinalities can help the query optimizer generate better and more robust query plans. However, existing methods fail to account for the high skewness of real data and produce loose upper bounds. In this paper, we propose a new framework *BoundEst*, which designs an upper bound formula that accounts for the presence of outliers in the data distribution and introduces the DBSCAN clustering algorithm to identify these outliers. Moreover, we incorporate the learning-based model to learn the correlation between attributes. Given queries, *BoundEst* efficiently estimates tight upper bounds for join cardinalities by applying separate calculation methods to outliers and other values. We evaluate our approach on real-world datasets, and the results show that *BoundEst* generates effective estimates for query optimizer.

Keywords: Cardinality estimation · Join queries · Query optimization

1 Introduction

Cardinality estimation (CardEst) is an important component in the query optimizer. It aims to estimate the number of tuples in the query results or intermediate results, which helps the optimizer choose the optimal join order and physical operators. The accuracy of CardEst directly affects the quality of the generated execution plans. Unfortunately, the existing methods inadequately estimate the cardinality of join queries in terms of both efficiency and accuracy.

Traditional CardEst approaches [3,5,8,18] rely on attribute independence and join-key uniformity assumptions. For instance, they presume that any tuple from one table is equally likely to join with any tuple from another table. However, in real-world datasets, these assumptions are impractical, so the estimated cardinality error can reach several orders of magnitude. When multiple tables are joined, the error will be exponentially propagated.

Query-driven methods [6,13] transform CardEst into supervised learning, with the goal of learning a function that maps queries to cardinalities. Such approaches require executing an impractical number of queries in advance and using true cardinalities as labels to train the model. Additionally, the trained model does not generalize well to unseen queries.

Data-driven methods [10,12,20,22,24] analyze data with machine learning and build distributions over pre-joined tables. They capture all attribute correlations and handle joins accurately, but they are much less efficient and have larger models than traditional methods.

The above CardEst methods work well for single-table queries, but they face challenges for multi-table queries. They need to estimate the cardinality of all sub-queries, which can be inaccurate and time-consuming. Previous researches [1, 4,11,21] have shown that underestimation can result in costly query execution plans, and upper bounds of join cardinalities can help generate efficient query plans. These methods estimate the join upper bounds by bucketing. *PessEst* [4] calculates the upper bounds for join queries by maintaining minimal statistics. Still, it does not support filter predicates, which are too expensive to execute on each relation during optimization. *FactorJoin* [21] can handle filter predicates efficiently, but it produces looser upper bounds due to data skewness as well as *PessEst*.

Estimating tight upper bounds for join cardinalities is very challenging in highly skewed data due to outliers (i.e., items with extreme attribute value frequencies). They are few but have a large impact on the results. In previous works, outliers were treated with equal importance, which affects the accuracy of the estimation. We will describe this in detail in Sect. 3.2. In this work, we develop a novel CardEst framework called *BoundEst* that estimates tight upper bounds of join cardinalities. Specifically, we utilize the Density-Based Spatial Clustering of Applications with Noise (DBSCAN) clustering algorithm [2] to detect outliers in the data distribution and perform separate calculations for these outliers and other values. To improve efficiency, we leverage learning-based methods to capture attribute correlation accurately and avoid executing filter predicates during optimization. We use Bayesian networks [22] for our implementation because they are easy to train and can produce fairly accurate results quickly.

In summary, our contributions are as follows:

(1) We design a new framework *BoundEst* that accounts for the presence of outliers in the data distribution and introduces the DBSCAN clustering algorithm to identify these outliers.
(2) We propose a formula for computing tight upper bounds for join cardinalities that apply separate calculation methods to outliers and other values.
(3) We conduct experiments on real-world datasets to demonstrate the advantages of our framework.

The paper is organized as follows. We define the problem and introduce the related work in Sect. 2. In Sect. 3, we design *BoundEst* framework to efficiently estimate join cardinalities with tight upper bounds. Section 4 shows the experimental analyses, and Sect. 5 concludes the whole paper.

2   Preliminaries

In Sect. 2.1, we present the problem definition of CardEst, and in Sect. 2.2, we provide an overview of existing methods.

2.1   Problem Definition

Given a database instance D with m tables $T_1, T_2, \ldots, T_m$, a query Q consists of two parts: join conditions among the selected tables and a set of base-table filter predicates. Then the cardinality of query Q can be denoted as $|Q| = |\delta(T_1) \bowtie \delta(T_2) \bowtie \ldots \bowtie \delta(T_m)|$, where $\delta(T_i)$ denotes the filters applied to table T_i and it can be empty set if table T_i is not touched by Q. Similarly, its subqueries can be denoted in the same form. The goal of the CardEst problem is to estimate the result size of query Q and its subqueries without executing the query.

2.2   Related Work

CardEst has been extensively studied by both academic and industrial communities. Open-source and commercial DBMSs mainly use histogram [3,5] in PostgreSQL and sampling [16,17,28] in MySQL. With the prosperity of machine learning (ML), we witness a proliferation of learned methods for CardEst in the recent years [6,10,12,20,22,25–27,29].

Traditional Methods. Histogram-based methods employ individual histograms for each column to approximate the distribution of data. The fundamental concept of sampling methods is to execute queries on a small dataset that is sampled from the original data, allowing for the estimation of the overall distribution. Many subsequent variants have been proposed to enhance their performance. Histogram-based variants focus on constructing multi-dimensional [3,5] histograms to capture the correlation between attributes. Sampling-based variants include index-based methods [16] and random walk based methods [17,28]. Sketch-based methods [4,21] have also been proposed for CardEst.

Learned Query-Driven Methods. Such methods try to learn a model to map each featured query to its cardinality directly. Some approaches leverage more complex models, such as deep neural networks (DNNs) [13], to enhance the estimation accuracy.

Learned Data-Driven Methods. The methods build the joint distribution $P(A, B, \ldots, N)$ on the original data, rendering their models query-independent. Prior research has employed various learning-based models to represent P, including deep auto-regression model [13,24], as well as probabilistic graphical models such as Bayesian networks [19,22], Sum-Product Networks (SPN) [12], and FSPN [23].

3 Estimating Join Cardinalities with Tight Upper Bounds

Given the CardEst problem definition, we now introduce the methodology of *BoundEst*. In Sect. 3.1, we describe how to calculate upper bounds of join cardinalities. Then, we discuss the impact of outliers on the accuracy of CardEst in Sect. 3.2. Section 3.3 introduces the workflow of *BoundEst*. Finally, we present a formula for estimating join cardinalities with tight upper bounds in Sect. 3.4.

3.1 Estimate Join Cardinalities with Upper Bounds

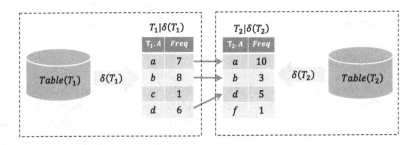

Fig. 1. A query Q join two tables on the join condition $T_1.A = T_2.A$ and filter predicates $\delta(T_1)$ and $\delta(T_2)$.

As shown in Fig. 1, a query Q joins tables T_1 and T_2 on the join condition $T_1.A = T_2.A$ and applies filter predicates $\delta(T_1)$ and $\delta(T_2)$. After filtering, the intermediate tables $T_1|\delta(T_1)$ and $T_2|\delta(T_2)$ contain only records that satisfy $\delta(T_1)$ and $\delta(T_2)$, respectively. The join operation then matches the values of the join-keys $T_1.A$ and $T_2.A$ from these intermediate tables. For example, the value a occurs 7 times in $T_1|\delta(T_1)$ and 10 times in $T_2|\delta(T_2)$, resulting in $7 \times 10 = 70$ occurrences of a in the join result. The true cardinality of query Q can be computed as $7 \times 10 + 8 \times 3 + 6 \times 5 = 124$. This procedure can be expressed as a statistical equation in Eq. 1, where $D(T_1.A)$ denotes the domain of all unique values of $T_1.A$.

$$
\begin{aligned}
|Q| &= \sum_{x \in \mathcal{D}(T_1.A)} P_{T_1}(T_1.A = x \mid \delta(T_1)) * |\delta(T_1)| * P_{T_2}(T_2.A = x \mid \delta(T_2)) * |\delta(T_2)| \\
&= \sum_{x \in \mathcal{D}(T_1.A)} P_{T_1}(T_1.A = x \wedge \delta(T_1)) * |T_1| * P_{T_2}(T_2.A = x \wedge \delta(T_2)) * |T_2|
\end{aligned}
$$

(1)

In a column of attributes, each value may appear with different frequencies. We refer to the maximum frequency among them as MaxFreq. Motivated by PessEst [4], we can derive an upper bound of the query Q by multiplying the size of $T_1|\delta(T_1)$ with the MaxFreq in table $T_2|\delta(T_2)$, i.e., $(7+8+1+6) \times 10 = 220$

(larger than true cardinality 124). Similarly, another upper bound can be derived as $(10 + 3 + 5 + 1) \times 8 = 152$ (larger than 124). Therefore, we can further transform Eq. 1 into Eq. 2 and derive an upper bound for join cardinality, where $|\delta(T_1)|$ is the size of intermediate table $A|\delta(T_1)$ and $MaxFreq\,(T_2.A \mid \delta(T_2))$ is the maximum value frequency of $T_2.A$ in $B \mid \delta(T_2)$.

$$
\begin{aligned}
|Q_1| &= \sum_{x \in \mathcal{D}(T_1.A)} P_{T_1}\,(T_1.A = x \wedge \delta(T_1)) * |T_1| * P_R\,(T_2.A = x \wedge \delta(T_2)) * |T_1| \\
&\leq \sum_{x \in \mathcal{D}(T_1.A)} P_{T_1}\,(T_1.A = x \wedge \delta(T_1)) * |T_1| * MaxFreq\,(T_2.A \mid \delta(T_2)) \\
&= |\delta(T_1)| * MaxFreq\,(T_2.A \mid \delta(T_2))
\end{aligned}
$$
(2)

The other upper bound can be derived in the same way, i.e., Eq. 3.

$$
|Q_2| \leq |\delta(T_2)| * MaxFreq\,(T_1.A \mid \delta(T_1))
$$
(3)

Finally, we obtain an upper bound for the query as shown in Eq. 4.

$$
\begin{aligned}
|Q| \leq min(&|\delta(T_1)| * MaxFreq\,(T_2.A \mid \delta(T_2)), \\
&|\delta(T_2)| * MaxFreq\,(T_1.A \mid \delta(T_1)))
\end{aligned}
$$
(4)

Based on the aforementioned procedure, for any given query, we are able to derive its corresponding upper bound. Importantly, the upper bound depends solely on the size of a single intermediate table (i.e., the filtered tables) and the MaxFreq of the join-keys and is independent of any other factors. This means we can completely use single-table cardinality estimation methods to calculate upper bounds for multi-table join queries.

3.2   How Outliers Affect Join Upper Bound Estimation

When calculating upper bounds, it is necessary to perform bucketing calculations to enhance the efficiency of estimation. In cases where data is uniformly distributed, bucketing calculations yield highly accurate results. However, real-world data displays high skewness, and the inclusion of outliers can notably affect the accuracy of estimation.

The distribution and bucketing of table T on join attribute $T.A$ are presented in Fig. 2. The outliers in this distribution are the attribute values with frequencies of 1, 1000, and 2000 respectively. When performing a self-join on attribute $T.A$, the true cardinality is 5010003. According to the formula proposed in Sect. 3.1, we calculate the upper bound of the join cardinality in each bucket. The upper bound for bin_1 is $50 \times 49 = 2450$, which is 48 more than the true cardinality of $1 \times 1 + 49 \times 49 = 2402$. The upper bound for bin_2 is $100 \times 50 = 5000$, which matches the true cardinality of $50 \times 50 + 50 \times 50 = 5000$. The upper bound for bin_3 is $3051 \times 2000 = 6102000$, which is 1099399 larger than the true cardinality of $51 \times 51 + 1000 \times 1000 + 2000 \times 2000 = 5002601$.

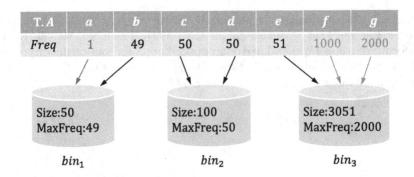

Fig. 2. The distribution and bucketing of table T on join attribute $T.A$.

This analysis reveals that treating all data equally in skewed data distributions leads to relatively loose upper bounds due to outliers. Therefore, it is essential to detect and isolate the outliers in the data. In the CardEst problem, outliers primarily refer to attribute values with exceptionally low or high frequencies that deviate from the uniform distribution. While the impact of low-frequency items on the estimation upper bounds is negligible, dealing with them incurs additional time and space costs. Consequently, our focus in this study is on individually addressing high-frequency outliers, as they exert a substantial influence on the estimation results.

Query: SELECT * FROM T_1, T_2 WHERE $T_1.A = T_2.A$ AND $\delta(T_1)$ AND $\delta(T_2)$;

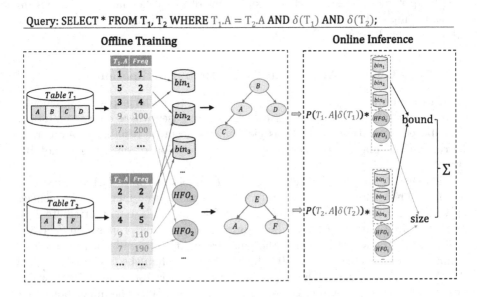

Fig. 3. The workflow overview of *BoundEst*.

3.3   Workflow

The workflow overview is shown in Fig. 3. *BoundEst* contains two phases: offline training and online inference. In the offline training phase, the model utilizes the DBSCAN clustering algorithm [2] to detect outliers in the data and learns the correlation between attributes. During the online phase, our framework derives tight upper bounds of join cardinalities.

Offline Training. Given a DB instance, *BoundEst* first analyzes its DB schema and tables to get all possible join-keys. When a join relationship exists between two join-keys, they are considered semantically equivalent and form an equivalent join group. For each equivalent join group, we identify and separate high-frequency outliers. Then we bin the domains of relatively evenly distributed join-keys, i.e., bin_i. For each outlier, we assign a separate bucket to it, i.e., HFO_i (High-Frequency Outlier). Based on the data tables and the binned join-keys, *BoundEst* builds models to learn the correlation between attributes within each table. We can incorporate any single-table CardEst method that can compute conditional probability into our framework.

Bayesian Network is a classical probabilistic graphical model that decomposes the joint distribution into multiple local conditional probability distributions by utilizing conditional independence. The directed edges in the network represent the dependencies between nodes, with each node maintaining a low-dimensional conditional probability distribution capturing the dependency between itself and its parent nodes. In this way, Bayesian Network not only reduces the complexity of model construction but also reduces the complexity of inference computations.

Online Inference. For a given query Q, *BoundEst* first parses its join graph and constructs single-table queries that handle filter predicates. Next, we use the trained Bayesian networks to estimate single-table queries and obtain the conditional probabilities of high-frequency outliers and remaining join-keys on the filter predicates of Q. We then compute the outliers and the values with relatively uniform distributions separately. Finally, we derive a tight upper bound for the query Q cardinality.

3.4   Tighten Upper Bounds of Join Cardinalities

This section focuses on tightening the upper bounds of join CardEst. Initially, we partition the join-keys into bins. We then perform separate calculations for high-frequency outliers and other values in each bucket to achieve precise estimations. Specifically, we accurately calculate the size of the join results for outliers, whereas, for the remaining values, we estimate upper bounds.

Offline Training. We use the DBSCAN clustering algorithm [2] to detect k high-frequency outliers in an equivalent join group. As shown in Fig. 3, we then bin the values and assign a separate bucket for each outlier. Therefore, in the subsequent Bayesian network learning, each HFO bucket learns the conditional probability of the corresponding outlier. The remaining values are distributed into $n - k$ buckets using the heuristic greedy binning selection algorithm [21] that optimizes the variance in value counts across all tables. Then, *BoundEst* records the MaxFreq of the join-keys in each bucket to compute upper bounds during online inference.

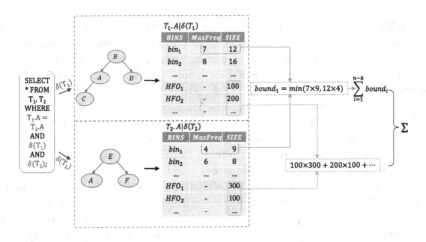

Fig. 4. An example of how to estimate upper bounds for queries.

Online Inference. As shown in Fig. 4, for a given query Q, the Bayesian networks estimate the conditional probability of join-keys to get the size of tuples satisfying filter predicates in each bucket. Next, *BoundEst* performs separate computations for the high-frequency outliers and the remaining values. Based on Sect. 3.1, we estimate the upper bound (i.e., $bound_i$) for each bucket containing relatively evenly distributed values and then sum them up.

As for the high-frequency outliers, since each HFO bucket only stores one value during the Bayesian network learning (e.g., $X_i = a$), after applying filter predicates, the HFO bucket contains the conditional probability of that outlier, i.e., $P(X_i = a \mid \delta(T_1))$. By multiplying this probability by the size of the relation, we obtain $size(X_i = a \mid \delta(T_1))$. Finally, accurate results for the outliers are obtained by multiplying the size of the value for each join-key in the equivalent join group (i.e., $100 \times 300 + 200 \times 100 + \ldots$). The tight upper bound for the join cardinality is derived by adding the estimation results of the high-frequency outliers and the relatively evenly distributed parts. We formally represent the aforementioned procedure in Eq. 5, where JK represents the join-keys in the equivalent group.

$$|Q_{tight}| = \sum_{i=1}^{n-k} bound_i + \sum_{j=1}^{k} \prod_{JK \in join\text{-}keys} size(JK) \qquad (5)$$

BoundEst. considers the substantial influence of high-frequency outliers in the data distribution on the estimation results. These outliers are identified using clustering algorithms and treated separately from other values in both offline training and online inference stages. Our framework employs the DBSCAN algorithm for outlier identification. The DBSCAN algorithm is sensitive to the number of clusters, so we must adjust suitable neighborhood distance thresholds for different datasets to identify their high-frequency outliers. Moreover, our approach is more capable of estimating tighter upper bounds on datasets with skewed data distributions, while its effect is less pronounced on datasets with uniform data distributions.

4 Experimental Evaluation

In this section, we experimentally evaluate the performance of our proposed *BoundEst* for solving the CardEst problem. We first introduce the experimental setting in Sect. 4.1. Then, we present experimental results in Sect. 4.2.

4.1 Experimental Setting

Datasets. We evaluate the performance of our framework on two benchmarks: STATS-CEB [9] and JOB-light [15], with their statistics compared in Table 1.

Table 1. Comparison of STATS-CEB and JOB-light benchmark

Category	STATS-CEB	JOB-light
# of table	8	6
# of columns in each table	$3-10$	$2-12$
# of rows in each table	$10^3 - 4 \cdot 10^5$	$3 \cdot 10^6 - 6 \cdot 10^7$
# of query	146	70
# of join templates	70	23
true cardinality range	$200 - 2 \cdot 10^{10}$	$9 - 9 \cdot 10^9$

Baselines. We compare our framework with the representative and competitive CardEst methods.

1) *PostgreSQL* refers to the histogram-based CardEst method used in the well-known DBMS PostgreSQL. We use PostgreSQL version 13.1 in experiments.
2) *TrueCard* uses true cardinality for given queries with no estimation latency. It represents the optimal CardEst performance.
3) *BayesCard* [22] is fundamentally based on BN, which models the dependence relations among all attributes on denormalized join tables as a directed acyclic graph.

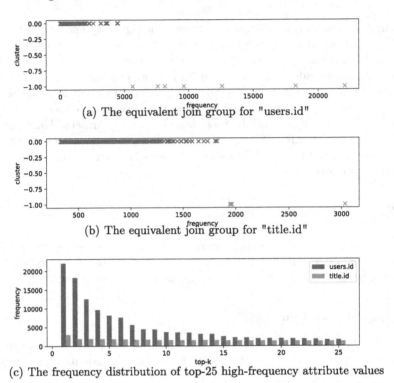

(a) The equivalent join group for "users.id"

(b) The equivalent join group for "title.id"

(c) The frequency distribution of top-25 high-frequency attribute values

Fig. 5. The value frequency distributions for two equivalent join groups.

4) *DeepDB* [12] based on SPN [7], approximates joint probability by recursively decomposing it into local and simpler PDFs.
5) *NeuroCard* [24] is built upon a deep auto-regression model. It is only designed for datasets with tree-structured join schema and it does not conform with STATS-CEB.
6) *PessEst* [4] is bound-based method developed for multi-join queries. For filtered join queries, it needs to build sketch bound online.
7) *FactorJoin* [21] is the state-of-the-art CardEst method. It uses statistics from single tables to calculate upper bounds.

In experiments, we test *DeepDB*, *PessEst*, and *BayesCard* methods using the optimally-tuned model parameters, as used in [9]. Both *FactorJoin* and *Bound-Est* use bucketing method, and we set the bin size to 200 for both datasets. For queries involving multiple equivalent join groups, *BoundEst* utilizes mutual information [14] to capture the correlations between them. Mutual information is a measure of dependence or "mutual dependence" between two random variables.

High-Frequency Outliers. Figure 5 shows the value frequency distributions for two equivalent join groups: *users.id* in the STATS dataset and *title.id* in the IMDB dataset. Figures 5(a) and 5(b) display the high-frequency outliers detected by the DBSCAN clustering algorithm within two equivalent join groups. Each group contains 7 and 4 high-frequency outliers, respectively. These outliers are values that deviate significantly from the rest of the data. Therefore, we allocate 7 HFO buckets for one group and 4 HFO buckets for the other in experiments. Figure 5(c) shows the frequency distribution of top-25 high-frequency attribute values. The *users.id* group has some values with very high frequencies, while the *title.id* group has a more even frequency distribution. The STATS dataset also contains another equivalent join group for *post.id*, but it does not have any shared values. Therefore, we do not set any HFO bucket for this group.

Environment. We run all of our experiments on a Ubuntu system with an Intel(R) Core(TM) i7-9700 CPU, 31.2GB DDR4 main memory, and 370GB SSD. To evaluate our framework and the baselines in an end-to-end setting, we integrated them into the query optimizer of PostgreSQL 13.1. We executed each query three times for each method, using the sub-plan query cardinalities estimated by each method as input for PostgreSQL.

4.2 Experimental Results

Our evaluations mainly focus on the end-to-end query performance inside the query optimizer, as well as the relative error between the estimated cardinalities and the true cardinalities. We summarize the overall average performance of STATS-CEB benchmark in Table 2, with the relative end-to-end improvement over *PostgreSQL* shown in the last column, i.e., (*PostgreSQL time − method time*) / *PostgreSQL time*. In addition, we show the relative errors between the methods' estimates and the true cardinalities (*estimate / true*) for all sub-queries in Fig. 6. We also show the overall average performance and the relative errors of the JOB-light benchmark in Table 3 and Fig. 7.

Table 2. Overall average performance of CardEst algorithms on STATS-CEB.

Methods	End-to-end time	Execute + plan time	Improvement
Postgres	8126.11 s	8125.89 s + 0.22 s	−
TrueCard	5525.78 s	5525.56 s + 0.22 s	32.00%
BayesCard	6955.15 s	6931.07 s + 24.08 s	14.41%
DeepDB	6101.48 s	5987.16 s + 114.32 s	24.92%
PessEst	5698.15 s	5628.21 s + 69.74 s	29.88%
FactorJoin	5618.76 s	5594.08 s + 24.68 s	30.86%
Ours	**5577.33 s**	**5554.51 s + 22.82 s**	**31.37%**

Performance on STATS-CEB. As shown in Table 2, *BoundEst* achieves the best end-to-end performance with the query runtime of 5577.33 s, and it outperforms the state-of-the-art method *FactorJoin* with an end-to-end execution time

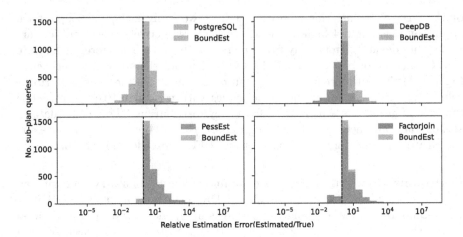

Fig. 6. Relative estimation errors on STATS-CEB.

Table 3. Overall average performance of CardEst algorithms on JOB-light.

Methods	End-to-end time	Execute + plan time	Improvement
Postgres	2108.37 s	2108.08 s + 0.29 s	–
TrueCard	1676.99 s	1676.70 s + 0.29 s	20.46%
BayesCard	**1735.33 s**	**1733.26 s + 2.07 s**	**17.69%**
DeepDB	1846.69 s	1827.95 s + 18.74 s	12.41%
NeuroCard	2208.94 s	1980.71 s + 228.23 s	−4.77%
PessEst	1998.47 s	1849.65 s + 148.82 s	5.21%
FactorJoin	1743.91 s	1740.42 s + 3.49 s	17.29%
Ours	1736.97 s	1733.73 s + 3.24 s	17.62%

of 5618.76 s. Meanwhile, it is close to the optimal performance of the true cardinality 5525.78 s. When compared with the *PosgreSQL*, *BayesCard* and *DeepDB*, *BoundEst* has significantly better query execution time, indicating that our estimates are much more effective at generating high-quality query plans. In terms of pure execution time, *PessEst* also shows relatively good performance. Furthermore, the planning time of our framework is much better than *DeepDB* and *PessEst* because the model of *DeepDB* is complex and inefficient at estimation. *PessEst* needs to execute filter predicates online during estimation, resulting in high planning latency.

In Fig. 6, in terms of estimation accuracy, *PostgreSQL* and *DeepDB* severely underestimate the cardinalities. *PessEst* generates the upper bounds and never underestimates. *BoundEst* exhibits fewer cases of underestimation compared to *FactorJoin* and outputs the upper bounds for almost sub-queries. Meanwhile, the marginal underestimates are very close to the true cardinalities, which can still help query optimizers generate effective plans. Compared with *PessEst* and *FactorJoin*, *BoundEst* tightens the upper bounds and obtains more accurate estimations, which makes it perform better in end-to-end time.

Performance on JOB-Light. As shown in Table 3, the end-to-end execution time of *BoundEst* is 1736.97 s, showcasing comparable performance to the state-of-the-art method *BayesCard*. *BayesCard* with an end-to-end execution time of 1735.33 s, is only 1.64 s faster than *BoundEst*. As methods for estimating upper bounds, *PessEst* spends too much time on processing filter predicates, resulting in high plan latency. In contrast, *BayesCard*, *FactorJoin*, and *BoundEst* can generate accurate estimates in a short plan time. *NeuroCard* performs the worst among these methods, with even lower performance than *PostgreSQL* itself when taking into account planning time.

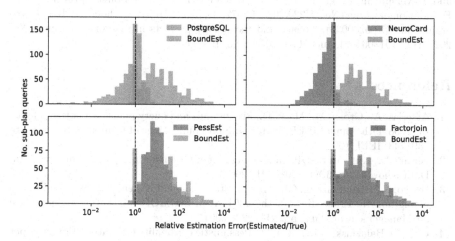

Fig. 7. Relative estimation errors on JOB-light.

In Fig. 7, *PostgreSQL* also suffers from the problem of underestimation. Besides, *NeuroCard* is also subject to underestimating cardinalities. These results indicate that underestimation often leads to worse execution plans. As for methods based on upper bounds, *BoundEst* has more queries whose estimated values are equal to the true cardinality than *PessEst* and *Factor-Join*. Compared with them, *BoundEst* does not significantly improve the accuracy of the estimated cardinality. This is due to the data distribution of the IMDB dataset, where it is difficult to find high-frequency outliers that are common in an equivalent join group. Specifically, the frequency of the join-key $movie\_info.movie\_id = 1197283$ is 2937, while the frequency of the join-key $cast\_info.movie\_id = 1197283$ joined to it is only 5.

5 Conclusion

In this paper, we proposed *BoundEst*, an efficient framework for estimating join cardinalities with tight upper bounds. Our method takes into account the impact of outliers on query results. We calculated the high-frequency outliers accurately and estimated the upper bounds for the remaining part. Our experiments showed that *BoundEst* generated effective estimates for the query optimizer. Specifically, on the STATS-CEB benchmark, *BoundEst* showed a 31.37% improvement in end-to-end time compared to *PostgreSQL*, outperforming the state-of-the-art methods. On the JOB-light benchmark, *BoundEst* achieved a 17.62% improvement in end-to-end time, which was comparable to the performance of the state-of-the-art methods.

Acknowledgement. The work is partially supported by the National Key Research and Development Program of China (2020YFB1707901), National Natural Science Foundation of China (Nos. U22A2025, 62072088, 62232007), Ten Thousand Talent Program (No. ZX20200035), Science and technology projects in Liaoning Province (No. 2023JH2/ 101300182), and 111 Project (No. B16009).

References

1. Atserias, A., Grohe, M., Marx, D.: Size bounds and query plans for relational joins. In: 2008 49th Annual IEEE Symposium on Foundations of Computer Science, pp. 739–748. IEEE (2008)
2. Birant, D., Kut, A.: St-DBScan: an algorithm for clustering spatial-temporal data. Data Knowl. Eng. **60**(1), 208–221 (2007)
3. Bruno, N., Chaudhuri, S., Gravano, L.: Stholes: a multidimensional workload-aware histogram. In: Proceedings of the 2001 ACM SIGMOD International Conference on Management of Data, pp. 211–222 (2001)
4. Cai, W., Balazinska, M., Suciu, D.: Pessimistic cardinality estimation: tighter upper bounds for intermediate join cardinalities. In: Proceedings of the 2019 International Conference on Management of Data, pp. 18–35 (2019)
5. Deshpande, A., Garofalakis, M., Rastogi, R.: Independence is good: dependency-based histogram synopses for high-dimensional data. ACM SIGMOD Rec. **30**(2), 199–210 (2001)
6. Dutt, A., Wang, C., Nazi, A., Kandula, S., Narasayya, V., Chaudhuri, S.: Selectivity estimation for range predicates using lightweight models. Proc. VLDB Endow. **12**(9), 1044–1057 (2019)
7. Gens, R., Pedro, D.: Learning the structure of sum-product networks. In: International Conference on Machine Learning, pp. 873–880. PMLR (2013)
8. Gunopulos, D., Kollios, G., Tsotras, V.J., Domeniconi, C.: Selectivity estimators for multidimensional range queries over real attributes. VLDB J. **14**, 137–154 (2005)
9. Han, Y., et al.: Cardinality estimation in DBMS: a comprehensive benchmark evaluation. arXiv preprint arXiv:2109.05877 (2021)
10. Hasan, S., Thirumuruganathan, S., Augustine, J., Koudas, N., Das, G.: Deep learning models for selectivity estimation of multi-attribute queries. In: Proceedings of the 2020 ACM SIGMOD International Conference on Management of Data, pp. 1035–1050 (2020)

11. Hertzschuch, A., Hartmann, C., Habich, D., Lehner, W.: Simplicity done right for join ordering. In: CIDR (2021)
12. Hilprecht, B., Schmidt, A., Kulessa, M., Molina, A., Kersting, K., Binnig, C.: DeepDB: learn from data, not from queries! arXiv preprint arXiv:1909.00607 (2019)
13. Kipf, A., Kipf, T., Radke, B., Leis, V., Boncz, P., Kemper, A.: Learned cardinalities: estimating correlated joins with deep learning. arXiv preprint arXiv:1809.00677 (2018)
14. Kraskov, A., Stögbauer, H., Grassberger, P.: Estimating mutual information. Phys. Rev. E **69**(6), 066138 (2004)
15. Leis, V., Gubichev, A., Mirchev, A., Boncz, P., Kemper, A., Neumann, T.: How good are query optimizers, really? Proc. VLDB Endow. **9**(3), 204–215 (2015)
16. Leis, V., Radke, B., Gubichev, A., Kemper, A., Neumann, T.: Cardinality estimation done right: index-based join sampling. In: CIDR (2017)
17. Li, F., Wu, B., Yi, K., Zhao, Z.: Wander join: online aggregation via random walks. In: Proceedings of the 2016 International Conference on Management of Data, pp. 615–629 (2016)
18. Muralikrishna, M., DeWitt, D.J.: Equi-depth multidimensional histograms. In: Proceedings of the 1988 ACM SIGMOD International Conference on Management of Data, pp. 28–36 (1988)
19. Tzoumas, K., Deshpande, A., Jensen, C.S.: Lightweight graphical models for selectivity estimation without independence assumptions. Proc. VLDB Endow. **4**(11), 852–863 (2011)
20. Wu, P., Cong, G.: A unified deep model of learning from both data and queries for cardinality estimation. In: Proceedings of the 2021 International Conference on Management of Data, pp. 2009–2022 (2021)
21. Wu, Z., Negi, P., Alizadeh, M., Kraska, T., Madden, S.: FactorJoin: a new cardinality estimation framework for join queries (2023)
22. Wu, Z., Shaikhha, A., Zhu, R., Zeng, K., Han, Y., Zhou, J.: Bayescard: revitilizing Bayesian frameworks for cardinality estimation. arXiv preprint arXiv:2012.14743 (2020)
23. Wu, Z., et al.: FSPN: a new class of probabilistic graphical model. arXiv preprint arXiv:2011.09020 (2020)
24. Yang, Z., et al.: Neurocard: one cardinality estimator for all tables. arXiv preprint arXiv:2006.08109 (2020)
25. Yi, P., Li, J., Choi, B., Bhowmick, S.S., Xu, J.: Flag: towards graph query auto-completion for large graphs. Data Sci. Eng. **7**(2), 175–191 (2022)
26. Yin, H., Gao, H., Wang, B., Li, S., Li, J.: Efficient trajectory compression and range query processing. World Wide Web **25**(3), 1259–1285 (2022)
27. Yu, T., et al.: Zebra: a novel method for optimizing text classification query in overload scenario. World Wide Web **26**(3), 905–931 (2023)
28. Zhao, Z., Christensen, R., Li, F., Hu, X., Yi, K.: Random sampling over joins revisited. In: Proceedings of the 2018 International Conference on Management of Data, pp. 1525–1539 (2018)
29. Zhu, R., et al.: Flat: fast, lightweight and accurate method for cardinality estimation. arXiv preprint arXiv:2011.09022 (2020)

A Multi-level Network with Multi-feature Clause Pair Graph for Emotion Cause Pair Extraction

Kai Kang[1], Guozheng Rao[1,3,4], Li Zhang[2]([✉]), Qing Cong[3], and Xin Wang[1,4]

[1] College of Intelligence and Computing, Tianjin University, Tianjin 300350, China
{kasalocc,rgz,wangx}@tju.edu.cn
[2] School of Economics and Management, Tianjin University of Science
and Technology, Tianjin 300457, China
zhangli2006@tust.edu.cn
[3] School of New Media and Communication, Tianjin University,
Tianjin 300072, China
chf@tju.edu.cn
[4] Tianjin Key Laboratory of Cognitive Computing and Applications,
Tianjin 300350, China

Abstract. Emotion-cause pair extraction (ECPE) is an emerging task which aims to extract all emotion clauses and corresponding cause clauses in a document. Most current methods use an end-to-end framework with two auxiliary tasks: emotion extraction and cause extraction, to solve this problem. However, these methods ignore the inter-task interaction among the three tasks and independently extract the features of emotion, cause, and emotion-cause pair. At the same time, most approaches do not fully utilize phrase-level information that is important for the ECPE task. To address these problems, we propose a novel multi-level network with multi-feature clause pair graph to extract different information from the document and generate multi-task features simultaneously. We use word-level and phrase-level network to capture the word and phrase information. At the clause-level, we construct a multi-feature clause pair graph to model the relationship between emotion, cause, and pair. Experimental results show that our method outperforms other baselines on a standard ECPE corpus.

Keywords: Emotion-cause pair extraction · Graph convolutional network · Temporal convolutional network

1 Introduction

Emotion Cause Extraction (ECE) is a subtask of emotion analysis, which has attracted more and more attention in the field of emotion analysis in recent years. The goal of the ECE task is to identify corresponding cause clause for a given emotion in the document [1]. ECE tasks are clearly defined but have a drawback. Before predicting the cause clause, the sentiment in the document

X. Song et al. (Eds.): APWeb-WAIM 2023, LNCS 14331, pp. 452–465, 2024.
https://doi.org/10.1007/978-981-97-2303-4_30

needs to be manually annotated in advance. The cost of annotation is very high, which greatly limits its application in real-world scenarios. To address this issue, Xia and Ding [2] propose a new task called emotional cause Pair Extraction (ECPE), which aims to extract all emotion-cause pairs consisting of emotion clauses and corresponding cause clauses from a document. Figure 1 shows an example of the ECPE task. The input is a document of nine clauses. Clauses c2, c4 and c8 contain emotion with the emotion expressions "sad", "anger" and "happy". The emotion c2 has one cause c1 ("Thinking of her little teddy hanging in the balance"), the emotion c4 has one cause c3 ("When told reporters about her encounter with dog abductors") and the emotion c8 has one cause c7 ("With this clue"). So the expected output is a set of valid emotion-cause pairs: (c2, c1), (c4, c3), (c8, c7). Intuitively, ECPE is more challenging because the clause classification task and the pair matching task need to be done simultaneously.

c1: Thinking of her little teddy hanging in the balance (cause)
c2: Ms. Yuan could not help feeling distressed (emotion)
c3: When told reporters about her encounter with dog abductors (cause)
c4: She was overcome with anger (emotion)
c5: The owner of a pet shop told Yuan he had heard something like this before
c6: and suggest she go to the dog walk over at Bag Park
c7: With this clue (cause)
c8: Ms. Yuan was thrilled (emotion)
c9: and hurried to the dog's place to ask

Fig. 1. An example of the ECPE task

For ECPE, a two-step approach was first proposed by Xia and Ding [2]. However, in this approach, the prediction of pairs relies on the prediction of emotions in the first step and that may lead to the problem of error propagation. To solve this problem, the previous work uses end-to-end methods with multi-task learning. Two features are extracted from each clause in the document for emotion extraction and cause extraction, and then a pairing matrix is generated by enumerating all possible clause combinations to select valid pairs for emotion cause pair extraction.

Although these methods achieve good performance, they still has the following drawbacks: First, they do not take the interactive relations between three tasks into consideration and extract the features of emotion, cause and emotion-cause pair independently without any direct contraction. In fact, if a clause is identified as an emotion or cause clause, it is more likely to appear in the emotion-cause pair and vice versa. Besides, these methods mainly focus on the clause-level relationship without using phrase-level information in clauses that contribute significantly to emotion or cause semantics.

To address these shortcomings, we proposed a multi-level network with multi-feature clause pair graph (MLNPG). By constructing a clause pair graph, we focus on clause-level information and model the interaction among emotion, cause and emotion-cause pair. For the internal information in clauses, we design word-level and phrase-level network to take the words, phrases and position information into account. Specially, we put the above sections into a multi-level network to further improve the performance of the ECPE task.

Our paper makes three main contributions:

- We propose a multi-level network with multi-feature clause pair graph (MLNPG) to extract document features from word-level, phrase-level, and clause-level.
- We construct a multi-feature clause pair graph to model the relationship among emotion, cause and pairs and use DAGNN to simultaneously generate the above multi-task feature in a joint feature framework at clause-level.
- Experimental results show that our method outperforms other baselines on a standard ECPE corpus. Furthermore, ablation experiments are performed to verify the effectiveness of the components in our model.

2   Related Work

2.1   Emotion Cause Extraction

The emotion cause extraction (ECE) task aim to extract the cause behind the given emotion. Lee et al. [3] first gave the definition of this task and formulated it as a word-level cause labeling problem. Early works of ECE focused on rule-based methods and machine learning methods. Chen et al. [4] redefined the ECE task as a clause-level extraction problem because the clause is the most appropriate unit to extract causes occurring in document. Gui et al. [5] constructed a Chinese emotion cause dataset base on SINA city news, which has become a benchmark corpus for the ECE task. Recent studies are mainly focused on deep learning methods and can be divided into two main categories: position-insensitive models and position-aware models. Position-insensitive models do not consider position information. For example, Li et al. [6] proposed a CNN-based model with co-attention mechanism. Yan et al. [7] import relevant knowledge paths from ConceptNet to enhance the connection between two clause nodes in a graph. In contrast, position-aware models usually use position information of the clause which is proved to be important in ECE task. For example, Ding et al. [8] incorporate content, relative position, global label and dynamic predict the cause clause by recording the predictions of the previous clauses. Xia et al. [9] propose a RNN-Transformer Hierarchical Network which use Transformer to capture the correlations between multiple clauses and classify them simultaneously.

2.2   Emotion-Cause Pair Extraction

ECE tasks have a drawback that the emotion in the document needs to be given in practice, which greatly limits its application in the actual scene. In this case, Xia and Ding [2] define a new task called Emotion-Cause Pair Extraction (ECPE) and propose a two steps model which extracts the emotion and cause first separately and then pairs and screens them. However, this two-step framework suffers from the error propagation from the first step to the second step. To solve this problem, many major works use end-to-end framework. Ding et al. design a two-dimensional matrix to represent emotion-cause pairs and use 2D-Transformer and its two variants to capture the interactions between them [10]. After that, they propose a sliding window mechanism [11] and achieve a substantial improvement. Yang et al. [12] propose a recurrent synchronization network that explicitly models the interaction among different tasks. Shi et al. [13] and Wang et al. [14] optimize end to end method by using different features in the document. On the other hand, some work transform the ECPE task into a unified sequence labeling problem. For example, Fan et al. [15] propose a transition-based model to transform the task into a procedure of parsing-like directed graph construction. Wu et al. [16] propose a sequence labeling based mo to predict whether two clauses constitute a cause pair by marking the relationship between them.

As graph models have made breakthrough progress in a variety of tasks, and they have also been applied to extract relations between clauses. Chen et al. [17] construct a pair graph convolutional network to capture dependency among local neighborhood candidate pairs. Wei et al. [18] use graph attention to model the inter-relationships between the clauses and design a ranking mechanism to extract emotion-cause pair. Chen et al. [19] and Yu et al. [20] construct a variety of different graphs to extract different features in the document. However, these typical methods mainly focus on extracting emotion, cause and pair features independently without using the relationship between them.

3   Methodology

3.1   Task Definition

Given a document consisting of N clauses $D = (c_1, c_2, ..., c_N)$ and the i-th clause has n words $c_i = (w_1^{c_i}, w_2^{c_i}, ..., w_n^{c_i})$. ECPE task aims to extract a set of emotion-cause pair in D:

$$P = \{..., (c_i, c_j), ...\} \quad (1 \leq i, j \leq N) \tag{1}$$

where c_i represent the emotion clause and c_j represent its corresponding cause clause in D.

3.2   An Overview of MLNPG

The overview architecture of MLNPG is shown in Fig. 2, the model consists of three parts: 1) Word-level network. We use Bert [21] to encode clauses in order to obtain word-level features. 2) Phrase-level network. We use Temporal Convolutional Networks (TCN) [22] to extract phrase-level information composed of multiple words, and then adopt position information to pair corresponding two clauses. 3) Multi-feature clause pair graph network. We construct a clause-level graph containing emotion, cause, and emotion-cause pair nodes to obtain the inter-relationship between different nodes. Finally, we apply Deep Adaptive Graph Neural Network (DAGNN) [23] to jointly encode three task features and make predictions.

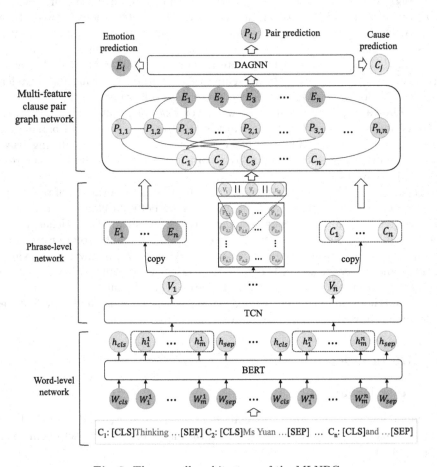

Fig. 2. The overall architecture of the MLNPG

3.3   Word-Level Network

Given a document consisting of n clauses $D = (c_1, c_2, ..., c_n)$ and each clause $c_i = (w_1^i, w_2^i, ..., w_m^i)$ consists of m words. We feed D as input into BERT [21]:

$$Input = ([CLS], w_1^1, w_2^1, ..., [SEP], ..., [CLS], w_1^n, w_2^n, ..., [SEP]) \qquad (2)$$

where w_j^i is the j-th token of the i-th clause in the document; [CLS], [SEP] are special BERT tokens. The we can obtain hidden representations after word-level encoder:

$$H = (h_{[CLS]}, h_1^1, h_2^1, ..., h_{[SEP]}, ..., h_{[CLS]}, h_1^n, h_2^n, ..., h_{[SEP]}) \qquad (3)$$

where $H \in R^{|I| \times d}$, d is the dimension of the hidden states; h_j^i denotes the hidden representation of token w_j^i.

3.4   Phrase-Level Network

Since some key phrases composed of multiple words are crucial to the whole sentence, we propose a Temporal Convolutional Network (TCN) based phrase-level network as the existence of convolution kernels can extract local phrase information. Compared with classical recurrent networks such as LSTM and GRU, the architecture of TCN is simpler and clearer as it can capture long-term dependencies better without retaining all the historical information. We fed word-level feature representation H into TCN to obtain phrase-level feature representation:

$$(V_1, V_2, ..., V_n) = TCN(H) \qquad (4)$$

where $V_i \in R^{n \times d}$ is the hidden representation of the i clause after averaging the output features after TCN. n is the number of clauses, d is the dimension of the hidden states.

After that, we obtain the emotion-cause pair by pairing the emotion clause and the cause clause. Since two clauses in an emotion-cause pair may occur within a specific distance (see Table 1), most of the cause clauses appear around the emotion clauses, and the probability of two clauses becoming an emotion-cause pair is 98.5% when the distance between them is less than 3. So we only pair two clauses whose distance between is less than K, where K is a hyperparameter. For a clause c_i as an emotion clause, the matching pair is:

$$P_{i,[j-K:j+K]} = \{P_{i,j-K}, ..., P_{i,j-1}, P_{i,j}, P_{i,j+1}, ..., P_{i,j+K}\} \qquad (5)$$

where $P_{i,j}$ is a emotion-cause pair consist of emotion clause c_i and cause clause c_j. To obtain the representations of pairs, we concatenate the features of corresponding two clauses and add a learnable relative position embedding:

$$P_{i,j} = W[V_i; V_j] + b + r_{i,j} \qquad (6)$$

where W, b are learnable parameters; $r_{i,j}$ is relative position embedding which is calculated as follows:

$$r_{i,j} = position\_embedding(pos_i - pos_j) \qquad (7)$$

where pos_i, pos_j are the position of c_i, c_j in the document, $position\_embedding$ is a embedding layer.

3.5 Multi-feature Clause Pair Graph Network

In order to extract the relationship between emotion, cause and emotion-cause pair, we construct a multi-feature clause pair graph to propagate the information between different nodes and join the multi-task features at the clause level. All nodes can be formulated as follows:

- **Emotoin node s(E)**: the emotion feature representation of the clause, updated with the graph and finally used in the emotion extraction subtask.
- **Cause nodes (C)**: the cause feature representation of the clause, updated with the graph and finally used in the cause extraction subtask.
- **Pair nodes (P)**: the pair feature representation of the emotion-cause pair, updated with the graph and finally used in the emotion-cause pair extraction task.

These three kinds of nodes contain the feature information of emotion, cause and pair respectively, and the information between nodes can be transmitted to each other by edges. All edges can be formulated as follows:

- **Emotion edges**: edges are formed between adjacent emotion nodes. In this way, we can capture the relationship between different emotion clauses.
- **Cause edges**: edges are formed between adjacent cause nodes. In this way, we can capture the relationship between different cause clauses.
- **Pair edges**: edges are formed between pair nodes and corresponding emotion nodes and cause nodes. In this way, we can capture the relationship between emotion, cause and pair.
- **Self-loop edges**: it can help each node to keep its feature in the process of interaction.

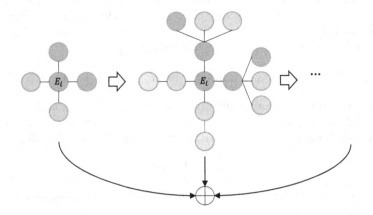

Fig. 3. Interaction of emotion, cause and pair features under DAGNN

Then we use Deep Adaptive Graph Neural Network (DAGNN) to aggregate the features from neighbors. Compared with GCN, DAGNN can be deeper, which helps it capture information from large and adaptive receptive fields. In our clause pair graph, with the increase of the number of graph network layers, nodes can obtain more interactive information, as shown in Fig. 3. For node E_i, it can only capture interaction information between pair nodes and adjacent emotion nodes at the first layer. At the second layer, it can aggregate further emotion nodes, cause nodes and pair nodes.

For the input $X = (E, C, P)$ in the graph, the node update as:

$$Z = MLP(X) \tag{8}$$

$$H_i = \hat{A}Z \tag{9}$$

$$H = stack(Z, H_1, ..., H_k) \tag{10}$$

$$S = \sigma(Hs) \tag{11}$$

$$\tilde{S} = reshape(S) \tag{12}$$

$$X_{out} = softmax(squeeze(\tilde{S}H)) \tag{13}$$

where $Z \in R^{c \times d}$ is the feature after applying an MLP network; c is the number of nodes; $H_i \in R^{d \times n}$ denotes the representations obtained after i layer of DAGNN; $\hat{A} = \tilde{D}^{-\frac{1}{2}}\tilde{A}\tilde{D}^{-\frac{1}{2}}$; $\tilde{A} = A + I$; k is the depth of the model; s is a trainable projection vector. Also we use *stack*, *reshape* and *squeeze* to rearrange the data dimension.

3.6   Prediction

After obtaining the final node representation, we use a MLP to obtain the predictions of emotional-cause pair extraction task:

$$\hat{y}_{ij} = \sigma(W_p p_{ij}^{final} + b_p) \tag{14}$$

where W_p and b_p are learnable parameters. Correspondingly, the binary cross entropy loss is utilized as loss of ECPE task:

$$\mathcal{L}_p = -\sum_i^n \sum_j^n y_{ij} \log(\hat{y}_{ij}) \tag{15}$$

where $\hat{y}_{ij}$, y_{ij} are predict label and truth label respectively.

Similarly, for the two auxiliary tasks: emotion extraction and cause extraction, we compute the probability as follows:

$$\hat{y}_i^e = \sigma(W_e e_i^{final} + b_e) \tag{16}$$

$$\hat{y}_j^c = \sigma(W_c c_j^{final} + b_c) \tag{17}$$

where W_e, b_e, W_c, b_c are learnable parameters. They have the corresponding loss:

$$\mathcal{L}_e = -\sum_i^n y_i^e \log(\hat{y}_i^e) \tag{18}$$

$$\mathcal{L}_c = -\sum_j^n y_j^c \log(\hat{y}_j^c) \tag{19}$$

where $\hat{y}_i^e$, y_i^e are predict label and truth label of emotion clause; $\hat{y}_j^c$, y_j^c are predict label and truth label of cause clause. Thus, the final loss function L was as follows:

$$\mathcal{L} = \mathcal{L}_p + \mathcal{L}_e + \mathcal{L}_c \tag{20}$$

4 Experiments

4.1 Dataset and Metrics

We use the Chinese benchmark dataset released by Xia and Ding [2], which is constructed from the SINA city news [5]. The detail of this corpus is shown in Table 1. It contains 1945 documents. Among them, 1746 documents have one emotion-cause pair, 177 documents have two emotion-cause pairs, and only 22 documents have more than two emotion-cause pairs. We use the 10-fold cross-validation as the data split strategy and precision (P), recall (R), and F1-score as evaluation metrics. Furthermore, during the emotion-cause pair prediction, we can also get emotions and causes prediction and these two subtasks are also evaluated.

Table 1. The detail of the Chinese ECPE corpus (Xia and Ding, 2019).

Item	Number	Percentage
document	1945	–
with one pair	1746	89.77%
with two pair	177	9.10%
over two pair	22	1.13%
pairs	2167	–
distance 0	511	23.58%
distance 1	1342	61.93%
distance 2	224	10.34%
distance > 2	90	4.15%

4.2 Experimental Settings

We use the BERT as the word encoder and the dimension of token embeddings is 768. During training, we use the Adam optimizer to update all parameters. Learning rate and warmup rate are set to 2e−5 and 0.1 respectively. Dropout is applied for networks to reduce overfitting and the dropout rate is set to 0.2. Finally, we set the mini-batch size to 4 and the training epoch to 30. Our experimental environment is Pytorch-1.9.0, ubuntu-18.0.4, 14 vCPU Intel(R) Xeon(R) Gold 6330 CPU and RTX 3090 (24 GB) GPU.

4.3   Baselines

In order to evaluate the performance of our model, we make a comparison with the following model using BERT.

- **ECPE-2D** [10]: uses a 2D transformers to model the interactions of different emotion-cause pairs.
- **TransECPE** [15]: propose a transition-based model to transform the task into a procedure of parsing-like directed graph construction
- **PairGCN** [17]: constructs a pair graph convolutional network to model dependency relations among local neighboring candidate pairs.
- **RANKCP** [18]: uses a graph attention network to model interactions between clauses and selects emotion-cause pairs by the ranking mechanism.
- **ECPE-MLL** [11]: extracts emotion-cause pairs based on sliding window multi-label learning.
- **RSN** [12]: propose a recurrent synchronization network that explicitly models the interaction among the EE, CE and ECPE tasks.
- **PTN** [16]: propose a sequence labeling based model to predict whether two clauses constitute a cause pair by marking the relationship between them.

4.4   Experimental Results

Table 2 shows the results of the emotion-cause pair extraction (ECPE) task and two subtasks: emotion clause extraction (EE) and cause clause extraction (CE). Our model shows a clear advantage over previous works. For the ECPE task, our model obtained F1 improvements of 1.12% and precision improvements of 1.08% compared to the previous best models ECPE-MLL and TransECPE. For the cause extraction task, our model also obtained F1 improvements of 1.56% and precision improvements of 1.59% compared to the previous best models PTN and PairGCN. Although our model is not the best in emotion clause extraction, our model performs better in cause identification. In fact, we can clearly see that cause extraction is more difficult than emotion extraction. Thus compared to emotion extraction which already works well, the improvement on the cause extraction helps us much more. We believe that the improvement of our model is due to the phrase-level information and the interaction information among emotion, cause and pair.

4.5   Ablation Study

Ablation studies are conducted to verify the effectiveness of our model. First, we design a set of ablation experiments for different components of our model. Table 3 shows the results of the ablation studies:

Table 2. Performance of our models and baselines. P, R and F1 denote precision, recall and F1-measure respectively. The best performance is in bold, the second-best performance is underlined

Model	E-C Pair Extraction			Emotion Extraction			Cause Extraction		
	P	R	F1	P	R	F1	P	R	F1
ECPE-2D	72.92	65.44	68.99	86.27	**92.21**	89.10	73.36	69.34	71.23
TransECPE	77.08	65.32	70.72	88.79	83.15	85.88	78.74	66.89	72.33
PairGCN	76.92	67.91	72.02	88.57	79.58	83.75	79.07	69.28	73.75
RANKCP	71.19	**76.30**	73.60	91.23	89.99	**90.57**	74.61	77.88	76.15
ECPE-MLL	77.00	72.35	74.52	86.08	91.91	88.86	73.82	**79.12**	76.30
RSN	76.01	72.19	73.93	86.14	89.22	87.55	77.27	73.98	75.45
PTN	76.40	72.40	74.30	85.09	91.59	88.19	74.87	77.90	76.31
MLNPG (ours)	**78.19**	73.35	**75.64**	90.61	87.07	88.78	**80.66**	75.33	**77.87**

- **w/o TCN**: to demonstrate the effectiveness of phrase information, we remove the TCN and the model performance decreases by 1.24% of the F1 value which indicates the phrase-level feature plays an important role.
- **w/o Pos**: to demonstrate the effectiveness of position information, we remove position embedding, it means that we simply concatenate the emotion and cause to obtain the pair feature and ignore the position information. The model performance decreases by 1.92% of the F1 value in ECPE task which proves the importance of position information.
- **w/o DAGNN**: to demonstrate the effectiveness of the multi-feature clause pair graph, we remove the whole graph as well as the graph algorithm and the model performance decreases by 2.59% of the F1 value in ECPE task. It proves the clause pair graph can obtain the clause-level relationship between emotion, cause and pair.

Table 3. The results of ablation study on the benchmark corpus for different components. The best performance is in bold, the second-best performance is underlined

Model	E-C Pair Extraction			Emotion Extraction			Cause Extraction		
	P	R	F1	P	R	F1	P	R	F1
MLNPG	**78.19**	**73.35**	**75.64**	**90.61**	87.07	88.78	**80.66**	75.33	**77.87**
-w/o TCN	77.27	71.81	74.40	90.44	86.39	88.35	79.43	73.74	76.44
-w/o Pos	74.51	73.13	73.72	89.59	87.25	88.36	78.95	**75.66**	77.21
-w/o DAGNN	75.21	71.08	73.05	90.33	**87.59**	**88.92**	77.41	72.93	75.06

In order to study the effect of different edges in the multi-feature clause pair graph. We design a set of ablation experiments and the results in Table 4 show that removing different edges has a certain impact on the performance of the emotion-cause extraction task.

Table 4. The results of ablation study on the benchmark corpus for different components. The best performance is in bold, the second-best performance is underlined

Model	E-C Pair Extraction			Emotion Extraction			Cause Extraction		
	P	R	F1	P	R	F1	P	R	F1
MLNPG	**78.19**	**73.35**	**75.64**	90.61	<u>87.07</u>	**88.78**	**80.66**	**75.33**	**77.87**
-w/o emo edge	77.80	72.16	<u>74.84</u>	<u>90.84</u>	86.68	88.68	<u>80.37</u>	74.56	<u>77.32</u>
-w/o cau edge	76.85	<u>72.87</u>	74.75	89.32	**87.19**	88.21	79.20	<u>75.10</u>	77.04
-w/o pair edge	<u>78.09</u>	70.74	74.20	**91.95**	85.85	<u>88.77</u>	79.94	72.58	76.05

When removing the emotion-emotion edge, cause-cause edge and emotion-cause-pair edge, the F1 value decreases by 0.80%, 0.89% and 1.44% respectively. Although the performance of the emotion extraction subtask does not change significantly, the F1 value of the cause extraction subtask decreases by 0.55%, 0.83% and 1.82%, respectively. At the same time, it can be seen that the influence of removing pair edges is slightly larger than removing emotion or cause edges. This may be because the pair edge has two edges connecting the emotion, pair and cause and remove this edge will remove the information interaction between them.

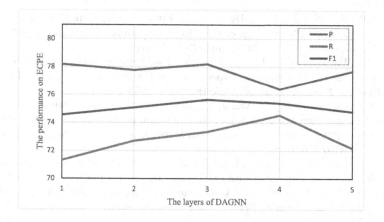

Fig. 4. Results of DAGNN with different depths

We also examine the effects of different numbers of DAGNN layer. As shown in Fig. 4. In this work, the best results are achieved when the number of DAGNN layers is 3. Compared with GCN, the number of layers of DAGNN can be deeper, this makes better use of the information of domain nodes, but too deep layers will still cause gradient disappearance problem.

5   Conclusion

In this paper, we propose a multi-level network with multi-feature clause pair graph (MLNPG) to extract different information from word-level, phrase-level and clause-level. Specially, at the clause-level, we construct a multi-feature clause pair graph to generate emotion, cause and pair features simultaneously and use the DAGNN framework to capture the relationship between them. The experiments on the Chinese benchmark corpus show that our model achieves state-of-the-art performance, and the ablation experiments demonstrate the effectiveness of our methods.

References

1. Gui, L., Xu, R., Wu, D., Lu, Q., Zhou, Y.: Event-driven emotion cause extraction with corpus construction. In: Social Media Content Analysis: Natural Language Processing and Beyond, pp. 145–160. World Scientific (2018)
2. Xia, R., Ding, Z.: Emotion-cause pair extraction: a new task to emotion analysis in texts. In: Proceedings of the 57th Annual Meeting of the Association for Computational Linguistics, pp. 1003–1012 (2019)
3. Lee, S.Y.M., Chen, Y., Huang, C.R.: A text-driven rule-based system for emotion cause detection. In: Proceedings of the NAACL HLT 2010 Workshop on Computational Approaches to Analysis and Generation of Emotion in Text, pp. 45–53 (2010)
4. Chen, Y., Lee, S.Y.M., Li, S., Huang, C.R.: Emotion cause detection with linguistic constructions. In: Proceedings of the 23rd International Conference on Computational Linguistics (Coling 2010), pp. 179–187 (2010)
5. Gui, L., Wu, D., Xu, R., Lu, Q., Zhou, Y.: Event-driven emotion cause extraction with corpus construction. In: Proceedings of the 2016 Conference on Empirical Methods in Natural Language Processing, pp. 1639–1649 (2016)
6. Li, X., Song, K., Feng, S., Wang, D., Zhang, Y.: A co-attention neural network model for emotion cause analysis with emotional context awareness. In: Proceedings of the 2018 Conference on Empirical Methods in Natural Language Processing, pp. 4752–4757 (2018)
7. Yan, H., Gui, L., Pergola, G., He, Y.: position bias mitigation: a knowledge-aware graph model for emotion cause extraction. In: Proceedings of the 59th Annual Meeting of the Association for Computational Linguistics and the 11th International Joint Conference on Natural Language Processing (Volume 1: Long Papers), pp. 3364–3375 (2021)
8. Ding, Z., He, H., Zhang, M., Xia, R.: From independent prediction to reordered prediction: integrating relative position and global label information to emotion cause identification. In: Proceedings of the AAAI Conference on Artificial Intelligence, vol. 33, pp. 6343–6350 (2019)
9. Xia, R., Zhang, M., Ding, Z.: RTHN: a RNN-transformer hierarchical network for emotion cause extraction. arXiv preprint arXiv:1906.01236 (2019)

10. Ding, Z., Xia, R., Yu, J.: ECPE-2D: emotion-cause pair extraction based on joint two-dimensional representation, interaction and prediction. In: Proceedings of the 58th Annual Meeting of the Association for Computational Linguistics, pp. 3161–3170 (2020)
11. Ding, Z., Xia, R., Yu, J.: End-to-end emotion-cause pair extraction based on sliding window multi-label learning. In: Proceedings of the 2020 Conference on Empirical Methods in Natural Language Processing (EMNLP), pp. 3574–3583 (2020)
12. Chen, F., Shi, Z., Yang, Z., Huang, Y.: Recurrent synchronization network for emotion-cause pair extraction. Knowl.-Based Syst. **238**, 107965 (2022)
13. Shi, J., Li, H., Zhou, J., Pang, Z., Wang, C.: Optimizing emotion–cause pair extraction task by using mutual assistance single-task model, clause position information and semantic features. J. Supercomput. 1–20 (2022)
14. Wang, B., Ma, T., Lu, Z., Xu, H.: An end-to-end mutually interactive emotion-cause pair extractor via soft sharing. Appl. Sci. **12**(18), 8998 (2022)
15. Fan, C., Yuan, C., Du, J., Gui, L., Yang, M., Xu, R.: Transition-based directed graph construction for emotion-cause pair extraction. In: Proceedings of the 58th Annual Meeting of the Association for Computational Linguistics, pp. 3707–3717 (2020)
16. Wu, Z., Dai, X., Xia, R.: Pairwise tagging framework for end-to-end emotion-cause pair extraction. Front. Comp. Sci. **17**(2), 172314 (2023)
17. Chen, Y., Hou, W., Li, S., Wu, C., Zhang, X.: End-to-end emotion-cause pair extraction with graph convolutional network. In: Proceedings of the 28th International Conference on Computational Linguistics, pp. 198–207 (2020)
18. Wei, P., Zhao, J., Mao, W.: Effective inter-clause modeling for end-to-end emotion-cause pair extraction. In: Proceedings of the 58th Annual Meeting of the Association for Computational Linguistics, pp. 3171–3181 (2020)
19. Chen, S., Mao, K.: A graph attention network utilizing multi-granular information for emotion-cause pair extraction. Neurocomputing **543**, 126252 (2023)
20. Yu, J., Liu, W., He, Y., Zhong, B.: A hierarchical heterogeneous graph attention network for emotion-cause pair extraction. Electronics **11**(18), 2884 (2022)
21. Kenton, J.D.M.W.C., Toutanova, L.K.: BERT: pre-training of deep bidirectional transformers for language understanding. In: Proceedings of NAACL-HLT, pp. 4171–4186 (2019)
22. Lea, C., Flynn, M.D., Vidal, R., Reiter, A., Hager, G.D.: Temporal convolutional networks for action segmentation and detection. In: Proceedings of the IEEE Conference on Computer Vision and Pattern Recognition, pp. 156–165 (2017)
23. Liu, M., Gao, H., Ji, S.: Towards deeper graph neural networks. In: Proceedings of the 26th ACM SIGKDD International Conference on Knowledge Discovery and Data Mining, pp. 338–348 (2020)

CCBTC: A Blockchain-Based Covert Communication Scheme over Bitcoin Transactions

Rundong Wang[1,2], Bohao Li[1], Wei Ren[1,2(✉)], and Jie He[2]

[1] School of Computer Science, China University of Geosciences, Wuhan, China
weirencs@cug.edu.cn
[2] Guangxi Key Laboratory of Machine Vision and Intelligent Control, Wuzhou University, Wuzhou, China

Abstract. Blockchain can be looked at as a new channel for covert communication. Since blockchain provides a tamper-proof distributed ledger, it is possible to employ blockchain to provide persistent and anonymous covert communications. This persistence includes the persistence of one-way message distribution and the persistence of multi-party communications. In this paper, we propose a covert communication method based on Bitcoin transactions (BTC). The method converts the secret message to a decimal value via ASCII code and utilizes the decimal transaction amount as the message carrier. The decimal value of the secret message can be divided into multiple parts, to shorten the communication delay and to decrease the transaction fee, which is sent to a designated destination account. Only the recipient of the transaction who possesses the correct key can decode the transaction amount to get the secret message, and this message is persistent in the blockchain. We evaluated different ways of dividing methods, to explore the way that fewer bitcoin transactions are consumed. Besides, the sender can always update his address after each transmission to further hide the recipient of the message by unlinking. We have verified the feasibility and efficiency of the method through iterative experiments on the Bitcoin test network, which justified the applicability of the scheme.

Keywords: Blockchain · Bitcoin Transactions · Covert Communication · Anonymous Communication · Transaction Amount

1 Introduction

In recent years, with the rapid development of computer and network technology, the security of transmission of modern network information has gradually become a concern. As many encryption algorithms are inefficient and can be broken, they cannot effectively guarantee the confidentiality and integrity of information transmission. As a result, covert communication methods for secret information have come into the limelight. Blockchain, as a distributed database,

X. Song et al. (Eds.): APWeb-WAIM 2023, LNCS 14331, pp. 466–480, 2024.
https://doi.org/10.1007/978-981-97-2303-4_31

provides a new way to achieve open, transparent, and tamper-evident transactions between multiple parties without the need for a central trust authority, and can be employed as a new public communication channel with trust among untrusted users.

With the development of blockchain, bitcoin trading has also become one of the widely used trading methods nowadays. Bitcoin is a cryptocurrency with currency issuance, transaction, and account management functions implemented open source on a P2P network with the help of cryptographic techniques [1]. Bitcoin uses blockchain technology to implement a decentralized digital currency system and transaction data can be looked at as a new communication channel. Thus, the blockchain can provide new ideas for information hiding and covert communication schemes due to its characteristics of being less tamperable, anonymous, and decentralized, enabling secret information embedded under the guise of Bitcoin transactions [2,3].

In the distributed, decentralized digital currency Bitcoin, the transaction system provides a natural vehicle for information transfer. Past research has explored how to use Bitcoin for covert communication, such as [4], but most of them require to modify the Bitcoin protocol or are tracked [5] and detected easily [6,7], which limits its practical applications. In this paper, we propose a covert communication scheme based on Bitcoin transactions without tampering with any BTC pr. Roughly speaking, our proposed scheme takes the transaction amount as the original carrier, converts the secret information into decimal numbers via ASCII codes, and embeds the secret information into the transaction amount using the four-bit numerical partitioning and multiplication method, thus realizing the covert communication of the information. Also, the security of the covert communication can further be enhanced by updating the account address of the sender after each covert communication between the two parties of the transaction. In addition, the proposed scheme does not require modification of the Bitcoin protocol, which can effectively achieve the purpose of secret information security transmission.

Contributions. In this paper, our main contributions are listed as follows:

- We propose a covert communication scheme based on Bitcoin transactions. The proposed scheme uses the transaction amount as the carrier, divides the transaction amount into multiple parts using the four-bit bit splitting method, and embeds the information using the multiplication method so that both parties of a Bitcoin transaction only need to consume a small portion of the transaction fee to complete the covert communication of secret information.
- We propose an empirical structure in concrete hiding. After repeated experiments on the Bitcoin test network, we have found the most suitable numerical break processing method (i.e., four-bit bit break) that can consume less Bitcoin transaction fees in covert communication and verified the feasibility of the scheme. In addition, the covert communication scheme also shows high efficiency in actual bitcoin transactions and has good practicality.
- We propose to change the destination address by updating the recipient address. In this covert communication solution, we update the transaction

address after each transaction to prevent leakage of transaction records and to ensure more effective covert and secure communication.

Organization. In Sect. 2, we describe the related work of covert communication based on blockchain. In Sect. 3, the covert communication scheme is explained. We perform the corresponding communication experiments in Sect. 4. Finally, in Sect. 5, our main work is concluded.

2 Related Work

Traditional covert information transmission methods embed covert information in pictures, videos or wireless channels [8], and used network multimedia transmission mechanisms to achieve the purpose of covert information delivery. However, existing information hiding methods and network covert channels have problems such as low bandwidth, easy detection and targeted blocking, which limit the use of covert information transmission. How to securely transmit information over public non-dedicated (covert) channels without being detected by adversaries has become a major challenge plaguing the scientific community. With the development of blockchain technology and bitcoin trading, covert communication of secret information with the help of the bitcoin blockchain has become a popular field.

In 2008, Nakamoto proposed the concept of Bitcoin blockchain. The Bitcoin blockchain has decentralisation, information immutability, wide information dissemination, information anonymity characteristics and frequent transaction characteristics. The above characteristics of the blockchain provide new ideas and solutions for the integration of cryptography and information hiding technology. A simple way to perform covert communication in blockchain is to write arbitrary content using the OP_RETURN field in the Bitcoin protocol [9].

Tian et al. [10] speak of target dynamic tags embedded in the OP_RETURN field to enable the identification of special transactions and the transmission of secret information. However, such an approach is vulnerable to detection by eavesdroppers, and thus the OP_RETURN field may be analysed, allowing information leakage. To address this issue, Partala et al. [11] combined steganography with blockchain to implement a covert communication scheme using the LSB algorithm to embed secret information into Bitcoin's address. However, due to the limitations of steganography, the scheme can only hide one bit of secret information for one transaction, making the capacity and efficiency of steganographic communication limited. Subsequently, Li et al. [12] constructed a new blockchain-based model for network covert communication, demonstrating the model's resistance to interference, receiver anonymity and multi-line communication. Castiglione et al. [13] analyzed the impact of information segmentation and encoding methods, and then, Cao et al. [14] designed a new scheme for encoding covert data on the chain to achieve secret information hiding.

In fact, due to its open and public nature, there are some threats to the privacy of blockchain users in these schemes [15]. In numerous scholarly studies, most have attempted to address these issues by increasing anonymity, for example, by unlinking payments from their sources [16], using zero-knowledge proofs [17] or blind signatures [18]. Typically, these schemes all require the application of encryption to protect stored application data and are not efficient in communication.

In order to solve the challenge of covert communication in blockchain and effectively ensure the security and efficiency of communication, this paper proposes and designs a covert communication scheme with transaction amount as the carrier in the framework of bitcoin transactions. The carrier is thus easy to obtained and the volume of covert communications is low, which is very hard to be aware due to the overwhelmingly large volume of normal transactions.

3   Proposed Scheme

3.1   Design Framework

Our proposed scheme uses the bitcoin transaction amount as the carrier, and encodes the secret message and embeds it into the transaction amount. When two parties conduct a bitcoin transaction, the sender of the transaction sends the BTC amount with the secret information embedded to the receiver. The receiver of the transaction can then decode the value of the transaction amount to obtain the hidden information, thus enabling the communication of the secret message between the two parties. The whole scheme can be simplified into three main steps: encoding message, creation and sending of the transaction, and decoding and extraction of the message. The specific symbolic representation is shown in Table 1 and the scheme is illustrated in Fig. 1.

Table 1. Notations.

Symbols	Description
M	Secret message
$\text{Enc}(M)$	Message encoding: Transforming M into encoded form
$\text{Decimal}(M)$	Converting the encoded information to decimal form
$\text{Amount}(M)$	Converting decimal codes to transaction amounts
Transaction	Bitcoin transaction creation based on hidden information
Broadcasting	Broadcast transactions to the Bitcoin network
Transaction Receipts	Receipts or records of bitcoin transactions
$\text{Dec}(\text{Amount}(M))$	Decoding of transaction amounts into binary encoded form
$\text{Dec}(\text{Binary}(M))$	Decoding binary codes into their original encoded form

Message Encoding. The original message M is encoded by the encoding function $\text{Enc}(M)$ to obtain the encoded result. The encoded result is converted to the decimal form $\text{Decimal}(M)$, which is expressed as a decimal value. The result in decimal form is further processed as the amount of the transaction $\text{Amount}(M)$.

Creation and Sending of Transactions. Create a Bitcoin transaction using the transaction amount Amount(M) and then broadcast the created transaction to the Bitcoin network so that other nodes can receive it.

Decoding and Extraction. After Broadcasting, the transaction is recorded by the miner node in Transaction Receipts, including information such as the amount of the transaction. The amount of the transaction is extracted from the Transaction Receipts Amount(M). The extracted Amount(M) is decoded by the Dec(Amount(M)) operation, which converts it back to the decimal form Decimal(M). The result in the decimal form is again decoded Dec(Decimal(M)) to obtain the hidden original message M.

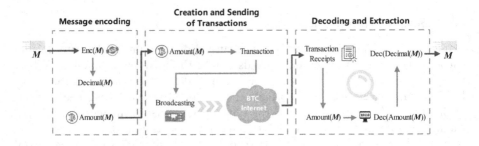

Fig. 1. Structure of the covert communication scheme on BTC.

3.2   Message Encoding

Before embedding the secret message, we need to encode the hidden message and convert it to a decimal value. Here, suppose the equation is Enc(M), where M is the message to be hidden, and we have used ASCII encoding to complete the encoding of the message.

In computers, data is represented in the form of binary numbers, both for storage and for arithmetic, such as a-z upper and lower case letters, 0–9 numbers, and some common symbols (e.g. !, $, &, etc.). In order for users to communicate with each other properly, it is necessary to unify the specific symbols corresponding to binary numbers and to follow the same encoding rules, so ISO (International Organization for Standardization) designed the ASCII code to unify the binary representation of each character.

ASCII (American Standard Code for Information Interchange) is an encoding standard for representing characters by mapping each character to a unique decimal value. In a decimal ASCII string, each character is represented as a decimal value and is separated by a comma or other separator. For example, the string "hello" can be represented as "104, 101, 108, 108, 111", where each character corresponds to the ASCII codes 104 (h), 101 (e), 108 (l), 108 (l), and 111 (o), in that order. This encoding can be used for data transmission or storage, especially in applications related to ASCII characters, such as text processing, network communication, and file transfer. It provides a simple way to represent characters, making it easier to convert and process between characters.

In our proposed scheme, $\text{Decimal}(M) = \text{Enc}(M)$, where Decimal is ASCII decimal encoding and $\text{Amount}(M) = \text{Decimal}(M)$. Then, it is multiplied by the smallest unit of BTC (Satoshi), 1 Satoshi = 0.00000001 BTC, and $\text{Amount}(M)$ is used as the transaction amount. It is important to note that in Bitcoin's transaction mechanism, each transaction needs to meet the minimum transaction amount in order to ensure the concealment of the information and the reversibility of the recipient after receiving it.

3.3 Creating and Sending of Transactions

Bitcoin transactions are the process of transferring ownership of bitcoins in the Bitcoin network. In our hidden message transfer scheme, we use bitcoin transactions to transfer hidden messages. The process of creating and sending bitcoin transactions is described in detail below.

3.3.1 Transaction Creating

Creating a transaction in the bitcoin network requires the following steps:

1. Input Settings: Select the bitcoin input to be used for payment. An input usually points to the output of a previous bitcoin transaction, proving that the sender has the right to use the bitcoin.

2. Output Settings: Set the output of the transaction. In our approach, we use the message as a hidden message via ASCII decimal encoding, which corresponds to the payment amount. Each output contains a specific payment amount and a payee address.

3. Signature: Sign the transaction to prove that the sender has the right to use the input and to ensure the integrity and security of the transaction.

4. Transaction Creation: Combines the input, output, and signature information to construct the bitcoin transaction data structure.

When creating a transaction, the detailed algorithmic process is described in Algorithm 1.

3.3.2 Transactions Sending

Transaction sending is the process of broadcasting a constructed transaction to the Bitcoin network so that it is included in the bitcoin blockchain by the miner nodes. The following are the key steps:

1. Connect to the Bitcoin Network: Establish a connection to the bitcoin network by using the bitcoin node client software (Bitcoin Core) or a third-party library.

Algorithm 1. Secret Message Hiding and Sending Algorithm

Input: Secret message M, Sender A's username and password $user_A, password_A$,
 Address of receiver B $address_B$, Bitcoin Core node's host $host_A$ and port $port_A$
Output: True (Make a transaction)
 1: **function** $Hiding\_M(M, user_A, password_A, address_B, host_A, port_A)$
 2:    Define segment length n
 3:    Establish RPC connection $RPC\_connection(user_A, password_A, host_A, port_A)$
 4:    Define bitcoin unlock wallet time $time_A$
 5:    $walletpassphrase(password_A, time_A)$ //Unlock the wallet
 6:    $getwalletinfo(), getbalance()$ //Read wallet information
 7:    $message\_dec \leftarrow Enc(M)$
 8:    **while** $i < length(message\_dec) - 1$ **do**
 9:       $segment\_M = Decimal(message\_dec(i : i + n))$
10:       $payment \leftarrow segment\_M*Decimal(0.00000001)+Decimal(0.00000294)$
11:       $sendtoaddress(address_A, address_B)$ //Send transaction amount
12:       $sleep(t)$ //Pause for a period of time t
13:    **end while**
14:    **return True**
15: **end function**

2. Broadcast the Transaction: Using the Bitcoin Network Interface, broadcast the constructed transaction data to the Bitcoin Network. This way, other nodes on the network will receive the transaction.

3. Confirmation and Recording: Once a transaction is verified by a miner node in the network and included in a bitcoin block, the transaction is confirmed and recorded on the blockchain. This indicates that the transaction has been completed and that the hidden information has been passed through the bitcoin network.

It is important to note that the process of transaction confirmation takes some time in the Bitcoin network, as it is necessary to wait for the miner node to pack the transaction into the block. The confirmation time of a transaction depends on the level of network congestion and the speed at which the miner node is working.

With the above steps, we can encode the hidden message as a payment amount and create a bitcoin transaction to send it to the bitcoin network. The receiver can extract the hidden message by using the decoding algorithm and the confirmation records of the bitcoin network.

3.4   Decoding and Extracting

Once the transaction is confirmed and recorded on the Bitcoin blockchain, the recipient can reduce the payment amount with hidden communication in the transaction amount Amount(M) to the ASCII string Dec(Amount(M)) by decoding algorithm, and the hidden information can be extracted by the method I = Dec(Decimal(M)), which converts the payment amount to the corresponding character.

Algorithm 2. Secret Message Extracting Algorithm

Input: Receiver B's username and password $user_B$, $password_B$, Bitcoin Core node's host $host_B$ and port $port_B$

Output: Secret message M

1: **function** $Extracting(user_A, password_A, address_B, host_A, port_A)$
2:     Establish RPC connection $RPC\_connection(user_B, password_B, host_B, port_B)$
3:     Transaction amount list $TA \leftarrow getwalletinfo(), getbalance()$ //Read wallet
4:     Number of transactions $N \leftarrow$ count(TA)
5:     **for** $i = 1$ **to** N
6:          Amount per transaction $X \leftarrow TA(i)$
7:          Splicing $Str +\!= X$
8:     **end for**
9:     $M \leftarrow$ Dec(Str)
10:    **return True**
11: **end function**
12: **Tips:** TA is an array containing multiple transactions.

4   Experiments and Analysis

4.1   Experiment Setting

Here, the experimental configuration and environment for the covert communication experiment is as follows:

- **Host Computer Configuration:** Legion Y7000P2020H; CPU: Intel(R) Core(TM) i7-10875H 2.30 GHz; Memory: 16G; Hard Disk: 512G; Operating System: Windows 11 Home Edition
- **Experimental Environment:** Bitcoin Core; Bitcoin Test Network; Windows 11 platform
- **Programming Language:** Python

In this experiment, we used Bitcoin Core, the open source software for Bitcoin, as the client for the experiment, and the Bitcoin Test Network (testnet) as the experimental environment. Bitcoin Core is a complete Bitcoin node software that runs on a computer and connects to other nodes in the Bitcoin network. The Bitcoin Test Network is an environment used for development and testing purposes, and is completely isolated from the main bitcoin network. The Bitcoin Test Network uses a blockchain and transaction structure similar to that of the main Bitcoin network, but it uses test coins that have no real financial value. Therefore our experiments can realistically and effectively simulate real bitcoin transactions.

And we use Python programs to interact with Bitcoin's RPC (Remote Procedure Call) API, using the functions and interfaces provided by the API to create and send Bitcoin transactions. This allows for an automated transaction creation and sending process, and allows for real-time viewing of transaction information through the Bitcoin Core client, increasing efficiency and scalability.

4.2 Experiment Design

Firstly, we used the bitcoin client to register two unused wallets, Alice and Bob, representing the two parties communicating with each other. Two wallets created an original address OA_0 and OB_0, respectively, and received a certain number of test coins.

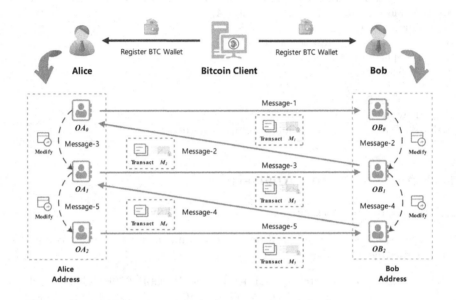

Fig. 2. Experiment trading process.

As shown in the Fig. 2, at the beginning of the covert communication, Alice first creates and sends a transaction containing secret information to Bob, and after Bob receives the transaction, he can decode the transaction amount and extract the secret information hidden in the payment amount. When Bob returns the message, Bob needs to create a new address OB_1 and transfer the message containing the hidden information from the address where he just received Alice's transaction to the new address OB_1, then Bob sends a transaction containing the hidden information to Alice's address OA_0 through the address OB_1, and extracts the hidden information from the transaction in a uniform way. The hidden information hidden in the payment amount. This completes the transmission of a mutual communication between Alice and Bob.

In this similar way, each message transfer is a transfer, each transfer, generate a new address, the old address transfer to the new address (themselves to themselves), the new address transfer each other is to carry out covert message communication.

Transaction Analysis. It is worth mentioning that Bitcoin has a minimum transaction amount limit for each transfer, where the Bitcoin Core client sends a minimum transaction amount of not less than 0.00000294, which requires us to ensure that each transaction amount is greater than the value of the change, so we have designed a method to make each transaction meet this condition. For example, I need to send a "hello world!" to the other party, so the decimal code can be transformed into "104101108108111032119111114108100033". Because of the minimum transaction limit and the amount of bitcoin held, we set the parameter n, which means that this string is cut and sent in parts. Since the minimum transaction amount limit has three values, we set $4 \leq n \leq 8$.

When n = 4, every group of four digits will be cut, respectively 1041, 0110, 8108, 1110, 3211, 9111, 1141, 0810, 0033, a total of nine groups, each group of digits then multiplied by the minimum bitcoin unit 0.00000001BTC, which becomes the transaction amount, but we can find some of them do not meet the transaction amount greater than the minimum transaction amount of 0.00000294. For example, $0033 * 0.00000001 = 0.00000033 \leq 0.00000294$.

Thus, we add 0.00000294 to the amount of the transaction after the multiplication operation, so that the minimum transaction amount limit is met. Also, consider that the first time each transaction is not transferred directly to the other party, but first to the new address of that wallet, and then to the other party via the new address. This results in two transactions, and Bitcoin charges a fee for each transaction. The following equation for calculating the amount of the transaction is defined here:

$$ActTA = TA + F \tag{1}$$

$$F = TD * RValue \tag{2}$$

where, in Eq. (1), TA denotes the transaction amount, F denotes the handling fee and $ActTA$ is the actual transaction amount transferred out. In Eq. (2), TD denotes the transaction data volume and $Rvalue$ is the current recommendation value.

Due to the fact that the handling fees charged are not always the same and that the new address requires a transfer amount to be transferred in order to effect the transfer out. In addition, a second transfer to the other party requires a fee, the current recommended value of which can be defined by yourself. Therefore, to set the fee for each transfer, we consider here adding a fee value to the amount of the first transfer, in order to prevent the amount of the first transfer to the new address being insufficient for the second transfer, or causing too much to be deducted as a transaction fee, resulting in information being lost.

4.3 Experiment Implementation

The recommended value of the handling fee is fixed at 0.00001000BTC/kvB for our experiments, so we set it at n = 4 and then add an extra 0.00001000BTC with the first transaction to be transferred to the new address, this amount is carried out in the own transfer itself, and even after deducting the transaction fee for the second transaction, the remaining amount remains in that wallet.

After encoding the "hello world!" message using the four-bit segmentation method, there are nine segments and the transactions are forwarded as follows (Table 2):

Table 2. Trading results for "hello world!" messages.

Message No.	Message Segments	Transaction Amount	First Trading	Second Trading	Extracted Message
1	1041	0.00001041	0.00002335	0.00001335	1041
2	0110	0.00000110	0.00001404	0.00000404	0110
3	8108	0.00008108	0.00009402	0.00008402	8108
4	1110	0.00001110	0.00002404	0.00001404	1110
5	3211	0.00003211	0.00004505	0.00003505	3211
6	9111	0.00009111	0.00010405	0.00009405	9111
7	1141	0.00001141	0.00002435	0.00001435	1141
8	0810	0.00000810	0.00002104	0.00001104	0810
9	0033	0.00000033	0.00001327	0.00000327	0033

Where, the amount of the first transaction is sent by Alice from address OA_0 to the new address OA_1, and the amount of the second transaction is sent by Alice from address OA_1 to address OB_0. The amount received by B is therefore the amount of the second transaction, and each amount is then decoded to obtain the corresponding hidden information. The total amount received by Bob is the sum of the second forwarded amount of **0.00027321 BTC**, without taking into account the processing fee, after the **four-bit segmentation method (n = 4)**.

In a message transfer, which consists of multiple transactions being sent, the latency of the Bitcoin network and the presence of fluctuating transmission rates can lead to sending errors and inconsistent sequences. In order to ensure that each transaction is received in the same order as it is sent, and that the correct message is decoded, we wait for a period of time after each transaction is sent. After repeated experimental tests in this environment, we chose to pause for 5 s after each transaction was sent to ensure that each transaction was sent and received correctly. From the experimental results, it is clear that the proposed covert communication scheme has good feasibility and practicality.

4.4   Experiment Results and Analysis

To verify the universality and practicality of the proposed scheme, we vary the content and length of the hidden messages to perform complexified covert communication experiments. The effect of covert communication in each case is also analysed for different segment lengths (n values).

In the evaluation of the experimental results, the main evaluation metrics we use include the amount of secret messages delivered (Messages size) and

the amount of transactions consumed (Transaction Amount), i.e. the number of
BTC. The message size can be expressed in terms of the number of characters
used to deliver a cryptic message. For example: "hello world!" is a cryptic message
forwarded once, Messages size = 12, spaces are also included. The larger
the Messages size value, the more messages are delivered, and the lower the
Transaction Amount spent, the more efficient the communication is, the lower
the overhead and the better the results. The Table 3 is a simulated conversation.

Table 3. Simulated conversation.

Alice: Hey Bob!
Bob: Hi Alice!
Alice: How's it going?
Bob: Good, you?
Alice: Not bad, thanks! Anything interesting happening?
Bob: Just the usual, nothing too exciting. How about you?
Alice: Same here, nothing out of the ordinary. Just going with the flow
Bob: That's cool. Sometimes it's nice to have some peaceful days

The first sentence is sent by Alice to Bob and is represented by Message1,
the second sentence is replied by Bob and is represented by Message2, and so
on. Each of the eight dialogues was coded as follows Table 4.

Table 4. Message encoding situation.

Message Label	Message Coding
Message1	1072101121032066111098033
Message2	0721050320651081050991010033
Message3	0721111190391150321...160321031111105110103063
Message4	0711111111000440321211111117063
Message5	0781111160320980097...112112101110105110103063
Message6	0741171151160321 16...11111711160321211111117063
Message7	0830971091010321041...04101032102108111119046
Message8	0841040971160391150...171080321000097121115046

The covert communication of messages is carried out with segment lengths
(n values) taken as 4, 5, 6, 7 and 8 respectively. The cumulative value of the
transaction amount to be forwarded to the other party was counted without
considering the handling fee. The detailed results are shown in Fig. 3:

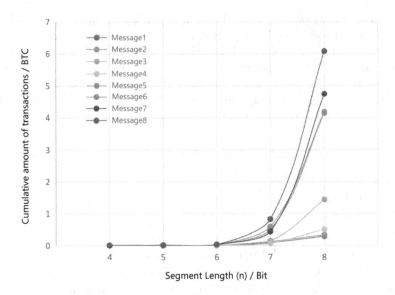

Fig. 3. The amount of BTC consumed by messages at different values of n.

As can be seen from Fig. 3, when the segmentation length is 4 (n = 4), the proposed scheme can use a very small amount of BTC for one message hidden transmission. When the amount of BTC transaction increases, it will increase the handling fee to a certain extent, and such a segmentation method will cause excessive communication consumption, reduce the efficiency of communication, and is not practical. In Fig. 4, we further compare and analyse the impact of each segmentation length on the amount of BTC consumed.

The following relationships can be obtained from the analysis in Fig. 4:

$$AvgTA_{n=5} = 7.2798AvgTA_{n=4}, \quad AvgTA_{n=6} = 24.0675AvgTA_{n=4}$$
$$AvgTA_{n=7} = 530.7185AvgTA_{n=4}, \quad AvgTA_{n=8} = 4244.902AvgTA_{n=4} \tag{3}$$

In Eq. (3), $AvgTA$ denotes the average BTC consumed by the exchange under sending the same message, and n is the segment length. Based on our experimental results, we can analyse that the same amount of messages n = 5 consumes on average about 7.3 times more bitcoins than n = 4, n = 6 is 24.0675 times more than n = 4, n = 7 is 530.7185 times more, and n = 8 reaches 4244.902 times more. The case of n = 4 is shown to be optimal and the best consumption.

Therefore, after processing the secret message through the segmentation method (at n = 4), it can achieve the purpose of covert communication with minimum BTC consumption. In the verification of the experimental results, it is also shown that the proposed covert communication scheme based on Bitcoin transactions is well feasible and efficient. The experimental results also show that our approach is able to handle different messages and lengths efficiently. We also examine the performance of this approach in the presence of network congestion, and the results show that message reception is guaranteed even when the network is congested.

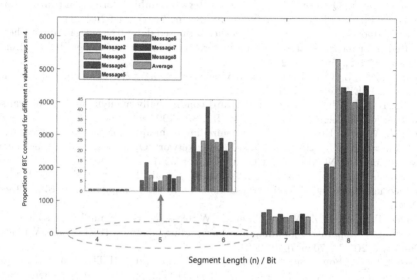

Fig. 4. Proportion of BTC consumed for different n values versus n = 4.

5   Conclusion

In this paper, we propose and validate a new method of covert communication based on Bitcoin transactions. The scheme uses the payment amount as the information carrier, adopts the four-bit segmentation method (n = 4) and product form to embed secret information, and realizes efficient covert communication. Furthermore, in tests of different segmentation methods, message delivery when n = 4 consumes far less BTC than other segmentation methods. In passing the "hello world!", When using the four-bit segmentation method (n = 4), only 0.00027321BTC was consumed. Repeated experiments have proved that the scheme is feasible and has good concealment and practicability. It is expected to promote more application areas of Bitcoin trading.

Acknowledgements. The research was financially supported by the Knowledge Innovation Program of Wuhan - Basic Research (No. 2022010801010197), the Opening Project of Nanchang Innovation Institute, Peking University (No. NCII2022A02), Guangxi Key Laboratory of Machine Vision and Intelligent Control (No. 2022B11), and National Natural Science Foundation of China (61972366, 61961036, 62162054).

References

1. Nakamoto S.: Bitcoin: a peer-to-peer electronic cash system. Decentralized Bus. Rev. (2008)
2. Zhang, T., Wu, Q., Tang, Z.: Research on steganographic information transmission based on bitcoin blockchain. J. Netw. Inf. Secur. **7**(01), 84–92 (2021)
3. Liu, Y., Hao, X., Ren, W., et al.: A blockchain-based decentralized, fair and authenticated information sharing scheme in zero trust internet-of-things. IEEE Trans. Comput. **72**(2), 501–512 (2022)
4. Gao, F., Zhu, L., Gai, K., et al.: Achieving a covert channel over an open blockchain network. IEEE Netw. **34**(2), 6–13 (2020)
5. Li, L., Liu, J., Chang, X., et al.: Toward conditionally anonymous Bitcoin transactions: a lightweight-script approach. Inf. Sci. **509**, 290–303 (2020)
6. Han, J., Woo, J., Hong, J.W.K.: Oversampling techniques for detecting bitcoin illegal transactions. In: 2020 21st Asia-Pacific Network Operations and Management Symposium (APNOMS), pp. 330–333. IEEE (2020)
7. Al Jawaheri, H., Al Sabah, M., Boshmaf, Y., et al.: Deanonymizing Tor hidden service users through Bitcoin transactions analysis. Comput. Secur. **89**, 101684 (2020)
8. Zheng, T.X., Yang, Z., Wang, C., et al.: Wireless covert communications aided by distributed cooperative jamming over slow fading channels. IEEE Trans. Wireless Commun. **20**(11), 7026–7039 (2021)
9. Bartoletti, M., Pompianu, L.: An analysis of bitcoin OP_RETURN metadata. In: Brenner, M., et al. (eds.) FC 2017. LNCS, vol. 10323, pp. 218–230. Springer, Cham (2017). https://doi.org/10.1007/978-3-319-70278-0_14
10. Tian, J., Gou, G., Liu, C., Chen, Y., Xiong, G., Li, Z.: DLchain: a covert channel over blockchain based on dynamic labels. In: Zhou, J., Luo, X., Shen, Q., Xu, Z. (eds.) ICICS 2019. LNCS, vol. 11999, pp. 814–830. Springer, Cham (2020). https://doi.org/10.1007/978-3-030-41579-2_47
11. Partala, J.: Provably secure covert communication on blockchain. Cryptography **2**(3), 18 (2018)
12. Li, Y., Ding, L., Wu, J., et al.: Research on a new network covert channel model in blockchain environment. J. Commun. **40**(5), 67–79 (2019)
13. Castiglione, A., De Santis, A., Fiore, U., et al.: An asynchronous covert channel using spam. Comput. Math. Appl. **63**(2), 437–447 (2012)
14. Cao, H., Yin, H., Gao, F., et al.: Chain-based covert data embedding schemes in blockchain. IEEE Internet Things J. **9**(16), 14699–14707 (2020)
15. Xiao, R., Ren, W., Zhu, T., et al.: A mixing scheme using a decentralized signature protocol for privacy protection in bitcoin blockchain. IEEE Trans. Dependable Secure Comput. **18**(4), 1793–1803 (2019)
16. Miers, I., Garman, C., Green, M., et al.: Zerocoin: Anonymous distributed E-Cash from Bitcoin. In: 2013 IEEE Symposium on Security and Privacy, pp. 397–411. IEEE (2013)
17. Sasson, E.B., Chiesa, A., Garman, C., et al.: Zerocash: decentralized anonymous payments from Bitcoin. 2014 IEEE Symposium on Security and Privacy, pp. 459–474. IEEE (2014)
18. Heilman, E., Baldimtsi, F., Goldberg, S.: Blindly signed contracts: anonymous on-blockchain and off-blockchain bitcoin transactions. In: Clark, J., Meiklejohn, S., Ryan, P.Y.A., Wallach, D., Brenner, M., Rohloff, K. (eds.) FC 2016. LNCS, vol. 9604, pp. 43–60. Springer, Heidelberg (2016). https://doi.org/10.1007/978-3-662-53357-4_4

Reliability Scheduling Algorithm
for Heterogeneous Multi-verified Time Systems

Fang Liu[1,2], Xing Gao[2], Di Cheng[3], Min Peng[1], and Yanxiang He[1(✉)]

[1] School of Computer Science, Wuhan University, Wuhan, China
{liufangfang,pengm,yxhe}@whu.edu.cn
[2] Wuhan City College, Wuhan, China
[3] College of Computer Science, Wuhan University of Science and Technology, Wuhan, China
chengdi@wust.edu.cn

Abstract. Heterogeneous multi-core systems are widely used in various real-time scheduling systems because of their high performance and low energy consumption. With the continuous progress of integrated circuit process technology, heterogeneous multi-core real-time systems have improved the power density of processors and the sensitivity of circuits, but they also lead to an increasing probability of system failure, and it is an important and difficult problem to improve the reliability of the whole system and ensure the normal operation of the system. In this paper, we propose the reliability algorithm BFSA (Backward-Forward Scheduling Algorithm), which aims to maximize the reliability of the system under the condition of satisfying the real-time constraint. The BFSA scheduling algorithm schedules tasks backward and forward after one copyless scheduling according to the deadline to obtain the final result of scheduling. The experimental results show that this algorithm maximizes the reliability of BFSA under different different constraints of three categories of real-time conditions, which effectively improves the reliability of the system.

Keywords: Heterogeneous Multicore · Real-time System · Reliability · Scheduling Algorithm

1 Introduction

The execution capabilities of each processor in a heterogeneous multi-core system are different, and tasks can be assigned to suitable processors for execution, which brings great advantages to the execution of tasks. However, along with the advantages of energy saving, high execution efficiency, and good performance, the failure rate of the system is also increasing, especially for systems that are critical to safety, which will have real-time and reliability requirements, such as: critical Patient monitoring system, aerospace system, unmanned system. Therefore, an extremely urgent need is to explore how to ensure the reliability of the system under the constraints of real-time conditions.

System failure may occur due to electromagnetic interference [1], hardware failure, human error, frequent temperature changes, excessive temperature [2], etc. The greater the probability of a system failure, the less reliable the system is. Faults are generally divided into transient faults and permanent faults. For transient failures, the system can recover on its own, but certain tasks may not be performed successfully. A permanent failure, however, will prevent all tasks on the failed processor from successfully executing. Since transient faults occur more frequently, this paper studies how to ensure system reliability under transient faults.

There are various methods to improve system reliability, such as: task migration [3], task backup, and reducing task execution time [4, 5]. The most typical scenario is task backup. Task backup is divided into active backup and passive backup. Active backup is to actively copy the task during the task scheduling process and schedule its copies on different processors to ensure the successful execution of the task; passive backup is to detect whether there is an error in the execution of the task during the scheduling process. When an error occurs, the task is executed again to ensure the reliability of the system. Since passive backup requires an error detection mechanism and requires time to adjust, under the condition of real-time scheduling, the system generally does not have time to adjust tasks after an error is detected. Therefore, in this paper, we choose the scheduling scheme of active replication. The active backup scheme belongs to time redundancy [6], and scheduling using this scheme will inevitably lead to excessive scheduling length. Therefore, how to improve the reliability of the system under the constraint of scheduling length is an urgent problem to be explored.

In recent years, the problem of system reliability has attracted the attention of various researchers [7–11]. Han et al. [7] proposed a multi-criteria optimization scheme to improve reliability and reduce energy consumption. But he works on periodic tasks and requires the use of error detection mechanisms. Liu et al. [8] studied scheduling problems with similar constraints, however, he studied permanent errors. Xie et al. [9] proposed a scheme to improve reliability under time constraints, however, he studied scheduling on homogeneous multi-core processors. Han et al. [10] used the active backup method and used a linear method to find the processor with the most reliability and the least energy consumption and the corresponding processor frequency for scheduling, but did not consider the time constraint. Youness et al. [11] combined a list scheduling algorithm with a simulated annealing algorithm to improve scheduling time and system reliability, however, based on an on-chip multi-core system.

In this paper, we consider the scheduling problem of reliability under heterogeneous multi-core real-time systems, and our aim is to maximize the reliability of heterogeneous multi-core systems under real-time constraints and task dependency constraints. This is NP's. To solve this problem, we propose a novel heuristic scheduling algorithm-forward-backward traversal combined scheduling algorithm, referred to as FBSA.

The main contributions of this paper are as follows:

1. We propose the BSA scheduling algorithm, which considers the barrel principle. Under the time constraint, tasks are backed up according to reliability from small to large, which improves the reliability of low-reliability tasks and improves the reliability of applications.

2. On the basis of BSA, a BFSA scheduling algorithm is proposed, which combines backward scheduling and forward scheduling, makes full use of the time gap in the scheduling process, and reschedules the primary and backup copies of the task on the most and next most reliable processors.
3. The experimental results show that the reliability of the application has been improved to a certain extent by using the BFSA scheduling algorithm.

We conduct extensive experiments using random applications and real applications, comparing our algorithm with the Heterogeneous Earliest-Finish-Time algorithm (HEFT) [12], the fast functional safety verification2 algorithm (FFSV2) [13], and find that our Algorithms are better and more reliable.

The rest of the paper is organized as follows: Sect. 2 reviews related work. Section 3 describes the model and problem statement. Section 4 presents an example of motivation. Section 5 explains our algorithm. The Sect. 6 analyzes the experimental results. The Sect 7 summarizes the full text.

2 Related Work

In recent years, researchers and scholars have proposed many solutions to improve system reliability [10, 14–18, 20–22]. [10, 14, 15] all focus on the improvement of reliability in heterogeneous multi-core systems with transient errors. Xie et al. [22] proposed that the ERRM algorithm should minimize the number of redundant copies under the requirement of reliability. However, none of them takes into account time constraints. Wang et al. [17] proposed a replication-based scheduling algorithm with the goal of maximizing system reliability. By comparing with the task reliability threshold to calculate the number of replicas for each task, Zhang et al. [18] proposed a new dual-objective genetic algorithm BOGA, which pursues low energy consumption and high system reliability. But they mainly focus on communication reliability. Niu et al. [20] proposed a task partition strategy to adaptively reserve recovery space for real-time tasks, which significantly improves the reliability of the system and solves the problem of insufficient system reliability. Kumar et al. [21] proposed an active replication-based framework that satisfies reliability and time constraints while minimizing energy consumption. Similarly, they both focus on periodic tasks.

Some use other methods to improve system reliability [13, 19]. Huang et al. [19] simultaneously consider energy consumption, reliability and scheduling completion time, use parameters to balance scheduling completion time and reliability, and use parameter l to balance scheduling completion time and energy consumption, achieving that the three are better than most in most cases Compare algorithms. Xie et al. [13] proposed the FFSV2 algorithm to migrate tasks to the most reliable processor to achieve maximum reliability under time constraints, but he did not consider task backup.

Many task replication based reliability enhancement methods have been proposed [8, 23–26]. Broberg et al. [23] proposed a dynamic fault-tolerant model and task scheduling scheme to ensure the reliability of system reservations through task replication, in order to shorten the scheduling time under the requirements of reliability. Liu et al. [8] adopted the pre-scheduling scheme, under the time constraint, backed up the entire application to find the maximum number of fault tolerances in the system and improve the reliability of the system, but they studied permanent errors. Haque et al. [24] propose statically and dynamically adaptive solutions to achieve maximum reliability goals with minimum energy consumption. Roy et al. [25] deliberately delay the backup task and dynamically cancel the execution of the backup task when the main task is successfully completed, in order to achieve the optimal effect of reliability and energy consumption. Likewise, they all need to use fault detection mechanisms to detect errors.

Different from existing research, this paper considers time constraints, dependencies constraints between tasks, and error constraints, and adopts task backup technology to maximize the reliability of the system by combining backward scheduling and forward scheduling.

3   Model and Problem Statement

In this section we detail the system model, task model, error model and state the definition of the problem.

3.1   System Model

The real-time heterogeneous multi-core system considered in this paper contains p interconnected processors, and the set of processors is denoted as $P = \{P_1, P_2, ..., P_p\}$ Processors can communicate with each other, but different processors communicate at different speeds. An example of the processor architecture diagram is shown in Fig. 1. Three heterogeneous processors are connected through a bus. Each processor mainly consists of a processing unit (PU), a local memory (LM), and a data transfer unit (DTU). The PU is the command and control center, the LM is used to temporarily store data, and the DTU transfers data between the processor and the memory. Different processors can be connected through a bus or the like. We assume that there is no contention for inter-processor communication and that inter-processor communication capabilities are symmetric. In other words, if the communication cost of processor p1 to p2 is $d(p_1, p_2)$, then the communication cost of processor 2 to processor 1 is $d(p_2, p_1) = d(p_1, p_2)$.

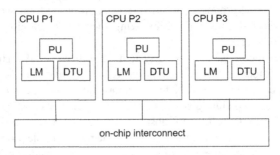

Fig. 1. Example of processor architecture

Table 1 shows examples of inter-processor communication capabilities, The 2 in the second row and first column of the table represents the cost of unit data transfer between processors P2 and P1. Similarly, the unit data communication cost of P1 to P2 in the first row and two columns in the table is also 2.

Table 1. Communication cost between processors

P	P_1	P_2	P_3
P_1	0	2	3
P_2	2	0	1
P_3	3	1	0

3.2 Task Model

A distributed application uses DAG, $G = <V, E>$. V is the set of tasks in G, denoted as $V = \{n_1, n_2, ..., n_n\}$, and n_n represents the nth task of G. E is the set of communication overhead between tasks. As shown in Fig. 2, the fixed point in the figure is the task, and the number on the edge of the connection between tasks is the amount of data that needs to be transmitted between tasks. There are dependencies between tasks in our model, in other words, if there is an edge from task n_i to task n_j, denoted by e_{ij}, then task n_j can start executing after task n_i is completed. If task n_i is assigned to processor P_u and task n_j is assigned to processor P_v, when u = v, when tasks are assigned to the same processor, the communication overhead is negligible, and the communication time is assumed to be $u \neq v$, the communication cost between tasks is $c(n_i, n_j) = d(P_u, P_v) \times e_{ij}$.

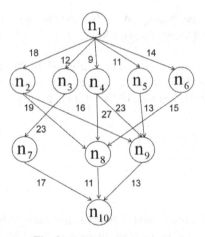

Fig. 2. DAG model example

In Fig. 2, there is an edge from n_1 to n_2, then n_1 is the predecessor of task n_2, and n_2 is the successor of n_1. We define the predecessor of task n_i as pre_i, and the successor set of task n_i as suc_i. The predecessor of n_1 is empty, and the successor is $\{n_2, n_3, n_4, n_5, n_6\}$. Assuming that the DAG has only one exit task and one entry task, if the DAG has multiple exit or entry tasks, add a zero-weight task as a virtual exit or entry task to ensure that all DAGs have only one entry and exit task. The entry task of the DAG is defined as n_{entry}, and the exit task is n_{exit}.

Generally speaking, the same processor performs different tasks and consumes different time. Due to the heterogeneity of processors, the execution time of the same task on different processors is different. Task n_i is allocated to the processor P_u, and the execution is defined as w(i, u). As shown in Table 2, the execution time of task n_1 on the three processors is w(1, 1) = 14, w(1, 2) = 16, w(1, 3) = 9.

Table 2. Task execution schedule on the processor

Task	P_1	P_2	P_3
n_1	14	16	9
n_2	13	19	18
n_3	11	13	19
n_4	13	8	17
n_5	12	13	10
n_6	13	16	9
n_7	7	15	11
n_8	5	11	14
n_9	18	12	20
n_{10}	21	7	16

This article involves task backup. We define the primary copy task of task v_i as n_i, the backup copy as n_i^k, and k represents the kth backup copy. Primary and backup copies are the exact same task.

The execution order of tasks is from the entry task to the exit task. The difference between the execution end time of the exit task and the execution start time of the entry task is the scheduling completion time of the entire application, which can be expressed as:

$$sl = ft(n_{exit}) - st(n_{entey})$$ (1)

3.3 Reliability Model

Random hardware failures occur randomly in the life cycle of hardware and are unpredictable and follow a probability distribution. Generally speaking, the transient failures of DAG-based distributed applications follow a Poisson distribution [26].

The reliability at unit time t is expressed as

$$R = e^{\lambda t}$$ (2)

When the execution time of task n_i on the processor P_u is w(i, u), the reliability is shown in the following formula, where λ_u is the error rate of the processor P_u per unit time.

$$rel(n_i, P_u) = e^{\lambda_u w(i,u)}$$ (3)

When the task has a backup copy, the reliability calculation rule is as follows: dup is the set of processors scheduled by the primary copy of the task and the backup copy

$$R(n_i) = 1 - \prod_{u \in dup, 1 \le u \le p} (1 - rel(n_i, P_u)) \tag{4}$$

The reliability of the entire distributed application is

$$R_{DAG} = \prod_{i=1}^{n} R(n_i) \tag{5}$$

3.4 Problem Statement

The problem scenario we want to solve is: given a DAG = <V, E>, a heterogeneous system contains p processors and n tasks, the time constraint is D, the error probability of the processor is in order to obtain the maximum reliability of the system as possible.

4 A Motivation Example

In this section we use a motivated example to demonstrate the effectiveness of our proposed algorithm. The task model, system model, and error model are described in Sect. 3. To facilitate settlement, we assume that the communication cost of a task to transfer a unit of data between different processors is 1. The communication relationship between tasks is shown in Fig. 3, and the computing overhead of tasks on all processors is shown in Table 2. Figure 4 shows four scheduling timing diagrams, namely earliest completion time (HEFT) scheduling timing diagram, fast functional safety verification 2 (FFSV2) scheduling timing diagram, backward traversal scheduling timing diagram (BSA), backward Front-traversal Combined Scheduling Moment Graph (BFSA). The BFA is the first part of the BFSA. Rounded rectangles represent tasks, white ones represent primary copies, and light gray ones represent backup copies.

Figure 3 is the scheduling result of HEFT. The task is scheduled on the processor with the earliest completion time, and the scheduling length is 80. (b) is the scheduling result of FFSV2, where tasks n_3 and n_7 are migrated from processor P_3 to processor P_1. Task n_9 is migrated from processor P_2 to processor P_1. Task n_6 is migrated from processor P_2 to processor P_3 with a schedule length of 98. (c) is the scheduling result of the intermediate process of BFSA. Tasks n_3 and n_7 are backed up on processor P_1, task n_9 is backed up on processor P_2, and task n_6 is backed up on processor P_3, , and the scheduling length is 100. (d) is the final scheduling result of BFSA. Compared with (c), task n_3 is scheduled from processor P_3 to processor P_2, and task n_8 is no longer scheduled on processor P_3, but on processor P_1. The length is 98. BFSA uses backup to improve the reliability of low-reliability tasks, reuses gaps to improve the reliability of certain tasks, and ultimately improves the reliability of the system.

BFSA improves the reliability of the system. Assuming that the errors of the three processors are $\lambda_1 = 0.002$, $\lambda_2 = 0.005$, and $\lambda_3 = 0.004$, the deadline D = 100, the reliability of HEFT is 0.63954, the reliability of FFSV2 is 0.74453, and the reliability of BSA is 0.78665. The reliability of BFSA is 0.82386. HEFT is time-redundant, FFSV2 does not take full advantage of the gap time on the processor, BFSA can improve the reliability of low reliability tasks, and make full use of the time gap.

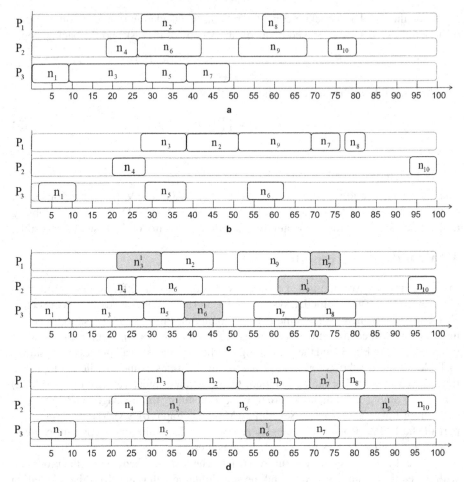

Fig. 3. An example of motivation; (a) HEFT scheduling time chart, the reliability is 0.63954, and the scheduling length is 80. (b) FFSV2 scheduling time chart, the reliability is 0.74453, and the scheduling length is 98. (c) BSA scheduling time chart, the reliability is 0.78665, and the scheduling length is 100. (d) BFSA scheduling time chart, the reliability is 0.82386, and the scheduling length is 98. $D = 100$.

5   The Proposed Algorithm

This section proposes a backward-forward scheduling algorithm (BFSA for short), which solves the problem of insufficient reliability in heterogeneous multi-core real-time systems. For a better understanding, first, we introduce some basic concepts, and then introduce our proposed algorithm in detail.

5.1 Basic Concepts of Backward Scheduling

The scheduling order of tasks in the process of backward scheduling is consistent with HEFT. Tasks are scheduled in descending order of their rank values. The larger the rank value, the higher the priority of the task. The calculation formula of rank is as follows:

$$\text{rank}(n_i) = \begin{cases} \overline{w(n_i)}, & \text{if } (n_i = n_{exit}) \\ \overline{w(n_i)} + \max\limits_{n_j \in suc(n_i)} \left\{ \overline{c(n_i, n_j)} + \text{rank}(n_j) \right\}, & \text{if } (n_i \neq n_{exit}) \end{cases} \tag{6}$$

$\overline{w(n_i)}$ is the average execution time of the task on all processors, $\overline{c(n_i, n_j)}$ is the average communication time between task n_i and task n_j, and $\overline{d}$ is the average unit cost of the task to transfer data between different processors.

$$\overline{w(n_i)} = \frac{\sum_j^p w(n_i, P_j)}{p} \tag{7}$$

$$\overline{c(n_i, n_j)} = e_{ij} \times \overline{d} \tag{8}$$

$$\overline{d} = \frac{\sum_{i=1}^p \sum_{j=1}^p d(P_i, P_j)}{p^2 - p} \tag{9}$$

Define $EST(n_i, P_m)$ as the earliest start time of task n_i on processor p_m, and $EFT(n_i, P_m)$ as the earliest completion time, which can be expressed as:

$$EST(n_i, P_m) = \begin{cases} 0, & \text{if } (n_i = n_{entry}) \\ \max\{dpre(n_i, P_m), \ avail(P_k)\}, & \text{if } (n_i \neq n_{entry}) \end{cases} \tag{10}$$

$avail(P_m)$ is the earliest available time of processor p_m, and $dpre(n_i, P_m)$ is the time when task n_i is assigned to processor p_m to receive all the data passed by the predecessor.

$$dpre(n_i, P_m) = \max \left\{ \max\limits_{\substack{0 \leq c \leq copy \\ n_j \in pre(n_i)}} \left(EFT\left(n_j^c, P\left(n_j^c\right)\right) + c(n_j, n_i) \right) \right\} \tag{11}$$

copy is the number of copies of a single task

$$EFT(n_i, P_m) = EST(n_i, P_m) + w(n_i, P_m) \tag{12}$$

5.2 Basic Concepts of Forward Scheduling

When traversing forward, the earliest start time of a task is defined as:

$$EST(n_i, P_m) = \begin{cases} 0, & if\ (n_i = n_{entry}) \\ \max_{n_j \in pre(n_i)} \{AFT(n_j) + c(n_j, n_i)\}, & if\ (n_i \neq n_{entry}) \end{cases} \qquad (13)$$

$AFT(n_j)$ is the actual completion time of task n_j.
The latest end time of the task is defined as:

$$\begin{cases} LET(n_i, P_m) = LET(G), & if\ (n_i = n_{exit}) \\ LET(n_i, P_m) = \min_{n_j \in suc(n_i)} \{AST(n_j) - c(n_i - n_j)\}, & if\ (n_i \neq n_{exit}) \end{cases} \qquad (14)$$

$AST(n_j)$ is the actual start time of task n_j.

During the scheduling process, the task n_i has a certain time slot on the processor P_m as $slot = \{k[0], k[1]\}$. If one of the following two conditions is satisfied, the task can be scheduled in this time slot.

When $LET(n_i, P_m) >= k[1]$ then $LET(n_i, P_m) = k[1]$, if $k[0] <= LET(n_i, P_m) - w(n_i, P_m)$, then the actual end time can be expressed as: $AFT(n_i, P_m) = k[1]$.

When $LET(n_i, P_m) < k[1]$, if $k[0] <= LET(n_i, P_m) - w(n_i, P_m)$, then the actual end time can be expressed as: $AFT(n_i, P_m) = LET(n_i, P_m)$.

The actual start time of the task can be expressed as:

$$AST(n_i, P_m) = AFT(n_i, P_m) - w(n_i, P_m) \qquad (15)$$

5.3 Backward-Forward Scheduling Algorithms

In this section, we introduce the algorithm BFSA. Its basic idea is as follows: First, we use the HEFT algorithm to schedule all tasks. If the scheduling length is less than or equal to the given deadline, then the application is schedulable, otherwise, it is not schedulable. Secondly, we sort all tasks according to reliability from small to large, and when the scheduling length does not exceed the deadline, the task copies are added to the scheduling process in turn. Finally, schedule from back to front, select an appropriate time slot, and reschedule the tasks on the processor in turn.

Algorithm 1 BSA.
Input:A DAG={V,E},a deadline D,$P = \{P_1, P_2, \ldots, P_p\}$;
Output:DAG scheduling result
1:Sort the tasks in BSAList in descending order of rank;
2:call HEFT scheduling algorithm,get HEFT scheduling length sl;
3:The tasks are listed in RelSeq in ascending order of reliability;
4:Task to_delete;
5:while(sl<=D)
6:$task \leftarrow Rel Seq.out()$;
7:task.isBackup=1,to_delete=task;
8:for (each t in BSAList)
9:for m=1 to p do
10:calculate $EFT(t, P_m)$,Store in ascending order in eftList;
11:end for
12:$EFT(t, P_m) \leftarrow$ eftList.out(),schedule task t on processor P_m;
13:if(t.isBackup)
14:$EFT(t, P_m) \leftarrow$ eftList.out(),schedule task t on processor P_m;
15:end if
16:end for
17:end while
18:Return the result of scheduling before copy to_delete is added.

In Algorithm 1, we sort the tasks in descending rank order in line 2–3, and use the HEFT scheduling algorithm to schedule tasks, and sort tasks in ascending order of reliability. Lines 5–17, when the scheduling length of the application is less than the time constraint, the task replicas are added to the scheduling process in ascending order of reliability, and the primary replicas of all tasks are scheduled on the processor with the earliest and next earliest completion time. When the scheduling length exceeds the time constraint, the loop ends. In line 18, remove the copy that was added last, and return the scheduling result before adding the last copy.

In Algorithm 2, the first line determines that the scheduling order of tasks is in ascending order of rank value; in lines 2–19, the while loop reschedules all tasks; in lines 4–8, the BSA scheduling result is compared with the current scheduling task id The same primary copy record is deleted. If there is a backup copy of the id task, the scheduling record is also deleted. Lines 9–14, in descending order of the starting value of the processor's gap, traverse all the gaps, and schedule tasks on the processor gap that meets the scheduling time requirements and has the greatest reliability. If the task has a backup copy, the backup copy is scheduled on the processor that meets the scheduling time requirement and has the next highest reliability.

Algorithm 2 BFSA.
Input:BSA scheduling result BSAList
Output:Reliability of the application
1:Store the tasks in BFSAOrderList in ascending rank order;
2:while(there are tasks in BFSAOrderList)
3:task←BFSAOrderList.out();
4:for(t in BSAList)
5:if(task.id==t.id)
6:BSAList.remove(t);//Delete the primary and backup copies together
7:end if
8:end for
9:for m=1 to p do
10:double met=$AFT(t,P_m) - AST(t,P_m)$;
11:if(met==$w(t,P_m)$)
12:compute$Re\,l\,(task,P_m)$,store in RelList;
13:end if
14:end for
15:schedule task on the most reliable processor in RelList;
16:if(task.isBackup)
17:schedule tasks on the next most reliable processor in RelList;
18:end if
19:end while
20:Return the reliability of the application.

6 Experimental Results and Analysis

In this section, we evaluate the effectiveness of our proposed algorithm BFSA, and we compare the BFSA algorithm with two classic algorithms, HEFT and FFSV2. HEFT is a typical static list scheduling algorithm that has been shown to have good performance on task scheduling problems. FFSV2 migrates tasks to more reliable processors under time constraints, effectively improving the reliability of applications. We use randomly generated applications and real-world applications to test the effectiveness of our proposed algorithm. The experimental parameters and result analysis are given below.

6.1 Experimental Parameters

In our experiments, we use artificially synthesized tasks [2] to validate our proposed algorithm, and the parameters used in the task generation process are as follows:

n: the total number of tasks generated;
CCR: the ratio of communication data volume and computing time, generally
              defined as the average communication data volume divided by the average
              computing time;
outDegree: the number of successor tasks of a task;
inDegree: the number of predecessor tasks of a task;
p: the number of processors;
λ: The error rate of the processor;

The range of n is from [10, 60], the increment is 10; the CCR is selected from {0.1, 0.5, 1, 2, 5}; the range of in-degree and out-degree is in [0, 4]; the calculation time range of the task is [1, 20]; the average communication time of the task is equal to the CCR multiplied by the average execution time of the task; we simulate a 4-core processor; the error rate of the task is in the range of $\left[10 \times 10^{-3}, 15 \times 10^{-3}\right]$; for the convenience of calculation, we will The inter-unit data communication cost is set to 1.

The parameter evaluated is the reliability of the system.

6.2 Randomly Generate Application Results

In this section, we test HEFT, FFSV2, and BFSA using randomly generated applications to verify the effectiveness of our algorithm.

Figure 4 shows the comparison chart of the scheduling results of three algorithms with the number of tasks from 10 to 50 and the interval of 10.The CCR is fixed at 1, the number of tasks is getting larger and the deadline is getting bigger and bigger, and the reliability of BFSA is increasing compared to HEFT and FFSV2.

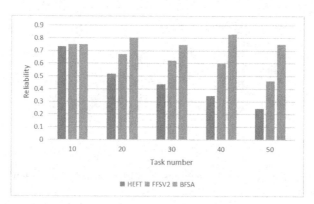

Fig. 4. Random task scheduling results. w = 15, CCR = 1, D = (w + w * CCR) * n/4

Figure 5 shows the changes in the reliability of the three algorithms as the deadline increases when the number of tasks is fixed at 60, CCR = 1, and w = 15. In this test example, the scheduling length of HEFT is 228, we set the deadline as 228, 258, 288, 318, 348, 378, and the interval is 30. The impact of the test time constraint on reliability can be seen from the figure, as the deadline increases, the BFSA's. Compared with the other two algorithms, the reliability is significantly increased.

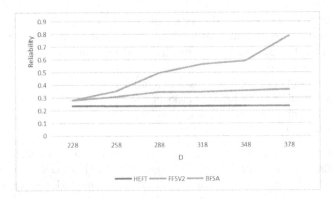

Fig. 5. Random task scheduling results. w = 15, CCR = 1, n = 60

6.3 Real World Application Results

Gaussian elimination: In the Gaussian elimination experiment, MS is used to represent the size of the matrix, and the relationship between MS and the total number of tasks n is: $n = \frac{MS^2 + MS - 2}{2}$.

Figure 6 shows the scheduling results of HEFT, FFSV2 and BFSA when MS goes from 5 to 17, the increment is 3, w = 15, CCR = 1, D = (w + w * CCR) * MS/2. It can be seen from the figure that under the same conditions, the reliability of FFSV2 is higher than that of HEFT, and the reliability of BFSA is higher than that of FFSV2.

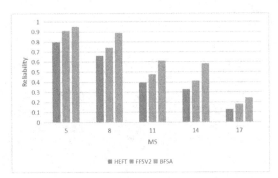

Fig. 6. Gaussian elimination scheduling results. w = 15, CCR = 1, D = (w + w * CCR) * MS/2

Fast Fourier Transform: Use S to represent the size of the input vector of polynomial coefficients. The relationship between S and the number of tasks n is: $n = S \times \log_2 S$, S takes a value from 2 to 32, and the multiplier is 2.

Figure 7 shows the reliability comparison of HEFT, FFSV2 and BFSA algorithms when w-15, CCR = 1, deadline D = (w + w * CCR) * S * 5/2, and S takes different values. It can be seen that when S is equal to 2 and 4, the reliability of the three algorithms is similar, and the reliability of the three algorithms begins to widen from S = 8.

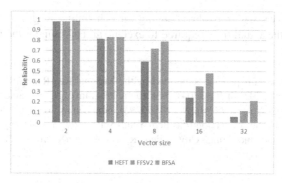

Fig. 7. Fast Fourier scheduling results. w = 15, CCR = 1, D = (w + w * CCR) * S * 5/2

Dynamic Molecular Coding: This application is also part of the experimental evaluation, in this experiment, the number of tasks is 40, the CCR is 0.1, 0.5, 1, 5, w = 15, D = (w + w * CCR) * S * 5 /2. Figure 8 shows the reliability results after the application is scheduled using the three algorithms: HEFT, FFSV2, and BFSA. It can be seen from the figure that under the same conditions, the reliability of BFSA is the largest, followed by FFSV2, and the reliability of HEFT is the smallest.

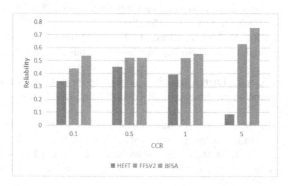

Fig. 8. Dynamic Molecular Coding results. w = 15, CCR = 1, D = (w + w * CCR) * S * 5/2

7 Summary

By combining backward and forward scheduling, this paper makes full use of the time gap in the scheduling process and improves the reliability of the system. In the process of backward scheduling, taking into account the barrel principle, active backup technology is used to back up tasks according to reliability from small to large under time constraints to improve system reliability. In the process of forward scheduling, the time gap in the scheduling process is fully utilized, and the task is migrated to the processor with high reliability, which further improves the reliability of the system.

In the future, we will consider reducing the scheduling length of the application, that is, reducing the actual time required to execute all tasks, under the requirements of reliability.

Acknowledgements. This paper is supported by the National Key Research and Development Program of China (No. 2021ZD0113304), the General Program of Natural Science Foundation of China (NSFC) (Grant No. 62072346, No. 61972293), Founded by Joint\&Laboratory on Credit Technology. Hubei Higher Education Excellent Youth Science and Technology Innovation Team Project (No. T2022060) and Wuhan City College Research Science Project (2023CYYYJJ01).

References

1. Niu, L., Zhu, D.: Reliable and energy-aware fixed-priority (m,k)-deadlines enforcement with standby-sparing. In: 2020 Design, Automation & Test in Europe Conference & Exhibition (DATE) (2020)
2. Zhou, J., Jin, S., Zhou, X., et al.: Resource management for improving soft-error and lifetime reliability of real-time MPSoCs. IEEE Trans. Comput. Aided Des. Integr. Cir. Syst. **38**, 2215-2228 (2019)
3. Namazi, A., Abdollahi, M., Safari, S., Mohammadi, S., Daneshtalab, M.: LRTM: life-time and reliability-aware task mapping approach for heterogeneous multi-core systems. In: 2018 11th International Workshop on Network on Chip Architectures (NoCArc), pp. 1–6 (2018). https://doi.org/10.1109/NOCARC.2018.8541223
4. Zhang, L., Li, K., Xu, Y., Mei, J., Zhang, F., Li, K.: Maximizing reliability with energy conservation for parallel task scheduling in a heterogeneous cluster. Inf. Sci. **319**, 113–131 (2015)
5. Lee, Y.C., Zomaya, A.Y.: Energy conscious scheduling for distributed computing systems under different operating conditions. IEEE Trans. Parallel Distrib. Syst. **22**(8), 1374–1381 (2011)
6. Roy, A., Aydin, H., Zhu, D.: Energy-aware primary/backup scheduling of periodic real-time tasks on heterogeneous multicore systems. Sustain. Comput. Inf. Syst. **29**(2), 100474 (2021)
7. Han, L., Gao, Y., Liu, J., et al.: Energy-aware strategies for reliability-oriented real-time task allocation on heterogeneous platforms. In: 49th International Conference on Parallel Processing, ICPP 2020 (2020)
8. Liu, J., Wei, M., Hu, W., et al.: Task scheduling with fault-tolerance in real-time heterogeneous systems. J. Syst. Archit. **90**, 23–33 (2018)
9. Xie, G., Zeng, G., Li, R.: Safety enhancement for real-time parallel applications in distributed automotive embedded systems: a stable stopping approach. IEEE Trans. Parallel Distrib. Syst. **31**(9), 2067–2080 (2020)
10. Han, Y., Liu, J., Hu, W., Gan, Y.: High-reliability and energy-saving dag scheduling in heterogeneous multi-core systems based on task replication. In: 2021 IEEE International Conference on Systems, Man, and Cybernetics (SMC), pp. 2012–2017 (2021). https://doi.org/10.1109/SMC52423.2021.9658608
11. Youness, H., Omar, A., Moness, M.: An optimized weighted average makespan in fault-tolerant heterogeneous MPSoCs. IEEE Trans. Parallel Distrib. Syst. **32**, 1933–1946 (2021)
12. Topcuoglu, H., Hariri, S., Wu, M.Y.: Performance-effective and low-complex task scheduling for heterogeneous computing. IEEE Trans. Parallel Distrib. Syst. **13**(3), 260–260 (2002)
13. Xie, G., Gang, Z., Yan, L., et al.: Fast functional safety verification for distributed automotive applications during early design phase. IEEE Trans. Ind. Electron. **65**, 4378–4391 (2017)

14. Han, Y., Hu, W., Liu, J., Gan, Y.: Energy-efficient scheduling algorithms with reliability goal on heterogeneous embedded systems. In: 2021 IEEE International Conference on Parallel & Distributed Processing with Applications, Big Data & Cloud Computing, Sustainable Computing & Communications, Social Computing & Networking (ISPA/BDCloud/SocialCom/SustainCom), pp. 555–562 (2021). https://doi.org/10.1109/ISPA-BDCloud-SocialCom-SustainCom52081.2021.00082

15. Xiao, X., Xie, G., Xu, C., et al.: Maximizing reliability of energy constrained parallel applications on heterogeneous distributed systems. J. Comput. Sci. **26**, 344–353 (2017)

16. Gupta, M., et al.: Reliability-aware data placement for heterogeneous memory architecture. In: 2018 IEEE International Symposium on High Performance Computer Architecture (HPCA), pp. 583–595 (2018). https://doi.org/10.1109/HPCA.2018.00056

17. Wang, S., Li, K., Mei, J., et al.: A reliability-aware task scheduling algorithm based on replication on heterogeneous computing systems. J. Grid Comput. **15**(1), 1–17 (2016)

18. Zhang, L., Li, K., Li, C., et al.: Bi-objective workflow scheduling of the energy consumption and reliability in heterogeneous computing systems. Inf. Sci. **379**, 241–256 (2017)

19. Huang, J., Li, R., Jiao, X., Jiang, Y., Chang, W.: Dynamic DAG scheduling on multiprocessor systems: reliability, energy, and makespan. IEEE Trans. Comput. Aided Des. Integr. Circ. Syst. **39**(11), 3336–3347 (2020). https://doi.org/10.1109/TCAD.2020.3013045

20. Niu, L.: Reliability-aware scheduling for periodic tasks requiring (m,k)-firm real-time data processing. In: 2021 IEEE 22nd International Conference on Information Reuse and Integration for Data Science (IRI), pp. 69–74 (2021). https://doi.org/10.1109/IRI51335.2021.00016

21. Kumar, N., Mayank, J., Mondal, A.: Reliability aware energy optimized scheduling of non-preemptive periodic real-time tasks on heterogeneous multiprocessor system. IEEE Trans. Parallel Distrib. Syst. **31**(4), 871–885 (2020). https://doi.org/10.1109/TPDS.2019.2950251

22. Xie, G., Zeng, G., Chen, Y., et al.: Minimizing redundancy to satisfy reliability requirement for a parallel application on heterogeneous service-oriented systems. IEEE Trans. Serv. Comput. **13**, 871–886 (2017)

23. Broberg, J., Ståhl, P.: Dynamic Fault Tolerance and Task Scheduling in Distributed Systems (2016)

24. Haque, M.A., Aydin, H., Zhu, D.: On reliability management of energy-aware real-time systems through task replication. IEEE Trans. Parallel Distrib. Syst. **28**(3), 813–825 (2017). https://doi.org/10.1109/TPDS.2016.2600595

25. Liu, F., Peng, M., Gao, X., et al.: A customized processor for convolutional neural networks based on RISC-V. J. Wuhan Univ. (Sci. Edn.) **69**(2), 147–155 (2023). https://doi.org/10.14188/j.1671-8836.2022.0215

26. Roy, A., Aydin, H., Zhu, D.: Energy-efficient primary/backup scheduling techniques for heterogeneous multicore systems. In: 2017 Eighth International Green and Sustainable Computing Conference (IGSC), pp. 1–8 (2017). https://doi.org/10.1109/IGCC.2017.8323569

Efficient Log Anomaly Detection Based on Dimension Reduction and Attention Aware TCN

Zhihao Xu[1], Yuliang Shi[1,2]([✉]), Zhiyuan Su[3], Li Song[4], Jianjun Zhang[4], Xinjun Wang[1], and Hui Li[1]

[1] School of Software, Shandong University, Jinan, China
shiyuliang@sdu.edu.cn
[2] Dareway Software Co., Ltd., Jinan, China
[3] Jinan Inspur Data Technology Co., Ltd., Jinan, China
[4] Shandong Agricultural Machinery Research Institute, Jinan, China

Abstract. The software system usually records important runtime information in the log for troubleshooting. Researchers mine large log data for anomalies. Many studies use log data to build deep-learning models for detecting system anomalies. Although progress has been made in log anomaly detection on high-performance computing platforms, it is still difficult to achieve real-time and accurate anomaly detection on mobile devices and Internet of Things devices, as these devices usually do not have high computational power. To solve the above limitations, we propose an efficient log anomaly detection based on dimension reduction and attention aware temporal convolutional network method, namely EfficientLog. The model achieves efficient and accurate log detection in two ways: (1) it reduces the communication cost between mobile devices and cloud computing platforms by reducing the log vector dimension through BERT-whitening, and (2) it detects log anomalies by using the attention aware temporal convolutional network to reduce model testing time and computational consumption. We evaluate the proposed method on two public datasets, and the experimental results show that Efficient-Log can outperform existing popular log-based anomaly detection methods in terms of detection accuracy and computational consumption.

Keywords: Anomaly detection · BERT-whitening · Temporal convolutional network · Deep learning

1 Introduction

With the advent of the era of big data, software systems are becoming more and more complex. Since these systems provide various services to a large number of users, a small problem may arise dissatisfaction and even significant financial losses. Therefore, accurate and timely anomaly detection is crucial. Large and complicated software-intensive systems such as online business systems and big data systems generate fault processing logs to handle specific errors. The

X. Song et al. (Eds.): APWeb-WAIM 2023, LNCS 14331, pp. 498–512, 2024.
https://doi.org/10.1007/978-981-97-2303-4_33

log is an indispensable data resource to record the information on service running time. Over the years, researchers have developed many automated methods to detect system anomalies. These works retrieve useful information from logs, which employed data mining and machine learning technology to analyze log data and detect system anomalies [1–3].

With the rapid development of deep learning, many studies have begun to utilize neural networks for anomaly detection. For example, Loganomaly [4] adopts the Word2Vec method to obtain the log data word vector and then uses the Long Short-Term Memory (LSTM) model to detect log anomalies. However, Word2Vec is unable to understand complex contexts, resulting in low accuracy of anomaly detection. Besides, LSTM is computationally intensive and time-consuming. NeuralLog [5] chooses to use the pre-trained language model BERT [6] to obtain log data word vectors and adopts the Transformer model to classify anomalies. However, BERT generates word vectors with high dimensionality, and Transformer, which is too large to deploy on edge devices, has so many parameters. Therefore, although progress has been made in log anomaly detection of neural networks on high-performance computing platforms [4,5,7–9], it is still difficult to achieve real-time and accurate anomaly detection on mobile devices and Internet of Things (IoT) devices. Time and memory overhead hinder its deployment on platforms with limited resources [10].

BERT-whitening [11] uses the whitening operation in traditional machine learning to enhance the isotropy of sentence representations and reduce the dimension of the sentence representation. By using it, the dimension of word vectors and the calculation consumption can be reduced. It is suitable for anomaly detection. Temporal convolutional network (TCN) [12] has stronger parallelism and significant advantages in terms of memory consumption and processing speed. For edge devices, TCN is easier to deploy. In addition, the attention mechanism can be used to understand complex semantics and polysemy of words and achieve higher precision anomaly detection results.

Inspired by the above methods, we propose EfficientLog, a novel log anomaly detection model with lightweight and accurate features, which can be deployed on resource-constrained devices. Different from the existing methods, EfficientLog uses BERT-whitening to reduce the word vector dimension of the pre-trained model ALBERT [13], and detects log anomalies by using the attention aware TCN. Due to the lightweight feature of TCN, it reduces the testing time and calculation consumption of the model. Because of the attention mechanism, it can understand complex semantics and polysemy of words. We evaluate the proposed method on two popular datasets. The experimental results show that EfficientLog is superior to the existing methods in terms of complexity and accuracy. The main contributions of this paper are as follows:

- We propose a novel anomaly detection method, EfficientLog, which reduces the dimension of word vectors by using the dimension-reducing method called BERT-whitening, improving efficiency and enhancing the isotropy of word vectors.

- We propose an attention aware TCN, which reduces the model testing time and computing power consumption. So this method can facilitate real-time and accurate anomaly detection for mobile devices or IoT devices.
- We evaluate the performance of EfficientLog on two public datasets, and the experimental results indicate that the accuracy of this method is better than the existing model with the reduction of the number of parameters, the size of the detection model, and the detection time.

2   Related Work

2.1   Log Representation Learning

Log data representation is usually divided into two steps, log parsing and log vector representation.

Log Parsing: There are two types of log parsing, which are parsing with the log parser and using the original log directly after preprocessing. Using a log parser is to convert log data into log templates through a specific parser. Among them, Drain [14] uses a heuristic parsing method that applies a fixed-depth tree structure to represent log messages and efficiently extracts common templates; LKE [15] applies the clustering method to log analysis and adopts a hierarchical clustering algorithm based on the weighted editing distance between paired log messages. However, the accuracy of log parsing highly affects the performance of log mining. Existing methods suffer from inaccurate log parsing and cannot handle Out-Of-Vocabulary (OOV) words well. Therefore, some researchers directly use raw logs to obtain structured log data through simple preprocessing [5,8].

Log Vector Representation: After obtaining the structured log data, some methods count the obtained log templates to generate a log count vector [1,2]. IM and ADR [16] find a linear relationship between log events and generate a log event sequence vector to represent the log data. NeuralLog uses Wordpiece [17] and pre-trained BERT to obtain a vector representation of log data with fixed dimensionality.

However, more factors need to be considered when deploying models on edge devices and mobile devices. Mobile devices cannot meet the needs of the high computational power of the model [18]. While the above methods have been verified effective on high-performance platforms, further research is needed in mobile device deployments.

2.2   Log Anomaly Detection

Existing log anomaly detection methods can be classified into machine learning methods and deep learning methods. In terms of traditional machine learning methods, Xu et al. [19] propose the Principal Component Analysis (PCA) method. They assume that there are different sessions in the log file, which can be easily identified by the session id attached to each log entry. Lou et al.

[3] propose IM, which mines linear relationships between log events from a log event count vector. Lin et al. [20] propose LogCluster, which applies clustering techniques to classify similar log sequences into the same category.

However, these machine learning methods lack robustness. The accuracy of anomaly detection will drop significantly when the logs contain noise caused by log message changes or regular updates, which limits their applicability in practical applications. The development of deep learning methods provides new solutions for log-based anomaly detection. Du et al. [7] propose DeepLog, which uses LSTM to learn the relationship between normal log sequences but cannot deal with noisy logs in time. Guo et al. [8] propose LogBERT, which uses the mask mechanism of a BERT model to randomly mask logs in the dialog window for training but does not learn the semantics of log data. Le et al. [5] propose NeuralLog to detect anomalies through a Transformer-based classification model, which can capture contextual information from log sequences. EdgeLog [21] proposes a lightweight log-based anomaly detection model based on a compressed TCN. However, most of the models proposed in the above methods are deep neural network models with long detection time, large models, and high computing power consumption, which are difficult to apply to edge devices with low computing power, small storage space, and real-time detection requirements.

We propose an attention aware TCN model that uses BERT-whitening dimensional reduction to effectively alleviate the problems of high dimensionality and large data volume of log data vectors. And the attention aware TCN solves the problems of multiple parameters and computational complexity of anomaly detection models. This method achieves satisfactory results in terms of detection efficiency and accuracy.

3   The Proposed EfficientLog

We propose EfficientLog, a log-based anomaly detection method that uses BERT-whitening to dimensionally clip the log embedding vectors generated by ALBERT and adopts a lightweight attention aware TCN model to detect anomalies. An overview of the approach is shown in Fig. 1(a). First, training log vector representation is performed on the cloud computing platform to obtain log vector data; log vector data are passed into the attention aware TCN to get a trained model and deploy this model to the mobile device. The mobile device uploads newly collected log data to the cloud computing platform to obtain log vector data. Finally, they are put into the model to detect anomalies.

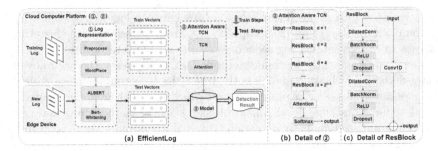

Fig. 1. Overview of EfficientLog and detailed pictures of key components.

3.1 Log Representation

This section focuses on log representation, which consists of three steps: data preprocessing, neural representation, and BERT-whitening dimension reduction. These three steps are all completed on the cloud computing platform. After uploading the new log data collected by the mobile device to the cloud computing platform, real-time log vector data generated by log representation for detection can be obtained.

Data Preprocessing. We first mark log messages as a set of word tokens. Use common separators (spaces, colons, commas, etc.) to split the log messages. Then, each uppercase letter is converted to lowercase and all non-character (operators, punctuation marks, and numbers) tokens are removed from the word set. This type of token is removed because it usually indicates a variable in the log message and has no information. So we can get the word character sequence $X = [x_0, x_1, x_2 \ldots, x_M]$, M is the length of character sequence.

Neural Representation. Each log message records a system event with its title and message content. The header contains fields determined by the logging framework, such as component and detail level. To preserve semantic information and capture the relationship between existing and new log messages, the presentation phase represents log messages in a vector format.

First, we use the WordPiece method for word segmentation, which has been shown in [5] to process OOV words more effectively and reduce vocabulary size. WordPiece first incorporates all characters and symbols into its basic vocabulary. It does not depend on the frequency of pairs, but chooses the one that maximizes the training data, maximizing the possibility of training data. It trains the language model from the basic vocabulary and selects the pair with the highest probability. The pair is added to the vocabulary, and the language model has trained again on the new vocabulary. These steps are repeated until the desired vocabulary is reached. For example, *PacketResponder*, a frequent occurrence in log data, is divided into more frequent subwords: *packet, respond, er*. The meaning of the OOV words is captured in this way.

Then we used the pre-trained language model ALBERT to obtain the log vector data. ALBERT employs two parameter reduction techniques: factorized embedding parameterization as well as cross-layer parameter sharing, removing a major obstacle to scaling pre-trained models. Both techniques significantly reduce the number of parameters in ALBERT without severely affecting performance, thus improving parameter efficiency. After tokenization, the set of words and subwords are passed to the ALBERT model and encoded as a vector representation with a fixed dimensionality. Since any words that do not appear in the vocabulary (i.e., OOV words) are decomposed into subwords, ALBERT can learn the representation vector of these OOV words based on the meaning of the set of subwords. In addition, the positional embedding layer allows ALBERT to capture the representation of words based on the context in the log message. ALBERT also contains a Self-Attention mechanism to efficiently measure the importance of each word in a sentence. Through the pre-training language model, we get $\mathbf{E} = [\mathbf{e_1} \ldots, \mathbf{e_N}], \mathbf{e_i} \in \mathbb{R}^D$, $\mathbf{E}$ is the log word vector set, N is the hyperparameter window size, D is the word vector dimension, at this time, the word vector dimension $D = 768$.

BERT-whitening. To obtain a vector representation of the log data, we use ALBERT. However, ALBERT has the problem of anisotropy [22]. At the same time, for mobile devices, the amount of high-dimensional vector data acquired by ALBERT is relatively large. To solve the above problems, this method adopts BERT-whitening, which cuts the vector dimension and reduces the communication overhead while solving the anisotropy problem.

It uses the whitening operation in traditional machine learning to enhance the isotropy of sentence representations and reduce the dimension of the sentence representation. BERT-whitening includes two steps: whitening transformation and dimension reduction.

BERT-whitening converts the mean value of sentence vectors to 0, and the covariance matrix to the identity matrix. Given a set of row vectors $\{\mathbf{e_i}\}_{i=1}^N$, BERT-whitening transforms them into $\{\tilde{\mathbf{e}}_i\}_{i=1}^N$:

$$\tilde{\mathbf{e}}_i = (\mathbf{e_i} - \mu)\mathbf{W}, \tag{1}$$

where the mean value $\mu = \frac{1}{N} \sum_{i=1}^N e_i$, $\mathbf{W}$ is solving the linear transformation of later. The original covariance matrix Σ represented as:

$$\Sigma = \frac{1}{N} \sum_{i=1}^N (\mathbf{e_i} - \mu)^T (\mathbf{e_i} - \mu). \tag{2}$$

After transformation of the covariance matrix $\tilde{\Sigma} = \mathbf{W}^T \Sigma \mathbf{W}$ and $\tilde{\Sigma} = I$, therefore,

$$\Sigma = (\mathbf{W}^\top)^{-1}\mathbf{W}^{-1} = (\mathbf{W}^{-1})^\top \mathbf{W}^{-1}. \tag{3}$$

Σ satisfies the following forms of SVD [23] is decomposed into:

$$\Sigma = U\Lambda U^T, \tag{4}$$

where U is the orthogonal matrix and Λ is the diagonal matrix, and the diagonal elements are positive. Therefore, setting $\mathbf{W}^{-1} = \sqrt{\Lambda}U^T$, we can obtain the solution:

$$\mathbf{W} = U\sqrt{\Lambda^{-1}}. \tag{5}$$

Elements in the diagonal matrix Λ derived from SVD have been sorted in descending order. We only need to keep the first n columns of $\mathbf{W}$ to achieve the dimensionality reduction. For details about the algorithm, see [11]. Finally, we get the log data vector $\{\widetilde{\mathbf{e}}_i\}_{i=1}^N$, and dimension is K.

3.2 Attention Aware TCN

TCN is based on two principles: the network produces an output of the same length as the input, and it cannot leak from the future to the past. To achieve the first point, TCN uses a 1D Full Convolutional Network (FCN) architecture. Specifically, each hidden layer has the same length as the input layer. And each hidden layer adds zero padding of length $(kernel\_size - 1)$ to keep the subsequent layers the same length as the previous layer. When it comes to the second principle, TCN uses causal convolution, where the output at time t is convolved only with the earlier elements at time t and in the previous layer. However, simple causal convolution is difficult to handle tasks with large amounts of data, so dilation convolution is used to increase the receptive field. The dilation convolution is calculated as follows:

$$F(s) = (\widetilde{e} *_d f)(s) = \sum_{i=0}^{k-1} f(i) \cdot \widetilde{e}_{s-d \cdot i}, \tag{6}$$

where d is the dilation factor, k is the filter size, and $s - d \cdot i$ represents the past direction. Finally, to reduce the complexity of the model and prevent overfitting, the Residual block (as shown in Fig. 1(c)) is used instead of convolutional layers.

Using the above three changes, compared with recurrent neural networks (RNN) and convolutional neural networks (CNN), TCN has stronger parallelism and significant advantages in terms of memory consumption and processing speed; it also has a flexible receptive field size, which can better control the memory size of the model; and it has a more stable gradient. Therefore, this paper decides to use TCN model for anomaly detection.

In the model proposed in this paper (as in Fig. 1(b)), we use the attention mechanism to replace the FCN layer in the original model. By using the attention mechanism, we can better understand the contextual meaning of the log data and get the feature weights of different word vectors to achieve higher accuracy anomaly detection. At the same time, by replacing the fully connected layer with the attention mechanism, the number of model parameters is reduced and the complexity of the model is simplified, which helps to deploy on mobile devices.

The attention model in this paper uses the query-key-value (QKV) model for attention score calculation. The input to the attention module is the sequence $\widetilde{\mathbf{E}} = \{\widetilde{\mathbf{e}}_i\}_{i=1}^N$. For the input, we map it into different vector spaces using the following formulas:

$$\mathbf{Q} = \mathbf{W}_q\widetilde{\mathbf{E}} \in \mathbb{R}^{K \times N},$$
$$\mathbf{K} = \mathbf{W}_k\widetilde{\mathbf{E}} \in \mathbb{R}^{K \times N}, \tag{7}$$
$$\mathbf{V} = \mathbf{W}_v\widetilde{\mathbf{E}} \in \mathbb{R}^{K \times N},$$

$\mathbf{W}_q \in \mathbb{R}^{K \times D}, \mathbf{W}_k \in \mathbb{R}^{K \times D}, \mathbf{W}_v \in \mathbb{R}^{K \times D}$ are the query-key-value linear mapping parameter matrices, respectively; $\mathbf{Q} = [\mathbf{q_0}, \ldots \mathbf{q_N}], \mathbf{K} = [\mathbf{k_0}, \ldots, \mathbf{k_N}], \mathbf{V} = [\mathbf{v_0}, \ldots, \mathbf{v_N}]$ are the matrices consisting of query vectors, key vectors and value vectors. For each query vector $\mathbf{q}_t \in \mathbf{Q}$, the corresponding output vector can be obtained using the following formula:

$$\mathbf{h_t} = \sum_{i=0}^{N} \alpha_{ti}\mathbf{v_i} = \sum_{i=0}^{N} \text{SoftMax}\big(s(\mathbf{q_t}, \mathbf{k_i})\big)\mathbf{v_i}, \tag{8}$$

which α_{ti} is the attention weight, $h_t \in \mathbf{H}$ is the weighted output vector based on the attention weight; SoftMax is the attention score function; This paper calculates the attention score in the form of scaled dot product, which can be abbreviated as follows:

$$H = \mathbf{V}\text{SoftMax}(\frac{\mathbf{K^T Q}}{\sqrt{D}}), \tag{9}$$

where $\mathbf{H} = [\mathbf{h_0}, ..., \mathbf{h_N}]$ is the sequence of attention mechanism output vectors. All vectors in the above output sequence are spliced to obtain the final output vector:

$$\mathbf{o}_{sum} = \Sigma_{i=0}^{N}\mathbf{h_i}. \tag{10}$$

After training, a trained anomaly detection model is obtained for real-time anomaly detection of mobile devices or IoT devices.

3.3 Anomaly Detection

By training attention aware TCN on a cloud computing platform, we deploy the obtained model on mobile devices for lightweight and efficient anomaly detection. We pass the log vector data transferred from the cloud into the trained model, thus enabling local real-time detection of online logs. The training task is carried out as binary classification, the loss is calculated based on binary cross entropy and trained using the Adam optimizer:

$$\text{Loss} = -\frac{1}{N} \sum_{i=1}^{N} y_i \cdot \log(\text{p}(y_i)) + (1 - y_i) \cdot \log(1 - \text{p}(y_i)), \tag{11}$$

where y_i is binary label 0 or 1, and $p(y)$ is the probability that the output belongs to y label.

4 Experiments

We conduct comparison experiments with the popular methods on two publicly available datasets to verify the superiority of the model in terms of accuracy and complexity. Through the experiments, we focus on answering following four research questions:

- RQ1: Is EfficientLog more accurate compared to baseline models?
- RQ2: Is EfficientLog more efficient than baseline models?
- RQ3: How does different dimension setting affect EfficientLog?
- RQ4: Have the key designs played a positive role?

4.1 Datasets

We use two of the most widely used datasets to evaluate EfficientLog, Hadoop Distributed File System (HDFS) [19] dataset and Blue Gene/L supercomputer (BGL) [24]. HDFS dataset contains 11,175,629 log messages collected from a Hadoop Distributed File System on the Amazon EC2 platform. Each session identified by block ID in the HDFS dataset is labeled as normal or abnormal. BGL dataset contains 4,747,963 log messages collected from the Blue Gene/L supercomputer at Lawrence Livermore National Labs. The specific data are shown in Table 1.

Table 1. Details of Log Datasets

Dataset	# Log Messages	# Anomalies	# Test Log	
			Normal	Anomalous
HDFS	11,175,629	575,061	111,709	3,304
BGL	4,747,963	348,460	889,412	60,161

4.2 Baselines

We compare our model with eight anomaly detection methods for log data, including three traditional machine learning methods (LR [1], IM [3] and SVM [2]) and five deep learning based anomaly detection methods (DeepLog [7], Log-Anomaly [4], PLELog [9], NeuralLog [5] and LogGD [25]). Specially, Deeplog, Log-Anomaly and PLELog are based on the recurrent neural network method to realize anomaly detection. NeuralLog uses Transformer for efficient detection. LogGD uses graph neural networks to convert log data into graph data and obtain accurate detection results.

4.3 Implementation and Evaluation Metrics

We run training experiments on a Linux server with two NVIDIA RTX3090 and 64G memory and edge devices are tested using NVIDIA JTX2. We achieve EfficientLog based on Python 3.6.15 and Tensorflow 2.6.0. For the parameters in EfficientLog, we set the dilated convolution rate to $1, 2, 4, 8$. We use an initial learning rate of 0.0001. In addition, we use Softmax as the activation function for the output layer and set the epoch to 20 and batch size to 1024. In the following experiments, the fixed-window size of BGL data is set to 20 logs and step-size is set to 1. The dimension of BGL is 300 and 32 for HDFS.

Evaluation Metrics: We evaluate the performance of our model and the baseline model using the Precision ($Precision = \frac{TP}{TP+FP}$), Recall ($Recall = \frac{TP}{TP+FN}$) and F1-Score ($F1 = 2 * \frac{Precision*Recall}{Precision+Recall}$), where TP is the correctly detected anomaly, FP is the falsely detected anomaly and FN is the falsely assigned normal. In addition, we use the number of parameters, model size, and detection time to measure the performance of the model in terms of efficiency and model storage space size. The number of parameters is related to the neural network kernel size and input vector dimension; the model size visualizes the size of the model resource occupation, and the detection time reflects the model's ability to handle large-scale log data.

4.4 Comparison with Baseline in Accurary(RQ1)

As shown in Table 2, the EfficientLog proposed in this paper achieves an F1 value of 0.997 on both datasets, which achieves a higher detection accuracy compared to the existing popular baseline models. We can find that the existing methods all perform better on the HDFS dataset than on the BGL, which is due to the fact that the BGL collected data over a longer time span compared to the HDFS and thus has more unstable data. More specifically, about 7.4% of log events in the BGL test set do not appear in its training set, while there are no such new log events on HDFS. For machine learning methods, SVM and LR convert log sequences into log count vectors, ignoring the temporal and semantic information of log sequences, resulting in low detection accuracy. Since IM uses the index of log templates to learn normal mode and abnormal mode, the accuracy is lower than that of SVM and LR. Because different templates may share the same semantics and exist log parser parsing errors.

Among deep learning methods, DeepLog and Loganomaly have the same problem as IM, only learning sequence information and then ignoring semantic information, and the results are also poor. Since Loganomaly proposes

Table 2. Results of Different Methods on Public Datasets

Model	BGL			HDFS		
	Pre	Rec	F1	Pre	Rec	F1
LR	0.131	0.932	0.230	0.950	0.921	0.935
SVM	0.971	0.303	0.462	0.991	0.942	0.966
IM	0.136	0.307	0.188	**1.000**	0.881	0.937
DeepLog	0.128	0.995	0.227	0.835	0.994	0.908
Loganomaly	0.136	0.970	0.239	0.886	0.961	0.922
PLELog	0.592	0.958	0.732	0.893	0.979	0.934
NeuralLog	0.989	0.983	0.986	0.977	0.990	0.988
LogGD	0.956	0.967	0.962	0.980	0.993	0.988
EfficientLog	**0.996**	**0.998**	**0.997**	0.996	**0.999**	**0.997**

Template2Vec, which is based on synonyms and antonyms to learn the information of log templates, the accuracy is relatively higher than DeepLog. PLELog learns the knowledge of historical anomalies through probabilistic token estimation and maintains immunity to unstable log data through semantic embedding, thus achieving higher detection accuracy on BGL dataset than the above models. LogGD can capture more expressive structure information from graphs than purely sequential relations between log events, leading to better identify the anomalous log sequences. NeuralLog achieves excellent detection results on both datasets because it uses a pre-trained language model to obtain semantic information of logs and uses Transformer to detect anomalies that can capture data contextual information. The proposed EfficientLog uses ALBERT and BERT-whitening to improve the efficiency and enhance the isotropy of word vectors. EfficientLog also uses attention aware TCN, which is more in line with the temporal characteristics of log data. Thus it can better capture temporal information and contextual relationships, leading to better results than NeuralLog.

4.5 Comparison with Baseline in Efficiency (RQ2)

As described in the previous section, EfficientLog achieves the most advanced results. We compare EfficientLog with DeepLog, NeuralLog, LogGD, Efficient-NoBW and EfficientLog-TF. Because the detection accuracy of machine learning is too low, the efficiency of machine learning methods is not compared in this experiment. Deeplog, Loganomaly and PLELog all use recurrent neural networks to detect anomalies, so we choose DeepLog with the lowest complexity for efficiency comparison. Efficientlog-NoBW refers to removing BERT-whitening parts of the models presented in this paper and EfficientLog-TF is EfficientLog using Transformer instead of TCN. These two models add in comparison in order to prove that dimensional reduction and attention aware TCN are helpful to improve model efficiency and reduce model complexity.

Table 3. Results Of Efficiency Experiments

Model	Count of Parameter		Model Size (KB)		Test Time (s)	
	HDFS	BGL	HDFS	BGL	HDFS	BGL
DeepLog	52,252	52,252	636	636	330	400
NeuralLog	31,516,258	31,516,258	123,146	123,146	69	556
LogGD	12,224,724	12,224,724	47,770	47,770	24	**28**
EfficientLog-NoBW	14,046	14,046	101	101	69	528
EfficientLog-TF	184,738	5,573,146	753	21,801	19	207
EfficientLog	**755**	**5,622**	**50**	**68**	**17**	187

As shown in Table 3, EfficientLog is much smaller than the other methods in terms of the number of parameters, model size, and detection time. Among them, the number of parameters of EfficientLog is only 1% of DeepLog and only one hundred-thousandth of NeuralLog. Though LogGD has the short test time, the number of parameters and the size of the model are much larger than EfficientLog. Efficient-NoBW's results prove that BERT-whitening is helpful to improve model efficiency and reduce model complexity. Due to EfficientLog-TF's Transformer model, too many parameters will cause the model to occupy too much space on edge devices. The number of parameters intuitively reflects the model complexity, which shows that the model complexity of EfficientLog is lower than other models, and the computational consumption is also much lower than other existing models, which is beneficial for porting to mobile devices and edge devices. For the model size, EfficientLog is smaller than the other five models and easier to use for edge devices with less storage space. Finally, the detection time indicates the efficiency of the method. As shown in the table, EfficientLog is the most efficient of the six, which shows that EfficientLog can handle large-scale log data quickly and efficiently.

4.6 Effects of Dimension on EfficientLog (RQ3)

In this section, we discuss how far we can reduce the dimensions of BGL and HDFS to achieve both detection accuracy and efficiency using BERT-whitening.

Results are shown in Fig. 2. For BGL, when the dimension is reduced too low, the model's performance in detecting anomalies is poor. This is because when using BERT whitening, SVD matrix decomposition is used for dimensionality reduction, retaining the first N dimensions with larger eigenvalues. When N is about small, fewer eigenvalues are retained, which may result in feature omission. When $N = 300$, the F1 value reaches its highest. Considering that the model size gap is small, we choose the dimension with higher F1 as the final use dimension. At this point, we chose to reduce the BGL dimension to 300. Similarly, HDFS achieves optimal performance at 32. Therefore, we choose to set the HDFS dimension to 32.

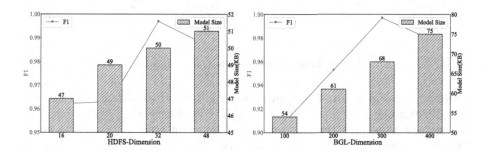

Fig. 2. Experiments on Dimensions of BGL and HDFS.

4.7 Ablation Study for Key Components (RQ4)

In this section, we first compare the effect of different pre-trained language models on the accuracy of the method and then compare whether the main components of the model play an active role.

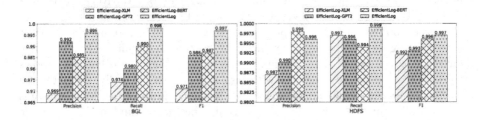

Fig. 3. Experiments on Pre-trained Language Models.

Experiments on Pre-trained Language Models: We replace the ALBERT model in EfficientLog with GPT2 [26], BERT, and Roberta [27] for experiments to evaluate the performance of log anomaly detection. For the GPT2 , BERT, and Roberta encoders, we use their basic models with 12 layers, 12 attention heads, and 768 hidden units. The Fig. 3 shows the results. We observe that the pre-trained models are able to understand the semantics of the log messages and achieve good results. Overall, ALBERT's performance is higher than the three pre-trained language models.

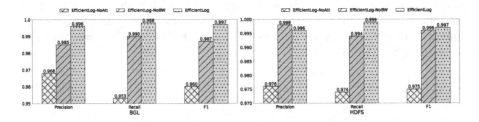

Fig. 4. Experiments on Main Components.

Experiments on Main Components: We design two variants of Efficient-Log, EfficientLog-NoAtt and EfficientLog-NoBW. EfficientLog-NoAtt refers to the method of leaving the rest unchanged and removing the attention layer. EfficientLog-NoBW refers to the direct input of vectors into the TCN without using BERT-whitening. As shown in Fig. 4, the two variants perform well on the dataset, illustrating that semantic information of log data can be captured using TCN. The EfficientLog-NoBW effect is not much different from the complete method, because BERT-whitening is more like a linear operation, which has little effect on the final vector representation, and is mainly used to shorten the detection time and model size. EfficientLog-NoAt is the worst of the three methods, and the EfficientLog ontology performs best because the attention mechanism can provide contextual information, capture global connections, and learn log semantic information more precisely.

5 Conclusion

In this paper, we design a log anomaly detection method based on dimensional reduction and attention aware TCN. The method uses the BERT-whitening algorithm to reduce the dimensionality of the semantic vectors generated by the pre-trained model ALBERT and improves the TCN by adding attention layer. After experiments on two public datasets, it is shown that the method not only achieves excellent detection performance, but also reduces the computing power consumption, detection time, and storage space occupation of edge devices, which is more suitable than popular baseline models. To sum up, this method is more suitable for deployment on mobile devices or edge devices.

Acknowledgements. This work is supported by the Key Research and Development Plan of Shandong Province (Major scientific and technological innovation project0) (2021CXGC010103).

References

1. Bodik, P., Goldszmidt, M., Fox, A., Woodard, D.B., Andersen, H.: Fingerprinting the datacenter: automated classification of performance crises. In: Proceedings of the 5th European Conference on Computer Systems, pp. 111–124 (2010)
2. Chen, M., Zheng, A.X., Lloyd, J., Jordan, M.I., Brewer, E.: Failure diagnosis using decision trees. In: International Conference on Autonomic Computing 2004. Proceedings, pp. 36–43. IEEE (2004)
3. Lou, J.G., Fu, Q., Yang, S., Xu, Y., Li, J.: Mining invariants from console logs for system problem detection. In: USENIX Annual Technical Conference, pp. 1–14 (2010)
4. Meng, W., et al.: LogAnomaly: unsupervised detection of sequential and quantitative anomalies in unstructured logs. In: IJCAI. vol. 19, pp. 4739–4745 (2019)
5. Le, V.H., Zhang, H.: Log-based anomaly detection without log parsing. In: 2021 36th IEEE/ACM International Conference on Automated Software Engineering (ASE), pp. 492–504. IEEE (2021)
6. Kenton, J.D.M.W.C., Toutanova, L.K.: BERT: pre-training of deep bidirectional transformers for language understanding. In: Proceedings of NAACL-HLT, vol. 1, p. 2 (2019)
7. Du, M., Li, F., Zheng, G., Srikumar, V.: DeepLog: anomaly detection and diagnosis from system logs through deep learning. In: Proceedings of the 2017 ACM SIGSAC Conference on Computer and Communications Security, pp. 1285–1298 (2017)
8. Guo, H., Yuan, S., Wu, X.: LogBERT: Log anomaly detection via BERT. In: 2021 International Joint Conference on Neural Networks (IJCNN), pp. 1–8. IEEE (2021)
9. Yang, L., et al.: Semi-supervised log-based anomaly detection via probabilistic label estimation. In: 2021 IEEE/ACM 43rd International Conference on Software Engineering (ICSE), pp. 1448–1460. IEEE (2021)

10. Chen, C., et al.: Deep learning on computational-resource-limited platforms: a survey. Mob. Inf. Syst. **2020**, 1–19 (2020)
11. Su, J., Cao, J., Liu, W., Ou, Y.: Whitening sentence representations for better semantics and faster retrieval. arXiv preprint arXiv:2103.15316 (2021)
12. Bai, S., Kolter, J.Z., Koltun, V.: An empirical evaluation of generic convolutional and recurrent networks for sequence modeling. arXiv preprint arXiv:1803.01271 (2018)
13. Lan, Z., Chen, M., Goodman, S., Gimpel, K., Sharma, P., Soricut, R.: ALBERT: a lite BERT for self-supervised learning of language representations. arXiv preprint arXiv:1909.11942 (2019)
14. He, P., Zhu, J., Zheng, Z., Lyu, M.R.: Drain: an online log parsing approach with fixed depth tree. In: 2017 IEEE International Conference on Web Services (ICWS), pp. 33–40. IEEE (2017)
15. Fu, Q., Lou, J.G., Wang, Y., Li, J.: Execution anomaly detection in distributed systems through unstructured log analysis. In: 2009 Ninth IEEE International Conference on Data Mining, pp. 149–158. IEEE (2009)
16. Zhang, B., Zhang, H., Moscato, P., Zhang, A.: Anomaly detection via mining numerical workflow relations from logs. In: 2020 International Symposium on Reliable Distributed Systems (SRDS), pp. 195–204. IEEE (2020)
17. Schuster, M., Nakajima, K.: Japanese and Korean voice search. In: 2012 IEEE International Conference on Acoustics, Speech and Signal Processing (ICASSP), pp. 5149–5152. IEEE (2012)
18. Perrucci, G.P., Fitzek, F.H., Widmer, J.: Survey on energy consumption entities on the smartphone platform. In: 2011 IEEE 73rd Vehicular Technology Conference (VTC Spring), pp. 1–6. IEEE (2011)
19. Xu, W., Huang, L., Fox, A., Patterson, D., Jordan, M.I.: Detecting large-scale system problems by mining console logs. In: Proceedings of the ACM SIGOPS 22nd Symposium on Operating Systems Principles, pp. 117–132 (2009)
20. Lin, Q., Zhang, H., Lou, J.G., Zhang, Y., Chen, X.: Log clustering based problem identification for online service systems. In: Proceedings of the 38th International Conference on Software Engineering Companion, pp. 102–111 (2016)
21. Chen, J., Chong, W., Yu, S., Xu, Z., Tan, C., Chen, N.: TCN-based lightweight log anomaly detection in cloud-edge collaborative environment. In: 2022 Tenth International Conference on Advanced Cloud and Big Data (CBD), pp. 13–18. IEEE (2022)
22. Ethayarajh, K.: How contextual are contextualized word representations? Comparing the geometry of BERT, ELMo, and GPT-2 embeddings. arXiv preprint arXiv:1909.00512 (2019)
23. Golub, G.H., Reinsch, C.: Singular value decomposition and least squares solutions. Linear Algebra **2**, 134–151 (1971)
24. Oliner, A., Stearley, J.: What supercomputers say: a study of five system logs. In: 37th Annual IEEE/IFIP International Conference on Dependable Systems and Networks (DSN 2007), pp. 575–584. IEEE (2007)
25. Xie, Y., Zhang, H., Babar, M.A.: LogGD: detecting anomalies from system logs by graph neural networks. arXiv preprint arXiv:2209.07869 (2022)
26. Radford, A., et al.: Language models are unsupervised multitask learners. OpenAI Blog **1**(8), 9 (2019)
27. Ruder, S., Søgaard, A., Vulić, I.: Unsupervised cross-lingual representation learning. In: Proceedings of the 57th Annual Meeting of the Association for Computational Linguistics: Tutorial Abstracts, pp. 31–38 (2019)

Author Index

Printed in the United States
by Baker & Taylor Publisher Services